Lecture Notes in Artificial Intelligence   5399

Edited by R. Goebel, J. Siekmann, and W. Wahlster

Subseries of Lecture Notes in Computer Science

T0135320

Luca Iocchi   Hitoshi Matsubara
Alfredo Weitzenfeld   Changjiu Zhou (Eds.)

RoboCup 2008:
Robot Soccer
World Cup XII

 Springer

Series Editors

Randy Goebel, University of Alberta, Edmonton, Canada
Jörg Siekmann, University of Saarland, Saarbrücken, Germany
Wolfgang Wahlster, DFKI and University of Saarland, Saarbrücken, Germany

Volume Editors

Luca Iocchi
Sapienza University, Roma, Italy
E-mail: iocchi@dis.uniroma1.it

Hitoshi Matsubara
Future University, Hakodate, Japan
E-mail: matsubar@fun.ac.jp

Alfredo Weitzenfeld
ITAM, México City, DF, México
E-mail: alfredo@itam.mx

Changjiu Zhou
Singapore Polytechnic, Singapore
E-mail: zhoucj@sp.edu.sg

Library of Congress Control Number: Applied for

CR Subject Classification (1998): I.2, C.2.4, D.2.7, H.5, I.5.4, J.4

LNCS Sublibrary: SL 7 – Artificial Intelligence

ISSN 0302-9743

ISBN 978-3-642-02920-2 Springer Berlin Heidelberg New York

Typesetting: Camera-ready by author, data conversion by Scientific Publishing Services, Chennai, India
Printed on acid-free paper   SPIN: 12629923   06/3180   5 4 3 2 1 0

Preface

The 12th annual RoboCup International Symposium was held during July 15–18, 2008 in conjunction with RoboCup 2008 Competitions and Demonstrations. The symposium represents the core meeting for the presentation and discussion of scientific contributions in diverse areas related to the main threads within RoboCupSoccer, RoboCupRescue, RoboCup@Home and RoboCupJunior. Its scope encompassed, but was not restricted to, research and education activities within the fields of artificial intelligence and robotics.

A fundamental aspect of RoboCup is promoting science and technology among young students and researchers, in addition to providing a forum for discussion and excitement about Robotics with practitioners from all over the world. Since its first edition in 1997 in Nagoya, the RoboCup Competitions and Symposium have attracted an increasing number of researchers and students from all the world and today it is a major event in robotics worldwide.

Due to its interdisciplinary nature and the exploration of various and intimate connections of theory and practice across a wide spectrum of different fields, the symposium offered an excellent opportunity to introduce new techniques to various scientific disciplines. The experimental, interactive and benchmark character of the RoboCup initiative created the opportunity to present, learn and evaluate novel ideas and approaches with significant potential. If promising, they are then rapidly adopted and field-tested by a large (and still strongly growing) community.

For the 12th RoboCup International Symposium we received 91 paper submissions, covering a variety of areas. The papers were carefully reviewed by the International Program Committee, which selected 36 oral and 20 poster presentations. Each paper was reviewed by at least three Program Committee members and the final decisions were made by the Co-chairs.

In addition to the paper and poster presentations, we had three invited speakers:

- Jean-Guy Fontaine, from the Italian Institute of Technology (IIT), Genoa, Italy, whose talk was about "Tele-operation and Tele-cooperation with Humanoids"
- Tianmiao Wang, from Beijing University of Aeronautics and Astronautics, Institute of Robotics, Beijing, China, who talked about "The Advanced Robotics R&D and Progress of the 863 High-Tech Program in China"
- Sven Behnke, from Rheinische Friedrich-Wilhelms-Universität, Institut für Informatik VI, Bonn, Germany, who talked about "Designing a Team of Humanoid Soccer Robots"

We would like to take this opportunity to thank all the authors for their contributions, the Program Committee members and the additional reviewers for their hard work, which had to be completed in a short period of time, the invited speakers for their participation, and the Chinese local organizers for making the RoboCup Symposium 2008 a successful event.

December 2008

Luca Iocchi
Hitoshi Matsubara
Alfredo Weitzenfeld
Changjiu Zhou

Organization

Symposium Co-chairs

Luca Iocchi   Sapienza University, Rome, Italy
Hitoshi Matsubara   Future University, Hakodate, Japan
Alfredo Weitzenfeld   ITAM, Mexico
Changjiu Zhou   Singapore Polytechnic, Singapore

International Symposium Program Committee

Acosta Calderon Carlos Antonio	Singapore Polytechnic, Singapore
Akin H. Levent	Bogazici University, Turkey
Almeida Luis	University of Aveiro, Portugal
Amigoni Francesco	Politecnico di Milano, Italy
Baltes Jacky	University of Manitoba, Canada
Behnke Sven	University of Bonn, Germany
Birk Andreas	Jacobs University Bremen, Germany
Bonarini Andrea	Politecnico di Milano, Italy
Bredenfeld Ansgar	Fraunhofer IAIS, Germany
Brena Ramon	Tecnologico de Monterrey, Mexico
Burkhard Hans-Dieter	Humboldt University, Germany
Caglioti Vincenzo	Politecnico di Milano, Italy
Cardeira Carlos	Technical University of Lisbon, Portugal
Carpin Stefano	University of California, Merced, USA
Cassinis Riccardo	University of Brescia, Italy
Chen Xiaoping	University of Science and Technology Hefei, China
Chen Weidong	Shanghai Jiaotong University, China
Coradeschi Silvia	Orebro University, Sweden
Costa Anna	University of Sao Paulo, Brazil
Dias M. Bernardine	Carnegie Mellon University, USA
Eguchi Amy	Bloomfield College, USA
Farinelli Alessandro	University of Southampton, UK
Förster Alexander	IDSIA, Switzerland
Frontoni Emanuele	Università Politecnica delle Marche, Italy
Gini Giuseppina	Politecnico di Milano, Italy
Grisetti Giorgio	Freiburg University, Germany
Guangming Xie	Peking University, China
Hong Dennis	Virginia Tech, USA
Hu Lingyun	Tsinghua University, China
Indiveri Giovanni	University of Lecce, Italy
Iocchi Luca	Sapienza University of Rome, Italy

Jahshan David	University of Melbourne, Australia
Jamzad Mansour	Sharif University of Technology, Iran
Karlapalem Kamal	International Institute of Information Technology, India
Kenn Holger	Microsoft EMIC, Germany
Kimura Testuya	Nagaoka University of Technology, Japan
Kleiner Alexander	Freiburg University, Germany
Kraetzschmar Gerhard	Bonn-Rhein-Sieg University of Applied Sciences, Germany
Kuwata Yoshitaka	NTT Data Corporation, Japan
Levi Paul	University of Stuttgart, Germany
Levy Simon	Washington and Lee University, USA
Lima Pedro	Instituto Superior Técnico, Portugal
Lopes Gil	Welding Engineering Research Centre Cranfield, UK
Matsubara Hitoshi	Future University, Hakodate, Japan
Matteucci Matteo	Politecnico di Milano, Italy
Mayer Gerd	University of Ulm, Germany
Menegatti Emanuele	University of Padova, Italy
Monekosso Dorothy	Kingston University, UK
Nakashima Tomoharu	Osaka Prefecture University, Japan
Nardi Daniele	Sapienza University of Rome, Italy
Noda Itsuki	Natl Inst of Advanced Industrial Science/Tech, Japan
Nomura Tairo	Saitama University, Japan
Ohashi Takeshi	Kyushu Institute of Technology, Japan
Pagello Enrico	University of Padova, Italy
Paiva Ana	INESC-ID and Instituto Superior Técnico, Portugal
Parsons Simon	City University of New York, USA
Pirri Fiora	Sapienza University of Rome, Italy
Polani Daniel	University of Hertfordshire, UK
Reis Luis Paulo	University of Porto, Portugal
Restelli Marcello	Politecnico di Milan, Italy
Ribeiro Carlos	Technological Institute of Aeronautics, Brazil
Ribeiro Fernando	Minho University, Portugal
Röfer Thomas	University of Bremen, Germany
Rojas Raul	Free University of Berlin, Germany
Ruiz-del-Solar Javier	Universidad de Chile, Chile
Rybski Paul E.	CMU, USA
Santos Vitor	Universidade de Aveiro, Portugal
Schiffer Stefan	RWTH Aachen University, Germany
Shiry Saeed	Amirkabir University of Technology, Iran
Sorrenti Domenico G.	University of Milan - Bicocca, Italy
Sridharan Mohan	Texas Tech University, USA
Takahashi Tomoichi	Meijo University, Japan
Takahashi Yasutake	Osaka University, Japan

Tawfik Ahmed	University of Windsor, Canada
van der Zant Tijn	Rijksuniversiteit Groningen, The Netherlands
Velastin Sergio	Kingston University, UK
Verner Igor	Technion - Israel Institute of Technology, Israel
Visser Ubbo	TZI, University of Bremen, Germany
Wagner Thomas	University of Bremen, Germany
Weitzenfeld Alfredo	Institute Tecnologico Autonomo de Mexico, Mexico
Williams Mary-Anne	University of Technology, Sydney, Australia
Wisse Martijn	Delft University of Technology, The Netherlands
Wisspeintner Thomas	Fraunhofer IAIS, Germany
Wotawa Franz	Technische Universität Graz, Austria
Wyeth Gordon	University of Queensland, Australia
Wyeth Peta	University of Queensland, Australia
Zell Andreas	University of Tuebingen, Germany
Zhao Mingguo	Tsinghua University, China
Zhou Changjiu	Singapore Polytechnic, Sinagpore
Ziparo Vittorio A.	Sapienza University of Rome, Italy

Additional Reviewers

Berger Ralf	Humboldt University, Germany
Calisi Daniele	Sapienza University of Rome, Italy
Certo João	University of Porto, Portugal
Dornhege Christian	Freiburg University, Germany
Gasparini Simone	Politecnico di Milano, Italy
Goehring Daniel	Humboldt University, Germany
Marchetti Luca	Sapienza University of Rome, Italy
Mellmann Heinrich	Humboldt University, Germany
Pellegrini Stefano	Sapienza University of Rome, Italy
Randelli Gabriele	Sapienza University of Rome, Italy
Ronny Novianto	University of Technology, Sydney, Australia
Selvatici Antonio	University of Sao Paulo, Brazil
Silva Valdinei	University of Sao Paulo, Brazil
Stanton Christophe	University of Technology, Sydney, Australia
Stückler Jörg	University of Bonn, Germany
Valero Alberto	Sapienza University of Rome, Italy
Werneck Nicolau	University of Sao Paulo, Brazil

Table of Contents

Papers with Poster Presentation

A Robust Speech Recognition System
for Service-Robotics Applications

Masrur Doostdar, Stefan Schiffer, and Gerhard Lakemeyer

Knowledge-based Systems Group
Department of Computer Science 5
RWTH Aachen University, Germany
doostdar@kbsg.rwth-aachen.de,
{schiffer,gerhard}@cs.rwth-aachen.de

Abstract. Mobile service robots in human environments need to have versatile abilities to perceive and to interact with their environment. Spoken language is a natural way to interact with a robot, in general, and to instruct it, in particular. However, most existing speech recognition systems often suffer from high environmental noise present in the target domain and they require in-depth knowledge of the underlying theory in case of necessary adaptation to reach the desired accuracy. We propose and evaluate an architecture for a robust speaker independent speech recognition system using off-the-shelf technology and simple additional methods. We first use close speech detection to segment closed utterances which alleviates the recognition process. By further utilizing a combination of an FSG based and an N-gram based speech decoder we reduce false positive recognitions while achieving high accuracy.

1 Introduction

Speech recognition is a crucial ability for mobile service robots to communicate with humans. Spoken language is a natural and convenient means to instruct a robot if it is processed reliably. Modern speech recognition systems can achieve high recognition rates, but their accuracy often decreases dramatically in noisy and crowded environments. This is usually dealt with by either requiring an almost noise-free environment or by placing the microphone very close to the speaker's mouth. Although we already assume the latter, all requirements for a sufficiently high accuracy cannot always be met in realistic scenarios.

Our target application is a service-robotics domain, in particular, the ROBOCUP-@HOME league [1], where robots should assist humans with their everyday activities in a home-like environment. Any interaction with a robot has to be done in a natural fashion. That is to say, instructions issued to the robot may only be given by means of gestures or natural spoken language. An important property of the domain, especially at a competition, is the high amount of non-stationary background noise and background speech. A successful speech recognition system in ROBOCUP@HOME must be able to provide robust speaker-independent recognition of mostly command-like sentences. For one, it is important that commands given to the robot are recognized robustly. For another, spoken language not directed to the robot must not be matched to an instruction

L. Iocchi et al. (Eds.): RoboCup 2008, LNAI 5399, pp. 1–12, 2009.

for the robot. This is a non-trivial task in an environment with a high amount of background noise. That is why most teams use a head mounted microphone for their speech recognition. Still, it is not easy to determine which audio input is actually addressed to the robot and which one is not. This is even more so, since within a competition there usually is a person that describes to the audience what is currently happening in the arena via loudspeakers. The words used for the presentation often are very similar if not even the same used to instruct the robot. This complicates the task of robust speech recognition even more.

We propose an architecture that tackles the problem of robust speech recognition in the above setting. It comprises two steps. First, we use a threshold based close speech detection module to segment utterances targeted at the robot from the continuous audio stream recorded by the microphone. Then, we decode these utterances with two different decoders in parallel, namely one very restrictive decoder based on finite state grammars and a second more lenient decoder using N-grams. We do this to filter out false positive recognitions by comparing the output of the two decoders and rejecting the input if it was not recognized by both decoders.

The paper is organized as follows. In Section 2, we describe some basics of common speech recognition systems. Then we propose our architecture and discuss related work. We go into detail about our close speech detection in Section 3 and the dual decoding in Section 4. After an experimental evaluation in Section 5, we conclude in Section 6.

2   Foundations and Approach

We first sketch properties of common statistical speech recognition systems. Then we propose a system architecture that tries to combine the features of different approaches to tackle the problems present in our target domain.

2.1   Statistical Speech Recognition

Most statistical speech recognition systems available today use hidden Markov models (HMMs). For a given vector of acoustical data x they choose a sequence of words w_{opt} as best-hypothesis by optimizing

$$w_{opt} = \arg\max_{w} p(w|x) = \arg\max_{w} \frac{p(x|w) \cdot p(w)}{p(x)} = \arg\max_{w} p(x|w) \cdot p(w). \quad (1)$$

Here $p(w|x)$ is the posterior probability of w being spoken, the fundamental Bayes' decision rule is applied to it. The constant normalization probability $p(x)$ can be omitted for the maximization. $p(x|w)$ denotes the probability of observing the acoustical data x for the assumption of w being spoken. This is given to the recognizer by the acoustic-model. $p(w)$ denotes the probability of the particular word-sequence w occurring. This prior probability is provided to the recognizer by the so-called language-model. The language model can either be represented by a grammar, e.g. a *finite state grammar* (FSG), or by means of a statistical model, mostly in form of so-called *N-grams* that provide probabilities for a word dependent on the previous (N-1) words. Common speech-recognizers use 3-grams, also called *TriGrams*.

Standard statistical speech recognition systems process a given speech utterance time-synchronously. Each time-frame, possible sub-word-units (modeled by HMM-states) and word-ends are scored considering their acoustical probability and their language model probability. Most of the possible hypotheses score considerably worse than the best hypothesis at this time frame and are pruned away. In the search for a best hypothesis information about possible near alternatives can be kept to allow for useful post-processing. For each time-frame, the most probable hypotheses of words ending at that frame are stored along with their acoustical scores. This information can be appropriately represented by a directed, acyclic, weighted graph called *word-graph* or *word-lattice*. Nodes and edges in the graph denote words, their start-frames and their acoustic likelihoods. Any path through the graph, starting at the single start-node and ending in the single end-node, represents a hypothesis for the complete utterance. By combining the acoustical likelihoods of the words contained along this path with the language model probability we obtain the score $p(x|w) \cdot p(w)$. The so-called N-*best list* contains all possible paths through the word-graph that were not pruned in the search, ordered by their score.

The language model used in searching for hypotheses largely influences the performance of a speech recognition system. Thus, it is crucial to choose a model appropriate for the particular target application to achieve sufficiently good results. On the one hand, FSG-based decoders perform good on sentences from their restricted grammar. On the other hand, they get easily confused for input that does not fit the grammar used. This can lead to high false recognition rates. N-gram based language models are good for larger vocabularies, since utterances do not have to follow a strict grammar.

2.2 Approach

For our target application we are confronted with a high amount of non-stationary background noise including speech similar to the vocabulary used to instruct the robot. Only using an FSG-based decoder would lead to high false recognition rates. We aim to eliminate false recognitions with a system that exploits the properties of different language models described above. In a first step, we try to segment utterances that are potential speech commands issued to the robot from the continuous audio stream recorded by the robot's microphone. Then, we decode those utterances using two decoders in parallel, one FSG-based and a second TriGram based one to combine the benefits of both. An overview of our system's architecture is shown in Figure 1.

Segmentation of close speech sections
We employ a module that is supposed to segment *close speech sections* from the continuous live stream, i.e. sections where the main person speaks closely into the microphone. We call this *close speech detection* (CSD). Doing this provides us with two advantages. First, with the (reasonable) presupposition that the speech to be recognized, we call it *positive speech*, is being carried out close to the microphone, we can discriminate it from other (background) speech events that are not relevant and thus may cause false recognitions. These false recognitions would be wrongly matched to a speech command for the robot. Second, the performance of speech recognition engines like SPHINX [2] increases considerably, if the speech input occurs in closed utterances instead of a continuous stream. Furthermore, we are able to reduce the computational

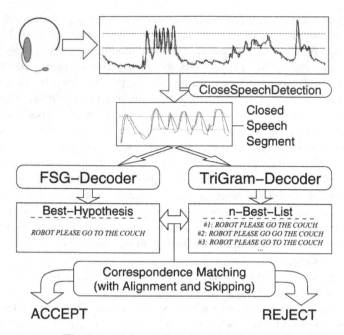

Fig. 1. Architecture of our dual decoder system

demands for speech recognition if we only process input of interest instead of decoding continuously. This is especially useful for mobile robotic platforms with limited computing power.

Multiple decoders

As already mentioned, different types of decoders exhibit different properties we would like to combine for our application. For one, we are interested in the high accuracy that very restricted FSG decoders provide. But we cannot afford to accept a high rate of falsely recognized speech that is then probably matched to a legal command. For another, TriGram-based decoders are able to reliably detect words from a larger vocabulary and they can generate appropriate hypotheses for utterances not coming from the grammar, i.e. that are not positive speech. However, that comes with the drawback of an increased error rate in overall sentence recognition. Still, a sentence at least similar to the actual utterance will very likely be contained in the N-best list, the list of the n hypotheses with the highest score. By decoding with an FSG and a TriGram decoder in parallel we seek to eliminate each decoders drawbacks retaining its benefits. We can look for similarities between the FSG's output and the N-best list of the trigram decoder. This allows detecting and rejecting *false positive recognitions* from the FSG decoder.

In principle, any automatic speech recognition system that provides the ability to use both, FSG and N-gram based decoding with N-best list generation, could be employed within our proposed architecture. We chose to use SPHINX 3 because it is a freely available open source software, it is under active development with good support, it is flexible to extend, and it provides techniques for speaker adaption and acoustic model generation. For an overview of an earlier version of the SPHINX system we refer to [2].

2.3 Related Work

For speech detection, also referred to as speech activity detection (SAD), endpoint detection, or speech/non-speech classification, speech events have to be detected and preferably also discriminated against non-speech events on various energy levels. There has been work on this problem in the last decades, some of which also employs threshold-approaches like an earlier work of Rabiner detecting energy pulses of isolated words [3], and [4]. One of the main differences to our approach is that they dynamically adapt the threshold to detect speech on various energy levels. For our application, however, it is more preferable to use a static threshold since the environmental conditions may vary but the characteristics of the aural input of interest do not. Furthermore, we dynamically allocate the distance allowed between two spoken words and we apply simple smoothing of a signal's energy-value sequence. For a more general solution to the problem of speech activity detection threshold-based approaches often do not work since they are not robust enough on higher noise levels. More robust approaches use, for example, linear discriminant analysis (LDA) like [5] and [6], or HMMs on Gaussian Mixtures [7]. Our aim is to use detection of close speech only as a pre-processing step before decoding. That is why we do not want to put up the additional costs for these more sophisticated approaches.

To improve the accuracy of speech-recognition systems on grammar-definable utterances while also rejecting false-positives, usually in-depth knowledge of the low-level HMM-decoding processes is required. There has been work on integrating N-grams and finite state grammars [8] in one decoding process for detecting FSG-definable sentences. They assume that the sentences to detect are usually surrounded by carrier phrases. The N-gram is aimed to cover those surrounding phrases and the FSG is triggered into the decoding-process if start-words of the grammar are hypothesized by the N-gram. To reject an FSG-hypothesis they consult thresholds on acoustical likelihoods of the hypothesized words. Whereas this approach requires integration with low-level decoding processes, our dual-decoder approach only performs some post-processing on the hypotheses of the N-best-list. In combination with the CSD front-end we achieve acceptable performance for our application without modifying essential parts of the underlying system. Instead of using two decoders in parallel, a more common method could be to use an N-gram language model in a first pass and to re-score the resulting word-graph or N-best list using an FSG based language model afterwards. However, independently decoding with an FSG-based decoder can be expected to provide higher accuracy for the best hypothesis than the best hypothesis after re-scoring a word-graph with an FSG language model. It would be promising, though, to combine a two-pass approach or our dual-decoder with a method for statistically approximating confidences [9] (in terms of posterior-probabilities) of hypothesized words given a word-graph. A reliable confidence measure would provide a good method for rejection with a threshold.

3 Close Speech Detection

Our approach to detect and segment sections of close speech from a continuous audio stream is quite simple. It makes use of the straightforward idea that sounds being produced close to the microphone exhibit considerably high energy levels. The *energy values* of an audio input are provided when working with speech recognition systems as they extract cepstral coefficients as features from the acoustic signals. The first value of the cepstral coefficients can be understood as the signal's logarithmic energy value. Close speech is detected by first searching for energy values that exceed some upper threshold. Then, we determine the start and the end-point of the segment. Therefore, we look in forward and backward direction for points where the speech's energy values fall below a lower threshold for some time. Note that this straightforward approach can only detect speech carried out close to the microphone. However, this is expressly aimed for in our application since it provides a simple and robust method to discriminate between utterances of the "legal speaker" and other nearby speakers as well as background noise.

3.1 Detailed Description

Examining a sequence of energy values, speech-segments are characterized by adjacent heaps (see Figure 2: E[t]). For our aim of detecting close speech segments we use two thresholds, namely T_{up} and T_{down}. The first threshold T_{up} mainly serves as a criterion for detecting a close-speech-segment when some energy values exceed it. Thus, T_{up} should be chosen so that for close speech segments some of the heaps are expected to exceed T_{up} while other segments do not.

After this initial detection, the beginning and the end of the speech segment have to be determined such that the resulting segment contains all heaps adjacent to the initial peak. Therefore, starting from the detection point, we proceed in forward and backward direction. We search for points where the energy value drops below the second threshold T_{down} and does not recover again within a certain amount of time-frames. We thereby identify the beginning and the end of the speech segment, respectively. T_{down} should be chosen largest so that still all heaps of a close speech section are expected to exceed it and lowest so that the energy-level of the background noise and most background sound events do not go beyond T_{down}. The amount of time-frames given for recovering again represents the maximal distance we allow between two heaps, i.e. between two consecutive words. We call this the *alive-time*. We further enhance this approach by smoothing the sequence of energy levels before processing it and by dynamically allocating the time to recover from dropping below T_{down}.

Smoothing the energy values. The energy-value-sequence is smoothed (cf. $sm(E, t)$ in Figure 2) to prevent punctual variations to take effect on the detection of speech-segments and the determination of the alive-time. We compute the smoothed values by averaging over the current energy-value and the three largest of the previous six energy-values, i.e. for an energy-sequence E and time-frame t we use the smoothing function:

$$sm(E, t) = \frac{1}{4} \cdot \max\{E[t] + E[t_1] + E[t_2] + E[t_3] \mid t_i \in \{t - 1, \ldots, t - 6\}\}.$$

Start/End-point detection. The amount of time-frames given to recover from falling below T_{down} is not fixed but is determined dependent on the height of the last heap's peak and the distance to this peak. Thus, the closer the peak of the last heap is to T_{down},

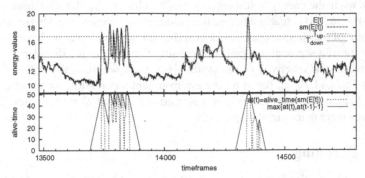

(a) Continuous stream with two utterances of interest:

bg-speech – **Oh, I forgot my cup!** – Should I go an get it for you? – **Yes-please!** – *bg-speech*

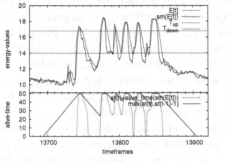

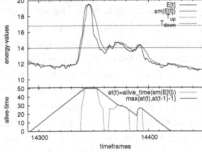

(b) First utterance "Oh, I forgot my cup!" (c) Second utterance "Yes please!"

Fig. 2. Segmentation of close speech segments

i.e. the less the confidence is for the last heap being produced by a close speech, the less time-frames we grant before the next heap must occur (cf. alive-time of right-most heap in Figure 2(c)). This helps to prevent that background sounds which intersect with the close-speech segment or directly succeed/precede them and which exceed T_{down} (like a nearby speech) cause the fixation of the start or end point of a close-speech segment to be postponed over and over again. For energy-values greater than T_{up} we assign the alive-time AT_{up}, for the value T_{down} we assign the alive-time AT_{down} and for values between T_{up} and T_{down} we calculate a time-value by linearly interpolating between AT_{up} and AT_{down}:

$$\text{alive\_time}(v) = \begin{cases} AT_{up} & , v \geq T_{up} \\ \frac{AT_{up} - AT_{down}}{T_{up} - T_{down}} \cdot (v - T_{down}) + T_{down} & , T_{down} < v < T_{up} \\ 0 & , v \leq T_{down} \end{cases}$$

After a close-speech segment is detected we proceed in forward/backward direction (see Figure 2(b) and Figure 2(c)). For each time-frame t, we compute the alive-time at as the maximum of the value associated with the smoothed energy-value $at(sm(E,t))$ and the alive-time value chosen in the previous time-frame minus one ($at(t-1) - 1$).

Obviously, when the energy-value-sequence falls below T_{down} the alive-time decreases each time-frame. If 0 is reached before the values recover again, we determine the start and end point of the close-speech-segment at that frame. As soon as we have determined the start point of a segment, we can start passing the input to the decoders. This drastically increases the reactivity of the system. For now, we manually define the actual thresholds T_{up} and T_{down} based on the environmental conditions at a particular site. We fixed AT_{up} at 50 time-frames (500 ms) and AT_{down} at 25 time frames (250 ms). These values were determined empirically.

4   Dual Decoding

As mentioned in Section 2.1, in statistical SR-systems the optimization of the posterior probabilities $p(w|x)$ is approached by maximizing the scores $p(x|w) \cdot p(w)$ where the likelihood $p(x|w)$ is given by an acoustical model and the prior probability $p(w)$ is provided by a language model. The set of utterances to recognize in our target application per task is quite limited and very structured. It can thus appropriately be defined by a grammar. Consequently, a language model based on a finite state grammar (FSG) seems most suitable. Even though we assume our CSD already filtered out some undesired input, we are confronted with a high rate of false positive recognitions of *out-of-grammar* (OOG) utterances which we cannot afford and have to take care of. Given an OOG utterance x, a restricted FSG-based decoder cannot come around to hypothesize x as an *in-grammar* (IG) sentence w (or prefixes of it), since the word-sequence probability for all other sentences $p(w')$ is 0 because they are not part of the grammar. This holds even if we suppose the acoustical probability $p(x|w)$ for an IG-sentence to be low. The acoustical probability mainly plays a decisive role for discrimination between different utterances w from within the language model. A TriGram-based language model contains many more possible utterances, hence a decoder using such a language model can also hypothesize those other sentences when it is given an utterance that is OOG (with respect to the FSG). Unfortunately, it cannot provide us with an accurate *best hypothesis* reliably enough. That is, the correct sentence w_x for a given IG-utterance x will not be the best hypothesis often enough. Otherwise, we could just stay with a single TriGram-based decoder for recognition. But we can utilize a TriGram language model to help rejecting OOG utterances hypothesized by the FSG-based decoder. For this the TriGram has to comprise a larger vocabulary than the specific grammar and provide not-too-low probabilities for appropriate combinations of these words, i.e. for OOG sentences.

Let us consider a false positive recognition where an OOG-utterance x is falsely recognized as an IG-sentence w by the FSG-based decoder. With an appropriate modeling of the TriGram we can assume it to provide OOG-sentences w' with significant probabilities within its language-model. We can also assume that the acoustical probability $p(x|w')$ for those w' corresponding to the actual utterance exceed the acoustical probability of each falsely hypothesized IG-sentence w considerably. So for the TriGram-based decoding process the comparatively low acoustical probability $p(x|w)$ causes some words of w to be pruned away around the corresponding time-frames they were hypothesized at by the FSG-decoder. Hence, the word-graph and thus the N-best list produced by the TriGram-based decoder will not contain the sequence w. On the other hand, given an IG utterance x and its sentence w_x, the comparatively high acoustical probability $p(x|w)$ (in combination with a still sufficiently high language probability) likely prevents that words of w_x are pruned in the decoding process. Therefore, the N-best-list will still contain w_x.

Consequently, we accept the hypothesis of the FSG-based decoder, if it can be matched with some hypothesis within the N-best list of the TriGram decoder. To not compare utterances from different instances of time, the matching also takes word-start-frames into consideration.

4.1 Hypothesis Matching

For the matching of the FSG-best hypothesis w_{FSG} with one of the N-best hypotheses of the TriGram w_n, we require that the words of w_{FSG} occur in the same order in w_n. The difference in the start-frames of the matched words shall not exceed a predefined maximal offset. For this, we simply iterate through the word-sequence w_n. If the current word of w_n matches with the current word of w_{FSG} (considering the maximal offset allowed), we proceed to the next word of w_{FSG}. If all words of w_{FSG} are processed, we accept the FSG's hypothesis. Within this matching, we always omit hypothesized filler-words like *SILENCE*. For some cases, we experienced that an additional heuristics can improve the acceptance rate. As so, for longer word-sequences w_{FSG} hypothesized by the FSG-based decoder it can be reasonable not to require that all words in w_{FSG} have to be matched on the N-best hypothesis compared to. There is a trade-off between good acceptance rate and good rejection rate when relaxing the matching. Since we only want to make sure that the FSG's hypothesis is not a false positive we argue that enough evidence is given if the FSG's and the TriGram's hypotheses have been similar enough. Therefore, we allow to skip some words of w_{FSG} during the matching dependent on the number of words of w_{FSG} (e.g. in our application we allow to skip 1 word if $|w_{FSG}| \geq 4$, skip 2 words if $|w_{FSG}| \geq 6$ ...). This can be incorporated very easily in our matching procedure we explained above.

5   Experimental Evaluation

To evaluate our approach we conducted several experiments on speech input recorded in the ROBOCUP@HOME environment during a competition. We use a freely available speaker independent acoustic-model for the SPHINX 3 speech engine build with the WSJ-corpus [10]. The FSG decoder was run with the specific grammar for the navigation task shown in Table 1.

The performance of the our dual-decoder systems is influenced by several parameters. We adjusted these in such a way, that the trade-off between higher acceptance-rate of IG-utterances and higher rejection-rates of OOG-utterances tends to a higher acceptance rate. That is because we expect to let pass a fairly low amount of OOG speech-like utterances in the close-speech detection step already.

5.1   Recognition Accuracy

To assess the overall recognition performance of our dual decoder system compared to a TriGram-only and an FSG-only system, we compiled a set of utterances that are legal

Table 1. Grammar for the navigation task

```
command = [ salut ] instruct TO THE location | STOP
salut     = ROBOT [ PLEASE ]              instruct = GO | NAVIGATE | DRIVE | GUIDE ME
location  = ARM CHAIR | PALM [TREE] | WASTE (BASKET | BIN) | TRASH CAN | UPLIGHT |
            REFRIGERATOR | FRIDGE | COUCH | SOFA | PLANT | BOOKSHELF | SHELF |
            (COUCH | SIDE | COFFEE | DINNER | DINNING) TABLE | [FRONT] DOOR | LAMP
```

commands of a particular task in the ROBOCUP@HOME domain. The FSG decoder is using the corresponding grammar of this specific task. The TriGram decoder in our dual decoder system uses a language model constructed from all tasks (excluding the task used for evaluating the rejection of OOG-utterances) of the ROBOCUP@HOME domain with an additional set of 100 sentences of general purpose English. To achieve best performance the TriGram-only decoder uses a language model constructed from navigation sentences only.

In our particular evaluation setup we fed 723 (20.6 minutes) correct commands from the navigation task (cf. Table 1) to all three decoders. Table 2 shows the accuracy and rejection results. For our dual-decoder, we consider an utterance successful if it is recognized correctly and accepted. The recognition rates are based on the FSG decoder while the rejection rates are based on the matching between the FSG's hypothesis and the first 25 entries of the TriGram's N-best list. In the TriGram-only case, we take the best hypothesis as the recognition output. The results indicate that using two decoders in parallel yields successful processing. Adding up 13.8% of falsely recognized commands and 8.6% of correctly recognized but rejected commands, we receive a total of 22.4% of unsuccessfully processed utterances in comparison to 30.7% of an TriGram-based decoder. The sentence-error rate (SER) is a more meaningful measure than the word-error rate (WER) here, because we are interested in the amount of sentences containing errors and not in the number of errors per sentence. The 10.2% of falsely recognized but accepted utterances are critical in the sense that they could have caused a false command to be interpreted. Depending on the application, hypotheses that differ from the reference spoken can still result in the same command, e.g. "ROBOT PLEASE GO TO THE REFRIGERATOR" yields the same command as "DRIVE TO FRIDGE". To give an idea about possible proportions, for the dual-decoder on our navigation task half of the potentially critical utterances (10.2%) are matched on the same command. For the TriGram-decoder this is the case for one fifth of the 30.7% of all sentence errors. 24.2% (overall) are OOG-sentences and thus are not matched to commands at all. To compare the processing speed, we also measured the real-time factor (RTF), i.e. the time it takes to process a signal of duration 1. We achieved an RTF of 1.16 for our dual-decoder

Table 2. Accuracy and rejection results of dual decoder for legal commands

(a) Dual decoder

	rejected	accepted	
recognized	8.6%	77.6%	86.2%
falsely recognized	3.6%	10.2%	13.8%
	12.2%	87.8%	

(b) single decoders

	WER	SER	RTF
TriGram-based	9.9%	30.7%	0.99
FSG-based	4.1%	13.8%	0.24

Table 3. Acceptance rates of false positive (FP) utterances and error rates on legal commands

Decoder	$FP_{accepted}$	Error rate on correct commands
Single (FSG only)	93.9%	13.8% (SER)
Single (TriGram only)	16.1%	30.7% (SER)
Dual (FSG+TriGram)	17.7%	13.8% (SER) + 8.6% (falsely rejected)

system on a Pentium M with 1.6 GHz. This is fast enough for our application, since we are given closed utterances by our CSD front-end and we only decode those. RTFs for the single-decoder systems (with relaxed pruning thresholds for best accuracy) on the same machine are listed in Table 2(b).

5.2 Rejection Accuracy

To assess the performance of our dual decoder system in rejecting OOG utterances (with respect to the FSG) we collected a set of utterances that are legitimate commands of the ROBOCUP@HOME domain (all tasks) but that do not belong to the specific task the FSG decoder is using. Please note that this is close to a worst case analysis since not all of the utterances that make it to the decoder stage in a real setup will be legal commands at all. In our particular case we took 1824 commands (44 minutes) from the final demonstration task and the manipulation task and fed those commands to an FSG-only system, a TriGram-only system, and our proposed dual decoder system. All three decoders had the same configuration as in the recognition setup above. The FSG decoder for the single case and within our dual-decoder system was using the navigation task grammar (cf. Table 1). As can be seen in Table 3, the single FSG decoder setup would have matched over 93% of the false positive utterances to valid robot commands. With our dual decoder approach, on the other hand, the system was able to reject more than 82% of those false utterances. With a TriGram-only decoder we would have been able to reject 84%, but this would have come with a prohibitive error rate of more than 30% for correct commands as shown in Table 3 and Table 2(b) already.

6 Conclusion

In this paper, we presented an architecture for a robust speech recognition system for service-robotics applications. We used off-the-shelf statistical speech recognition technology and combined two decoders with different language models to filter out false recognitions which we cannot afford for a reliable system to instruct a robot. The advantages of our system in comparison to more sophisticated approaches mentioned are as follows. It provides sufficiently accurate speech detection results as a front-end for ASR-systems. Our approach is computationally efficient and relatively simple to implement without deeper knowledge about speech recognition interiors and sophisticated classifiers like HMMs, GMMs or LDA. It is therefore valuable for groups lacking background knowledge in speech recognition and aiming for a robust speech recognition system in restricted domains.

As results are very promising so far, a future issue would be to examine the system's performance for far-field speech, that is not using a headset. We imagine this to be worthwhile especially when we integrate filter methods such as beam forming for on-board microphones with sound-source localization [11,12].

Acknowledgment

This work was supported by the German National Science Foundation (DFG) in the Graduate School 643 *Software for Mobile Communication Systems* and partly within the Priority Program 1125. Further, we thank the reviewers for their comments.

References

1. van der Zant, T., Wisspeintner, T.: Robocup x: A proposal for a new league where robocup goes real world. In: Bredenfeld, A., et al. (eds.) RoboCup 2005. LNCS, vol. 4020, pp. 166–172. Springer, Heidelberg (2006)
2. Huang, X., Alleva, F., Hon, H.W., Hwang, M.Y., Rosenfeld, R.: The SPHINX-II speech recognition system: an overview. Computer Speech and Language 7(2), 137–148 (1993)
3. Lamel, L., Rabiner, L., Rosenberg, A., Wilpon, J.: An improved endpoint detector for isolated word recognition. IEEE Trans. on Acoustics, Speech, and Signal Processing [see also IEEE Trans. on Signal Processing] 29(4), 777–785 (1981)
4. Macho, D., Padrell, J., Abad, A., Nadeu, C., Hernando, J., McDonough, J., Wolfel, M., Klee, U., Omologo, M., Brutti, A., Svaizer, P., Potamianos, G., Chu, S.: Automatic speech activity detection, source localization, and speech recognition on the chil seminar corpus. In: IEEE Int. Conf. on Multimedia and Expo, 2005 (ICME 2005), July 6, pp. 876–879 (2005)
5. Padrell, J., Macho, D., Nadeu, C.: Robust speech activity detection using lda applied to ff parameters. In: Proc. of the IEEE Int. Conf. on Acoustics, Speech, and Signal Processing (ICASSP 2005), March 18-23, vol. 1, pp. 557–560 (2005)
6. Rentzeperis, E., Stergiou, A., Boukis, C., Souretis, G., Pnevmatikakis, A., Polymenakos, L.: An Adaptive Speech Activity Detector Based on Signal Energy and LDA. In: 3rd Joint Workshop on Multi-Modal Interaction and Related Machine Learning Algorithms (2006)
7. Ruhi Sarikaya, J.H.L.H.: Robust Speech Activity Detection in the Presence of Noise. In: Proc. of the 5th Int. Conf. on Spoken Language Processing (1998)
8. Lin, Q., Lubensky, D., Picheny, M., Rao, P.S.: Key-phrase spotting using an integrated language model of n-grams and finite-state grammar. In: Proc. of the 5th European Conference on Speech Communication and Technology (EUROSPEECH 1997), pp. 255–258 (1997)
9. Wessel, F., Schlüter, R., Macherey, K., Ney, H.: Confidence measures for large vocabulary continuous speech recognition. IEEE Trans. on Speech and Audio Processing 9(3), 288–298 (2001)
10. Seymore, K., Chen, S., Doh, S., Eskenazi, M., Gouvea, E., Raj, B., Ravishankar, M., Rosenfeld, R., Siegler, M., Stern, R., Thayer, E.: The 1997 CMU Sphinx-3 English Broadcast News transcription system. In: Proc. of the DARPA Speech Recognition Workshop (1998)
11. Calmes, L., Lakemeyer, G., Wagner, H.: Azimuthal sound localization using coincidence of timing across frequency on a robotic platform. Journal of the Acoustical Society of America 121(4), 2034–2048 (2007)
12. Calmes, L., Wagner, H., Schiffer, S., Lakemeyer, G.: Combining sound localization and laser based object recognition. In: Tapus, A., Michalowski, M., Sabanovic, S. (eds.) Papers from the AAAI Spring Symposium, Stanford, CA, pp. 1–6. AAAI Press, Menlo Park (2007)

Intuitive Humanoid Motion Generation Joining User-Defined Key-Frames and Automatic Learning

Marco Antonelli[1], Fabio Dalla Libera[1], Emanuele Menegatti[1],
Takashi Minato[3], and Hiroshi Ishiguro[2,3]

[1] Intelligent Autonomous Systems Laboratory,
Department of Information Engineering (DEI), Faculty of Engineering,
University of Padua, Via Gradenigo 6/a, I-35131 Padova, Italy
[2] Department of Adaptive Machine Systems, Osaka University, Suita, Osaka,
565-0871 Japan
[3] ERATO, Japan Science and Technology Agency, Osaka University, Suita, Osaka,
565-0871, Japan

Abstract. In this paper we present a new method for generating humanoid robot movements. We propose to merge the intuitiveness of the widely used key-frame technique with the optimization provided by automatic learning algorithms. Key-frame approaches are straightforward but require the user to precisely define the position of each robot joint, a very time consuming task. Automatic learning strategies can search for a good combination of parameters resulting in an effective motion of the robot without requiring user effort. On the other hand their search usually cannot be easily driven by the operator and the results can hardly be modified manually. While the fitness function gives a quantitative evaluation of the motion (e.g. "How far the robot moved?"), it cannot provide a qualitative evaluation, for instance the similarity to the human movements. In the proposed technique the user, exploiting the key-frame approach, can intuitively bound the search by specifying relationships to be maintained between the joints and by giving a range of possible values for easily understandable parameters. The automatic learning algorithm then performs a local exploration of the parameter space inside the defined bounds. Thanks to the clear meaning of the parameters provided by the user, s/he can give qualitative evaluation of the generated motion (e.g. "This walking gait looks odd. Let's raise the knee more") and easily introduce new constraints to the motion. Experimental results proved the approach to be successful in terms of reduction of motion-development time, in terms of natural appearance of the motion, and in terms of stability of the walking.

1 Introduction

It is widely accepted that robots will become part in everyone's life in the near future and that their use will not be limited to manufacturing chains in factories. Emblems of this tendency are the humanoid robots, which structure and purpose

L. Iocchi et al. (Eds.): RoboCup 2008, LNAI 5399, pp. 13–24, 2009.

is very different from classic robotic arms. Small size humanoids have recently become more and more popular among robots, mainly for entertainment and research purpose. This is due to the strong decrease in their cost in the last few years. Small humanoids are usually less than 50 cm tall and actuated by low voltage servomotors (around 5V) at each of the joints. See [11], [12] or [13] for some examples. Though these robots are often simpler than big ones, the number of their degrees of freedom (d.o.f.) is very high. For instance Kondo's KHR-2 HV has 17 d.o.f while VStone's Robovie-M has 22. Because of the high dimensionality of the configuration space, generating motions for this kind of robots is a complex task.

In general, three major strategies are adopted. The first one is about generating the motion off line, and then replaying it on the robot. The second one is about calculating the motion online, so that all the information available can be employed to produce an optimal movement. The third approach, that is placed in between the previous two, consists in calculating the motion off line and then adjusting it online depending on the data coming from the sensors, for instance to assure stability. While of course online motion planning and generation usually allows a better movement, nowadays off line motion generation (with or without online correction) is widely employed, mainly due to the restrictions posed by the limited computing power available on board of the robot. Several approaches were proposed to generate the motion, for instance by splines [1], by truncate Fourier series [2] or by central pattern generators. Nevertheless, the most basic but, surprisingly, the most widespread way of realizing motions is still to specify a motion as a set of "frames", that is a set of time instants for which the position of each joint is provided. The position to be assumed by each motor at each time is usually obtained by a simple linear interpolation of the positions it must assume in the previous and in the following frames. Figure 1 shows a commercial motion editor based on this design principle, while [3] or [4] provide examples of recent papers where keyframe-based interfaces are used. As easily observable in all these examples, this kind of interfaces usually presents one slider for each of the joints, which allows to choose the rotation angle assumed by the servomotor at the frame being edited.

Some improvements were proposed, for instance [5] introduced an interface where the motion is represented by Fourier coefficients calculated from control points provided by the user and [6] introduced a system by which the user develops movements by intuitively pushing the robot parts instead of acting on sliders.

However, these approaches force the user to specify an exact configuration of the joints, usually obtained through a trial and error process. On the other hand many articles, like [7], [8], and [10] present results achieved specifying the motion in a parametric way and letting a genetic algorithms determine good parameter configuration. As a drawback when CPGs are employed the effect of the parameters on the motion becomes often difficult to identify. This means that the user cannot easily modify a resulting motion by directly varying the parameters or impose constraints on the motion. This work aims at merging

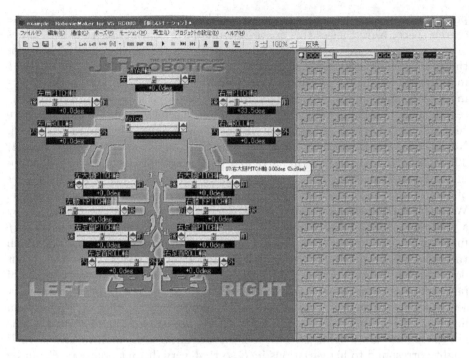

Fig. 1. Robovie Maker, a commercial motion development software developed by VS-tone

the classical, time consuming but easily understandable slider-based approach, with genetic algorithms that automatically search for a good motion. In the developed interface the user can specify a range of possible values instead of a fixed joint position and relationships that should be maintained between the joints. A genetic algorithm tries then to increase the value of a specified fitness function - that describes the quality of the motion - identifying good values that obey the constraints. If users notice bad features in the resulting motion, they can easily modify the constrains of the movement and run the genetic algorithm again. This paper is organized as follows. Section 2 introduces the advantages of the proposed approach and illustrates the ideas underlying the implementation. Section 3 describes how a walking pattern was obtained through the developed method and the performed tests. Section 4 presents the results of the experiment and is followed by a short section commenting these results. Finally,section 6 will summarize the presented concepts and discuss the future work.

2   Main Idea

A motion of a robot with n degrees of freedom can be defined by a function $f: \mathbb{R}^{\geq 0} \to M$, where $M \subseteq \mathbb{R}^n$ i.e. that given a real positive or null number, i.e. time, provides an n dimensional vector that specifies the position of each motor, that is some areas of M are not reachable, as the ones representing self collision positions. Discretization of this space could also be considered to reduce the search space. Nevertheless this space remains huge considering that the values of n is around 20 for small humanoids available on the market. Let us consider the space F of functions describing the movements of the robot, i.e. functions f mapping time to motor positions. More formally, F is the space of functions having $\mathbb{R}$ as domain and M as codomain. In this case too, it is possible to reduce the search space considering domain and codomain discretization. We can also note that the variation between consecutive positions is limited by the motor speed. Nevertheless the dimension of F is, by definition, exponentially higher than M. Therefore search algorithms (in particular local search algorithms like hill climbing, simulated annealing or genetic algorithms) can explore just a very little part of this space. Thus, strong constraints must be introduced in the search space. With classical interfaces the user needs to completely specify the f function by a trial and error process. When search algorithms are employed, instead, the f function is nearly completely defined and it depends on few parameters which values are determined by the search algorithm. In other terms, the search is not done directly in F, but in a p dimensional space P, where p is the number of parameters and $|P| << |F|$. The mapping between the points in P (parameter sets) and the points in F (trajectories in M) can be very complex and highly non linear, as it happens, for instance, when CPGs are employed and P dimensions are the connection weights [8]. This fact has advantages as small variations in P might correspond to big variations in F so that very distant points in F can be sampled. The drawback deriving from this technique is that the meaning of the parameters is difficult to understand so in many cases nearly no predictions can be done regarding consequences of changing the value of one of them.

Conversely the approach adopted in this research is to work directly on M so that the meaning of the parameters is straightforward. In details, the f function is defined as a linear interpolation of key-frames (points in M), that is the movement is given by linear transitions between consecutive motor configurations m_i. To make things even more easily understandable, each position m_i is given by a sum of basic activations of some joints a_j, where the role of each activation a_j is very simple, like "raise the left arm" or "bring the right leg backwards". An activation is then simply a point in M having most of the coordinates equal to zero (the joints that do not need to be moved for the intended purpose) and the few coordinates used for each basic joint activation are defined in terms of constants, parameters or sum of parameters. Of course it would be possible not to introduce the concept of activations a_j, and define directly the frames m_i, but, as will be clear in the experiment section, the introduction of basic activations allows their reuse in several frames and makes the process of defining frames nearly trivial. Thanks to the vector nature of activations and frames, each m_i can also be expressed as a sum of $k_{i,j} * a_j$, where each $k_{i,j}$ is a scalar value. Figure 2 provides an example on what composing "activations" to create "frames" means.

For each parameter appearing in the activations the user specifies the range in which its value should be used. This allows the user to easily constrain the

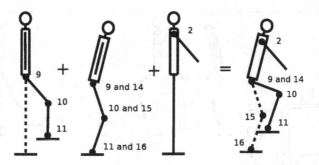

Fig. 2. Three "activations" (i.e. basic joint movements) are composed to create a "frame" (i.e. a posture at a certain time). Note that the distinctions between "activations" and "frames" is introduced just for clarity, since both of them are simple vectors describing the angle of each joint.

movement directly, which is often a big advantage. In fact when search algorithms are employed, if the parameter meaning is unknown the only way to constrain the final movement is to introduce penalizing factors in the score/fitness function. For instance, when developing a walking motion, an obvious fitness function would be the distance covered in a certain amount of time. The result of the algorithm could anyway be a motion where the robot moves fast by sliding its feet without lifting them and without falling. It would be then necessary to introduce a reward for lifting the feet in the fitness function, but deciding how to weight the covered distance and the feet lift is very difficult. With the approach taken in this paper, it is just necessary to choose a proper range for the parameters (actually, joint angles) in the activation a_j that has the role of lifting the foot. Certainly it is possible to provide many other examples regarding situations in which constraining the movement by using the fitness function is difficult. For instance, imagine to define a fitness function for a "human-like" gait. As it should be clear, instead, with the approach taken in this paper the user can employ an easy fitness function, run the genetic algorithm, observe the resulting motion and constrain the activations a_j (or introduce new ones) to correct the motion in the desired way. The genetic algorithm can then be run again and this process can be carried on until the motion determined by the genetic algorithm (and constrained by the user) is satisfactory. It is interesting to note the generality of this method, since no assumptions are made on the robot structure or on the desired movement. Therefore, even if the proposed approach had been applied to a very standard task, walking, we'd like to stress that the method we describe can be used for the realization of any kind of motion. We also feel that reporting a comparison of this method with other techniques specifically devised for the generation of walking motions is out of the scope of this paper. Finally it is worth to state that the proposed method aims at allowing the user to develop in an intuitive way and in short time a variety of motions, without the need to devise any mathematical description of the trajectories. Indeed, the maximization of the performance is not one of the purposes of this work.

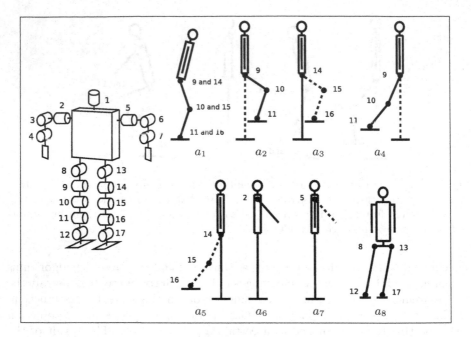

Fig. 3. Activations used to define the walking motion. The figure on the left shows the disposition of the Kondo's KHR-2 degrees of freedom, each one labeled with a number. The same numbers are used in the activation pictures to indicate the joints employed. Dashed lines represent the left part of the body, continuous ones the right side.

3   Realization of a Walking Motion

In our experiments a Kondo KHR-2HV was used. This is a small humanoid robot with 17 degrees of freedom, 34 cm tall and weighting about 1.3 Kg. For time reasons and to avoid hardware damages, preliminary tests were conducted using a simulator, namely USARSim. See [9] for a description of this simulator. A walking motion was developed for this robot using the proposed approach. In detail, first some "activations" a_i, that is basic set of joint modification used to create reusable features of postures were defined. Later, using these "activations", "frames" m_j,

Eight basic activations, also depicted in figure 3, were defined.

More precisely

- a_1 defines the basic posture of the robot during walking, consisting in slightly bending the knees (turning therefore the ankles too) and turning the hip joints so that the trunk is slightly bent forward. This activation defines three parameters, $p_{posture,thigh}$, $p_{posture,knee}$, $p_{posture,ankle}$ corresponding to the three joints mentioned (their values are equal for both legs).
- a_2 and a_3 define, respectively, the action of lifting up the right and the left foot. Two parameters, $p_{lift,thigh}$ and $p_{lift,knee}$ are defined and represent,

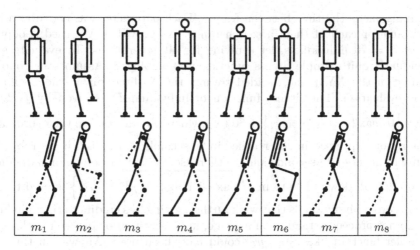

Fig. 4. Frames designed to have a walking motion

respectively, the thigh angle and the knee angle. The ankle is moved so that it maintains parallel to the ground (it is rotated by the opposite of the sum of the knee and thigh angles). For these two activation,as well as for the ones described hereafter the same parameter defines the position of the corresponding joints in the two legs.

- a_4 and a_5 are used to bring backwards, respectively, the right and the left leg, by turning the thigh and the knee. Also in this case, two parameters $p_{back,thigh}$ and $p_{back,knee}$ are defined for these joints while the ankle degree of freedom is used to keep the foot parallel to the ground.
- a_6 and a_7 provide the actions of rotating, respectively, the right and left shoulder joints to swing the arms. Each of these activations has just one parameter, $p_{armrotate,shoulder}$ which defines the shoulder angle.
- a_8 consists in rotating the thigh joints by an angle set by a parameter, $p_{swing,thigh}$, and the ankles by the opposite angle, so the legs keep parallel and the body moves in the frontal plane.

Given these activations, the definition of the frames was straightforward:

$$m_1 = a_1 + a_5 - a8 \qquad m_2 = a_1 + a_3 + \tfrac{1}{2}a_6 - a_8$$
$$m_3 = a_1 + a_4 + a_6 - \tfrac{1}{2}a_7 \quad m_4 = a_1 + a_4 + \tfrac{1}{2}a_6$$
$$m_5 = a_1 + a_4 + a_8 \qquad m_6 = a_1 + a_2 + \tfrac{1}{2}a_7 + a_8$$
$$m_7 = a_1 + a_5 - \tfrac{1}{2}a_6 + a_7 \quad m_8 = a_1 + a_5 + \tfrac{1}{2}a_7$$

For simplicity, in the current implementation the time between any two consecutive frames (m_i and m_{i+1}) is the same and is a parameter determined by the genetic algorithm (in other terms, the speed is a parameter too).

3.1 Genetic Algorithm

As understandable from the previous subsection, the genetic algorithm must determine 10 parameters, 9 of which are appearing in the activations and the remaining one is the time between frames. Each parameter was coded as a real value, and for each of them a search range was provided. We used a population of $S = 20$ individuals, let evolving for 50 generations. To evaluate each individual a quite simple fitness function was employed. Assume the robot is standing in the XY plane, headed toward the X direction at the beginning of the evaluation. The fitness function of individual P_i is given by $f(P_i) = \sqrt{\max\left\{0, \max\left\{0, x_T\right\}^2 - |y_T|\right\}} + 50 * s_T$, where x_T and y_T are the coordinates of the robot, expressed in cm, reached in 20 seconds and s_T is 0 if the robot felt down within these 20 seconds and 1 otherwise, i.e. it is a bonus for individuals that do not fall down. The term $\sqrt{\max\left\{0, \max\left\{0, x_T\right\}^2 - |y_T|\right\}}$ is used to give higher scores to individuals that walk further, giving less credits to the individual that do not proceed straight (on average, in 20 seconds) or proceed backwards. A simpler function like $x_T - |y_T|$ could have been used. Anyway in this case the evolution prefers individuals that just stand at the initial position over individuals that moved far but have $y_T > x_T$. The employed function instead gives an higher fitness to walking patterns that make the robot move forward, unless $y_T > x_T^2$, which could be considered as a side walk. For evolution between successive generations a ranking selection was chosen; in details, the Z individuals of the population are ordered by decreasing fitness. Let us denote by P_i the individual of the population appearing as the i-th in the ranking. The individuals $P_1 \ldots P_k$ with fitness equal to at least half the the fitness of P_1 are then identified (i.e. k, $1 \leq k \leq Z$, is determined) and only these are considered to produce the following generation. Tests were conducted applying both crossover and then mutation. More precisely, when taking individuals at random the probability of taking P_i is given by

$$\wp\left(P_i\right) = \frac{k - i + 1}{\sum_{j=1}^{k} j} = \frac{k - i + 1}{k * (k + 1)/2}$$

Firstly two individuals P_a and P_b were picked up at random, and the resulting individual P_c was obtained taking each gene from P_a with probability $\frac{f(P_a)}{f(P_a) + f(P_b)}$ and from P_b with probability $\frac{f(P_b)}{f(P_a) + f(P_b)}$. Mutation was then applied on P_c, and particularly each gene was modified (with probability $\wp_{mutation} = 1$) multiplying its value by a random number in the $[0.75, 1.25]$ range.

4   Experimental Results

The approach presented in this paper, with the activations and frames reported in the previous section, successfully led to a walking pattern on simulation. During the development of the motion we modified the motion constrains, and in detail the parameter ranges, three times. Initially $p_{lift,thigh}$ and $p_{lift,knee}$ were given a range of $[0, \pi/2]$ but the algorithm always chose very low values for these 2 parameters and the robot just slided the foot, giving a stable but

Table 1. Values of the parameters

Parameter	Range	Best,simulation	Best,real robot
$p_{posture,thigh}$	$\left[\frac{\pi}{2} - \frac{\pi}{16}, \frac{\pi}{2}\right]$	1.388	1.4801
$p_{posture,knee}$	$\left[\frac{\pi}{2} - \frac{\pi}{16}, \frac{\pi}{2}\right]$	1.428	1.4403
$p_{posture,ankle}$	$\left[\frac{\pi}{6} - \frac{\pi}{16}, \frac{\pi}{2} + \frac{\pi}{16}\right]$	0.3316	0.5195
$p_{lift,thigh}$	$\left[\frac{\pi}{8}, \frac{\pi}{3}\right]$	0.3927	0.3826
$p_{lift,knee}$	$\left[\frac{\pi}{8}, \frac{\pi}{4}\right]$	0.7854	0.2826
$p_{back,thigh}$	$\left[\frac{\pi}{8}, \frac{\pi}{4}\right]$	0.7854	0.2489
$p_{back,knee}$	$\left[\frac{\pi}{8}, \frac{\pi}{6}\right]$	0.1992	0.0981
$p_{armrotate,shoulder}$	$\left[\frac{\pi}{16}, \frac{\pi}{4}\right]$	0.6882	0.4961
$p_{swing,thigh}$	$\left[\frac{\pi}{32}, \frac{\pi}{8}\right]$	0.1470	0.2240
$time$	$[0.1, 0.4]$	117.2	140.3

not human like walking pattern. The lower value was then increased to $\pi/8$. Surprisingly after this modification the algorithm tended to select values close to the maximum of the range. The evolution tended also to prefer very fast and small steps over slower and longer ones. To account for this we increased the minimum time between frames from 10ms to 100ms. The maximum value of the lateral swing ($p_{swing,thing}$ parameter), instead, was decreased from $\pi/4$ to $\pi/8$. These modifications of the range show the advantage of the proposed method, i.e. the clear meaning of the parameters allows to constrain and improve the motion very easily. Table 1 reports the ranges of each parameter in the initial population, their values on the best individual obtained in simulation during 50 generations and the values for the best individual in the real world determined with the last 10 steps of evolution on the real robot. Angles are expressed in radians, times in seconds. Note that for all the individuals of the first generation each parameter is set to the medium value of its range. The fitness of the best individual found using the simulator is 181.38, corresponding to an individual proceeding for 131.38 cm in the X direction and 4.03 in the Y direction. On the real robot the distances reduced to 48.9 cm and 4.3 cm in the X and Y direction respectively. This is due to the increase of the inter-frame time and to the reduction of friction between the feet and the ground in the real world compared to the simulated case.

5   Discussion

Observing section 3 it is possible to understand how easy it is to define motions in terms of "activations" and "frames". Employing the method for the development of a walking pattern, a very important movement for the Robocup competition, we could verify that the approach is, as expected, less time consuming (at least for operator time) than directly setting each joint. In fact, while designing a walking pattern by a slider based interface took about eight hours, the design of frames and the adjustments on the range took less than three hours to the same user. The achieved speed is not very high, anyway the walking is very stable, in

Fig. 5. Snapshots of the walking pattern achieved on the real robot. The time between successiva images is 150 ms.

fact the robot walked for over seven minutes without falling. We strongly believe that inserting new frames to refine the motion improvements of the motion could be easily achieved. While these tests are of course too limited to completely prove the efficacy of the presented method, these preliminary results are encouraging and suggest this could be a good direction for further research.

6   Conclusions and Future Work

This work presented a simple approach to develop motions for small sized humanoid robots. Our method is structured as follows: the user provides a set of "activations" that indicate the joints required to execute a basic action. The joint positions can be specified in terms of relationships between the joints and possible ranges. By using weighted sums of these activations the human operator is able to build "frames", i.e. key positions to be assumed by the robot during the movement. Each parameters is bounded within a range also specified by the user. Once the genetic algorithm has detected a good set of parameters (that is, parameters that provide a high value for the fitness function) the user can verify if the motion is satisfactory. If this is not the case, instead of changing the fitness function (an often complex process) s/he can directly edit the ranges to modify the constraints on the motion in a straightforward way. This actually happened three times during a walking pattern used as a test-bed of the approach. Our approach is fast to implement, since it requires nothing more than writing a standard genetic algorithm (or any search algorithm, like policy gradient) and linear interpolation. The main drawback of this method is the time required by the genetic algorithm to calculate a good set of parameters. In the current implementation a single virtual robot is moved in the simulated world in real time. In order to have good evaluations of the fitness each individual is tested three times. The evaluation of each individual takes therefore one minute. The code has a client server architecture that allows to run multiple evaluations simultaneously. In the experimental setup two PCs were used. The time required to evolve for 50 generations was then 8 hours and 20 minutes, since it is possible to run just one instance of the simulator on one PC. Future work will therefore have the purpose of adapting the current system to a non-realtime, lightweight simulator to decrease the computation time.

Acknowledgments

We'd like to acknowledge Luca Iocchi for the fruitful discussions and Alejandro Gatto for his help in the realization of the experiments.

References

1. Ude, A., Atkeson, C., Riley, M.: Planning of joint trajectories for humanoid robots using B-spline wavelets. In: IEEE Conference on Robotics and Automation, San Francisco (2000)
2. Yang, L., Chew, C.M., Poo, A.N.: Adjustable Bipedal Gait Generation using Genetic Algorithm Optimized Fourier Series Formulation. In: Proceedings of the 2006 IEEE/RSJ International Conference on Intelligent Robots and Systems, Beijing (2006)
3. Wama, T., Higuchi, M., Sakamoto, H., Nakatsu, R.: Realization of tai-chi motion using a humanoid robot. In: Rauterberg, M. (ed.) ICEC 2004. LNCS, vol. 3166, pp. 14–19. Springer, Heidelberg (2004)
4. Baltes, J., McCann, S., Anderson, J.: Humanoid Robots: Abarenbou and DaoDan. In: RoboCup - Humanoid League Team Description (2006)
5. Mayer, N.M., Boedecker, J., Masui, K., Ogino, M., Asada, M.: HMDP:A new protocol for motion pattern generation towards behavior abstraction. In: RoboCup Symposium, Atlanta (2007)
6. Dalla Libera, F., Minato, T., Fasel, I., Ishiguro, H., Menegatti, E., Pagello, E.: Teaching by touching: an intuitive method for development of humanoid robot motions. In: IEEE-RAS International Conference on Humanoid Robots (Humanoids 2007), Pittsburg, USA (December 2007)
7. Yamasaki, F., Endo, K., Kitano, H., Asada, M.: Acquisition of Humanoid Walking Motion Using Genetic Algorithm - Considering Characteristics of Servo Modules. In: Proceedings of the 2002 IEEE InternationalConference on Robotics & Automation, Washington, DC (2002)
8. Inada, H., Ishii, K.: Behavior Generation of Bipedal Robot Using Central Pattern Generator(CPG) (1st Report: CPG Parameters Searching Method by Genetic Algorithm). In: Proceedings of the 2003 IEE/RSJ Intl. Conference on Intelligent Robots and Systems, Las Vegas, Nevada (2003)

9. Carpin, S., Lewis, M., Wang, J., Balakirsky, S., Scrapper, C.: USARSim: a robot simulator for research and education. In: Proceedings of the 2007 IEEE International Conference on Robotics and Automation (ICRA 2007) (2007)
10. Hebbel, M., Kosse, R., Nistico, W.: Modeling and Learning Walking Gaits of Biped Robots. In: Proceedings of the Workshop on Humanoid Soccer Robots of the IEEE-RAS International Conference on Humanoid Robots, pp. 40–48 (2006)
11. VStone's official homepage, http://wwsw.vstone.co.jp/
12. Kondo Kagaku's official homepage, http://www.kondo-robot.com/
13. Robonova's official homepage, http://www.hitecrobotics.com/

Landmark-Based Representations for Navigating Holonomic Soccer Robots

Daniel Beck, Alexander Ferrein, and Gerhard Lakemeyer

Knowledge-based Systems Group
Computer Science Department
RWTH Aachen
Aachen, Germany
{dbeck,ferrein,gerhard}@cs.rwth-aachen.de

Abstract. For navigating mobile robots the central problems of path planning and collision avoidance have to be solved. In this paper we propose a method to solve the (local) path planning problem in a reactive fashion given a landmark-based representation of the environment. The perceived obstacles define a point set for a Delaunay tessellation based on which a traversal graph containing possible paths to the target position is constructed. By applying A^* we find a short and safe path through the obstacles. Although the traversal graph is recomputed in every iteration in order to achieve a high degree of reactivity the method guarantees stable paths in a static environment; oscillating behavior known from other local methods is precluded. This method has been successfully implemented on our Middle-size robots.

1 Introduction

One fundamental problem which has to be addressed for a mobile robot is the problem of navigation and path planning while avoiding to collide with obstacles in the environment. In a dynamic environment a navigation scheme with the ability to rapidly incorporate changes in the environment into the navigation process is required to safely and quickly navigate to the given target location. The navigation problem is often divided into two problems, which are solved independently from each other: *path planning* and *collision avoidance*. Whereas collision avoidance is regarded as a local problem, the path planning problem is often examined at a more global scale. In contrast to local methods which solely plan on the basis of the information gained from the current sensor readings, global methods rely on a global map of the environment, say, a floor plan of a building. Usually, those maps cover an area that exceeds the limited perception range of the sensors by far. As a consequence thereof the path can be planned from the start to the target position and not only around the next obstacles as it is the case for the local setting. Due to this fact, local methods often suffer from oscillating behavior. Drawbacks of global methods are that they do not account for avoiding obstacles and need a global map of the environment. Examples of global path planning are methods using probabilistic road maps [1] or dynamic

L. Iocchi et al. (Eds.): RoboCup 2008, LNAI 5399, pp. 25–36, 2009.

programming-based approaches [2]; examples of local path planning (collision avoidance) are curvature-velocity methods [3] or potential fields [4,5].

In this paper we propose a method which combines features from both, local as well as global path planning. The obstacles around the robot as they are perceived by means of its sensors are entered into a local map. As such, our method is local. We employ a geometric, landmark-based representation of the environment as opposed to many collision avoidance algorithms which are density-based and make use of a grid map for the perceived obstacles. Based on a Delaunay tessellation over the obstacles we construct a traversal graph. This traversal graph is a topological representation of all paths through the perceived obstacles to the target position. On this space we apply A^* for finding a short and safe path. The safety of a path is accounted for by incorporating the distances between obstacles passed along the path into A^*'s cost function.

This way we obtain a path from the current position of the robot, around the detected obstacles to the target position, i.e., the search cannot get trapped in certain obstacle configurations—potential field methods, for example, get stuck in local minima. Nevertheless, we reach a degree of reactivity that is comparable to other collision avoidance schemes, which is due to the fact that the traversal graph is re-computed and a new path is searched for after every sensor update. This is feasible since the traversal graph can be efficiently constructed and its size is linear in the number of obstacles. Furthermore, it is shown that the re-computation does not lead to an oscillating behavior but yields stable paths in a static environment. These characteristics make this a very well-suited approach for robotic soccer applications where the number of obstacles is rather small and new obstacle information extracted from camera images usually arrive with 20 – 30 Hz. Since holonomic robot platforms are the de-facto standard in the Middle-size league we describe how a path through the traversal graph as it is found by A^* can be realized with such a robot. This means that, though path planning is involved, we keep up the reactivity of a local collision avoidance method, just like with potential fields. We show that the landmark-based representation is much more suited to scenarios like RoboCup and yield much smaller search problems. We compare the size of the path planning problems for landmark-based and density-based approaches and show that the branching factor as well as the solution length are smaller by an order of magnitude.

This paper is organized as follows. In Sect. 2 we discuss the large body of related approaches to the collision avoidance and path planning problem of mobile robots. In Sect. 3 we show our method in detail. We start with a concise description of the construction of the traversal graph and how A^* is applied on the traversal graph to find a short and safe path. We show the optimality of the calculated path and that it is stable. Then, we discuss in detail how a Bézier curve is constructed based on the previously calculated path, before we show how we combine this approach with a potential field method to avoid nearby obstacles. Sect. 4 shows experimental results of our method. We conclude with Sect. 5.

2   Related Work

The problems of path planning and collision avoidance are essential for mobile robotics and the techniques developed to solve those problems are also of great importance for a multitude of applications of industrial robots, for animating digital actors, and also for studying the problem of protein folding. Therefore, a rich body of work exists in this area. A comprehensive overview of the subject is given in the excellent textbooks [6] and [7].

Since we developed our approach with the robot soccer scenario in mind and implemented it on our Middle-size league robots we concentrate the discussion of related work on the robot soccer domain and, additionally, also some other landmark-based approaches which are related to ours. In the robot soccer domain, most of the methods rely on density-based represenations; the most popular kind of such representations are occupancy grid maps. For example, potential field methods [4] make, in general, use this kind of representation. At sample points, scattered over the environment, a force is computed which is the combination of the repelling forces of the obstacles and the attracting force of the target. The idea is that the resulting force then guides the robot to the target. In [8] an in-depth discussion of the limitations of potential field methods is presented. A major drawback is the problem of getting stuck in local minima. In our approach we compute a path to the target position and, therefore, are not affected by such problems. Nevertheless, potential field-based methods appeal due to their simplicity and high degree of reactivity which is achieved by constantly re-computing the forces. Various extensions to the basic potential field approach have been proposed. For instance, in [9], a collision avoidance method is presented which combines artificial potential fields with a simulated annealing strategy to circumvent the problem of getting stuck in local minima. A similar approach is followed in [10] and [11].

Other grid based methods rely on search alogrithms to find a path to the target as it was presented in [12] for path planning with a humanoid robot on 2.5 D grid and in [13]. In [14] the authors define a cellular automaton on top of the grid representation. Similar to value iteration approaches the cellular automaton computes a lowest cost path to the target. Especially in sparsely populated environments a uniform resolution of the grid cell size, as it is commonly used, provokes unnecessary computations since the path is computed with the same precision regardless of the size of the free space around the robot. A (partial) remedy for this problem was proposed in [13] where the resolution of the grid decreases with the distance to the robot.

Generally, landmark-based approaches are better suited for sparsely populated environments because the length of the path is dependent on the number of obstacles and not on the resolution of the grid. Although, sensor-based variants of those approaches exist they are mostly intended for scenarios where a global map is available. Besides probabilistic roadmap methods, landmark-based and geometrical methods have in commmon that they rely on some kind of cell decomposition to sub-divide the free space into faces. One such example is given in [15]. There, the Voronoi diagram around the obstacles is constructed and a path is defined as a sequence of edges in the Voronoi diagram. The Voronoi diagram is dual to the Delaunay tessellation and as such the approach mentioned above is similar to ours. With our approach, however, a trade-off between the riskiness of a path and its length can be implemented in a natural way which is not possible for Voronoi based approaches. This is because the obstacles are not part of the Voronoi diagram and in order to obtain the distances between the obstacles their position needs to be reconstructed from the given Voronoi diagram.

3   Triangulation-Based Path Planning

We now describe our landmark-based traversal graph in detail. With this representation and the cost function we use for planning, we found a good trade-off between shortness and the safety of the path. We prove that, for every point on the path, the calculated optimal path is stable given that the configuration of obstacles stays unchanged. Further, we show how a suchlike calculated path can be realized on a holonomic robot platform. In order to achieve the greatest possible reactivity, we moreover check if obstacles are very close to the robot and calculate, similar to the potential field methods, an avoidance course.

3.1   The Triangulation and the Traversal Graph

For the rest of the paper we assume that the robot is equipped with sensors that allows it to detect obstacles located in a circular perception field area around the robot using, say, an omnidirectional camera. The perceived obstacles are then represented in one of the following ways:

- Smaller obstacles (e.g., other robots) are represented by a point P denoting the obstacle's center and an associated radius r indicating the equidirectional extent of the obstacle. We write $\| P \| = r$ to denote the extension of obstacle P to be r.
- Obstacles which do not have an equi-directional extent (e.g., walls) are represented by a set of points and edges connecting the perceived contour points. Such points have no associated extent and the edges are explicitly marked as *not passable*. Generally, only the front-side contour can be perceived since no information about the extent of the obstacles on its rear side is available.

We construct a Delaunay tessellation for all such points which are perceived as obstacles. (See Fig. 1(a) for an example.)

Definition 1 (Delaunay Tessellation). *The Delaunay tessellation $DT(S)$ of a set S of points in the plane is a graph with the vertices $V_{DT(S)}$ and the edges $E_{DT(S)}$. It is obtained by connecting any two points $p, q \in S$ with a straight line segment such that a circle C exists which passes through p and q but no other point of S lies within the circle C.*

A face $f^{(i)}$ of the tessellation is defined by the three vertices $v_0^{(i)}, v_1^{(i)}, v_2^{(i)}$. The edge $e_j^{(i)}$ connects the vertices $v_{j-1}^{(i)}$ and $v_{j+1}^{(i)}$ (indices mod 3). A concise overview of Delaunay tessellations can be found in [16]. Due to the circumcircle criterion the triangulation as it is given by $DT(S)$ ensures that if the robot is located in the triangle $f^{(i)}$ the obstacles represented by $v_0^{(i)}$, $v_1^{(i)}$, $v_2^{(i)}$ are the next obstacles surrounding the robot. This implies that whenever the robot passes through any two obstacles it also crosses the edge in the triangulation $DT(S)$ between the vertices representing these two obstacles. Clearly, the safest way to pass through two obstacles is to drive right through the middle between them. This is why we choose the midpoints of the triangulation edges as way points. Then, the edges between the midpoints of a particular triangle represent, in a topological way, all possibilities to traverse the triangle in question. With this in mind, the problem of path planning now can be defined as finding a short and safe path through $DT(S)$ along those midpoints leading from the current position of the robot to the given target position. We thus construct a traversal graph $T(S)$ based on Delaunay tessellation $DT(S)$ over the set of points S representing the obstacles.

Algorithm 1 (Traversal Graph). *Let S be a set of points in the plane and $DT(S) = \langle V_{DT(S)}, E_{DT(S)} \rangle$ a Delaunay tessellation of S (as shown in Fig. 1(a)) with $F = \{f^{(0)}, \dots, f^{(m)}\}$ the set of faces of $DT(S)$ and $v_0^{(i)}$, $v_1^{(i)}$, $v_2^{(i)}$ the vertices of a face $f^{(i)}$ and $e_0^{(i)}$, $e_1^{(i)}$, and $e_2^{(i)}$ its edges as above. Let R, T be points in the plane denoting the robot's position and the target position, respectively. Let $T(S) = \langle V_{T(S)}, E_{T(S)} \rangle$ be the traversal graph of $DT(S)$. $V_{T(S)}$ and $E_{T(S)}$ are constructed as follows:*

1. *Add the midpoints of all edges in $E_{DT(S)}$ to $V_{T(S)}$. For all triangles $f^{(i)}$ in $DT(S)$ that neither contain R nor T add the edges connecting the midpoints of $f^{(i)}$'s edges to $E_{T(S)}$. In case, R or T are located inside $f^{(i)}$ add further edges between the respective point and the midpoints of the circumjacent triangulation edges $E_{T(S)}$. (See Fig. 1(b).)*
2. *If neither R nor T are enclosed by $DT(S)$, we extend it in the following way:*
   (a) *determine the set of outer edges of $DT(S)$, i.e., the edges in $E_{DT(S)}$ that are not shared by any two adjacent triangles;*
   (b) *consider all outer edges (cf. arrows in Fig. 1(b)) for which no direct connection between the respective midpoint and R or T exists. (A direct connection exists if the straight line segment between the midpoint and R or T does not intersect with another outer edge);*
   (c) *for each such outer edge e define line segments s_{v_1}, s_{v_2} starting in v_1, v_2, resp., which are incident to e. The s_{v_i} are angle bisectors of the outer angles between e and the neighboring outer edges and are of length $|s_{v_i}| = 2 \cdot (\|v_i\| + \|R\|)$. Construct three further line segments h_{v_1}, h_{v_2}, and h_m, all perpendicular to e. The h_{v_1}, h_{v_2} start in v_1, v_2, resp. ; h_m starts in the midpoint m of edge e; $|h_{v_i}| = |s_{v_i}|$ and $|h_m| = max\{|h_{v_i}|\}$. Fig. 1(c) illustrates this step;*
   (d) *let S' be S extended by the endpoints of these line segments. Re-establish $DT(S')$ and $T(S')$. Fig. 1(d) shows the resulting graph.*

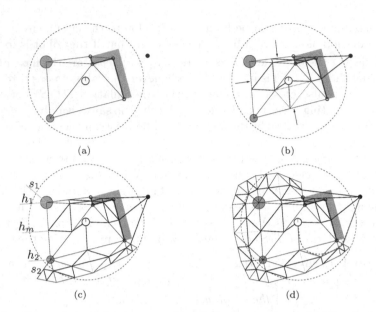

Fig. 1. An illustration of the construction of the traversal graph. The dashed circle around the robot depicts the perception border of the robot's sensors; the black dot on the right side is the target. The contour of non-circular obstacles is modeled by subdividing it into linear pieces which are represented by edges marked as not passable (cf. L-shaped obstacle to the right).

Every edge in $E_{T(S)}$ describes one possibility to traverse a triangle in $DT(S)$ by moving from the midpoint between two obstacles to the safest point to pass through the next two obstacles on the path. In order to also take into account paths that leave the area which is covered by the triangulation $DT(S)$ over an edge from which no direct path to the target exists, a construction as it is given in step 2 of the above construction algorithm is required. This construction guarantees that, topologically speaking, all possible paths through and also around the perceived obstacles can be represented. The traversal graph can be efficiently established. Note that the Delaunay tessellation can be established in $\mathcal{O}(n \log n)$ with $n = |S|$ (cf. [16]). The number of faces of $DT(S)$ and, consequently, also the size of $V_{T(S)}$ and $E_{T(S)}$ are in $\mathcal{O}(n)$. It can be shown that the whole algorithm runs in $\mathcal{O}(n)$.

3.2 Searching for a Short and Safe Path

The task is now to find a path between the robot R and the target T in the traversal graph $T(S)$. To find such a path, we use A^* for two reasons: (1) it is known to be optimal efficient with an admissible heuristic, and (2) even though the worst-case complexity is $\mathcal{O}(b^d)$, where b denotes the branching factor and d the depth of the search tree, we have in our case a branching factor of at most 2. To apply A^* we have to define an admissible heuristic and a cost function.

With the cost function we can influence the optimality criteria and furthermore prohibit edges with too few clearance.

Definition 2 (Heuristic and Cost Function). *For a Delaunay tessellation* $DT(S) = \langle V_{DT(S)}, E_{DT(S)} \rangle$ *and its traversal graph* $T(S) = \langle V_{T(S)}, E_{T(S)} \rangle$ *for a set* S *of points in the plane,* R *the robot's position, and* T *the position of the target point, define a heuristic function* $h : V_{T(S)} \to \mathbb{R}$ *as* $h(p) = dist(p, T)$ *as the straight line Euclidean distance between a vertex* p *in* $T(S)$ *and the target position* T. *The cost function* $g : V_{T(S)} \times \cdots \times V_{T(S)} \to \mathbb{R}$ *of a path* $p_0, \ldots, p_n$ *is defined as:*

$$g(p_0, \ldots, p_n) = \sum_{i=0}^{n-1} dist(p_i, p_{i+1}) + \sum_{j=1}^{n-1} clear(p_j)$$

with $clear(p_i) = \alpha \cdot ||e_i||^{-1} + \pi$, *where* $||e_i||$ *denotes the length of the triangulation edge* $e_i \in E_{DT(S)}$ *on which* $p_i \in V_{T(S)}$ *is located, and* π *is a term punishing edges which are too short for the robot to cross.* π *is defined as*

$$\pi = \begin{cases} \infty, & \text{if } ||e_i|| < ||R|| + ||v_{i-1}|| + ||v_{i+1}|| \\ & \text{or if } e_i \text{ is marked as not passable} \\ 0, & \text{otherwise} \end{cases}$$

with $||R||$ *denoting the robot's diameter,* $||v_{i-1}||$ *and* $||v_{i+1}||$ *the extent of the obstacles* $v_{i-1}, v_{i+1} \in V_{DT(S)}$ *incident to* e_i.

The result of the search will be a short and safe path through the graph that leads form the current position of the robot to the given target position.

Theorem 1. *Let* S *be a set of points in the plane, let* $DT(S) = \langle V_{DT(S)}, E_{DT(S)} \rangle$ *be a Delaunay tessellation of* S *with* $T(S) = \langle V_{T(S)}, E_{T(S)} \rangle$ *its traversal graph as constructed by Alg. 1. Let* $P^* = \langle p_0, \ldots, p_n \rangle$ *be the optimal path from* p_0 *to* p_n *in* $T(S)$ *according to the cost function* $f = h + g$ *as given in Def. 2, and let* R *be a point on the edge* $\langle p_i, p_{i+1} \rangle$ *of* P^*. *Let* $\tilde{P} = \langle p_{i+1}, \ldots, p_n \rangle \subset P^*$ *and* $P = \langle R, \tilde{P} \rangle$ *be a path starting in* R *and following the optimal way points thereafter. Then, there exists no path starting in* R *with lower costs than* P.

Proof. Assume the opposite, i.e., there exists a path $\tilde{P}' = \langle p'_{i+1}, \ldots, p'_m \rangle$ with $p'_m = p_n$ as depicted in Fig. 2 such that $P' = \langle R, \tilde{P}' \rangle$ is the optimal path from R to the target position, but $P' \not\subset P^*$. Since, for all points on the edges of the path $\langle p_0, \ldots, p_i \rangle$, a path continuing with $\langle p_{i+1}, \ldots, p_n \rangle$ after p_i has been chosen instead of one continuing with $\langle p'_{i+1}, \ldots, p'_m \rangle$, it must hold that: $g(p_i, \ldots, p_n) < g(p_i, p'_{i+1}, \ldots, p'_m)$. Under the assumption that the optimal path starting in R follows the way points $\langle p'_{i+1}, \ldots p'_m \rangle$, it must hold that $g(p_i, p'_{i+1}, \ldots, p'_m) - (d' - d'_\epsilon) < g(p_i, \ldots, p_n) - \epsilon$ because otherwise $\langle p'_{i+1}, \ldots, p'_m \rangle$ would not have been selected as the optimal path from R to the target. Since it holds for every triangle that each side of the triangle is longer than the absolute difference of the lengths of the other two sides, $d' - d'_\epsilon$ has to be less than ϵ. This is in contradiction to the initial assumption, hence the claim follows. $\square$

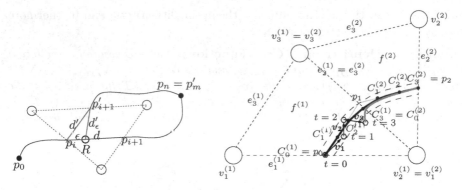

Fig. 2. Sketch of the proof of The-   **Fig. 3.** Bézier curve of the path
orem 1

Corollary 1. *The cost function of Def. 2 yields stable paths in* $T(S)$.

The corollary states that the decision which path to take does not oscillate for any point on the optimal path in the traversal graph $T(S)$. This makes the traversal graph a well-suited representation for the reactive path planning problem.

3.3   Realizing a Path on a Robot

After having found a path in $T(S)$, we have to realize this path on the robot. Let $P = \langle p_0, p_1, \ldots, p_{n-1}, p_n \rangle$ where $p_0 = R$ denotes the robot's location, the p_i denote the points on the path, and $p_n = T$ the target position. As the path given through the path points is very angled, we aim at smoothing it. To this end, we approximate the path by a set of cubic Bézier patches. The parametrized representation of a Bézier curve of degree n is given by $C(\tau) = \sum_{i=0}^{n} \binom{n}{i} (1 - \tau)^{n-i} \tau^i C_i$, where $0 \leq \tau \leq 1$ is the curve parameter, and the C_i are the control points of the curve. For our path approximation we make use of cubic Bézier patches with $n = 3$. Fig. 3 shows an example. The path through the faces f_1 and f_2 is given by the points p_0, p_1, and p_2 from which the control points $C_j^{(i)}$, $0 \leq j \leq 3$, $1 \leq i \leq 2$, of the two Bézier patches, can be constructed. The construction ensures that the curve consisting of several Bézier patches is C^2 continuous at the junctures. This property is very helpful for planning the robot's velocity. A detailed overview can be found, for example, in [17].

Now, for demonstrating how the robot drives along this curve, assume that, at time $t = 0$ the robot, located in $C_0^{(1)}$, is given the path through the points p_0, p_1, p_2. With these path points it derives the black curve given by the control points $C_j^{(1)}$. The robot now derives its next driving command by adding the difference quotient of the curve point at the next time step to its own movement vector. At time $t = 1$ the robot advanced to the depicted position. Its movement is described by its movement vector v_1, depicted in the figure. At this time step the robots checks if it diverted too much from the curve calculated at $t = 0$ due to slippage or other imprecisions in executing the previous motion command. The dashed lines left and right to the curve depict how much the robot might divert from the curve. To construct the black curve at $t = 1$, we must derive the first control point $C_0^{(1)}$ from where the robot started, as the Bézier curve is uniquely defined by its control points. This can be done by subtracting movement vector v_1 from the robot's current position. At $t = 2$, we again check whether our actual position is near the original curve, which was calculated based on p_0, p_1, p_2. Note that in order to regain $C_0^{(1)}$, we project back the vector sum $v_1 + v_2$ to time $t = 0$. At $t = 3$ the robot's location diverted too much from the curve in our example. Consequently, we have to compute a new curve (depicted with the dark gray curve in Fig. 3) to reach path point p_2 in face f_2.

Generally, it is only necessary to project the robot's position back to the last juncture since only this point is necessary to compute the current Bézier patch, i.e., after passing $C_3^{(1)}$ we do not have to go back to $p_0 = C_0^{(1)}$.

3.4   The Fallback: Collision Avoidance with Nearby Obstacles

Although the aspect of the path's safeness was addressed during the search, it might nevertheless happen, under certain circumstances, that the robot gets too close to one of the obstacles. This might be due to the high dynamics of the environment (an obstacles approaches the robot very fast) or motion errors due to, say, slippage such that the robot ends up at another position than the intended one. To be on the safe side, we avoid bumping into the obstacle by using an additional reactive collision avoidance scheme which borrows fundamental ideas from the potential field method. If the robot's distance to an obstacle is less than a specified safety clearance, we compute a repelling force vector keeping the robot away from the obstacle and an attracting force vector guiding the robot to the next way point on the optimal path. These force vectors are computed in the same fashion as the force vectors in the potential field method. The intention behind this is to steer towards the next way point while increasing the distance between the robot and the obstacle in order to continue on the previously computed path. In contrast to a pure potential field approach not all detected obstacles have to be taken into account but only the closest. Furthermore, the drawback of potential fields, i.e., getting stuck in local minima, cannot occur here since it is guaranteed that no other obstacle is located in the area between the obstacle and the next way point. Thus, we do not run into the risk of oscillating motion behaviors.

4   Evaluation

In comparison to grid-based approaches a considerable speed-up in the search for an optimal path is achieved with our triangulation-based method. The reason for this lies, in general, in the lower number of triangles compared to the number of cells, which have to be searched. Moreover, the number of triangles depends on the number of obstacles and their relative positions to each other in contrast

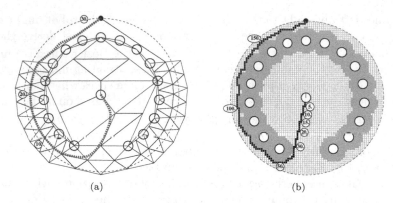

(a)   (b)

Fig. 4. Comparison of the traversal graph with a grid-based solution. Note: very small triangles are omitted in Fig. 4(a); length of path is correct, nevertheless.

to a non-adaptive sub-division of the area in cells with constant size.[1] Finally, the triangulation guides the search into the right direction in a natural way, i.e., it is not possible to stray too far from the optimal path.

In the following we give an example underlining this intuitive considerations. The example is a kind of a worst-case example which, nevertheless, is not unlikely to happen in reality. The situation is depicted in Fig. 4. The robot is surrounded by a number of obstacles arranged in an U-shaped fashion. The target is on the opposite side of the opening of the U such that the robot, at first, has to move away from the target to get out of the U-shape, and then needs to navigate around the obstacles to reach the target, eventually. For the following comparison we assume that perception radius is 4 m; the obstacles have a distance of 3 m to the robot's initial position; the robot has to keep a security distance of 50 cm to the obstacles; the cell size is 10×10 cm^2.

As can be seen in Fig. 4(a) the path from the initial position to the target traverses 30 triangles. The branching factor of the search tree is two (possible successors are the two other midpoints of the edges of the triangle the robot is entering) and consequently the size of the search tree is in $\mathcal{O}(2^{30})$. For the grid-based approach the search depth depends on the length of the path. The path leads around the obstacles which occupy the gray shaded cells. Note, the the security distance is added by the common method of obstacle growing, as depicted in Fig. 4(b). We now can estimate that the path, as it is depicted in black in the Fig. 4(b), traverses roughly 180 cells. Even if we assume that the branching factor is only four, i.e., no diagonal movements are allowed, the size of the search tree is in $\mathcal{O}(4^{180})$. Of course, these measures are both upper bounds which are hardly reached in practice—the effective branching factor of the search tree as it is generated by A* is in both cases smaller. Nevertheless, for the grid-based representation, nearly the whole inner of the U has to be searched before

---

[1] As described in Sect. 2, there also exist adaptive methods. Nevertheless, we compare our representation to non-adaptive grids as even an adaptive grid size does not remedy the basic problem of much longer paths.

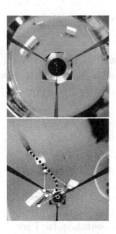
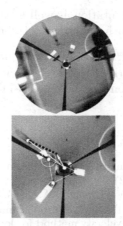

Fig. 5. (Top row from left to right): Omni-vision image of the scene, white boxes are the obstacles; undistorted image; the remaining images show a detailed view of the triangulation over the perceived obstacles, R denotes the robot, T the target. For the sake of clarity only the nearest obstacles are marked.

cells outside are taken into account. This shows that the search space is much larger than with our representation. Also, the length of the path gives a good impression of this fact.

Fig. 5 shows a real-world example. The top left image shows an omni-vision view of the scenario; the same view, though undistored this time, is shown in the middle. In the remaining images details of the undistorted views while navigating around the obstacles are depicted. The middle image in the bottom row pictures the nearby collision avoidance as it is mentioned in Sect. 3.4.

5   Conclusion

In this paper we presented a solution to the problem of reactive path planning. We introduce the traversal graph, a compact, landmark-based representation for sparsely populated environments like the robotic soccer domain. The traversal graph defines possible traversals of faces in a Delaunay tessellation leading through the midpoints of the Delaunay edges between two obstacles. In this graph we search for a path deploying A^* with a heuristics and a cost function which yields a good trade-off between shortness and safety of the path. This path can be proved to be stable, that is, the optimal path does not change for any point on the path, under the assumption that the environment is static. This is a major improvement over local path planning methods like potential fields which cannot guarantee non-oscillating behaviors. Finally, we compare our method with a standard grid-based local path planning approaches and show that the solution can be computed much faster as the search problem is much smaller. This allows us to search for a new path each time, a new sensor update arrives.

The compactness of the representation as well as the stable path property make it possible to apply reactive path planning to the robot.

Acknowlegments. This work was partially supported by the German Science Foundation (DFG) in the Priority Program 1125 and by the Bonn-Aachen International Center for Information Technology (B-IT). We thank the anonymous reviewers for their comments.

References

1. Canny, J.: Computing roadmaps of general semi-algebraic sets. The Computer Journal 36(5), 504–514 (1993)
2. Buhmann, J.M., Burgard, W., Cremers, A.B., Fox, D., Hofmann, T., Schneider, F.E., Strikos, J., Thrun, S.: The mobile robot rhino. AI Mag. 16(2), 31–38 (1995)
3. Simmons, R.: The curvature-velocity method for local obstacle avoidance. In: Proc. ICRA 1996. IEEE Computer Society Press, Los Alamitos (1996)
4. Khatib, O.: Real-time obstacle avoidance for manipulators and mobile robots. International Journal on Robotics Research 5(1), 90–98 (1986)
5. Borenstein, J., Koren, Y.: The vector field histogram - fast obstacle avoidance for mobile robots. IEEE Trans. on Robotics and Automation 3(7), 278–288 (1991)
6. LaValle, S.: Planning Algorithms. Cambridge University Press, Cambridge (2006)
7. Choset, H., Lynch, K.M., Hutchinson, S., Kantor, G.A., Burgard, W., Kavraki, L.E., Thrun, S.: Principles of Robot Motion: Theory, Algorithms, and Implementations. MIT Press, Cambridge (2005)
8. Koren, Y., Borenstein, J.: Potential field methods and their inherent limitations for mobile robot navigation. In: Proc. ICRA 1991, pp. 1398–1404 (1991)
9. Park, M.G., Jeon, J.H., Lee, M.C.: Obstacle avoidance for mobile robots using artificial potential field approach with simulated annealing. In: Proc. ISIE 2001, vol. 3, pp. 1530–1535 (2001)
10. Zhang, P.Y., Lü, T.S., Song, L.B.: Soccer robot path planning based on the artificial potential field approach with simulated annealing. Robotica 22(5), 563–566 (2004)
11. Zhu, Q., Yan, Y., Xing, Z.: Robot path planning based on artificial potential field approach with simulated annealing. In: Proc. ISDA 2006, pp. 622–627 (2006)
12. Gutman, J., Fukuchi, M., Fujita, M.: Real-time path planning for humanoid robot navigation. In: Proc. IJCAI 2005, pp. 1232–1238 (2005)
13. Behnke, S.: Local multiresolution path planning. In: Polani, D., Browning, B., Bonarini, A., Yoshida, K. (eds.) RoboCup 2003. LNCS, vol. 3020, pp. 332–343. Springer, Heidelberg (2004)
14. Behring, C., Bracho, M., Castro, M., Moreno, J.A.: An algorithm for robot path planning with cellular automata. In: Proc. ACRI 2000, pp. 11–19. Springer, Heidelberg (2000)
15. Sahraei, A., Manzuri, M.T., Razvan, M.R., Tajfard, M., Khoshbakht, S.: Real-Time Trajectory Generation for Mobile Robots. In: Basili, R., Pazienza, M.T. (eds.) AI*IA 2007. LNCS (LNAI), vol. 4733, pp. 459–470. Springer, Heidelberg (2007)
16. Aurenhammer, F., Klein, R.: Voronoi diagrams. In: Sack, J.R., Urrutia, J. (eds.) Handbook of Computational Geometry. Elsevier Science Publishers, Amsterdam (2000)
17. Foley, J.D., Phillips, R.L., Hughes, J.F., van Dam, A., Feiner, S.K.: Introduction to Computer Graphics. Addison-Wesley Longman Publishing Co., Inc., Amsterdam (1994)

Mutual Localization in a Team of Autonomous Robots Using Acoustic Robot Detection

David Becker and Max Risler

Technische Universität Darmstadt
Simulation, Systems Optimization, and Robotics Group
Hochschulstr. 10, 64289 Darmstadt, Germany
{becker,risler}@sim.informatik.tu-darmstadt.de

Abstract. In order to improve self-localization accuracy we are exploring ways of mutual localization in a team of autonomous robots. Detecting team mates visually usually leads to inaccurate bearings and only rough distance estimates. Also, visually *identifying* teammates is not possible. Therefore we are investigating methods of gaining relative position information *acoustically* in a team of robots.

The technique introduced in this paper is a variant of code-multiplexed communication (CDMA, code division multiple access). In a CDMA system, several receivers and senders can communicate at the same time, using the same carrier frequency. Well-known examples of CDMA systems include wireless computer networks and the Global Positioning System, GPS. While these systems use electro-magnetic waves, we will try to adopt the CDMA principle towards using *acoustic* pattern recognition, enabling robots to calculate distances and bearings to each other.

First, we explain the general idea of cross-correlation functions and appropriate signal pattern generation. We will further explain the importance of synchronized clocks and discuss the problems arising from clock drifts.

Finally, we describe an implementation using the Aibo ERS-7 as platform and briefly state basic results, including measurement accuracy and a runtime estimate. We will briefly discuss acoustic localization in the specific scenario of a RoboCup soccer game.

1 Introduction

Knowing exact relative positions of team mates in a team of autonomous robots can be a valuable information. For example, self-localization accuracy can be improved by including the team mates position beliefs, together with according measurements.

A typical situation in a RoboCup soccer game is a player, who tries to grab and shoot the ball while being attacked by opponent players. Usually, this leads to a bad localization accuracy, because the player is unknowingly being pushed without seeing any landmarks. A well localized team mate can support this player by localizing him and sending him a position belief via wireless communication.

L. Iocchi et al. (Eds.): RoboCup 2008, LNAI 5399, pp. 37–48, 2009.

Other applications of knowing relative team mate positions are several issues of tactical game play behavior, and special applications like passing.

Hereby we present an approach of measuring distances and bearings acoustically, using a speaker and a pair of microphones. Besides, the techniques described are platform independent. Measuring distances requires each robot to have a speaker and a microphone installed. Measuring bearings additionally requires a second microphone installed.

Measuring distances is done by measuring the time the sound travels from robot "A" to robot "B". If the sending of signals follows an exact timetable, both A and B know the point in time when A sends. (This usually requires the robots to have synchronized clocks, see section 3.2.) The main challenge for B is now to detect the *exact* point in time when A's signal reaches his microphone, even under noisy conditions. This signal detection technique will be the main topic of this paper. Our approach will allow several robots of a team to send their signals continuously and simultaneously.

While the calculation of bearings is inspired by the human sense of hearing, humans usually are unable to measure distances acoustically. However, there exist everyday life examples, like estimating the distance to a thunderstorm by counting the seconds between lightning and thunder.

Note, that the frequencies used in our approach are *human audible*, in a range between 500Hz and 1000Hz. This might be a downside of the approach when used in a quiet surrounding. The acoustic schemes might sound unpleasant for the human ear. However, the sound volume can be adapted to the level of noise in the environment.

2   State of Research

The *transmission of information* using acoustic signals has been investigated and also successfully used in a RoboCup competition (see [1] and [2]). Gaining *position information* acoustically in a team of robots is, however, a quite unexplored field of research.

We hereby discuss a wireless CDMA communication system based on acoustic signals. The approach mainly deals with pattern recognition using cross correlation functions. A lot of research has already been done regarding wireless CDMA communication (see [3]) and cross correlation functions in general. Well-known applications that use this technique include wireless network standards like WiFi, and also the Global Positioning System (GPS) [4]. Note, that all of these applications are based on electro-magnetic waves, while we aim to use acoustic waves.

There has been some research about gaining position information acoustically. See [5] for an approach for measuring distances between a single sender and a receiver using cross correlation. This is very similar to our approach, while we try to set up such a system for teams of autonomous robots.

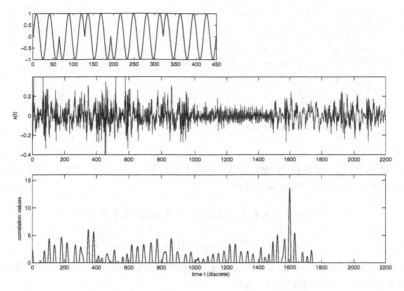

Fig. 1. Top: An example signal pattern p, Middle: A noisy input signal $x(t)$ containing a single pattern p, Bottom: The discrete correlation function $\rho_{x,p}(t)$ (positive range only), which correctly "reveals" the position of the signal pattern: $t_p = 1600$

2.1 Cross-Correlation-Functions

In the following the process of detecting a signal using cross correlation is described: A specific signal pattern is assigned to each sender, and all receivers know these patterns. After a sender "A" has sent his signal pattern, a receiver "B" calculates the cross correlation function of his input signal and A's specific signal pattern. The cross correlation function $\rho(t)$ of the discrete input signal $x(t)$ and A's discrete signal pattern $p(\tau)$ of length n is a similarity measure, given as

$$\rho_{x,p}(t) = \sum_{i=0}^{n} x(t+i) \cdot p(i).$$

B now assumes, that A's signal arrived at time $t_A = \text{argmax}_t\, \rho_{x,p}(t)$. Thus, the highest peak, reflecting the position of highest similarity, is chosen. To achieve reliable measurements, this peak should be clearly higher than any other correlation value. Fig 1 gives a simple example.

2.2 Code Generation

The selection of appropriate signal patterns is a crucial issue. Since several senders will send out their signal simultaneously, we have to make sure, that the receivers will not confuse the different patterns.

A signal pattern follows a binary code sequence, usually written as a sequence of ones and zeros. The elements of a binary code sequence are called "chips"

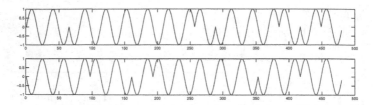

Fig. 2. Signal patterns of k_1 (top) and k_2 (bottom)

instead of "bits", since they are not part of any transmitted data. We give two 15-chip example code sequences here:

$$k_1 = (0, 0, 1, 1, 1, 1, 1, 1, 0, 1, 1, 1, 0, 1, 0),$$

$$k_2 = (1, 1, 1, 0, 0, 1, 0, 0, 0, 0, 0, 1, 1, 0, 0).$$

Fig 2 shows their corresponding signal patterns. As you can see, chips are modulated here using phase jumps.

Kasami [6] and Gold [7] gave algorithms to create sets of binary sequences with the following properties:

1. Every sequence of the set correlates badly with *any shifted copy* of itself.
2. Any two sequences of the set correlate badly with any shifted or non-shifted copy of each other.

"Correlating badly" means, that absolute correlation values are low. The process of correlating a function with itself is called "auto-correlation". Of course, the auto-correlation value of the *unshifted* sequence is maximal, reflecting maximum similarity, i.e., identity.

The two example code sequences, k_1 and k_2, are part of the same Kasami set of sequences. Their corresponding signal patterns are denoted with p_{k_1} and

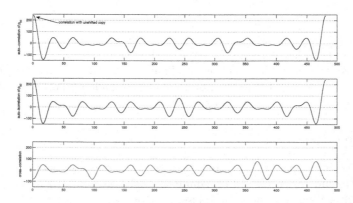

Fig. 3. Auto-correlation functions of p_{k_1} (top) and p_{k_2} (middle), and their cross-correlation function (bottom)

p_{k_2}. Fig 3 shows their auto-correlation functions and their cross-correlation function. As you can see, the two signal patterns indeed have good auto- and cross-correlation properties: The unshifted auto-correlation peak is at approx. 240, while all other correlation values are < 100. Longer code sequences yield even better correlation properties.

3 Acoustic Robot Detection

In this section we adopt the general strategies of cross-correlation functions and code generation towards using acoustic robot detection in a team of autonomous robots.

3.1 Codes and Modulation

Kasami [6] code sequences have very good auto- and cross-correlation properties, reaching the Welch Lower Bound [8]. A Kasami set consists of $2^{\frac{n}{2}}$ sequences, with n even. Each sequence has length $2^n - 1$. For $n = 4$, a set consists of four 15-chip-sequences. (k_1 and k_2, as introduced in Section 2.2, are part of a Kasami set of $n = 4$.) $n = 6$ yields a set of 8 sequences, with each of length 63. If more codes are required, one could either make codes longer or use Gold codes instead, which have slightly worse correlation properties while enabling much larger sequence sets.

Next, the process of *modulation* is discussed. The previous section already showed how chips can be modulated using phase jumps. This approach yields optimal results: The correlation value of different chips in two signals is the negative of the correlation value of two identical chips. This works well for electromagnetic waves. However, phase jumps in acoustic waves are difficult to produce accurately with an ordinary loudspeaker. Also, an acoustic waveform, when propagating through the air, will flatten in the regions of phase jumps. (Recall, that acoustic waves are pressure gradients.)

A second approach is to use frequency changes instead of phase jumps. Therefore, we use different frequencies for 0-chips and 1-chips. Obviously, 0-chips and 1-chips must have equal signal lengths, so using a higher frequency requires more sine cycles. The downside of this approach is, that the correlation value of two different chips is now zero, causing a bad overall correlation accuracy.

For a third approach, we return to the idea of phase jumps. If we use more than one sine cycle per chip, the effect of flattening around the phase jump declines. Instead of forcing explicit phase jumps, the waveform can be smoothed out by using short pauses of e.g. a half sine cycle. This approach almost yields the accuracy of the first approach.

The three different modulations of the sequence $k_3 = (1, 0, 1, 1, 0, 0, 0)$ are shown in Fig. 4.

When measuring distances while sending signal patterns continuously, the *length* of the code sequence is of importance. The measured distance is ambiguous since we use finite code lengths. To achieve definite measures, we have to make

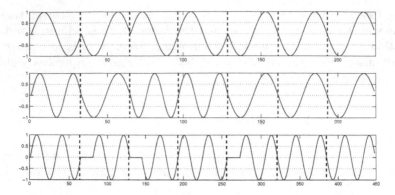

Fig. 4. Three approaches of code modulation: Using phase jumps (top), using frequency jumps (middle), using phase jumps with pauses (bottom)

sure that a signal pattern, when propagating as an electro-magnetic or acoustic waveform, is longer than the maximal possible distance between any receiver and sender. For example, assume the distance between two robots never exceeds 20 meters. Using acoustic waves, the minimum pattern length is

$$\chi_{min} = \frac{d_{max}}{c} = \frac{20m}{343\frac{m}{s}} \cong 58ms.$$

3.2   Clock Drift, Clock Synchronization and Distance Calculation

To measure distances, the robots clocks must run synchronized with high accuracy. For example, a clock error of $0.03ms$ equals a distance error of approx. $1cm$.

The clocks usually installed in PCs or mobile robots, respectively, are quartz clocks with very limited accuracy in terms of distance measuring. For example, a clock drift of one minute per month equals a drift of approx. $28ms$ per 20 minutes, which is a measured distance error of more than $9m$.

Hence, a single initial synchronization of the robots' clocks is insufficient. The clocks have to be re-synchronized repeatedly.

Fortunately, there exists an elegant way of synchronizing a pair of robots permanently. Because every sender in our scenario is also a receiver, we always have *two measures* of the *same* actual distance. As a result of the clock drift, these two measures will diverge. Taking the arithmetic average of the two measured distances gives us the *actual* distance.

Fig. 5 gives an example of a mutual distance calculation and synchronization between two robots R1 and R2. Initially, the two robots clocks run asynchronously. Each of the robots was switched on at local time 0. Each robot can determine the time when he sends or receives a signal (t_{S1}, t_{R1}, t_{S2}, t_{R2}). Assume, it takes time t_d for the acoustic waveform to travel from R1 to R2. t_d can

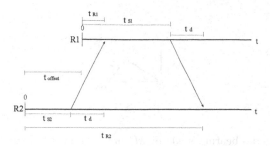

Fig. 5. Time view of the mutual localization and synchronization process

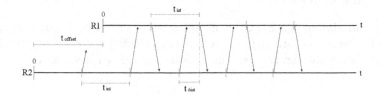

Fig. 6. Time overview: the green lines represent time interval edges

be determined as follows: R1 sends $(t_{R1} - t_{S1})$ to R2 via wireless LAN. Then R2 can calculate t_d using the equation

$$t_d = \frac{1}{2}[(t_{R2} - t_{S2}) + (t_{R1} - t_{S1})].$$

(R1 calculates t_d equivalently.)

R2 now also knows the local time when R1 sent out the signal. Since the sending of signals is done at regular time intervals t_{int}, R2 also knows the next sending times of R1 in *local time*:

$$t_{S1,i} = t_{R2} - t_d + i \cdot t_{int}, \text{ with } i = 0, 1, 2, ...$$

Therefore, R2 may calculate the distance to R1 only by having t_{R2} in the upcoming intervals.

While this synchronization procedure does not yield the main system time offset t_{offset}, it gives the relative time offset *between two adjacent time interval edges* of R1 and R2, $t_{\delta int}$. Note, that this suffices for measuring t_d. See Fig. 6 for an overview.

Altogether, a pair of robots can synchronize their clocks whenever they detect their signals coevally. For the meantime, they use the relative clock offset calculated at the last synchronization interval to perform a regular distance calculation.

Note, that considering the *relative* clock drift suffices here. There is no need to synchronize the clocks with civil time.

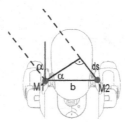

Fig. 7. Calculating the bearing angle α. M_1 and M_2 are the two microphones and b the distance between them.

3.3 Bearing Calculation

For the calculation of bearings, the incoming signals from both microphones are compared. The cross-correlation function is evaluated for each of the signals, yielding a distance difference ds. The bearing angle is then given as

$$\alpha = \arcsin(\frac{ds}{b}),$$

with b being the distance between the two microphones, see Fig. 7.

Note, that calculating bearings requires a high measurement accuracy for ds, see Section 4.3.

4   Implementation and Results

The four-legged Aibo ERS-7 was used as platform. The Aibo has a small speaker installed in its chest, and one microphone in each ear. The digital sound format we use for both input and output is 16bit at 16kHz sampling rate, which is the highest sound quality the Aibo offers. A RoboCup soccer game usually is a very noisy environment. When walking, however, the majority of noise comes from the Aibo itself.

In signal processing, the mean ratio of signal strength and noise is called *Signal to Noise Ratio (SNR)*. In a RoboCup soccer game, the SNR can be as low as 1:20.

4.1 Selecting Frequency and Code Length

Selecting an appropriate transmitting frequency is finding a trade-off between several issues, of which the most important ones are listed here:

– First of all, we need a high signal strength. The Aibos speaker is much louder at certain frequencies.
– Lower-frequency waves have better propagation properties, as they are prop-agate more linearly through the air than higher-frequency waves, which are more vulnerable to reflection and diffraction.

– Higher-frequency waves lead to narrower peaks in the cross-correlation function, resulting in a higher accuracy.

Experiments show, that frequencies in the range from $500Hz$ to $800Hz$ provide the best results.

Since there are five players in a four-legged-team, we use a Kasami-set of $n = 6$ yielding 8 code sequences, each of length 63. Using two full sine cycles per chip and a frequency of $f_0 = 750Hz$ gives a pattern of length

$$\chi = \frac{2}{f_0} \cdot 63 = 168ms.$$

The corresponding acoustic wave has a length of $343\frac{m}{s} \cdot 168ms \approx 57.6m$, which guarantees unambiguous distance measurements in the scenario of a RoboCup soccer game (see Section 3.1).

4.2   Clock Drift

The relative clock drift between two Aibos can be as large as $1ms$ per minute in the worst case, resulting in a distance measurement drift of up to $34cm$ per minute. To keep this error low, the robots need to synchronize their clocks once every two seconds, as described in Section 3.2. This delimits the maximum relative clock drift to $0.033ms$, which is equivalent to $e_{ClockDrift} \approx 1cm$ in distance.

4.3   Distance and Bearing Accuracy

The difference in time of a single sample equals a distance of about $e_{Resolution} = 2cm$ when using a 16kHz sampling rate. This error can be reduced by interpolation of neighboring correlation values.

Another error comes from the fact, that the Aibo's speaker and microphone reference points are not congruent when projected on the ground, depending on the head position. Two Aibos with the same orientation have an estimated maximum error of $e_{Orientation} \leq 4cm$, see Fig. 8. On the contrary, $e_{Orientation} = 0$ for two Aibos facing each other. Since the relative orientation is a priori unknown, we have to bargain for the maximum error.

Diffraction, reflection and interference lead to a third error. The original wave front interferes with diffracted and reflected wave fronts of lower amplitude, yielding small positive phase shifts in the resulting wave. The effect is predominantly noticeable in a signal of a microphone *aiming the opposite direction* of the signal source, because diffractions *at the Aibo's own chassis* seem to make up the biggest portion. Experiments show, that the resulting maximum error of this channel can be estimated by $e_{Interference} \leq 5cm$. The signal of a microphone facing the signal source is almost unaffected by diffraction and interference.

For distance measurements, we use the arithmetic average of the distances to the both microphones yielding a resulting maximum error of $e_{Interference} \cdot 0.5$ here.

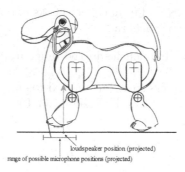

loudspeaker position (projected)

range of possible microphone positions (projected)

Fig. 8. Loudspeaker and microphone positions, projected onto the ground. The projected microphone position depends on the position of the head.

The overall *distance* measurement error can be estimated as

$$e_{Distance} \leq e_{ClockDrift} + e_{Resolution} + e_{Orientation} + 0.5 \cdot e_{Interference} \cong 9.5cm.$$

If the signal comes *from the side*, the effect of diffraction and interference yields very inaccurate bearing measurements. We can only make a qualitative statement for the signal source position in that case. The actual error also varies on nearby obstacles like walls or other robots, and therefore can hardly be estimated.

If the signal source is *in front of* the Aibo, within a range of about $\alpha \leq \pm 35°$, the effect of diffraction almost disappears, because the Aibo's chassis does not hamper the wave propagation. However, the resulting bearing measurement error still can be as large as $e_{Bearing} \approx \pm 10°$ due to obstacles.

4.4   Runtime Considerations

Evaluating the correlation function takes the majority of calculation time. At 16kHz sampling rate, a 32ms-frame equals 512 sampling points of sound data.

The evaluation of a *single* correlation value for the sound data portion received in a 32ms-frame needs 512 integer multiplications and 512 integer additions. We can use integer operations, because the signal pattern and the input signal as well are integer values. Note: Since we only look out for a single, distinct maximum peak, the actual, *exact* correlation values are not of any interest.

Finding a players signal correlation peak initially is costly, because we have to calculate all possible correlation values. In our setup, this means 2560 evaluations. Usually, this is too costly to be done in a single 32ms-frame and therefore is split into smaller steps. E.g., if the number of evaluations is limited to 40 for each 32ms-frame, the initial finding of a team mate's signal may take 2 seconds. If we lose a team mate's distance, e.g. when manually replaced by a referee, we know that the distance to that team mate maximally changes by approx. $12m$. Finding this team mate's signal pattern again will take less than half a second.

Once a player's signal was found, i.e. the maximum correlation peak was found, we can use this knowledge for the next time step. Assuming, that the distance to another team mate maximally changes by 20cm during the sending of a signal pattern (168ms), the correlation only needs to be done in the corresponding vicinity of the previously found peak position. We refer to this procedure as "distance tracking". Therefore, evaluating only approx. 20 correlation values per 32ms frame is sufficient to track the position of a team mate.

The overall sum of operations to be calculated in a 32ms-frame is given by

- 512 additions and 512 multiplications per evaluation
- 20 evaluations needed for distance tracking
- 2 channels to be evaluated (left and right microphone)
- 4 team mates to be evaluated coevally.

This multiplies up to $81,920$ integer additions and multiplications per 32ms-frame. Note, that additional calculation steps are required for array indexing, memory accesses and increasing of counter variables.

4.5 Experimental Results

In an experimental setup, two moving robots sent their signals simultaneously and continuously. The environment was a laboratory yielding a moderate environmental noise. For evaluating the acoustic distance measures, the relative distance was also recorded by a ceiling camera.

Fig. 9 shows the measured distances. The robots acoustic measurements are drawn in red, the camera measurements in green. (The plot is interrupted at times the ceiling camera was not able to recognize both robots.) As you can see, the distances calculated by the robots differ by less than 10cm from the camera measurements. The experiment took about two minutes.

A second experiment was set up in order to test bearing measurement quality. A robot moved around another still-standing robot on a circle of radius 1m. Please recall the denotations in Fig. 7. The values of ds were measured every 22.5° on the robots way along the circle. Fig. 10 shows these measurements ds against their corresponding theoretical values. As expected, interferences lead to larger absolute values. As you can see, ds becomes as large as $1.35 \cdot b$ for angles of $\pm 90°$. Obviously, ds can not get larger than b in theory. This tendency is experimentally reproducible. There might be a chance to increase bearing

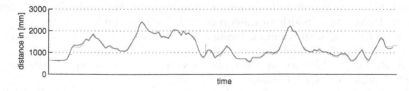

Fig. 9. Relative distances measured acoustically by the robots (red) and using a ceiling camera (green)

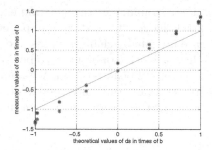

Fig. 10. Measured bearings between the two robots. red: sending the signal from behind, blue: from the front.

quality by using some correcting function. Overall, bearing measurements are rough, but still can be useful for several purposes.

5   Conclusions

A robot can measure distances and rough bearings to his team mates several times a second. The technique of using cross-correlation functions turns out to be very cost-efficient, while being robust against noise. However, to achieve good results in *very* noisy environments, the robots have to be equipped with adequate loudspeaker hardware.

The technique discussed is suitable for real-time applications, as computations can easily be spread over several time-frames. The overall process is expandable to larger teams of robots and also easily adoptable to other platforms or different scenarios.

References

1. Chen, S., et al.: The UNSW RoboCup 2001 sony legged league team report (2001)
2. Röfer, T., et al.: GermanTeam (2002),
   `http://www.tzi.de/kogrob/papers/GermanTeam2002.pdf`
3. Kohno, R., Milstein, L.B., Meidan, L.M.: Spread spectrum access methods for wireless communications. IEEE Communication Magazine 41 (1995)
4. Parkinson, B.: Global positioning system: Theory and applications. Technical report, American Institute of Aeronautics and Astronautics, Washington, D.C. (1996)
5. Girod, L.: Development and characterization of an acoustic rangefinder. Technical report, University of Southern California, Los Angeles (2000)
6. Kasami, T.: Weight distribution formula for some class of cyclic codes. Technical Report No. R-285, Univ. of Illinois (1966)
7. Gold, R.: Optimal binary sequences for spread spectrum multiplexing (corresp.). IEEE Transactions on Information Theory 13(4) (1967)
8. Welch, L.: Lower bounds on the maximum cross correlation of signals. IEEE Transactions on Information Theory 20(3) (1974)

Tracking of Ball Trajectories with a
Free Moving Camera-Inertial Sensor

Oliver Birbach[1], Jörg Kurlbaum[2], Tim Laue[1], and Udo Frese[1]

[1] Deutsches Forschungszentrum für Künstliche Intelligenz GmbH,
Sichere Kognitive Systeme, Enrique-Schmidt-Str. 5, 28359 Bremen, Germany
{oliver.birbach,tim.laue,udo.frese}@dfki.de
[2] Fachbereich 3 - Mathematik und Informatik, Universität Bremen,
Postfach 330 440, 28334 Bremen, Germany
jkur@informatik.uni-bremen.de

Abstract. This paper is motivated by the goal of a visual perception system for the RoboCup 2050 challenge to win against the human world-cup champion. Its contribution is to answer two questions on the subproblem of predicting the motion of a flying ball. First, if we could detect the ball in images, is that enough information to predict its motion precise enough? And second, how much do we lose by using the real-time capable Unscented Kalman Filter (UKF) instead of non-linear maximum likelihood as a *gold standard*? We present experiments with a camera and an inertial sensor on a helmet worn by a human soccer player. These confirm that the precision is roughly enough and using an UKF is feasible.

1 Introduction

1.1 Motivation: A Vision for 2050

In RoboCup's humanoid leagues, the limiting factor is often the robot hardware, in particular actuation. Current robots can walk and kick the ball, but are far away from running, jumping, or tackling. Comparing the official RoboCup vision to win against the human world champion in 2050 to the last 10 years of hardware development, 40 years to go sound realistic but not overly conservative.

As computer vision researchers, we look at the other end of the causal chain. How long is the road to a visual perception system that meets the RoboCup 2050 challenge? – Not too long, maybe a decade, was our conclusion in a recent analysis [1]. This idea is encouraging. But would it help, if we would build a vision system and then wait until 2050 to try it out? We propose an experiment without a robot instead. We mount a camera and an inertial sensor on a helmet worn by a human soccer player (Fig. 1) and ask, whether this setup and computer vision could provide enough information for a humanoid to take the human's place.

The paper is our first step towards this goal. It contributes a study on the subproblem of predicting a flying ball. The study focuses on the tracking part and hence uses manually processed images from real human soccer. We deliberately chose not to implement a full system because real-time vision of soccer scenes

L. Iocchi et al. (Eds.): RoboCup 2008, LNAI 5399, pp. 49–60, 2009.

Fig. 1. The proposed sensor setup with a wide angle camera and an inertial sensor

from a helmet camera is way beyond the scope of a single paper and instead of presenting incremental progress towards this, we rather want to definitely answer two basic questions on the tracking part: Is the information provided by a camera and an inertial sensor enough to predict a flying ball with the necessery precision for playing soccer? And, how much do we lose by using a real-time capable filter compared to the slow *gold standard* solution of non-linear maximum likelihood estimation? – In short, the answers will be positive: "Yes" and "not too much".

1.2   From Walking to Running

For localization and tracking, it is essential to know the camera pose relative to the ground. Today, it is usually obtained by forward kinematics since the robots have well defined contact with the ground. Once robots start to run and jump this will not be possible anymore. Instead, an additional sensor is needed that perceives the camera's motion, i.e. an inertial sensor measuring acceleration and angular velocity. The insight that motion without ground contact will come up in RoboCup sooner or later additionally motivates our experimental setup, where the camera motion has to be treated as free motion anyway, because "on a human", of course, no kinematic information is available.

1.3   Methods

To show how to track a ball and its observer state with this combined sensor setup, we will use well known methods and provide a quantitative evaluation.

Tracking algorithms exist in several variations. The family of Kalman filters is well understood, reliable and often used. You will mostly find its variations, the Extended Kalman Filter (EKF) [2] and the Unscented Kalman filter (UKF) [3], in RoboCup applications. Also, filters based on Monte-Carlo methods, such as the particle filter, or combinations of both [4] are popular.

In our case, the filter estimates the orientation of the free-moving camera. The orientation has three degrees of freedom, but suffers from singularity problems when parameterized with three variables, such as Euler angles. We therefore use quaternions as singularity-free representations and a special technique, the so-called embedding operator ⊞, for handling them in an UKF.

To evaluate the UKF, we compared it to maximum likelihood estimation using Levenberg-Marquardt as a *gold standard*. Furthermore, we evaluated the accuracy by comparing the calculated bouncing point from the tracker to the ground truth bouncing point obtained manually from the images.

1.4   Related Work

In RoboCup, tracking of balls was studied by Voigtländer et al. [5]. Their system recognizes balls in the Middle-Size League and tracks their state over time with a position error of $\leq 0.5m$. Their sensor setup consists of a perspective as well as a catadioptric camera. Within the Small-Size League Rojas et al. [6] developed a system to detect and track lifted kicks with a statically mounted overhead camera. After a couple of frames, this system allowed a reliable prediction of the ball's bouncing point. Another interesting work was done by Frese et al. [7]. There, the catch point for a robotic softball-catcher was determined by predicting the ball trajectory which was tracked using stereo vision and an EKF. Similarly, in [8] a system was developed that tracks a flying dart and repositions the dartboard to always hit the bull's eye.

Recently, even a commercial system has been introduced [9]. The so-called RoboKeeper employs cameras to track the ball. It intercepts the ball with a plastic plate rotated around the lower end by a single motor. The system is intended for public entertainment, so the plate shows the image of a goal keeper.

All these approaches either have a static camera or a camera fixed relative to the ground plane. In contrast, we are dealing with a freely moving observer. This requires to track the observer itself, making the problem much harder.

1.5   Outline

This paper is structured as follows. We introduce the proposed camera-inertial system and the measurements obtained from it. Then, the calibration of the alignment of the inertial-sensor and the camera will be discussed. The description of the system variables and the underlying model lead then to an overview of the UKF implementation. Finally, we present experimental results and a conclusion.

2   Experimental Setup

As mentioned above, we propose to use a combination of a monocular camera and an inertial sensor as the sensor setup for answering the question if the motion of flying balls can be predicted sufficiently.

2.1   Camera-Inertial Sensor System

In our setup, the monocular camera is a Basler A312fc (54 Hz) with a Pentax H(416)KP lens (86° wide-angle). Attached to this camera is a XSens MTx inertial

Table 1. Measurements from the sensor setup and their corresponding units

Accelerometer	Gyroscope	Monocular Camera				
Acceleration in $\frac{m}{s^2}$	Ang. vel. in $\frac{rad}{s}$	Ball (pixel)			Landmark (pixel)	
a	ω	x_b	y_b	r_b	$x_{l,n}$	$y_{l,n}$

sensor synchronized to the camera shutter. The camera provides information about the environment and the inertial sensor about the setup's motion (Tab. 1).

What information can these sensors provide? In principle, the pose of the camera can be obtained by integrating gyro and accelerometer measurements, however with accumulating error. To compensate this error, vision measurements of landmarks, e.g. lines on the field, are used. The camera provides the direction towards the ball from its image position. The image radius provides the depth, however with a large error since the ball is small. Interestingly, implicit depth information also comes from the effect of gravity over time. After a time of t the ball has fallen $\frac{1}{2}gt^2$ relative to motion without gravity. This distance is implicitly observed and, after some time, provides preciser depth than the ball radius.

One might also consider a stereo vision setup instead of a monocular one to obtain the distance by triangulating at the stereo baseline. However, we use the image radius of the ball implicitly triangulating at the ball diameter which is rather even a bit larger. So we expect no great benefit from stereo.

2.2 Sensor Calibration

Both sensors operate in different coordinate systems. For that, a transformation, mapping measurements between both, is needed. Our setup is not rotating too fast. Therefore, we can neglect the translation and the problem reduces to finding the rotational displacement. It is parameterized as a quaternion q, leading to the following mapping of a vector v_{cam} from camera to v_{iner} in inertial coordinates:

$$(0, v_{\text{iner}}) = q_{\text{iner}}^{\text{cam}} \cdot (0, v_{\text{cam}}) \cdot q_{\text{iner}}^{\text{cam}\ -1} \tag{1}$$

The rotational displacement $q_{\text{iner}}^{\text{cam}}$ is found by calibration. The idea here is to use a horizontal checkerboard calibration plate so both camera and inertial sensor know the direction "down" in their respective coordinate systems [10]. The camera parameters are computed by nonlinear least squares estimation following Zhang's method [11] which is extended to estimate the rotational displacement q with (1). The camera equations are described in Sec. 3.2. For our setup, the calibration had ≈ 0.2 pixels (camera) and $\approx 0.26°$ (inertial sensor) residual.

3   System Model

The flight of a ball observed by a camera-inertial sensor can be formally represented by a dynamical system. Within this system, the state of the ball and sensor changes as the ball flies and time progresses. The composition of the system's state and its variables is given in Tab. 2.

Table 2. Tabulated summary of the systems's state $\in S$

Ball		Sensor		
Position in m	Velocity in $\frac{m}{s}$	Position in m	Velocity in $\frac{m}{s}$	Orientation as quaternion
x_b	v_b	x_s	v_s	$q_{\text{world}}^{\text{iner}}$

3.1 Dynamic Equations

The evolution of the state, i.e. the motion, is given by ordinary differential equations. For the ball, these obey classical mechanics with gravitation and air drag.

$$\dot{x}_b = v_b \tag{2}$$

$$\dot{v}_b = g - \alpha \cdot |v_b| \cdot v_b, \quad \alpha = \frac{c_d A \rho}{2m} \tag{3}$$

where g represents gravity. The effect of air drag is determined by the factor α, where c_d is the drag coefficient, ρ is the density of air, A is the cross-sectional area and m the mass of the ball. Our football had a cross-sectional area of $A = 0.039 \ m^2$ and mass of $m = 0.43 \ kg$, resulting in $\alpha = 0.011 \ m^{-1}$.

Similarly, the motion of the sensor follows differential equations that incorporate the measurements from the inertial sensor (Tab. 1 and 2).

$$\dot{q}_{\text{world}}^{\text{iner}} = q_{\text{world}}^{\text{iner}} \cdot \frac{1}{2}(0, \ \omega) \tag{4}$$

$$\dot{x}_s = v_s \tag{5}$$

$$\dot{v}_s = g + q_{\text{world}}^{\text{iner}} \cdot (0, a) \cdot q_{\text{world}}^{\text{iner}-1} \tag{6}$$

Basically the inertial measurements are converted from body- to world coordinates and integrated once (q from ω) or twice (x from a).

3.2 Measurement Equations

The measurement equations model the measurements from the monocular camera, i.e. image position and radius of the ball and of point landmarks on the soccer field. We model the projection of these features from the state using a pin-hole camera model plus radial distortion. The first function simply projects a point x_{world} from a 3D-scene into the image plane at $(u, v)^T$:

$$\begin{pmatrix} X_{cam} \\ Y_{cam} \\ Z_{cam} \end{pmatrix} = q_{\text{world}}^{\text{cam}-1} \begin{pmatrix} 0 \\ x_{world} - x_s \end{pmatrix} q_{\text{world}}^{\text{cam}} \tag{7}$$

$$\begin{pmatrix} x \\ y \end{pmatrix} = \begin{pmatrix} X_{cam} \\ Y_{cam} \end{pmatrix} / Z_{cam} \tag{8}$$

$$\begin{pmatrix} u \\ v \end{pmatrix} = \begin{pmatrix} f_x \\ f_y \end{pmatrix} \cdot \begin{pmatrix} x' \\ y' \end{pmatrix} + \begin{pmatrix} u_0 \\ v_0 \end{pmatrix} \tag{9}$$

The first equation transforms the world point into camera space according to the quaternion $q_{\text{world}}^{\text{cam}} = q_{\text{world}}^{\text{imu}} \cdot q_{\text{imu}}^{\text{cam}}$ and the sensor's position x_s. Equation (8) is the actual perspective projection. The last equation transforms the result to pixels according to f_x, f_y (effective focal distance) and u_0, v_0 (image center).

The camera's radial distortion is considered by scaling $(x, y)^T$ according to

$$\begin{pmatrix} x' \\ y' \end{pmatrix} = \begin{pmatrix} x \\ y \end{pmatrix} \left(1 + \frac{a_2 r + a_3 r^2}{1 + b_1 r + b_2 r^2 + b_3 r^3} \right), \quad r^2 = x^2 + y^2. \tag{10}$$

To model the ball measurements, we project the ball from a 3D-scene as a circle into the image plane. For this, we use the introduced projection to calculate the position of four outer points of the ball state, which are computed from the always known ball diameter, on the image plane. After that, we recombine the projected points to center and radius by computing mean and std. deviation.

4   Unscented Kalman Filter (UKF) for Ball Tracking

For tracking the ball and sensor state over time, we use an UKF incorporating the preceding equations into the dynamic update and measurement update step.

4.1   Dynamic and Measurement Equations

The UKF's dynamic update step uses a numerical solution of (2), (3), (5), (6). For (4) an analytic solution is used. The camera measurements are integrated in the measurement update step using the perspective projection (7)-(10).

Kalman filters assume noisy processes. Therefore, the dynamic noise R and measurement noise Q need to be defined. We set (2) and (5) to zero noise assuming that every error in position comes from an error in velocity. The dynamic noise of the inertial sensor, affecting (4) and (6), was calibrated from a resting sensor. The dynamic noise of the ball (3) was tuned by looking at the recorded trajectory. This parameter is important since it models side wind and bended trajectories caused by ball spin. Based on these values, the measurement noise for the ball and landmark measurements were calibrated by maximizing their likelihood in an recorded series of a ball flight [12].

4.2   Quaternions in the UKF State

To track the state of the system described above, we use a customized UKF. While tracking x_s, v_s, x_b, v_b is straight-forward, the orientation q_{world}^{iner} poses a special problem. In Sec. 3, we modeled the orientation in the state space S using a unit quaternion to avoid singularities. Unfortunately, such a state cannot be treated as a vector $\mathbb{R}^4$ because of the constraint $|q_{world}^{iner}| = 1$. In mathematics such a structure is called a 3-manifold in $\mathbb{R}^4$. The filter doesn't know this structure and assumes to operate on a flat vector. Hence, after a measurement update, usually $|q_{world}^{iner}| \neq 1$ and the result is not an orientation anymore.

Our idea to solve this problem is to use the mean-state from the UKF as a reference and parameterize small deviations from that reference as a flat vector with a dimension corresponding to the dimension of the manifold (i.e. 3 for $q_{\text{world}}^{\text{iner}}$). As the deviations are small, this is possible without singularities.

The operation of applying such a small change to the state is encapsulated by an operator $\boxplus$ and the inverse operator $\boxminus$. They provide an interface for the UKF to access the state, which is then treated as a black-box datatype.

$$\boxplus \; : \; S \times \mathbb{R}^n \to S \tag{11}$$

$$\boxminus \; : \; S \times S \to \mathbb{R}^n \tag{12}$$

$$s_1 \boxplus (s_2 \boxminus s_1) = s_2 \tag{13}$$

$$s \boxplus (a + b) \approx (s \boxplus a) \boxplus b \tag{14}$$

To summarize the interaction between the UKF and the blackbox state datatype, dynamic and measurement equations access the state's internal structure as they depend on the concrete problem. However, the generic UKF equations for sigma point propagation and measurement update access the state only through $\boxplus$ and $\boxminus$ with a flat vector. Thereby, the state's internal structure, in particular the manifold structure underlying quaternions, is hidden from the UKF. Notably, this corresponds to a common implementation issue, that one would prefer the state to have named members sometimes and sometimes a flat vector.

As a result, in all the matrix computations of the UKF the quaternion part of the state $q_{\text{world}}^{\text{iner}}$ simply corresponds to 3 columns just as the vector part x_s, v_s, x_b, v_b. The only difference arises when the UKF finally applies the innovation to the state by $\boxplus$, where the vector part is a simple $+$ and the quaternion part is a more complex operation (multiplication with an angle-axis rotation).

This method is more elaborated in [13]. Its beauty lies in the fact that it is obtained from the classical UKF simply by replacing $+$ with $\boxplus$ and $-$ with $\boxminus$.

5 Experiments

We implemented the ball tracker as described above and held a series of experiments on a real soccer field. A human wearing the helmet with the camera-inertial sensor followed the trajectories of several ball flights which where initiated by another person. The data was recorded and the images were manually processed (confer the introduction why). The image center and radius of the ball and the image position of the field's landmark points were extracted. Landmarks were all intersections of two lines on the field, the lower end of the goal posts, the intersections of the goal post with the crossbar and the penalty spot.

5.1 Performance Compared to a Gold Standard

To evaluate the performance of the unscented Kalman filter, we implemented a nonlinear maximum likelihood estimator using the Levenberg-Marquardt (LM) algorithm [14]. It performs the same task and uses the same model as the Kalman filter. In contrast to the UKF, however, estimating the current state requires to estimate all past states as well. This makes maximum likelihood estimation

Fig. 2. Grid overlay of a homography between the image plane and the field plane. Observe the fine correspondence near the line in front of the goal.

computationally expensive and unsuitable for real-time tracking. The question we want to answer here is, whether the UKF is good enough or whether we will have to think about better real-time capable alternatives in the future.

5.2   Obtaining Ground Truth

To analyze the accuracy of the tracker, we wanted to compare the predicted bouncing points of the tracked ball states with the real bouncing point. This turned out to be not that easy as the ball leaves no permanent mark on field, making a manual measurement infeasible. However, at least temporarily raised dust is visible on the video. So for the evaluation here, the actual bouncing point was determined from the image position of the dust in the video using a homography between at least four landmarks on the field and their corresponding points in the image plane (correcting for distortion). If the dust was visible in several images, we took the mean of these results. While we have no quantitive value, the homography looks visually rather precise (Fig. 2).

5.3   Results

In order to show how well the ball flight observed by a moving camera-inertial sensor can be estimated and predicted, results from one fully filtered ball flight recorded during the experiments will be given. For this, a ball flying about 3 m (reaching an maximum height of 1.6 m) towards the observer from about 7 m away was chosen. The ball was flying almost spin-less and the impact of the side wind to the ball at this range of flight and speed was minimal. The performance of tracking and predicting the trajectory is obtained by comparing the predicted bouncing points of the tracked trajectory with the ground truth bouncing point.

Figure 3 shows the trajectory and velocity vectors tracked by the unscented Kalman filter and the LM algorithm from this flight. As can be seen, both methods calculate almost the same series of state estimates. But the tracked trajectories do not resemble the smoothness of the actual trajectory. Note that,

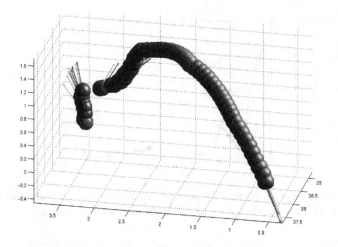

Fig. 3. Tracked trajectories and velocity vectors of the unscented Kalman filter and the Levenberg-Marquardt algorithm. As can be seen, both methods estimate almost the same set of states.

since we plotted the estimate over time, Fig. 3 does not show how the tracker believes the ball flew, but how the belief of the tracker on where the ball is changed. So the jerkiness in the beginning is due to new information arriving.

Another interesting behavior is the estimation of the velocity vectors pointing away from the actual flight direction. This behavior is not suprising since the monocular camera defines the ball distance by the ball diameter. Within the first few frames the diameter is 16 ± 1 pixels, leading to an expected error of 6.25% in the distance. So, initially estimating the distance is very difficult, the situation only gets better after the ball has been observed for some time.

5.4   Prediction Performance Compared to Ground Truth

To show how the prediction performs over time, Fig. 4 depicts the error between the prediction of the tracked ball state and the assumed bouncing points as a function of time. It can be clearly seen that the UKF and the LM method perform similar. The error of both methods decreases with time. Interestingly, the prediction performance improves in a wave-like shape, caused by the integer based image radius measurements. The peaks of the waves decrease, since the influence of the radius diminishes and the distance becomes more defined by gravity (Sec. 2.1).

A quantative analysis reveals that the estimation of ball states behind the initial ball position lead to very inaccurate ball predictions in the beginning. The error decreases significantly after 7 frames (or about 129 ms), reaching a surprisingly good estimate with a difference between 0.1 m and 0.4 m. Then, the accuracy decreases rapidly for a couple of frames before it reaches a next good estimate after 16 frames (297 ms) with an error between 0.1 and 0.15 m. From

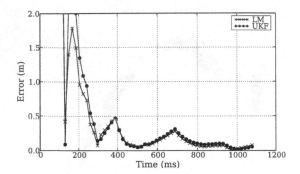

Fig. 4. Error between the predicted bouncing points computed from the estimated trajectory and the mean of the ground truth measurements over time

Fig. 5. Predicted trajectory of a flying ball observed by a moving camera-inertial sensor projected into image coordinates. See `http://www.informatik.uni-bremen.de/agebv/en/Vision2050`.

here, the estimates seem to be quite reliable and the prediction performance improves in the above mentioned wave-like shape.

Figure 5 shows a predicted trajectory of a ball from its tracked state projected into images coordinates. This image is part of a video visualizing the prediction performance within the recorded images from the experiments.

The results observed within this ball flight match the results from the two other ball flights that were evaluated. All three flights showed a suprisingly good predicted trajectory after $8 - 10$ frames. But all of these predictions only lasted for one or two frames, assuming that there is still a large amount of uncertainty at this point of the time during tracking. A couple of frames later, namely after $15 - 16$ frames ($278 - 296\ ms$), all examined flights show a quite accurate indication of the predicted bouncing point. This number of frames seem to be the point in time in which the distance is more defined by gravity than by the image radius.

For more information on our experiments and results on further scenes, see [12].

6 Conclusion

In this paper, we have shown how to track and predict the trajectories of flying balls using a freely moving monocular camera-inertial sensor. We have evaluated the precision by comparing the predicted bouncing points over time with the real bouncing point. The results indicate that after half the time of flight the error is about $0.3m$ which we judge to be enough for playing soccer. The real-time capable UKF and Levenberg-Marquardt, the *gold standard*, perform very similar. We can conclude that with such a setup, flying balls can be predicted roughly precise enough for playing soccer and the UKF is a feasible algorithm for that.

The next step is, of course, to replace the manual vision by computer vision. This is difficult because of the uncontrolled lighting and background on a real soccer field. Our idea [1] is to use, in turn, the tracking component as context for the ball detection. In real images, many things may look like a ball, e.g. a head, a white spot, or reflections, but few of them move according to the physics of a flying ball. So by treating the problem as a combined likelihood optimization with respect to both the physical trajectory model and visual appearance, we hope to obtain a reliable vision system.

Furthermore, we want to go away from predicting a single pass to tracking the ball during a full scene. Therefore, we need a multi-state tracker which not only tracks the state of a flying ball, but also a ball which is rolling or just lying.

References

1. Frese, U., Laue, T. (A) VISION FOR 2050 – The Road Towards Image Understanding for a Human-Robot Soccer Match. In: International Conference on Informatics in Control, Automation and Robotics (2008)
2. Welch, G., Bishop, G.: An Introduction to the Kalman Filter. Technical Report TR 95-041, Chapel Hill, NC, USA (1995)
3. Julier, S., Uhlmann, J.: A new extension of the Kalman filter to nonlinear systems. In: Int. Symp. Aerospace/Defense Sensing, Simul. and Controls, Orlando, FL (1997)
4. Kwok, C., Fox, D.: Map-based multiple model tracking of a moving object. In: Nardi, D., Riedmiller, M., Sammut, C., Santos-Victor, J. (eds.) RoboCup 2004. LNCS, vol. 3276, pp. 18–33. Springer, Heidelberg (2005)
5. Voigtländer, A., Lange, S., Lauer, M., Riedmiller, M.: Real-time 3D Ball Recognition using Perspective and Catadioptric Cameras. In: Proc. of the 3rd European Conference on Mobile Robots, ECMR (2007)
6. Rojas, R., Simon, M., Tenchio, O.: Parabolic Flight Reconstruction from Multiple Images from a Single Camera in General Position. In: Lakemeyer, G., Sklar, E., Sorrenti, D.G., Takahashi, T. (eds.) RoboCup 2006: Robot Soccer World Cup X. LNCS, vol. 4434, pp. 183–193. Springer, Heidelberg (2007)
7. Frese, U., Bäuml, B., Haidacher, S., Schreiber, G., Schaefer, I., Hähnle, M., Hirzinger, G.: Off-the-Shelf Vision for a Robotic Ball Catcher. In: Proceedings of the IEEE/RSJ International Conference on Intelligent Robots and Systems, Maui, pp. 1623–1629 (2001)
8. Kormann, B.: Highspeed dartboard (2008),
   http://www.treffsichere-dartscheibe.de/

9. 4attention GmbH. RoboKeeper web site (2006), http://www.robokeeper.com/
10. Lobo, J., Dias, J.: Relative pose calibration between visual and inertial sensors. In: ICRA Workshop InerVis (2004)
11. Zhang, Z.: A Flexible New Technique for Camera Calibration. IEEE Transactions on Pattern Analysis and Machine Intelligence 22(11), 1330–1334 (2000)
12. Birbach, O.: Accuracy Analysis of Camera-Inertial Sensor-Based Ball Trajectory Prediction. Master's thesis, Universität Bremen (2008)
13. Kurlbaum, J.: Verfolgung von Ballflugbahnen mit einem frei beweglichen Kamera-Inertialsensor. Master's thesis, Universität Bremen (2007)
14. Press, W.H., Teukolsky, S.A., Vetterling, W.T., Flannery, B.P.: Numerical Recipes in C: The Art of Scientific Computing, ch. 15.5. Cambridge University Press, New York (1992)

A Case Study on Improving
Defense Behavior in Soccer Simulation 2D:
The NeuroHassle Approach

Thomas Gabel, Martin Riedmiller, and Florian Trost

Neuroinformatics Group
Institute of Mathematics and Computer Science
Institute of Cognitive Science
University of Osnabrück, 49069 Osnabrück, Germany
{thomas.gabel,martin.riedmiller}@uos.de, floriantrost@gmx.net

Abstract. While a lot of papers on RoboCup's robotic 2D soccer simulation have focused on the players' offensive behavior, there are only a few papers that specifically address a team's defense strategy. In this paper, we consider a defense scenario of crucial importance: We focus on situations where one of our players must interfere and disturb an opponent ball leading player in order to scotch the opponent team's attack at an early stage and, even better, to eventually conquer the ball initiating a counter attack. We employ a reinforcement learning methodology that enables our players to autonomously acquire such an aggressive duel behavior, and we have embedded it into our soccer simulation team's defensive strategy. Employing the learned *NeuroHassle* policy in our competition team, we were able to clearly improve the capabilities of our defense and, thus, to increase the performance of our team as a whole.

1 Introduction

Over the years, RoboCup simulated soccer has turned out to be a highly attractive domain for applying machine learning and artificial intelligence approaches [1]. Beyond the scientific focus and the goal of beating the human soccer World Champion team in 2050, the annual RoboCup competitions also represent a highly competitive event. Consequently, if a team participating in a RoboCup tournament aims not just at exploring learning approaches in this application field, but also at using player behaviors, that were obtained with the help of some learning approach, during competitions, then there is a natural desire that also the AI-based player capabilities represent optimal or at least near-optimal behavior policies, so that their employment does not worsen the overall quality of playing.

This, of course, also holds true for a team's defense. Research in soccer simulation, however, has to the biggest part focused on offensive playing. To this end, mainly the tasks of passing and scoring [2,3] were addressed, as well as the

L. Iocchi et al. (Eds.): RoboCup 2008, LNAI 5399, pp. 61–72, 2009.

positioning of players not in ball possession [4,5]. Moreover, there were also studies that considered offensive playing in a holistic manner [6,7]. By contrast, the challenges of using learning approaches for a team's defensive capabilities have almost been neglected. A notable exception in this regard represents the work on positioning defending players [8]. Generally, defensive playing strategies comprise two core sub-tasks: On the one hand, there is the task of positioning players in free space so as to intercept opponent passes or to cover or mark opponent players. On the other hand, there must be a player that attacks the opponent ball leading player, interferes him, hassles him, and aims at bringing the ball under his control. While the paper mentioned before [8] focuses on the former task of positioning, the work at hand presents a case study concerning the latter: Using a reinforcement learning (RL) methodology, we aim at acquiring an aggressive player behavior which makes the learning agent interfere and hassle a ball leading opponent player effectively. Because the RL approach employed makes use of neural net-based value function approximation, we call the resulting behavior *NeuroHassle*, subsequently. The policy we obtained clearly surpasses any hand-coded approach for this task and improved our team's defensive strength significantly. With respect to the remarks made at the beginning of this section, a noteworthy property of the acquired hassling policy is its high degree of competitiveness which allowed us to integrate NeuroHassle into our competition team *Brainstormers* and to employ it successfully at RoboCup 2007 in Atlanta.

Section 2 provides some foundations on RL and on the soccer simulation environment. In Section 3, we introduce in detail our learning methodology and algorithm, whereas Section 4 evaluates the learning results empirically and discusses the integration of the acquired hassling policy with our competition team.

2   Basics

One reason for the attractiveness of reinforcement learning (RL) [9] is its applicability to unknown environments with unidentified dynamics, where an agent acquires its behavior by repeated interaction with the environment on the basis of success and failure signals. This situation is, also, given when an initially clueless soccer-playing agent is faced with the task of conquering the ball from an opponent ball leading player (of an adversary team regarding whose dribbling capabilities nothing is known). Accordingly, the usage of RL to tackle this learning problem is very promising. A more comprehensive review of our efforts on utilizing neural reinforcement learning methods in the scope of the RoboCup soccer simulation 2D League can be found in [6,10].

In each time step an RL agent observes the environmental state and makes a decision for a specific action, which may incur an immediate reward generated by the environment and, furthermore, transfers the agent to some successor state. The agent's goal is not to maximize its immediate reward, but its long-term, expected reward. To do so it must learn a decision policy π that is used to determine the best action in any state. Such a policy is a function that maps the current state $s \in S$ to an action a from a set of viable actions A and the goal is to learn the mapping $\pi : S \to A$ only on the basis of the rewards the agent gets from its environment. By repeatedly performing actions and observing resulting rewards, the agent tries to improve and fine-tune its policy. The respective reinforcement learning algorithm used specifies how experience from past interaction is used to adapt the policy. Assuming that a sufficient amount of states has been observed and rewards have been received, the optimal policy may have been found and an agent following that policy will behave perfectly in the particular environment.

Reinforcement learning problems are usually formalized using Markov Decision Processes (MDPs) [11], where an MDP is a 4-tuple $[S, A, r, p]$ with S as the set of states, A the set of actions the agent can perform, and $r : S \times A \to \mathbb{R}$ a function of immediate rewards $r(s, a)$ (also called costs of actions) that arise when taking action a in state s. Function $p : S \times A \times S \to [0, 1]$ depicts a probability distribution $p(s, a, s')$ that tells how likely it is to end up in state s' when performing action a in state s. Trying to act optimally, the agent needs a facility to differentiate between the desirability of possible successor states, in order to decide on a good action. A common way to rank states is by computing a state value function $V^\pi : S \to \mathbb{R}$ that estimates the future rewards that are to be expected when starting in a specific state s and taking actions determined by policy π from then on. In this work, our goal is to learn a value function for the hassling problem that we shall represent using multilayer neural networks. If we assume to be in possession of an optimal state value function $V^\star$, it is easy to infer the corresponding optimal behavior policy by exploiting that value function greedily according to $\pi^\star(s) := \arg\max_{a \in A}\{r(s, a) + \sum_{s \in S} p(s, a, s') \cdot V^\star(s')\}$.

The Soccer Server [12] is a software that allows agents in the soccer simulation 2D League to play soccer in a client/server-based style: It simulates the playing field, communication, the environment and its dynamics, while the clients – eleven agents per team – are permitted to send their intended actions (e.g. a parameterized kick or dash command) once per simulation cycle to the server. Then, the server takes all agents' actions into account, computes the subsequent world state and provides all agents with information about their environment. Therefore, while the reward function is unknown to the agent, in soccer simulation the transition function p (model of the environment) is given since the way the Soccer Server simulates a soccer match is known. In the hassling task, however, the situation is aggravated: The presence of an adversary whose next actions cannot be controlled and hardly be predicted makes it impossible to accurately anticipate the successor state. Hence, only a rough approximation of p, that merely takes into account that part of the state that can be directly be influenced by the learning agent, is available.

3   Learning a Duel Behavior: The NeuroHassle Approach

"Aggressive playing" is a collocation frequently used in today's press coverage of human soccer-playing. By aggressiveness it is usually referred to a player's willingness to interfere the opponent team's game build-up early and, in particular, to quickly and efficiently hassle and attack opponent ball leading players, which is often considered to be crucial for a team's success.

3.1 Outline of the Learning Approach

The Brainstormers' former approach for letting players duel with opponent ball leaders for the ball was a rather naive one: The player next to the opponent ball leading player simply moved towards the ball leader and towards the ball, respectively, in order to try to bring the ball into his kickable area. Needless to say that such a straightforward strategy is not difficult to overcome. Consequently, our players failed in conquering the ball in almost two thirds of all attempts – in particular when playing against teams with highly developed dribble behaviors.

A general strategy to hassle an opponent with the goal of conquering the ball is difficult to implement because

- the task itself is far beyond trivial and its degree of difficulty heavily depends on the respective adversary,
- there is a high danger of creating an over-specialized behavior that works well against some teams, but performs poorly against others, and
- duels between players (one without and the other with the ball in his possession) are of high importance for the team as a whole since they may bring about ball possession, but also bear some risk, if, for example, a defending player looses his duel, is overrun by the dribbling player, and thus opens a scoring opportunity for the opponent team.

To handle these challenges holistically, we decided to employ a neural reinforcement learning approach that allows our players to train the hassling of opponent ball leading players. The state space for the problem at hand is continuous and high-dimensional; we restricted ourselves to 9 state dimensions:

- distance d between our player and the opponent ball leading player
- velocity (v_x and v_y component) of our player
- absolute value v_{opp} of the opponent's velocity
- position (b_x and b_y component) of the ball
- our player's body angle α relative to the opponent's position
- opponent player's body angle β relative to his direction towards our goal[1]
- value of the strategic angle $\gamma = \angle GOM$ with G as position of our goal, O as position of the opponent, and M as the position of our player

There are three important remarks to be made concerning that state representation. First, to simplify the state representation the center of the coordinate system is placed in the center of our player and the abscissa cuts through our and the opponent player. Yet, by the two features listed last, also the position of the hassling situation on the playing field is taken into account. Second, these nine state features do not fully represent the actual state (e.g. the adversary's

---

[1] Most dribbling players are heading towards the opponent team's goal when being allowed to dribble freely without interference.

exact velocity vector is unknown and cannot be inferred), yet the degree of partial observability is kept low. And third, the high dimensionality of the problem space requires the use of value function approximation mechanisms if we aim at applying value function-based reinforcement learning. To this end, we rely on multilayer perceptron neural networks.

The learning agent is allowed to use $dash(x)$ and $turn(y)$ commands where the domains of bots commands' parameters ($x \in [-100, 100]$, $y \in [-180°, 180°]$) are discretized such that in total 76 actions are available to the agent at each time step.

3.2 Training Situations and Training Regions

We designed a specialized set of *training situations* S for the learning agent ($|S| = 5000$) which is visualized in Figure 1. It basically consists of two semicircles across which the opponent ball leading player is placed randomly, whereas our learning player resides in the center. While the semicircle that lies in the direction towards our goal (defensive direction) has a radius of $3.0m$, the one in the opposite (offensive) direction is larger ($5.0m$). The intention behind that design of starting situations is that, on the one hand, the ball leading opponent typically starts immediately to dribble towards our goal, whereas the learning agent must interfere and try to hinder him from making progress. On the other hand, the intended hassle behavior shall be primarily applied in situations where our player is closer to our goal or where the opponent ball leader has only a small head start. Regarding the remaining state space dimensions, the ball is always randomly placed in the ball leading player's kickable area with zero velocity, and the velocities of both players as well as their body angles are chosen randomly as well.

Moreover, we defined four *training regions* on the playing field as sketched in Figure 2. The midfield training region is situated at the center of the field, the central defensive region is halfway on the way towards our goal. Finally, there

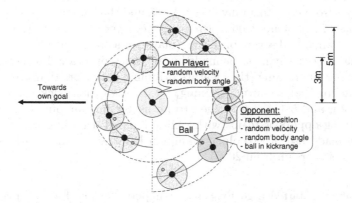

Fig. 1. General Training Scenario for Learning to Hassle Opponent Ball Leading Players

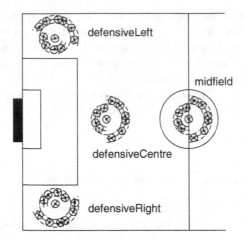

Fig. 2. Considered Sets of Training and Test Situations on the Playing Field

are a left and a right defensive region that are placed near the corners of the field with a distance of 25 meters to our goal. The idea behind this definition of different training and testing places is that dribbling players are very likely to behave differently depending on where they are positioned on the field. As a consequence, a duel for the ball may proceed very differently depending on the current position on the field. It is worth noting that the sets of situations used for the defensive left and right scenario are turned by approximately ±70° so that the smaller semicircle of starting positions for the ball leading opponent is oriented towards our goal.

3.3   A Note on Training Opponents

For training, of course, an adversary team is required. In the learning experiments we conducted, however, we did not employ an entire team, but just started a single player from each team involved[2]. We found that among the 16 published team binaries of RoboCup 2006 in Bremen, there were 2 binaries that were not functioning at all, 3 had severe problems (players crashing) with the player repositioning that necessarily must be made by the coach program during training, 6 were not usable for training (they feature dribbling behaviors that are seemingly inappropriate for the use during training, e.g. most of them kicked the ball just straight away), and 5 binaries seemed to be usable for our purposes. Since we preferred strong opponent teams with well developed dribbling capabilities, we additionally made use of two older champion binaries (teams STEP and UvA Trilearn in their version from 2005).

---

[2] Additionally, we started a goalkeeper as "support" for the learning player, since otherwise the ball leading player is sometimes mislead to shoot on the empty goal from a rather long distance.

3.4 Successful Episodes and Failures

A highly important issue concerns the question whether a single hassling episode, i.e. a single duel for the ball, was successful or not. Here, simply distinguishing between episodes during which the ball could be brought into the hassling player's kickable area or not, is not adequate – a more sophisticated differentiation is crucial for properly assessing the hassling capabilities acquired and is also essential for the use of a reinforcement learning approach.

After a careful analysis of the learning setting and of the way opponent ball leading players may behave during training, we found that the following five main outcomes of a training episode must be distinguished.

a) **Erroneous Episode.** A training episode can fail for many reasons. The dribbling player may lose the ball, the ball may go out of the field, a player may have not localized himself correctly and thus behave suboptimally, to name just a few. If any of these indications is perceived, then the corresponding episode is ended and abolished.

b) **Success.** A hassling episode can be considered successful, if the ball has been brought into the learning player's kickable area or if it has managed to position in such a manner that issuing a tackle command yields a successful tackle for the ball with very high probability.

c) **Opponent Panic.** It may also happen that the ball leading opponent player simply kicks the ball away (usually forwards), as soon as the learning agent has approached or hassled him too much, or if it simply considers his situation to be too hopeless to continue dribbling. Consequently, if an opponent issues such a kind of a panic kick, the episode under consideration may be regarded as a draw.

d) **Failure.** The hassling player fails, if none of the other cases has occurred. In particular, this means that the ball leading player has kept the ball in its kick range, or has even overrun the learning agent and escaped more than $7m$ from him, or has approached the goal such that a goal shot might become promising.

e) **Time Out.** We allocate a maximal episode duration of 35 time steps. If none of the cases mentioned before has occurred until that time, then the hassling episode is ended without a clear winner.

3.5 The Learning Algorithm

Temporal difference (TD) methods comprise a set of RL algorithms that incrementally update state value functions $V(s)$ after each transition (from state s to s') the agent has gone through. This is particularly useful when learning along trajectories $(s_0, s_1, \ldots, s_N)$ – as we do here – starting in some starting state s_0 and ending up in some terminal state $s_N \in G$. So, learning can be performed online, i.e. the processes of collecting (simulated) experience, and learning the value function run in parallel. Learning to hassle, we update the value function's estimates according to the $TD(1)$ update rule [13], where the new estimate for $V(s_k)$ is calculated as $V(s_k) := (1-\alpha) \cdot V(s_k) + \alpha \cdot ret(s_k)$ with

$ret(s_k) = \sum_{j=k}^{N} r(s_k, \pi(s_k))$ indicating the summed rewards following state s_k and α as a learning rate. Each time step incurs small negative rewards, a success goal state a large positive one, and the final state of a failure episode a large negative one.

We did also a number of tests concerning the usage of episodes that ended in some kind of a draw, i.e. in time-out episodes and in episodes where the adversary got into panic and kicked the ball straight away. Since an episode with a time-out is rather difficult to assess, we decided to not use those episodes during training. The probability of an opponent kicking the ball away when under high pressure, strongly depends on the respective opponent's dribbling behavior. For example, players from team TokyoTech frequently tend to perform panic kicks, even if the dribbling player is not yet under real pressure. Therefore, when training against such opponents it may be advisable to interpret panic kicks as failures, since, from the hassling player's perspective, the goal of bringing the ball into his control could not be achieved. As far as the learning results reported subsequently are concerned, we restricted ourselves to opponents with normal dribble behavior that do not panically and prematurely kick the ball away.

Due to the continuous, 9-dimensional state space to be covered a tabular representation of the state value function is impossible and instead the employment of a function approximator is necessary. For this task, we employ multi-layer perceptron neural networks with one hidden layer containing 18 neurons with sigmoidal activation function (hence, a 9:18:1-topology). We perform neural network training in batch mode: Repeatedly a number of training episodes is simulated and in so doing a set of representative states $\tilde{S} \subset S$ is incrementally built up where for each $s \in \tilde{S}$ we have an estimated value $V(s)$ calculated as mentioned above. We do not discard old training instances and invoke a retraining of the neural network each time $|\tilde{S}|$ has been incremented by 250 instances. Let the state value function approximation provided by the net be denoted as $\tilde{V}(s, w)$ where w corresponds to a vector of tunable parameters, i.e. the net's connection weights. Then, the actual training means determining w by solving the least squares optimization problem $min_w \sum_{s \in \tilde{S}} (\tilde{V}(s, w) - V(s))^2$. For the minimization we rely on the efficient back-propagation variant $RPROP$ [14].

4   Empirical Evaluation

To simplify training, the view restrictions imposed by the simulation environment (normally the Soccer Server provides agents only with information about objects in a restricted view cone) are turned off during learning, so that the learning agent obtains full state information.

4.1   Training to Hassle

Since the learning task is episodic, we do not use discounting. Further, we use a learning rate α of 1.0 and employ a purely greedy policy during training, since the neural network-based approximation of V in combination with the large sets

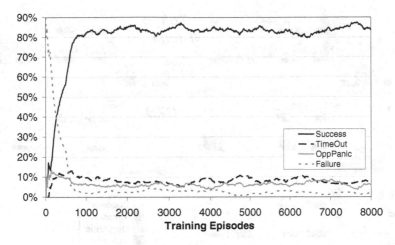

Fig. 3. Exemplary Learning Curve for Learning to Hassle (opponent during training: Wright Eagle, binary from RoboCup 2006)

of random episode start situations brings about a sufficient level of state space exploration. Figure 3 shows the learning curves for an example of a hassle learning experiment. Here, the neural network representing the value function from which a greedy policy is induced has been trained against the champion binary of WrightEagle (2006) for midfield training situations. Apparently, the initially clueless learning agent quickly improves its behavior and finally succeeds in successfully conquering the ball in more than 80% of all attempts. In particular, the number of failure episodes is extremely reduced (to less than 5%) which highlights the effectiveness of this learning approach.

4.2 Testing the NeuroHassle Performance

Being trained against a single opponent team, the danger arises that the resulting behavior is over-specialized, does not generalize very well, and performs poorly against other teams. For this reason, we trained our hassling policy against a selection of teams and for different sets of training situations (on different places on the field). Having obtained a larger number candidate hassle behaviors in this way, it is in principle possible to utilize the corresponding highly specialized hassling policy depending on the current situation. E.g., when playing against UvA Trilearn and given that the ball leading opponent is positioned at $(-36, 24)$, it is most likely that the hassling policy performing best is the one that has been acquired when training against UvA for the *defensiveLeft* training set.

Apart from that, we are aware that the effectiveness of the policy obtained will be influenced by the presence of further opponent teammates. Then, of course, the ball leading opponent may behave differently, for example, it may play a pass quite a time before it is seriously hassled. However, if the defensive strategy is customized to support the intended aggressive hassling of the ball leader, and

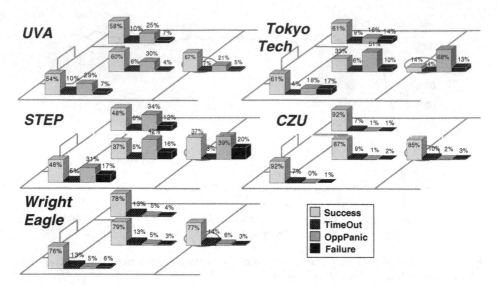

Fig. 4. Generalization Capabilities of the Acquired NeuroHassle Behavior against Various Opponent Teams and for Different Test Situation Sets

if it can be achieved that other adversary players are covered appropriately so that no passing opportunity arises (e.g. [8] focuses on that problem), then it is very likely that, during a real match, the ball leading player behaves similarly as during a training session.

In practice, unfortunately, the above-mentioned approach of employing a large ensemble of highly specialized policies (trained against various opponents and for varying position sets on the field, becomes quickly intractable: First, as discussed before (Section 3.3) it is impossible to acquire a specialized hassling behavior for every potential opponent since the binaries of most teams are not useful for training to hassle against them. And, of course, at times new teams appear for which no binary to be used as training partner is available at all. Second, each team changes from year to year. Hence, the risk of overfitting the hassling policy subject to a specific version of a certain team should not be underestimated.

Consequently, it has been our overall goal to create a hassling policy that generalizes very well over a number of teams (at least over those against which we could test and, in particular, against those teams that are still active). Figure 4 visualizes the performance of our NeuroHassle behavior when dueling against a selection of adversary teams and subject to test situations on different places on the playing field. Since a minimization of the share of failures is our main goal, one can conclude that the behavior acquired is quite effective. In any scenario considered the failure rates are not higher than 20%. Employing the learned NeuroHassle policy in our competition team *Brainstormers* was one of the crucial moves for winning the World Championships tournament RoboCup 2007 in Atlanta.

4.3   Integration into the Competition Team

Designing a strong team defense strategy represents an extremely challenging (and fragile) task when developing a soccer simulation team. It is well-known that in a defensive situation, i.e. when the ball is under control of the opponent team, the players' mission is to either (a) cover opponent players and block potential pass ways, or (b) to try to conquer the ball from the ball leader while simultaneously hindering him from dribbling ahead. Moreover, both tasks have to be assigned to the defending team's players in such a manner that no conflicts arise (e.g. no two players decide for hassling the ball leader in parallel while another opponent remains uncovered) and such that collaborative defense utility is maximized. Generally, it is advisable to select the player closest to the ball leading opponent for executing task (b), whereas the remaining players perform task (a), although a number of exceptional situations exist. A thorough discussion of this issue as well as of how to best solve task (a) is beyond the scope of this paper (see, for example, [8] for more details on these points).

We finish this section with providing some numbers on the utilization of the NeuroHassle policy during actual games against strong opponent teams. During the run of a standard game (6000 time steps), the players of our team start on average 66 hassling episodes. Therefore, even under the very conservative assumption that only about half of all hassle attempts are successful, we can draw the conclusion that the NeuroHassle behavior allows for conquering the ball at least 30 times per game. Furthermore, we determined that on average an opponent ball leading player is being hassled by one of our players during approximately 41.2% of the time he is in ball possession (ball within kick range). This corresponds to a NeuroHassle usage share of circa 14.8% in defensive situations during which the opponent team is building up its attack.

5   Summary

In this paper, we have reported on a comprehensive case study in RoboCup simulated soccer that aims at acquiring a defensive behavior which is meant to disturb an opponent ball leader and to reconquer the ball from him – a capability which is of substantial importance for team success. We developed a comprehensive set of training scenarios and pursued a reinforcement learning method based on neural value function approximation to obtain a good hassle policy for our *Brainstormers* soccer simulation 2D team. The empirical results of applying the acquired policy against various opponent teams were highly encouraging. Consequently, we also successfully applied the described NeuroHassle behavior during 2007's RoboCup World Championships tournament.

Generally, prior to new tournaments a retraining of the NeuroHassle policy must be performed so that it is up-to-date with recent changes introduced in opponent teams' strategies. Apart from that, in future work we also plan to let the agent train against different training opponents that are exchanged from time to time (during a single training session) so that the generalization capabilities can be further increased.

References

1. Kitano, H., Asada, M., Kuniyoshi, Y., Noda, I., Abd, H., Matsubara, E.O.: RoboCup: A Challenge Problem for AI. AI Magazine 18, 73–85 (1997)
2. Kyrylov, V., Greber, M., Bergman, D.: Multi-Criteria Optimization of Ball Passing in Simulated Soccer. Journal of Multi-Criteria Decision Analysis 13, 103–113 (2005)
3. Stone, P., Sutton, R., Kuhlmann, G.: Reinforcement Learning for RoboCup-Soccer Keepaway. Adaptive Bahvior 13, 165–188 (2005)
4. Dashtı, H., Aghaeepour, N., Asadi, S., Bastani, M., Delafkar, Z., Disfani, F., Ghaderi, S., Kamali, S.: Dynamic Positioning Based on Voronoi Cells (DPVC). In: Bredenfeld, A., Jacoff, A., Noda, I., Takahashi, Y. (eds.) RoboCup 2005. LNCS, vol. 4020, pp. 219–229. Springer, Heidelberg (2006)
5. Reis, L., Lau, N., Oliveira, E.: Situation Based Strategic Positioning for Coordinating a Team of Homogeneous Agents. In: Hannebauer, M., Wendler, J., Pagello, E. (eds.) ECAI-WS 2000. LNCS, vol. 2103, pp. 175–197. Springer, Heidelberg (2001)
6. Riedmiller, M., Gabel, T.: On Experiences in a Complex and Competitive Gaming Domain: Reinforcement Learning Meets RoboCup. In: Proceedings of the 3rd IEEE Symposium on Computational Intelligence and Games (CIG 2007), Honolulu, USA, pp. 68–75. IEEE Press, Los Alamitos (2007)
7. Kalyanakrishnan, S., Liu, Y., Stone, P.: Half Field Offense in RoboCup Soccer: A Multiagent Reinforcement Learning Case Study. In: Lakemeyer, G., Sklar, E., Sorrenti, D.G., Takahashi, T. (eds.) RoboCup 2006: Robot Soccer World Cup X. LNCS, vol. 4434, pp. 72–85. Springer, Heidelberg (2007)
8. Kyrylov, V., Hou, E.: While the Ball in the Digital Soccer Is Rolling, Where the Non-Player Characters Should Go in a Defensive Situation?. In: Proceedings of Future Play, Toronto, Canada, pp. 13–17 (2007)
9. Sutton, R.S., Barto, A.G.: Reinforcement Learning. An Introduction. MIT Press/A Bradford Book, Cambridge (1998)
10. Gabel, T., Riedmiller, M.: Learning a Partial Behavior for a Competitive Robotic Soccer Agent. KI Zeitschrift 20, 18–23 (2006)
11. Bertsekas, D.P., Tsitsiklis, J.N.: Neuro Dynamic Programming. Athena Scientific, Belmont, USA (1996)
12. Noda, I., Matsubara, H., Hiraki, K., Frank, I.: Soccer Server: A Tool for Research on Multi-Agent Systems. Applied Artificial Intelligence 12, 233–250 (1998)
13. Sutton, R.S.: Learning to Predict by the Methods of Temporal Differences. Machine Learning 3, 9–44 (1988)
14. Riedmiller, M., Braun, H.: A Direct Adaptive Method for Faster Backpropagation Learning: The RPROP Algorithm. In: Ruspini, H. (ed.) Proceedings of the International Conference on Neural Networks (ICNN), San Francisco, pp. 586–591 (1993)

Constraint Based Belief Modeling

Daniel Göhring, Heinrich Mellmann, and Hans-Dieter Burkhard

Institut für Informatik
LFG Künstliche Intelligenz
Humboldt-Universität zu Berlin, Germany
{goehring,mellmann,hdb}@informatik.hu-berlin.de

Abstract. In this paper we present a novel approach using constraint based techniques for world modeling, i.e. self localization and object modeling. Within the last years, we have seen a reduction of landmarks as beacons, colored goals, within the RoboCup domain. Using other features as line information becomes more important. Using such sensor data is tricky, especially when the resulting position belief is stretched over a larger area. Constraints can overcome this limitations, as they have several advantages: They can represent large distributions and are easy to store and to communicate to other robots. Propagation of a several constraints can be computationally cheap. Even high dimensional belief functions can be used. We will describe a sample implementation and show experimental results.

1   Introduction

Self localization and object tracking is crucial for a mobile robot. Especially when sensing capabilities are limited a short term memory about the surrounding is required. Thus modeling techniques have widely been researched in the past. Common approaches use Bayesian algorithms [5] as Kalman [6] or particle filters [2]. Under some circumstances, when sensor data is sparse and computational power is limited, those approaches can show disadvantages. Complex belief functions are hard to represent for Kalman filters which use Gaussians, particle filters don't have this limitation but need a high number for approximating the belief, resulting in high computational needs which often cannot be satisfied. We tackle this problem by using constraints for sensor data and belief representation. Constraint based modeling approaches have been proposed for localization in [8], or for slam map building in [7]. Constraint based approaches have several advantages: a) constraints are easy to create and to store. b) they have a high representational power, c) combining different constraints is computationally cheap. In this paper we discuss constraint propagation methods for solving navigation problems. The main difference to classical propagation is due to the fact that navigation tasks do always have a solution in reality.

1.1   Motivation

In many domains landmarks are very sparsely arranged. In RoboCup landmarks like beacons were more and more removed during the last years. Other sensor

L. Iocchi et al. (Eds.): RoboCup 2008, LNAI 5399, pp. 73–84, 2009.

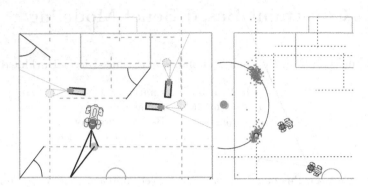

Fig. 1. Robots are localizing a ball relative to field lines, resulting in large distributions a) a robot is seeing a ball next to a line, resulting possible ball positions, b) particle based multi-agent ball localization, as described in [3]

data like field line information has to be used for self localization. We found out that seeing one field line results in a complex belief function which is hard to represent by a Gaussian or by a small set of samples as in Monte-Carlo approaches. Therefore we developed this constraint based representation method.

1.2   Outline

We will show how a constraint based localization can be implemented within the RoboCup Legged league. Furthermore we will compare the constraint based approach to a Monte-Carlo Particle Filter. We will use real robot sensor data and will discuss thereby how noisy and inconsistent sensor data can be considered for constraint localization.

2   Perceptual Constraints

A constraint C is defined over a set of variables $v(1), v(2), ..., v(k)$. It defines the values those variables can take:

$$C \subseteq Dom(v(1)) \times ... \times Dom(v(k))$$

We start with an example from the Legged league where the camera image of a robot shows a goal in front and the ball before the white line of the penalty area (Figure 2). It is not too difficult for a human interpreter to give an estimate for the position (x_B, y_B) of the ball and the position (x_R, y_R) of the observing robot. Humans can do that, regarding relations between objects, like the estimated distance d_{BR} between the robot and the ball, and by their knowledge about the world, like the positions of the goalposts and of the penalty line.

The program of the robot can use the related features using image processing. The distance d_{BR} can be calculated from the size of the ball in the image, or from the angle of the camera. The distance d_{BL} between the ball and the penalty line

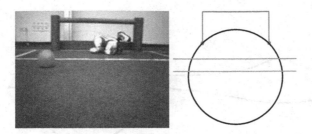

Fig. 2. Example from RoboCup (Four legged league): A robot is seeing a goal and the ball in front of a penalty line. The right picture shows the resulting robot positions represented by the periphery circle according to C_1, and the line of the Ball-Line-Constraint C_2.

can be calculated, too. Other values are known parameters of the environment: $(x_{Gl}, y_{Gl}), (x_{Gr}, y_{Gr})$ are the coordinates of the goalposts, and the penalty line is given as the set of points $\{(x, b_{PL}) | -a_{PL} \leq x \leq a_{PL}\}$. The coordinate system has its origins at the center point, the y-axis points to the observed goal.

The relations between the objects can be described by constraints. The following four constraints are obvious by looking to the image, and they can be determined by the program of the observing robot:

C_1: The view angle γ between the goalposts (the distance between them in the image) defines a circle (periphery circle), which contains the goal posts coordinates $(x_{Gl}, y_{Gl}), (x_{Gr}, y_{Gr})$ and the coordinates (x_R, y_R) of the robot:

$$\{(x_R, y_R) | \arctan \frac{y_{Gl} - y_R}{x_{Gl} - x_R} - \arctan \frac{y_{Gr} - y_R}{x_{Gr} - x_R} = \gamma\}$$

C_2: The ball lies in the distance d_{BL} before the penalty line. Therefore, the ball position must be from the set

$$\{(x_B, y_B) | x_B \in [-a_{PL}, a_{PL}], y_B = b_{PL} - d_{BL}\}$$

C_3: The distance d_{BR} between the robot and the ball defines a circle such that the robot is on that circle around the ball:

$$(x_B - x_R)^2 + (y_B - y_R)^2 = d_{BR}^2.$$

C_4: The observer, the ball and the left goal post are on a line:

$$\frac{x_R - x_B}{y_R - y_B} = \frac{x_B - x_{Gl}}{y_B - y_{Gl}}$$

The points satisfying the constraints by C_1 (for the robot) and by C_2 (for the ball) can be visualized immediately on the playground as in Figure 2.

The constraint by C_3 does not give any restriction to the position of the ball. The ball may be at any position on the playground, and then the robot has a

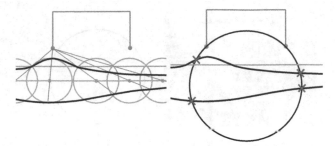

Fig. 3. Left: The picture shows the Constraint C_2 for the ball, some of the circles according to constraint C_5, some of the lines according to C_4, and the resulting two lines for C_6 (bold black lines). Right: Constraints according to C_7: The position of the robot is one of the four intersection points between the periphery circle (C_1) and the lines according to C_6.

position somewhere on the circle around the ball. Or vice versa for reasons of symmetry: The robot is on any position of the playground, and the ball around him on a circle. In fact, we have four variables which are restricted by C_3 to a subset of a four dimensional space. The same applies to constraint C_4.

The solution (i.e. the positions) must satisfy all four constraints. We can consider all constraints in the four dimensional space of the variables (x_B, y_B, x_R, y_R) such that each constraint defines a subset of this space. Then we get the following constraints:

$$C_1 = \{\arctan \frac{y_{Gl} - y_R}{x_{Gl} - x_R} - \arctan \frac{y_{Gr} - y_R}{x_{Gr} - x_R} = \gamma\} \tag{1}$$

$$C_2 = \{(x_B \in [-a_{PL}, a_{PL}], y_B = b_{PL} - d_{BL}\} \tag{2}$$

$$C_3 = \{(x_B - x_R)^2 + (y_B - y_R)^2 = d_{BR}^2\} \tag{3}$$

$$C_4 = \{\frac{x_R - x_B}{y_R - y_B} = \frac{x_B - x_{Gl}}{y_B - y_{Gl}}\} \tag{4}$$

Thus the possible solutions (as far as determined by C_1 to C_4) are given by the intersection $\bigcap_{1,\dots,4} C_i$. According to this fact, we can consider more constraints $C_5, \dots, C_n$ as far as they do not change this intersection, i.e. as far as $\bigcap_{1,\dots,n} C_i = \bigcap_{1,\dots,4} C_i$. Especially, we can combine some of the given constraints.

By combining C_2 and C_3 we get the constraint $C_5 = C_2 \cap C_3$ where the ball position is restricted to any position on the penalty line, and the player is located on a circle around the ball. Then, by combining C_4 and C_5 we get the constraint $C_6 = C_4 \cap C_5$ which restricts the positions of the robot to the two lines shown in Figure 3 (left).

Now intersecting C_1 and C_6 we get the constraint C_7 with four intersection points as shown in Figure 3 (right). According to the original constraints C_1 to C_4, these four points are determined as possible positions of the robot. The corresponding ball positions are then given by C_2 and C_4.

To find the real positions, we would need additional constraints from the image, e.g. that the ball lies between the robot and the goal (which removes one of the lines of C_6), and that the robot is located on the left site of the field (by exploiting perspective).

3 Formal Definitions of Constraints

We define all constraints over the set of all variables $v(1), v(2), ..., v(k)$ (even if some of the variables are not affected by a constraint). The domain of a variable v is denoted by $Dom(v)$, and the whole universe under consideration is given by

$$U = Dom(v(1)) \times \cdots \times Dom(v(k))$$

For this paper, we will consider all domains $Dom(v)$ as (may be infinite) intervals of real numbers, i.e. $U \subseteq \mathbb{R}^k$.

Definition 1. *(Constraints)*

1. *A **constraint** C over $v(1), ..., v(k)$ is a subset $C \subseteq U$.*
2. *An assignment β of values to the variables $v(1), ..., v(k)$, i.e. $\beta \in U$, is a* **solution** *of C iff $\beta \in C$.*

Definition 2. *(Constraint Sets)*

1. *A **constraint set** $\mathcal{C}$ over $v(1), ..., v(k)$ is a finite set of constraints over those variables: $\mathcal{C} = \{C_1, ..., C_n\}$.*
2. *An assignment $\beta \in U$ is a **solution** of $\mathcal{C}$ if β is a solution of all $C \in \mathcal{C}$, i.e. if $\beta \in \bigcap \mathcal{C}$.*
3. *A constraint set $\mathcal{C}$ is **inconsistent** if there is no solution, i.e. if $\bigcap \mathcal{C}$ is empty.*

The problem of finding solutions is usually denoted as solving a constraint satisfaction problem (CSP) which is given by a constraint set $\mathcal{C}$. By our definition, a solution is a point of the universe U, i.e. an assignment of values to all variables. For navigation problems it might be possible that only some variables are of interest. This would be the case if we are interested only in the position of the robot in our example above. Nevertheless we had to solve the whole problem to find a solution.

In case of robot navigation, there is always a unique solution of the problem in reality (the positions in the real scene). This has an impact on the interpretation of solutions and inconsistencies of the constraint system (cf. Section 4.1).

The constraints are models of relations (restrictions) between objects in the scene. The information can be derived from sensory data, from communication with other robots, and from knowledge about the world – as in the example from above. Since information may be noisy, the constraints might not be as strict as in the introductory example from Section 2. Instead of a circle we get an annulus for the positions of the robot around the ball according to C_3 in the example. In general, a constraint may concern a subspace of any dimension (e.g. the whole penalty area, the possible positions of an occluded object, etc.). Moreover, constraints need not to be connected: If there are indistinguishable landmarks, then the distance to such landmarks defines a constraint consisting of several circles. Further constraints are given by velocities: changes of locations are restricted by the direction and speed of objects.

4   Algorithms

In principle, many of the problems can be solved by grid based techniques. For each grid cell we can test if constraints are satisfied. This corresponds to some of the known Bayesian techniques including particle filters.

Another alternative are techniques from constraint propagation. We can successively restrict the domains of variables by combining constraints. We will discuss constraint propagation in the following subsection, later we will present experimental results for this approach.

4.1   Constraint Propagation

Known techniques (cf. e.g. [1] [4]) for constraint problems produce successively reduced sets leading to a sequence of decreasing restrictions

$$U = D_0 \supseteq D_1 \supseteq D_2, \supseteq \ldots$$

Restrictions for numerical constraints are often considered in the form of k-dimensional intervals $I = [a, b] := \{x | a \leq x \leq b\}$ where $a, b \in U$ and the $\leq$-relation is defined componentwise. The set of all intervals in U is denoted by $\mathcal{I}$. A basic scheme for constraint propagation with

- A constraint set $\mathcal{C} = \{C_1, ..., C_n\}$ over variables $v(1), ..., v(k)$ with domain $U = Dom(v(1)) \times ... \times Dom(v(k))$.
- A selection function $c : \mathbb{N} \to \mathcal{C}$ which selects a constraint C for processing in each step i.
- A propagation function $d : 2^U \times \mathcal{C} \to 2^U$ for constraint propagation which is monotonously decreasing in the first argument: $d(D, C) \subseteq D$.
- A stop function $t : \mathbb{N} \to \{true, false\}$.

works as follows:

Definition 3. *(Basic Scheme for Constraint Propagation, BSCP)*

- *Step(0) Initialization: $D_0 := U$, $i := 1$*
- *Step(i) Propagation: $D_i := d(D_{i-1}, c(i))$.*
- *If $t(i) = true$: Stop.*
- *Otherwise $i := i + 1$, continue with Step(i).*

We call any algorithm which is defined according to this scheme a BSCP-algorithm.

The restrictions are used to shrink the search space for possible solutions. If the shrinkage is too strong, possible solutions may be lost. For that, backtracking is allowed in related algorithms.

Definition 4. *(Locally consistent propagation function)*

1. *A restriction D is called* **locally consistent w.r.t. a constraint** C *if*

$$\forall d = [d_1, ..., d_k] \in D \quad \forall i = [1, ..., k] \exists d' = [d'_1, ..., d'_k] \in D \cap C \ : d_i = d'_i$$

   i.e. if each value of a variable of an assignment from D can be completed to an assignment in D which satisfies C.
2. *A propagation function $d : 2^U \times C \to 2^U$ is* **locally consistent** *if it holds for all D, C: $d(D, C)$ is locally consistent for C.*
3. *The* **maximal locally consistent** *propagation function $d_{maxlc} : 2^U \times C \to 2^U$ is defined by $d_{maxlc}(D, C) := Max\{d(D, C) | d$ is locally consistent$\}$.*

Since the search for solutions is easier in a more restricted search space (as provided by smaller restrictions D_i), constraint propagation is often performed not with d_{maxlc}, but with more restrictive ones. Backtracking to other restrictions is used if no solution is found.

For localization tasks, the situations is different: We want to have an overview about **all** possible poses. Furthermore, if a classical constraint problem is inconsistent, then the problem has no solution. As already stated, for localization problems always exists a solution in reality (the real poses of the objects under consideration) so we must be careful not to loose solutions.

Definition 5. *(Conservative propagation function)*
   A propagation function $d : 2^U \times C \to 2^U$ is called **conservative** *if $D \cap C \subseteq d(D, C)$ for all D and C.*

Note that the maximal locally consistent restriction function d_{maxlc} is conservative. We have:

Proposition 1. *Let the propagation function d be conservative.*

1. *Then it holds for all restrictions $D_i : \bigcap C \subseteq D_i$.*
2. *If any restriction D_i is empty, then there exists no solution, i.e. $\bigcap C = \emptyset$.*

If no solution can be found, then the constraint set is inconsistent. There exist different strategies to deal with that:

- enlargement of some constraints from C,
- usage of only some constraints from C,
- computation of the best fitting hypothesis according to C.

As already mentioned above, n-dimensional intervals are often used for the restrictions D, since the computations are much easier. Constraints are intersected with intervals, and the smallest bounding interval can be used as a conservative result. Examples are given in Fig. 4.

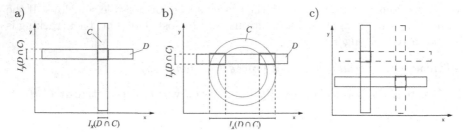

Fig. 4. Constraint propagation with intervals D for a) two rectangular constraints C b) a rectangular and a circular constraint C, resulting in a constraint consisting of two rectangular areas. *Intervals of Projection* w.r.t. $C \cap D$ are illustrated. c) Two constraints consisting of two boxes each are intersected with each other, resulting constraints depicted as bold red squares.

While local consistency is the traditional approach (to find only some solutions), the approach with conservative intervals is more suited for localization tasks because it can be modified w.r.t. to enlarging constraints during propagation for preventing from inconsistency.

Now we want to present a constraint propagation scheme. The stop condition compares the progress after processing each constraint once.

Algorithm 1. Constraint Propagation with Minimal Conservative Intervals, MCI-algorithm

---

> **Input:** constraint set $\mathcal{C} = \{C_1, ..., C_n\}$ with variables $\mathcal{V} = \{v_1, ..., v_k\}$
> over domain U and a time bound T
> **Data:** $D \leftarrow U$, $s \leftarrow 1$, $D_{old} \leftarrow \emptyset$
> **Result:** minimal conservative k-dimensional interval D

1 **while** $s < T$ & $D \neq D_{old}$ **do**
2 $\quad$ $D_{old} \leftarrow D$;
3 $\quad$ **foreach** $C \in \mathcal{C}$ **do**
4 $\quad\quad$ **foreach** $v \in \mathcal{V}$ **do**
5 $\quad\quad\quad$ $D(v) \leftarrow I_v(D \cap C)$;
6 $\quad\quad$ **end**
7 $\quad\quad$ $D \leftarrow D(v_1) \times \cdots \times D(v_n)$;
8 $\quad$ **end**
9 $\quad$ $s \leftarrow s + 1$;
10 **end**

Looking closer to the possible intersections of constraints (e.g. to the intersection of two circular rings or to the intersection of a circular ring with an rectangle like in Fig. 4a), the sets $D \cap C$ might be better approximated by sets of intervals instead of a single interval (see Fig. 4 b)). Thus, the algorithm was extended for implementation this way: The input and the output for each step are sets of intervals, and all input intervals are processed in parallel. For such purposes the propagation function d of the BSCP could be defined over sets as well. As in other constraint propagation algorithms, it might lead to better propagation results if we split a given interval to a union of smaller intervals. In many cases, when using more constraints, the restrictions end up with only one of the related intervals anyway.

Using Odometry data. When the robot moves, in self-localization it shifts the constraint boundaries into to movement direction. The odometry noise results in an enlargement of the shifted constraints to pay tribute to slippery ground, collisions and walking noise. The appropriate constraint enlargement was found experimentally.

4.2 Handling Inconsistencies

When performing real robot experiments, we realized that many constraints were based on noisy sensor data, resulting to inconsistencies. We dealt with that in the following way: We distinguished sensor data that was consistent to the current belief from inconsistent data, calling it the inconsistent data ratio IDR. When IDR was low, updating the belief used consistent sensor data only, resulting in a stable localization. But when new sensor data became inconsistent, e.g. as in kidnapped robot problem, resulting in a very large IDR, we added that inconsistent data to the belief constraints as well, enabling the robot belief to converge to the new position. For deciding what to do we used a threshold value.

5   Experimental Results

In our experiments within the RoboCup soccer domain (see section 2), we compared an implementation of a Monte-Carlo particle filter (MCPF) with the constraint based algorithm described above. We had our focus on calculation time and on localization accuracy.

We used constraints given by fixed objects like goalposts, flags and field lines identified in the images by the camera of the robot. The creation of the related constraints was done as follows: distances to landmarks are defined by circular rings, where only the distances derived from the vision system of the robot and the standard deviation of the measurement error have to be injected. Constraints given by observed field lines are defined by a set of rectangles and angles (Fig. 5 a)), the distances and the horizontal bearings are sufficient to define these constraints. All this can be done automatically. An example for constraints generated from lines and their propagation is given in Fig. 5 b).

During our first experiments we let a simulated robot move on a predefined path. Then we compared the modeled position with the ground truth position and calculated the localization error. Furthermore we measured in every time step the calculation time. As reference algorithm we used a Monte-Carlo particle filter.

a)   b)

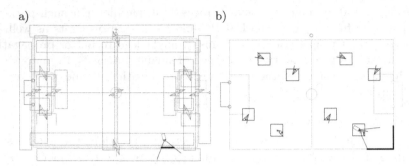

Fig. 5. Simulated robot situated on a soccer field. Bold black lines depict the line segment seen by the robot. a) Gray boxes illustrate a constraint generated from only one seen line segment. b) Two constraints are generated from perceived lines (not in the figure), black boxes depict the resulting constraint after propagation of the two constraints.

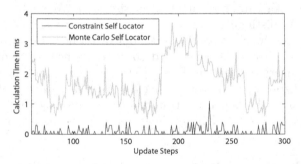

Fig. 6. Calculation time for one modeling step on a 1.5 GHz processor. Gray line: Monte Carlo particle filter, using 100 samples. Black line: Calculation time per step using the constraint based algorithm.

The time measurement data showed that the constraint based algorithm (MCI) algorithm works about 5-10 times faster than the particle filter (see Fig. 6). It also showed that the calculation time for the particle based approach is varying much more than for the constraint based approach.

In a further experiment we measured the localization accuracy for both approaches (Fig. 8). Most of the time the accuracies were comparable. Sometimes the constraint based approach was more sensible to noisy sensor data, which resulted in slightly jumping positions, as Fig. 8 b) shows. In future work we will investigate how the position can be more stabilized over time.

In another experiment we investigated more ambiguous data (i.e. when only few constraints are available as in Fig. 5). In this case, the constraint based approach provided a much better representation of all possible positions (all those positions which are consistent with the vision data). The handling of such cases

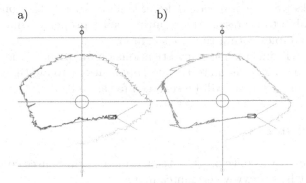

Fig. 7. Localization accuracy experiment. A robot is walking on the field in a circle (simulated) a) Monte-Carlo Particle filter based localization, the straight reference line is shown as well under the modeled localization trace. b) Constraint based localization.

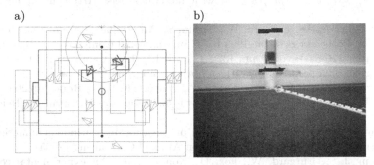

Fig. 8. Real robot experiment: a) The constraints generated from scene b) Recognized flag and line depicted. The two bold rectangles (left) show that image data leaves two possible position areas.

is difficult for particle filters because many particles are necessary for representing large belief distributions. Related situations may appear for sparse sensor data and for the kidnapped robot problem.

6   Conclusion

Constraint propagation techniques are an interesting alternative to probabilistic approaches. They could help for better understanding of navigation tasks at all and permit the investigation of larger search spaces employing the constraints between various data. Therewith, the many redundancies provided by images and other sensor data can be exploited better.

We presented an algorithm for constraint propagation and discussed some differences to classical constraint solving techniques. In our experiments, the algorithm outperformed approaches like particle filters with regard to calculational needs. The localization quality of both algorithms was comparable when many landmarks were seen. In case of fewer landmarks only the constraint based approach was able to represent the resulting belief properly, which was caused by the lack of samples within the particle filter.

Future work will include more investigations on algorithms and further comparisons with existing Bayesian techniques, applications to multiple target tracking with non-unique targets will be searched for as well.

Acknowledgments

Program code used was developed by the GermanTeam. Source code is available for download at http://www.germanteam.org

References

1. Davis, E.: Constraint propagation with interval labels. Artificial Intelligence 32 (1987)
2. Dellaert, F., Fox, D., Burgard, W., Thrun, S.: Monte carlo localization for mobile robots. In: Proceedings of the 1999 IEEE International Conference on Robotics and Automation (ICRA), vol. 2, pp. 1322–1328. IEEE, Los Alamitos (1999)
3. Göhring, D.: Cooperative object localization using line-based percept communication. In: Visser, U., Ribeiro, F., Ohashi, T., Dellaert, F. (eds.) RoboCup 2007: Robot Soccer World Cup XI. LNCS (LNAI), vol. 5001, pp. 53–64. Springer, Heidelberg (2008)
4. Goualard, F., Granvilliers, L.: Controlled propagation in continuous numerical constraint networks. In: ACM Symposium on Applied Computing (2005)
5. Gutmann, J.-S., Burgard, W., Fox, D., Konolige, K.: An experimental comparison of localization methods. In: Proceedings of the 1998 IEEE/RSJ International Conference on Intelligent Robots and Systems (IROS). IEEE, Los Alamitos (1998)
6. Kalman, R.: A new approach to linear filtering and prediction problems. Transactions of the ASME - Journal of Basic Engineering 82, 35–45 (1960)
7. Olson, E., Leonard, J., Teller, S.: Fast iterative alignment of pose graphs with poor initial estimates. In: International Conference on Robotics and Automation, 2006. ICRA 2006. Proceedings 2006 IEEE (2006)
8. Stroupe, A., Martin, M., Balch, T.: Distributed sensor fusion for object position estimation by multi-robot systems. In: Bredenfeld, A., Jacoff, A., Noda, I., Takahashi, Y. (eds.) Proceedings of the 2001 IEEE International Conference on Robotics and Automation (ICRA 2001) (2001)

Explicitly Task Oriented Probabilistic Active Vision for a Mobile Robot[*]

Pablo Guerrero, Javier Ruiz-del-Solar, and Miguel Romero

Department of Electrical Engineering, Universidad de Chile
{pguerrer,jruizd}@ing.uchile.cl

Abstract. A mobile robot has always uncertainty about the world model. Reducing this uncertainty is very hard because there is a huge amount of information and the robot must focus on the most relevant one. The selection of the most relevant information must be based on the task the robot is executing, but there could be several sources of information where the robot would like to focus on. This is also true in robot soccer where the robot must pay attention to landmarks in order to self-localize and to the ball and robots in order to follow the status of the game. In the presented work, an explicitly task oriented probabilistic active vision system is proposed. The system tries to minimize the most relevant components of the uncertainty for the task that is been performed and it is explicitly task oriented in the sense that it explicitly considers a task specific value function. As a result, the system estimates the convenience of looking towards each of the available objects. As a test-bed for the presented active vision approach, we selected a robot soccer attention problem: goal covering by a goalie player.

1 Introduction

A mobile robot that is performing a task must make decisions based on information of the state of its environment and itself, often called the *world model*. Unfortunately, a mobile robot has always some degree of uncertainty in its world model. This uncertainty comes from the lack of information and the noise in the sensor measurements and the actions results. One of the key questions in mobile robotics is how to measure, reduce, and handle this uncertainty. Since this uncertainty may have a tremendous negative impact on the performance of the task that the robot is executing, one important issue is how the robot can select actions to reduce it. Active vision basically consists of executing control strategies with the purpose of improving the perception performance by focusing on the most relevant parts of the sensor information. An active vision system may control physical variables, such as gaze direction, camera parameters, etc., and/or the way data is processed such as the *region of interest* (ROI), vision methodology and parameters, resolution, etc. The relationship between perception and action has been studied in several fields apart from robotics such as neurosciences, psychology and cognitive science. For instance, experiments have showed that

---

[*] This research was partially supported by FONDECYT (Chile) under Project Number 1090250.

L. Iocchi et al. (Eds.): RoboCup 2008, LNAI 5399, pp. 85–96, 2009.

the task that a human is executing has a strong influence on his/her gaze selection [1]. This suggests that the selection of the most relevant information must be based on the task the robot is executing.

In the robot soccer domain, an autonomous robot player must estimate the state of a complex and dynamic world: it must pay attention to landmarks in order to self-localize and to the ball and robots in order to follow the status of the game. The process of estimating the absolute *pose*[1] of the robot itself from *landmark*[2] observations is known as *localization*, while the process of estimating a mobile object position from observations of the object is known as *object tracking*. A robot soccer player must perform localization and ball tracking processes simultaneously, which is a problem because these two processes require the observation of different objects to correct the odometry and observation noise.

In this work we present an explicitly task oriented approach to the active vision problem, which tries to reduce the more relevant components of the uncertainty in the world model, for the execution of the current task the robot is executing. The presented approach is explicitly task oriented because it intends to minimize the variance of a task specific *value function*[3]. The output of the system is a selected object which is the most convenient to focus on. The focus on the selected object could be subsequently achieved by means of any of the control actions already mentioned. The proposed approach is generally applicable for problems where the following assumptions hold: (i) there is an independence between the actions necessary to accomplish the task and those necessary to direct the gaze or select the vision parameters, (ii) there is a world modeling stage which uses a Bayesian filter to estimate the world state, (iii) the task been executed has a value function defined, and (iv) the system knows a priori which objects it can observe and the state belief bring some information about where they can be observed. For problems where the former independence is not present (for example, robots without neck movements), the proposed approach is not generally applicable. In such cases, an approach that considers the interdependence between actions and observations (for example POMDP's) may be better suited. Besides, even in a system where that independence holds, sometimes it is convenient to modify the body position in order to improve the perceptions. Handling these situations is beyond the scope of this paper. Regarding the state estimation, the presented system assumes that the world modeling stage is based on a Kalman filter (KF) or any of its nonlinear variants (EKF, UKF, etc.), and thus, the belief is Gaussian. However, the extension for world modeling stages based on particle filters (PF), and thus on sample-based representations of the belief, is simple. Finally, the existence of a value function is held by a wide variety of applications, since it is difficult to even make decisions without a measurement of the convenience of each state.

We test our general approach on a specific application taken from robot soccer: goal covering by a goalie player. Robot soccer is a very interesting platform for testing robotic methodologies and algorithms since it presents a challenging environment

---

[1] The *pose* of the robot is defined as its absolute position and orientation.
[2] A *landmark* is defined as an observable object whose absolute position is fixed and known.
[3] A *value function* is defined as a function of the world state that increases with the degree of convenience of the world state for the execution of the current task.

and complex tasks [2]. This problem makes the selected application interesting for testing the proposed approach for gaze direction selection.

The structure of the paper is as follows. First, in section 2, the related work is reviewed. Then, section 3 presents the general statement of the proposed approach, while section 4 studies a particular case taken from the robot soccer problem. In section 5, experimental results are presented and discussed. Finally, in section 6, conclusions from the presented work are drawn.

2 Related Work

The basic ideas of the active vision paradigm were introduced under different names such as: *active vision* [3], *active perception* [4], and *animate vision* [5]. The contribution of this paradigm is that it does not consider perception as a passive process but as one that could conveniently influence the control actions. A huge amount of literature has been developed regarding active vision but it is beyond the scope of this paper to survey it, so we will focus on the most pertinent parts of it.

There are works where the control actions explicitly aim to reduce uncertainty in a probabilistic fashion, but they do it "blindly" (they focus on reducing the belief entropy), and thus, they are not generally optimal from the task performance point of view. The controlled variables in these works include, for example, robot movements and/or sensor direction [6][8], camera parameters [7]. Furthermore, they are not general from the task point of view since their goal is always the uncertainty reduction itself. From our point of view, the uncertainty reduction is a means but not an end. While the optimality from the task performance point of view for continuous beliefs is not achievable using digital computers, because integrals over the belief cannot be calculated, this work presents a suboptimal approximation which approximates the continuous beliefs by discrete ones.

In the RoboCup domain, there are some published papers on active vision. One of them [9] uses the localization estimate to change the way the vision system looks for the fixed objects in the image. As a result, the mean time required to process an image is reduced with no relevant change in the detection performance. Another work builds a decision tree for decision making and gaze selection [11]. The active vision system consists of maximizing the information gain in terms of which action should be selected, which is an advantage of this system, since it is oriented to the action selection itself. However, from our point of view, the direct mapping between observations and actions (without a state estimator) and the use of a decision tree, is not able to cope with the complexity and dynamism of real mobile robot applications (including robot soccer), and then the applicability of this kind of solutions is restricted. Finally, there is a particular work [10] which also uses a value function for the active vision system but: (i) the decision that is made in that case is whether the robot should move or observe, which we believe is control decision which can have very weaker effects and less applicability than the one of in which object to focus on, since for complex tasks several objects may bring relevant information, and (ii) it is restricted to a particular application: the navigation task with sample based beliefs of the robot localization.

3 Proposed Approach

3.1 Basic Definitions

In time step k, the *world state*, $\mathbf{x}_k$, is the set of the relevant variables of the world model (robot localization, other objects position, etc.) for the robot to make decisions. Then, we can define the *state belief*, or simply the *belief*, as the probability density function (pdf) of $\mathbf{x}_k$. For any task that the robot is performing, we should be able to define a *value function* $V_k(\mathbf{x}_k)$, which tells the robot how convenient is the state $\mathbf{x}_k$ for the accomplishment of this task. Depending on the complexity of the task, $V_k(\mathbf{x}_k)$ might be obtained from simple ad-hoc heuristics or calculated from popular methods like *Markov Decision Processes* or *Reinforcement Learning*. Note that the function $V_k(\mathbf{x}_k)$ is, in general, dependent on k^4. In the case when $V_k(\mathbf{x}_k)$ does not depend on k, the subscript can be omitted. The proposed approach assumes that the robot control system has at least three functionalities: (i) *vision*, where the camera images are processed in order to get the observation $\mathbf{z}_k$, (ii) *world modeling*, where $\mathbf{z}_k$ is used to calculate the belief, and (iii) *decision making*, where the belief is used to select the action $\mathbf{u}_{k+1}$ that the robot will execute in the next instant, in order to minimize the variance of the value function: $\mathrm{var}(V_{k+1}(\mathbf{x}_{k+1}))$.

The world modeling and decision making modules make use of the knowledge of two probabilistic models: the *process model*, $\mathbf{x}_{k+1} = f(\mathbf{x}_k, \mathbf{u}_k, \mathbf{w}_k)$, which models the effect of the robots actions in the state transitions, and the *observation model*, $\mathbf{z}_k = h(\mathbf{x}_k, \mathbf{u}_k, \mathbf{v}_k)$, which models the relationship between the current state and the current observation. Both models consider the existence of zero-mean random noises $\mathbf{w}_k$ and $\mathbf{v}_k$ with respective covariances $\mathbf{Q}_k$ and $\mathbf{R}_k$. Also, the Jacobians of f and h functions are respectively noted as $\mathbf{F}_k$ and $\mathbf{H}_k$.

In many mobile robotics applications, the action at time k, $\mathbf{u}_k$, can be decomposed into two parts: one, $\mathbf{u}_k^{act}$, that influences f and possibly h, and other, $\mathbf{u}_k^{sense}$, that only influences h. For example, in a legged robot with a camera mounted on a mobile neck, whose neck movements does not influence the world state, $\mathbf{u}_k^{act}$ corresponds to the movements of the legs, and $\mathbf{u}_k^{sense}$ corresponds to the movements of the neck and the vision parameters selection. Defining $f(\mathbf{x}_k, \mathbf{u}_k) = f(\mathbf{x}_k, \mathbf{u}_k, 0)$, the former condition is equivalent to $f(\mathbf{x}_k, \mathbf{u}_k) = f(\mathbf{x}_k, \mathbf{u}_k^{act})$. In this case, the selection of $\mathbf{u}_k^{act}$ may be performed before that of $\mathbf{u}_k^{sense}$, and then the selected $\mathbf{u}_k^{act}$ may be known in the $\mathbf{u}_k^{sense}$ selection process. The presented approach consists of selecting $\mathbf{u}_k^{sense}$ given $\mathbf{u}_k^{act}$ known, and assuming the active vision module is not able to modify $\mathbf{u}_k^{act}$. Since

---

[4] An example of tasks in which V_k depends on k is when the task has a finite time horizon, because V_k might depend on how much time is still available for the task execution.

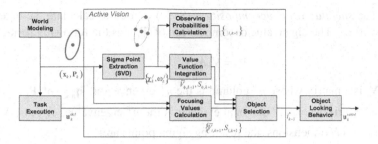

Fig. 1. Block diagram of the proposed approach for active vision (see explanations in sections 3.2-3.6)

$\mathbf{u}_k^{sense}$ cannot directly influence the future states, its main goal is to reduce the uncertainty of the state. To reduce that uncertainty in the most relevant components from the task point of view, we have selected to minimize the variance of the value function as an optimality criterion. Note that if the active vision system uses the maximization of the expected value function as an optimality criterion it will aim to be as optimistic as possible about the future values, but it does not ensure the reduction of the uncertainty.

The proposed system is designed to have as an input a Gaussian representation, $N(\overline{\mathbf{x}}_k, \mathbf{P}_k)$, of the belief. However, the system can be easily adapted to receive sample-based belief representations. The approach is based on the assumption that, given the *predicted state* $\overline{\mathbf{x}}_{k+1}^-$, each object i will have a possibly different observation noise covariance $\mathbf{R}_{k+1}^i$[5] and naturally a different observation model jacobian $\mathbf{H}_{k+1}^i$, evaluated in the next predicted state. For each of the objects that the robot is able to look at, the variance of $V_{k+1}(\mathbf{x}_{k+1})$, given that the object will be seen in the next instant, is estimated. Then, the object with the lowest value function variance is selected, and an *object focusing behavior* executes the respective control actions, such as directing the robot gaze towards this object and/or setting the vision ROI in the object. A block diagram of the proposed active vision system may be seen in figure 1. In the next subsections, the system modules will be described.

3.2 Sigma Point Extraction

In order to make the integrals over the state space computable, the Gaussian pdf coming from the world modeling stage is transformed into a *sigma-point* pdf. A sigma point pdf is a set $\{\chi_i, \omega_i\}$ of weighted points which are sampled, from a source pdf, following some deterministic pattern which conserves the pdf's two first moments. The sigma points are calculated using the eigenvalue decomposition of the covariance

---

[5] Here both $\mathbf{H}_{k+1}^i$ and $\mathbf{R}_{k+1}^i$ are at most function of i and $\overline{\mathbf{x}}_{k+1}^-$. If they are also a function of the observation $\mathbf{z}_{k+1}$, the expected observation $\mathbf{h}^i(\overline{\mathbf{x}}_{k+1}^-)$ of the ith object can be considered.

matrix. The *singular value decomposition* (SVD) is used to calculate the eigenvalues decomposition[6]. The eigenvalue decomposition of $\mathbf{P}_k$ has the following form:

$$\mathbf{P}_k = \mathbf{V}_k \mathbf{\Lambda}_k \mathbf{V}_k^T \tag{1}$$

Where $\mathbf{V}_k$ is a matrix whose n^{th} column is the nth eigenvector, $\mathbf{q}_{n,k}$, of $\mathbf{P}_k$, and $\mathbf{\Lambda}_k$ is a diagonal matrix whose n^{th} diagonal element is the nth eigenvalue, $\lambda_{i,k}$, of $\mathbf{P}_k$. If the state space has N dimensions, the resulting sigma points are:

$$\chi_k^j = \begin{cases} \overline{\mathbf{x}}_k & j = 0 \\ \overline{\mathbf{x}}_k + \sqrt{(N+\eta)\lambda_{j,k}}\,\mathbf{q}_{j,k} & j \in [1,N] \\ \overline{\mathbf{x}}_k - \sqrt{(N+\eta)\lambda_{j-N,k}}\,\mathbf{q}_{j-N,k} & j \in [N+1,2N] \end{cases} \tag{2}$$

We have set the scaling parameter η to 1. The corresponding weights are:

$$\omega_k^j = \begin{cases} \omega_0 = \dfrac{\eta}{(N+\eta)} & j = 0 \\ \dfrac{1}{2(N+\eta)} & \sim \end{cases} \tag{3}$$

3.3 Observing Probabilities Calculation

Let us define $\kappa_{i,k}$ and $\varphi_{i,k}$ as the events: (i) "at time k, the robot focus on the ith object" and (ii) "at time k, the robot observes the ith object", respectively. Note that from our definition of the problem (where $\mathbf{u}_k^{sense}$ does not influence $\mathbf{u}_k^{act}$), $\kappa_{i,k}$ does not influence $\mathbf{x}_k$, and thus $p(\mathbf{x}_k|\kappa_{i,k}) = p(\mathbf{x}_k)$. Then we can define $\alpha_{i,k}$ as the probability of $\varphi_{i,k}$ conditioned on $\kappa_{i,k}$:

$$\alpha_{i,k} = p(\varphi_{i,k}|\kappa_{i,k}) = \int_{\mathbf{x}_k \in \mathbf{X}} p(\varphi_{i,k}|\mathbf{x}_k, \kappa_{i,k}) p(\mathbf{x}_k) d\mathbf{x}_k \tag{4}$$

We can approximate the former integral using the sigma point representation $\{\chi_k^j, \omega_{j,k}\}$ of the pdf of $\mathbf{x}_k$ calculated as explained in section 3.2:

$$\alpha_{i,k} \approx \sum_{j \in [0,2N]} p(\varphi_{i,k}|\mathbf{x}_k = \chi_k^j, \kappa_{i,k}) \omega_{j,k} \tag{5}$$

The event $\varphi_{i,k}$ can be seen as the joint occurrence of three independent events: (i) $\nu_{i,k}$, "the ith object is inside the field of view of the camera", (ii) $\vartheta_{i,k}$, "the ith object is not been occluded by other objects", and (iii) $\rho_{i,k}$, "the camera and the vision module

---

[6] Since the covariance matrix is symmetric and positive semi-definite, SVD and the eigenvalue decomposition are equivalent.

with their physical limitations and current parameters are able to percept the ith object". Then,

$$p(\varphi_{i,k}|\mathbf{x}_k,\kappa_{i,k}) = p(\nu_{i,k}|\mathbf{x}_k,\kappa_{i,k})p(\vartheta_{i,k}|\mathbf{x}_k,\kappa_{i,k})p(\rho_{i,k}|\mathbf{x}_k,\kappa_{i,k}) \qquad (6)$$

$p(\nu_{i,k}|\mathbf{x}_k,\kappa_{i,k})$ can be calculated using the space configuration of the objects and the robot. $p(\vartheta_{i,k}|\mathbf{x}_k,\kappa_{i,k})$ can be estimated using both the robot pose and the object position plus some estimation of the probability that another object is between them. Finally, $p(\rho_{i,k}|\mathbf{x}_k,\kappa_{i,k})$ can be estimated as a function of the distance between the ith object and the robot, and the parameters of the camera and the vision module.

3.4 Value Function Integration

Let us define ϕ_k as the event "no object is observed at instant k". Then, given ϕ_k, $\mathbf{P}_k$ will suffer no correction and thus the expected value of the state pdf will remain unaltered. For the calculation of the variances, we will use the well known property:

$$\text{var}(X) = S - E(X)^2 \; ; \; S = E(X^2) \qquad (7)$$

The means of the value function and its square can be calculated as:

$$\overline{V}_{\phi,k} = \int_{\mathbf{x}_k \in X} V_k(\mathbf{x}_k)p(\mathbf{x}_k|\phi_k)d\mathbf{x}_k \; ; \; S_{\phi,k} = \int_{\mathbf{x}_k \in X} V_k(\mathbf{x}_k)^2 p(\mathbf{x}_k|\phi_k)d\mathbf{x}_k \qquad (8)$$

Again, we can approximate the integrals using the sigma point representation $\{\chi_{j,k},\omega_{j,k}\}$ of the pdf of $\mathbf{x}_k$ that was calculated as detailed in section 3.2.:

$$\overline{V}_{\phi,k} \approx \sum_{j\in[0,2N]} V_k(\chi_k^j)\omega_{j,k} \; ; \; S_{\phi,k} \approx \sum_{j\in[0,2N]} V_k(\chi_k^j)^2 \omega_{j,k} \qquad (9)$$

3.5 Focusing Values Calculation

Let us define a *candidate object* as one that the robot can focus on by only modifying $\mathbf{u}_k^{sense}$. For every candidate object, we can define its *focusing value variance* as the value function variance after focusing on it:

$$\sigma_{i,k}^V = \text{var}(V_k(\mathbf{x}_k)|\kappa_{i,k}) \qquad (10)$$

For every candidate object, its focusing value variance is estimated, following the procedure shown in Figure 2 and detailed afterwards, in order to test the utility of focusing on that object.

Perception Simulation (i^{th}* object).* In order to estimate $p(\mathbf{x}_{k+1}|\varphi_{i,k+1})$, a perception $\mathbf{z}_{k+1}^i$ of the ith object is simulated as the expected one. The predicted state $\overline{\mathbf{x}}_{k+1}^-$ and

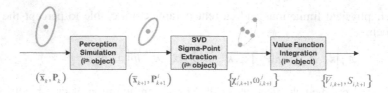

Fig. 2. The stages of the focusing values calculation process

covariance $\mathbf{P}_{k+1}^{-}$ should come from a prediction stage. As previously stated, $\mathbf{u}_{k}^{act}$ is selected before $\mathbf{u}_{k}^{sense}$, and thus, it is known in advance.

$$\overline{\mathbf{x}}_{k+1}^{-} = f\left(\overline{\mathbf{x}}_{k}, \mathbf{u}_{k}^{act}\right); \quad \mathbf{P}_{k+1}^{-} = \mathbf{F}_{k+1}\mathbf{P}_{k}\mathbf{F}_{k+1}^{T} + \mathbf{Q}_{k+1} \tag{11}$$

The correction stage of the state estimate has the following form:

$$\overline{\mathbf{x}}_{k+1}^{i} = \overline{\mathbf{x}}_{k+1}^{-} + \mathbf{K}_{k+1}^{i}\left(\mathbf{z}_{k+1}^{i} - \mathbf{h}^{i}\left(\overline{\mathbf{x}}_{k+1}^{-}\right)\right) \tag{12}$$

Where $\mathbf{K}_{k+1}^{i}$ is the Kalman gain given $\varphi_{i,k+1}$:

$$\mathbf{K}_{k+1}^{i} = \mathbf{P}_{k+1}^{-}\left(\mathbf{H}_{k+1}^{i}\right)^{T}\left(\mathbf{H}_{k+1}^{i}\mathbf{P}_{k+1}^{-}\left(\mathbf{H}_{k+1}^{i}\right)^{T} + \mathbf{R}_{k+1}^{i}\right)^{-1} \tag{13}$$

Simulating the observation as the expected one means $\mathbf{z}_{k+1}^{i} = \mathbf{h}^{i}\left(\overline{\mathbf{x}}_{k+1}^{-}\right)$, and then the corrected state estimate is independent of i:

$$\overline{\mathbf{x}}_{k+1}^{i} = \overline{\mathbf{x}}_{k+1}^{-} = \overline{\mathbf{x}}_{k+1} \tag{14}$$

Thus, the state estimate remains unaltered and only the state covariance is corrected. We can calculate the state covariance given $\varphi_{i,k+1}$:

$$\mathbf{P}_{k+1}^{i} = \left(\mathbf{I} - \mathbf{K}_{k+1}^{i}\mathbf{H}_{k+1}^{i}\right)\mathbf{P}_{k+1}^{-} \tag{15}$$

Sigma Point Extraction (i^{th} object). Following the procedure detailed in section 3.2, a sigma point representation $\left\{\chi_{i,k+1}^{j}, \omega_{i,k+1}^{j}\right\}$ is calculated from the Gaussian representation $\left(\overline{\mathbf{x}}_{k+1}, \mathbf{P}_{k+1}^{i}\right)$ of $p\left(\mathbf{x}_{k+1} | \varphi_{i,k+1}\right)$.

Value Function Integration (i^{th} object). Following the procedure detailed in section 3.4, the desired means are approximated using the sigma point representation $\left\{\chi_{i,k+1}^{j}, \omega_{i,k+1}^{j}\right\}$ of $p\left(\mathbf{x}_{k+1} | \varphi_{i,k+1}\right)$:

$$\overline{V}_{i,k+1} \approx \sum_{j\in[0,2N]} V_{k+1}\left(\chi_{i,k+1}^{j}\right)\omega_{i,k+1}^{j}; \quad S_{i,k+1} \approx \sum_{j\in[0,2N]} V_{k+1}\left(\chi_{i,k+1}^{j}\right)^{2}\omega_{i,k+1}^{j} \tag{16}$$

3.6   Object Selection

Finally, the object i_{k+1}^{*} to which the robot will focus on is selected,

$$i_{k+1}^{*} = \arg\min_{i} \sigma_{i,k+1}^{V} \tag{17}$$

Applying the total probabilities theorem, we get the following equation:

$$p(\mathbf{x}_k | \kappa_{i,k}) = p(\mathbf{x}_k | \varphi_{i,k}, \kappa_{i,k}) p(\varphi_{i,k} | \kappa_{i,k}) + p(\mathbf{x}_k | \sim \varphi_{i,k}, \kappa_{i,k}) p(\sim \varphi_{i,k} | \kappa_{i,k})$$ (18)

We assume that given $\kappa_{i,k}$ no other object is observed, then,

$$p(\mathbf{x}_k | \sim \varphi_{i,k}, \kappa_{i,k}) = p(\mathbf{x}_k | \phi_k, \kappa_{i,k})$$ (19)

Besides, given ϕ_k or $\theta_{i,k}$, $\kappa_{i,k}$ brings no additional information about $\mathbf{x}_k$, then,

$$p(\mathbf{x}_k | \kappa_{i,k}) = p(\mathbf{x}_k | \varphi_{i,k}) \alpha_{i,k} + p(\mathbf{x}_k | \phi)(1 - \alpha_{i,k})$$ (20)

Now, from equations (10), (7), (9), (16), and (20),

$$\sigma_{i,k+1}^V = \left(S_{i,k+1} \alpha_{i,k+1} + S_{\phi,k+1} (1 - \alpha_{i,k+1}) \right) - \left(\overline{V}_{i,k+1} \alpha_{i,k+1} + \overline{V}_{\phi,k+1} (1 - \alpha_{i,k+1}) \right)^2$$ (21)

4   Case Study: Goal Covering by a Goalie Player

With the purpose of showing the applicability of the proposed approach the robot soccer problem of goal covering by a goalie player is studied in this section. One key task that a goalie must follow is the *goal covering* task, which consists of maintaining its own position between the ball and the own goal (see fig. 3).

For the goal covering task, $V(\mathbf{x}_k)$ is independent of the instant k, and is defined as:

$$V(\mathbf{x}_k) = 1 - \frac{\beta_k}{\pi}$$ (22)

Where the *free goal angle* β_k corresponds to the maximum angle, defining an origin in the position of the ball, in which the own goal is not obstructed (see figure 3). Our selected $\mathbf{u}_k^{act}$ for this task consists of positioning the robot over the bisector of

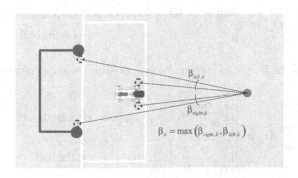

Fig. 3. Geometry of the goal covering task

the goal angle from the ball. Then, the potential attacker has two equal free goal angles to each side of the robot. The goalie must select then, to which distance d^* from the goal it should position itself. There is always a minimum distance $d_{\beta_k=0}^{\min}$ from the goal where the goalie can position to make $\beta_k = 0$. We will call $d_{zone}^{\max}$ the maximum distance to the goal where the goalie is still inside its goal zone. Then, we will select $d^* = \min\left(d_{\beta_k=0}^{\min}, d_{zone}^{\max}\right)$ to add the restriction that the goalie should keep inside its goal zone. Note that both $d_{\beta_k=0}^{\min}$ and $d_{zone}^{\max}$ depend on the absolute pose $\mathbf{G}_k = \left(G_{x,k}, G_{y,k}, G_{\theta,k}\right)$ of the goalie, and position $\mathbf{B}_k = \left(B_{x,k}, B_{y,k}\right)$ of the ball.

Then, for the goal covering task, the state of the world may be defined as $\mathbf{x}_k = \left(\mathbf{G}_k, \tilde{\mathbf{B}}_k\right)$, where $\tilde{\mathbf{B}}_k = \left(\tilde{B}_{x,k}, \tilde{B}_{y,k}\right)$ is the position of the ball with respect to the goalie. This definition of $\mathbf{x}_k$, which uses $\tilde{\mathbf{B}}_k$ instead of $\mathbf{B}_k$, is convenient from the world modeling point of view, since $\mathbf{G}_k$ may be estimated using observations of the existent landmarks and $\tilde{\mathbf{B}}_k$ may be estimated using observations of the ball. From these definitions, the localization and ball tracking processes are independent each other, which is not a requisite but is convenient for the simplicity of the problem formulation. Note that for the calculation of β_k, it is necessary to calculate:

$$\mathbf{B}_k = Rot\left(G_{\theta,k}\right)\tilde{\mathbf{B}}_k + \left(G_{x,k}, G_{y,k}\right) \tag{23}$$

Where $Rot(\theta)$ the rotation matrix in angle θ. We will define the observation of the ith object as its relative to the robot position in polar coordinates, i.e., $\mathbf{z}_k^i = \left(\mathbf{z}_{r,k}^i, \mathbf{z}_{\theta,k}^i\right)$. Then, the observation model for the ith object is:

$$\mathbf{h}^i(\mathbf{x}_k) = \left(\sqrt{\left(\tilde{O}_{x,k}^i\right)^2 + \left(\tilde{O}_{y,k}^i\right)^2}, \arctan\left(\tilde{O}_{y,k}^i / \tilde{O}_{x,k}^i\right)\right) \tag{24}$$

With

$$\tilde{\mathbf{O}}_k^i = \left(\tilde{O}_{x,k}^i, \tilde{O}_{y,k}^i\right) = \begin{cases} Rot\left(-G_{\theta,k}\right)\left(\mathbf{O}_k^i - \left(G_{x,k}, G_{y,k}\right)\right) & i \text{ is a landmark} \\ \tilde{\mathbf{B}}_k & i \text{ is the ball} \end{cases} \tag{25}$$

Where, $\mathbf{O}_k^i$ is the fixed and known absolute position of the ith object. Then,

$$\mathbf{H}_k^i = \frac{\partial \mathbf{h}^i}{\partial \mathbf{x}_k}(\mathbf{x}_k) = \begin{cases} \left[\left[\frac{\partial \mathbf{h}^i}{\partial \mathbf{G}_k}(\mathbf{G}_k)\right] \begin{matrix} 0 & 0 \\ 0 & 0 \end{matrix}\right] & i \text{ is a landmark} \\ \begin{bmatrix} 0 & 0 & 0 \\ 0 & 0 & 0 \end{bmatrix} \left[\frac{\partial \mathbf{h}^i}{\partial \tilde{\mathbf{B}}_k}(\tilde{\mathbf{B}}_k)\right] & i \text{ is the ball} \end{cases} \tag{26}$$

The calculation of the Jacobians is not detailed for space restrictions. In the next section, the results of simulated experiments for this application are presented.

5 Results and Discussion

In this section, we will show the applicability of the presented active vision method by testing it in the goal covering task. In this particular implementation, the active vision system acts only on the gaze direction.

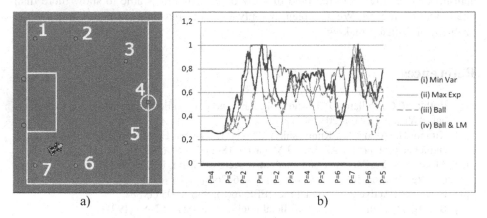

a)   b)

Fig. 4. Performed experiment: a) initial position of the goalie and cyclic positions (1 to 7) of the ball. b) Evolution of $V(\mathbf{x}_k)$ in the first experiment cycle, for the presented method and the compared heuristics (See posterior description of the (i)-(iv) indexes). In the horizontal axis, P=j represents an instant when the ball is on point j.

The experiment showed in this section has been performed in a high level simulator, which simulates the world modeling and decision making process and takes into account process and observation noises and observation probabilities. Only the goalie and the ball are inside the field. The goalie executes the goal covering task for a period of time of around 1 minute, while the ball rolls cyclically (4 repetitions) among 7 fixed positions. Figure 4.a shows the initial position of the goalie and the intermediate positions of the ball. Figure 4.b shows the evolution of $V(\mathbf{x}_k)$ over time.

The selected performance measures are the average and standard deviation of the actual value, $V(\mathbf{x}_k)$, evaluated in the actual state (unknown for the robot). The presented method (i) is compared to: (ii) the same method using the maximum expectation of $V(\mathbf{x}_k)$ as the optimality criterion; and two heuristics: (iii) looking always to the ball, and (iv) alternating for fixed periods (2 seconds) from the ball to a randomly selected landmark. The average and standard deviation of the actual values are respectively: 0.67 ± 0.14 for the presented method (i), 0.66 ± 0.15 for optimality criterion (ii), 0.57 ± 0.16 for heuristic (iii), and 0.62 ± 0.19 for heuristic (iv).

6 Conclusions

We have presented a probabilistic and explicitly task oriented active vision system which is able to select in which object the robot should focus. The applicability of the system is very wide from the point of view of the executed task. A particular robot soccer application, the goal covering task, is selected for showing this applicability. Preliminary results show that the presented method has a better performance in this task that some selected heuristics. As a future work, we plan to allow the system to consider the possibility of focusing in more than one object with the same sensing action, for example when the field of view of the camera is able to show more than one object of interest. We also plan to apply a similar reasoning structure to solve the problem of decision making.

References

[1] Rothkopf, C., Ballard, D., Hayhoe, M.: Task and context determine where you look. Journal of Vision 7(14), 1–20 (2007)

[2] Kitano, H., Asada, M., Kuniyoshi, Y., Noda, I., Osawa, E., Matsubara, H.: RoboCup: A challenge problem for AI. The AI Magazine 18(1), 73–85 (1997)

[3] Aloimonos, Y.: Purposive and Qualitative Active Vision. In: Proc. IAPR Int'l. Conf. Pattern Recognition, ICPR 1990, vol. 1, pp. 346–360 (1990)

[4] Bajcsy, R.: Active Perception. Proc. IEEE 76(8), 996–1005 (1988)

[5] Ballard, D.: Animate Vision. Artificial Intelligence 48(1), 57–86 (1991)

[6] Fox, D., Burgard, W., Thrun, S.: Active markov localization for mobile robots. Robotics and Autonomous Systems 25, 195–207 (1998)

[7] Denzler, J., Brown, C., Niemann, H.: Optimal Camera Parameter Selection for State Estimation with Applications in Object Recognition. In: Radig, B., Florczyk, S. (eds.) DAGM 2001. LNCS, vol. 2191, pp. 305–312. Springer, Heidelberg (2001)

[8] Porta, J.M., Terwijn, B., Kröse, B.: Efficient Entropy-Based Action Selection for Appearance- Based Robot Localization. In: Proc. IEEE Int. Conf. on Robotics and Automation, Taipei (2003)

[9] Stronger, D., Stone, P.: Selective Visual Attention for Object Detection on a Legged Robot. In: Lakemeyer, G., Sklar, E., Sorrenti, D.G., Takahashi, T. (eds.) RoboCup 2006: Robot Soccer World Cup X. LNCS, vol. 4434, pp. 158–170. Springer, Heidelberg (2007)

[10] Fukase, T., Yokoi, M., Kobayashi, Y., Ueda, R., Yuasa, H., Arai, T.: Quadruped Robot Navigation Considering the Observational Cost. In: Birk, A., Coradeschi, S., Tadokoro, S. (eds.) RoboCup 2001. LNCS, vol. 2377, pp. 350–355. Springer, Heidelberg (2002)

[11] Mitsunaga, N., Asada, M.: Visual Attention Control by Sensor Space Segmentation for a Small Quadruped Robot Based on Information Criterion. In: Birk, A., Coradeschi, S., Tadokoro, S. (eds.) RoboCup 2001. LNCS, vol. 2377, pp. 154–163. Springer, Heidelberg (2002)

An Incremental SLAM Algorithm with Inter-calibration between State Estimation and Data Association

Xiucai Ji, Hui Zhang, Dan Hai, and Zhiqiang Zheng

College of Mechatronics Engineering and Automation,
National University of Defense Technology, 410073 Changsha, China
{jxc_nudt,zhanghui_nudt,haidan,zqzheng}@nudt.edu.cn

Abstract. In most SLAM (Simultaneous Localization and Mapping) approaches, there is only unilateral data stream from data association (DA) to state estimation (SE), and the SE model estimates states according to the results of DA. This paper focuses on the reciprocity between DA and SE, and an incremental algorithm with inter-calibration between SE and DA is presented. Our approach uses a tree model called *correspondence tree* (CT) to represent the solution space of data association. CT is layered according to time steps and every node in it is a data association hypothesis for all the measurements gotten at the same time step. A best-first search with limited back-tracking is designed to find the optimal path in CT, and a state estimation approach based on the least-squares method is used to compute the cost of nodes in CT and update state estimation incrementally, so direct feedback is introduced from the SE to DA. With the interaction between DA and SE, and combining with tree pruning techniques, our approach can get accurate data association and state estimation for on-line SLAM applications.

Keywords: SLAM, data association, state estimation, least-squares, backtracking search.

1 Introduction

The simultaneous localization and mapping (SLAM) problem requires an autonomous robot, moving in an unknown environment, to incrementally derive a map of the environment only from its relative observations of the environment, and then simultaneously determine its own position in the map. A solution to SLAM has been seen as a 'holy grail' for mobile robotics community as it provides a mean to make a robot truly autonomous [1-3].

The feature-based SLAM problem is consisted of a continuous and a discrete component [4]. The continuous component is the *state estimation* (SE) problem, which estimates the location of individual features in the environment and the position the robot relative to these features. The discrete part is the *data association* (DA) problem [5], which is the problem of determining the correspondence relationships between observed features and landmarks in the map built thus far. Data association is very

L. Iocchi et al. (Eds.): RoboCup 2008, LNAI 5399, pp. 97–108, 2009.

important to the consistent map construction in SLAM, since incorrect data association may make the state estimation divergent, and cause the entire SLAM process fail.

Some approaches have been put forward to solve the data association problem in SLAM, and most of them are classical methods for tracking [5, 6]. The *nearest neighbor* (NN) algorithm is the simplest [7]. But in complex environment with cluttered features, it may accept wrong data association hypotheses, and cause the state estimation divergent. Many methods such as the joint compatibility test (JCBB) [8], the integer programming method [9] and the graph theoretic approach [10] have been presented to overcome this shortcoming. These methods make incremental maximum likelihood decisions for all the features gotten at the same time step, and could get the local optimal data association. More sophisticated algorithms called the multiple-hypothesis method, such as the multi-hypothesis Kalman filter [2] and FastSLAM [11], generate many data association hypotheses, and later determine which is the best as more observes arrive. These algorithms are more robust, but need more computational overhead. Reference [4] presented a lazy data association algorithm, which searches the interpretation tree with backtracking to revise past data association. This method can find the global optimal data association, but it need calculate the inverse of high-dimensional matrixes.

All above algorithms lose sight of the interaction between data association and state estimation. There is only data flow from the data association model to the state estimation process, and the state estimation process has no direct influence to the he data association model. In this paper we focus on the inter-calibration between data association and states estimation, and introduce direct feedback from the state estimation process to the data association model. The data association problem is modeled by a tree structure, called *correspondence tree* (CT), and nodes in it are features extracted at-a-time with a data association hypothesis. An incremental state estimating approach based on the least-squares method is used, and its least-squares residual is used as the cost of a node in CT. A best-first backtracking search strategy to CT is designed to search the best data association. Because the state estimation process provides feedback to the data association model, and at the same time past data association can be revised with backtracking, it's guaranteed that the global optimal data association is found. Additionally, since only a path of CT is expanded at most time, our method has moderate computational cost.

2 Formulation of SLAM

In this section we present the formulation of the SLAM problem with unknown data association. This paper uses the same mathematical framework and notations as that adopted in [7]. We are interested in the maximum a posterior (MAP) estimation for the entire robot trajectory $X \triangleq \{x_i\}$, $i \in 1...M$ and the map of landmarks $L \triangleq \{l_j\}$, $j \in 1...N$, given the measurements $Z \triangleq \{z_k\}$, $k \in 1...K$ and the control inputs, $U \triangleq \{u_i\}$, with unknown data association $C \triangleq \{c_k\}$. If $c_k = n$, the measurement z_k corresponds to landmark l_n. Let $\Theta \triangleq (X,L)$, and then the SLAM problem with unknown data association can be formulated as the following optimization problem:

$$\{\Theta^*, C^*\} \triangleq \underset{\Theta, C}{\arg \max}\, P(\Theta \mid Z, U, C) = \underset{X, L, C}{\arg \max}\, P(X, L \mid Z, U, C). \tag{1}$$

In (1), the solution space of C is discrete, while 's is continuous. Direct method solving (1) is to solve the sub-problem $\max P(\mid Z, U, C)$ for every value of C by traversing the whole solution space of C. This method is impractical, because the solution space of C will grow exponentially with the increas of measurements and landmarks. Most approaches solve (1) in two steps. Firstly, calculating a sub-optimal value of C by solving the following equation:

$$C^* \approx \underset{C}{\arg \max}\, P(\Theta \mid Z, U, C). \tag{2}$$

This is the *data association* problem. And then a *state estimation* problem with known data association is solved:

$$\Theta^* = \underset{\Theta}{\arg \max}\, P(\Theta \mid Z, U, C^*). \tag{3}$$

3 Data Association

3.1 Correspondence Tree

Every value of C represents a correspondence relationship between Z and L. Let c^i denotes the data association of the measurements z^i gotten at time step i, and $C^t \triangleq \{c^i\}$, $Z^t \triangleq \{z^i\}$, $i=1...t$. The solution space of C^t can been represented as a *correspondence tree* (CT) of t levels, as shown in Fig.1. The tree has a *null* root, level i

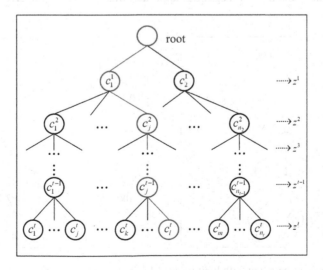

Fig. 1. An example of correspondence tree. c_j^i is the jth possible data association for the observations z^i at time step i, and n_i is the number of possible data associations for the observations z^i.

corresponds to measurement z^i and every node c^i corresponds to a data association hypothesis of z^i then a path from the root to a node in level i corresponds to a data association hypothesis of Z as shown by red lines in Fig.1.

The CT tree is layered according to time steps, so it is more preferable than the interpretation tree [12] to represent the data association problem in SLAM. The incremental maximum likelihood data association approaches, such as the joint compatibility test [8] and the graph theoretic approach [10], choose the best data association of z^t based on the data association of z^{t-1} and then freeze it forever, that is

$$\hat{c}^t = \arg\max_{c^t} P(\Theta \mid Z, U, \hat{C}^{t-1}, c^t). \tag{4}$$

In nature, these methods are *hill-climbing* searching to CT, which is a greedy tree searching method, and can only find a local optimal solution. Multiple-hypothesis methods actually are *breadth-first* searching to CT, so they could find the global optimal solution. But because these methods need to expand and save the whole or most parts of CT, their computation complex is very high.

3.2 Limited Backtracking Data Association

From above analysis we know that tree searching methods can be used to solve the data association problem. We take

$$\Gamma(c^t) = \min - \log P(\Theta \mid Z, U, C^t) \tag{5}$$

as the evaluation function to calculate the cost of node c^t. In fact, this function calculates the cost of path C^t to which c^t belongs. The cost presents the desirability of expanding node c^t, and the node with the least cost will be expanded first. Fig. 2 illustrates the basic idea of our searching strategy: our approach maintains an H-layered path above a

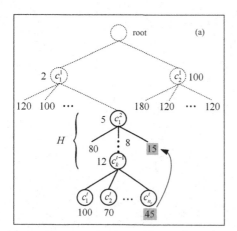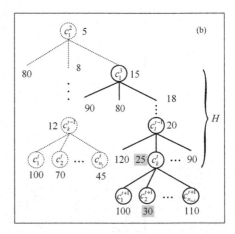

Fig. 2. A example of best-first with limit backtracking search. (a) At time step t, a backtracking appears; (b) When the node with least cost is in level t, the searching at time step t stops, and the H-layered path (*real line*) is saved. At time step $t+1$, the node with least cost is in level t is expand first.

node in the deepest level at the moment, at the same time the entire frontier of the path is also saved. At time step t, after new measurements z^t arrives, the node with the least cost in level t-1 will be expanded first to produce all of its children nodes, all children nodes are assessed and their costs are saved. Then all frontier nodes in the H-layered path are compared: if the frontier node with the least cost is among those new children nodes, the searching stops, the children node with the least cost will be expanded first at the time step $t+1$, the new H-layered path with its frontier are memorized and other nodes in CT are abandoned (dashed nodes in Fig. 2); If a frontier node not in level t has a lower cost than the least one of those children nodes, the algorithm will back-track to this frontier node and expand it, and such procedure will go on until the frontier node with least cost is in level t. Then the frontier node will be expanded first at the time step $t+1$, the new H-layered path its frontier above the node are memorized, and other nodes in CT are also abandoned.

The reasons why a path of H layer is kept are that:

- Since the cost of nodes increase monotonically with the depth of CT, if we maintain the whole optimal path, when CT is expanded deep enough, it's possible that the cost of an early frontier node is lower than the current frontier node of the optimal path, and then meaningless backtracking will appear;
- It's shown that good data association for a measurement can be obtained in a few time steps, and in some circumstance it only needs 2 time steps to get a near optimum solution [6], so we needn't keep the whole path, and for most circumstance, H=5 is enough.

In fact, our approach is a best-first search with limit backtracking. It's well known that in a limited tree, if the cost of nodes increase monotonically with the depth of the path which it belongs to, best-first searching with backtracking is certain to find the global optimal path. In addition, our searching method needn't expand the whole CT, and only needs to keep one path at most time, so it has moderate computational cost.

4 SLAM as a Least-Squares Problem

According to the Bayesian networks model, the SLAM problem with known data association C' can be formulated by the following equation [12]:

$$P(X,L,Z,U \mid C') = P(x_0)\prod_{i=1}^{M} P(x_i \mid x_{i-1},u_i)\prod_{k=1}^{K} P(z_k \mid x_{i_k},l_{c_k}), \qquad (6)$$

where $P(x_0)$ is a prior on the initial state, $P(x_i|x_{i-1},u_i)$ is the robot *motion model*, and $P(z_k \mid x_{i_k},l_{c_k})$ is the *landmark measurement model*. The MAP estimate Θ^* can be obtained by minimizing the negative log of the joint probability of (6):

$$\Theta^* = \arg\max_\Theta P(X,L \mid Z,U,C') = \arg\min_\Theta -\log P(X,L,Z,U \mid C'). \qquad (7)$$

It's assumed that the motion and measurement noises are Gaussian. The motion model is defined by $x_i = f_i(x_{i-1},u_i)+\omega_i$, where $\omega_i \sim N(0,\Lambda_i)$. The landmark measurement model is defined by $z_k = h_k(x_{i_k}, l_{j_k}) + v_k$, where $v_k \sim N(0, \Sigma_k)$. Combining them with (7), leads to the following least-squares problem:

$$\Theta^* = \arg\min_{\Theta} \left\{ \sum_{i=1}^{M} \| x_i - f_i(x_{i-1}, u_i) \|_{\Lambda_i}^2 + \sum_{k=1}^{K} \| z_k - h_k(x_{i_k}, l_{j_k}) \|_{\Sigma_k}^2 \right\} \qquad (8)$$

where

$$\| e \|_{\Sigma}^2 \triangleq e^T \Sigma^{-1} c = (\Sigma^{-T/2} e)^T (\Sigma^{-T/2} e) = \| \Sigma^{-T/2} e \|_2^2 \qquad (9)$$

is squared Mahalanobis distance given a covariance matrix Σ.

Linearizing the non-linear items in (8), we can obtain the following linear problem:

$$\delta\Theta^* = \arg\min_{\Theta} \left\{ \sum_{i=1}^{M} \| F_i^{i-1} \delta x_{i-1} + G_i^i \delta x_i - a_i \|_{\Lambda_i}^2 + \sum_{k=1}^{K} \| H_k^{i_k} \delta x_{i_k} + J_k^{j_k} \delta l_{j_k} - c_k \|_{\Sigma_k}^2 \right\}, \qquad (10)$$

where F_i^{i-1}, $H_k^{i_k}$, and $J_k^{j_k}$ are Jacobians of f_i and h_k, $G_i^i = I$, and a_i and c_k are odometry and observation measurement prediction error. According to (9), we can eliminate Λ_i and Σ_k from (10), and obtain a standard linear least-squares problem:

$$\theta^* = \arg\min_{\theta} \| A\theta - b \|_2^2. \qquad (11)$$

Please see [12] and [13] for more detail derivation.

This standard least-squares problem can be solved by the QR matrix factorization of the Jocobian $A \in \mathbb{R}^{m \times n}$:

$$A = Q \begin{pmatrix} R \\ 0 \end{pmatrix}, \qquad (12)$$

where $Q \in \mathbb{R}^{m \times m}$ is orthogonal, and $R \in \mathbb{R}^{n \times n}$ is upper triangular. Multiplying with the orthogonal matrix Q^T doesn't change the norm:

$$\| A\theta - b \|_2^2 = \| Q^T A\theta - Q^T b \|_2^2 = \| R\theta - d \|_2^2 + \| e \|_2^2, \qquad (13)$$

so the solution of $R\theta = d$ is the least-squares solution of (11), where $[d, e]^T \triangleq Q^T b$, and $\| e \|_2^2$ is the residual of the least-squares problem (11):

$$\| e \|_2^2 = \min \| A\theta - b \|_2^2 = \min - \log P(\Theta | Z, U, C'). \qquad (14)$$

According to (5), we take $E \triangleq \| e \|_2^2$ as the cost of the final node of the path C'. So there will be direct interaction and reciprocal promotion between data association and state estimation: the state estimation provides direct feedback to select the best data association, and the best data association will lead to more precise state estimation.

5 Incremental SLAM Algorithm

For real-time application, it needs the SLAM algorithm rapidly calculate the costs of nodes i.e. the residual of the least-squares problem (11). In this section, we present such an incremental SLAM algorithm.

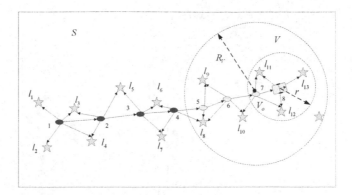

Fig. 3. The illustration of variables set V and S. Vo is the center of V, R_V is its radius, and r is the maximum perception radius of sensors.

5.1 Variable Ordering

We divide the variables in $\Theta=\{X, L\}$ into a static set S and a variable set V:

$$\Theta = S \bigcup V \tag{15}$$

The set V includes all variables that close to the robot's current position. For example, in Fig. 3, $V=\{l_8, l_9, l_{10}, l_{11}, l_{12}, l_{13}, x_5, x_6, x_7, x_8\}$, and $S=\{l_1, l_2, l_3, l_4, l_5, l_6, l_7, x_1, x_2, x_3, x_4\}$. Let V_o denote the center of V, which is one of the robot positions on its trajectory, and R_V denote the radius of V. When the robot's current position is far away from V_o:

$$\| \hat{x}_t - V_o \| > R_V - r , \tag{16}$$

we set $\hat{x}_t$ as the new center of V, where r is the maximum perception radius of sensors. All variables whose distance from the new V_o is smaller than R_V will compose the new V, and other variables left compose S. In our approach, R_V is chosen carefully so that V includes all variables related to the saved H-layered path of CT up to current time step.

For the SLAM problem, the Jacobian matrix A in (11) is sparse [12]. The QR factorization of A will lead to non-zero fill-ins for matrix R. However, such fill-ins can be avoided through column reordering of A. Since finding an optimal ordering is NP-hard, various approximate algorithms have been developed, and among them the *approximate minimum degree* (AMD) works well [14]. In our algorithm, A is ordered in the following way:

- S is put in the front of A and V follows:

$$A = \begin{bmatrix} A_S & A_V \end{bmatrix}, \tag{17}$$

- Every time V_o changes, we use AMD to column order A_S and A_V for the new S and V.
- At time steps that V_o isn't changed, the current pose and new landmarks are put into V, and ordered at the end of A_V. The current pose is in front of new landmarks.
- We row order A according to time, and early equations are ordered on the top. At every time steps, the motion model equations are ordered ahead of measure-ments and old landmarks' measurements are ordered ahead of those of new landmarks.

Fig. 4 depicts the row and column ordering of the Jacobian matrix A in the SLAM scene shown in Fig. 3, and the side number is the row order of A, the 16th and 17th row are the equations relating to variables in V and S, and are the links between V and S.

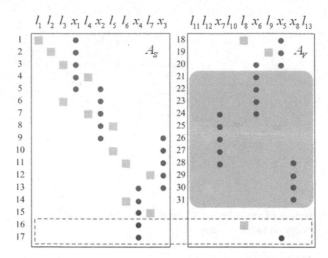

Fig. 4. The row and column ordering of the Jacobian matrix A in the SLAM scene shown in Fig. 3. On the top is the column order of variables and the side number is the row order. The 16th and 17th row are the equations relating to variables in V and S. The shadowed part is related to the saved H-layered path of CT.

The above column ordering will reduce fill-ins to R, and the row ordering will facilitate the backtracking search of our data association algorithm. Because we row order A according to time steps, equations that related to the saved H-layered path of CT is at the bottom of A. For example, the shadowed part in Fig. 4 is related to a 3-layered path.

5.2 Incremental State Estimation

Since V includes all variables related to the saved H-layered path of CT up to current time step, the calculation of the costs of nodes in the backtracking search to CT only relates to variables inside V. Assuming that the optimal data association up to time step t has been found, and after QR factorization, the equations only including variables in V is $R_V^t \delta V^t = d_V^t$, with a least-squares residual $E^t = \| e^t \|_2^2$, i.e. the cost of the last node in the saved optimal path of CT, where $R_V^t = Q_V^t A_V^t$.

At time step $t+1$, the last node of the saved optimal path is first expanded according to z^{t+1}'s association hypotheses. Let c' be one children nodes. According to c', we divide z^{t+1} into two parts: measurements from old landmarks and others from new landmarks. Following the variables ordering rules in section 5.1, we can get the following equations:

$$
\begin{pmatrix}
R_V^t & 0 & 0 \\
F & G & 0 \\
J_{old} & H_{old} & 0 \\
0 & H_{new} & J_{new} \\
0 & 0 & 0
\end{pmatrix}
\begin{pmatrix}
\delta V^t \\
\delta x_{t+1} \\
\delta L_{new}
\end{pmatrix}
=
\begin{pmatrix}
d_V^t \\
b_x \\
b_{Lold} \\
b_{Lnew} \\
e^t
\end{pmatrix}.
\tag{18}
$$

We denote (18) as $A_v^{t+1} \delta V^{t+1} = b^{t+1}$. It's easy to verify that new landmark measurements have no influence to the cost of c'. The QR factorization of the coefficient matrix in (18) can be obtained by eliminating lower diagonal non-zeroes of F, G, J_{old} and H_{old} row-by-row through Givens rotations [13], and then we will get

$$\begin{pmatrix} \bar{R}_v^t & T \\ 0 & \bar{R}_x \\ 0 & 0 \end{pmatrix} \begin{pmatrix} \delta V' \\ \delta x_{t+1} \end{pmatrix} = \begin{pmatrix} \bar{d}_v^t \\ \bar{d}_x \\ e^{t+1} \end{pmatrix} \tag{19}$$

Thus the least-squares residual of (18) is

$$E^{t+1} = \| e' \|_2^2 + \| e_{t+1} \|_2^2 = E^t + \| e_{t+1} \|_2^2 . \tag{20}$$

Apparently $E^{t+1} \geq E^t$, and the costs of nodes in CT increase monotonically, so our algorithm can find the global optimal path of CT. Equation (20) shows that the cost of children nodes can be incrementally computed from its parent's cost. After the optimal node in level $t+1$ is founded, the estimation update of variables in V can be obtained by back-substitution [13].

Above computation is enclosed in V, so our algorithm has $O(n^2)$ time complex, where n is the size of V. Since the size of V is limited and relatively small, our algorithm has a moderate computation cost.

5.3 Pruning

In our approach the following techniques are adapted to prune those branches in CT with little chance to be expanded:

- A range W around the current robot position is set. Landmarks out of the range will not be considered in the expanding of nodes in CT. W is decided by $R_W = r + \rho$, where R_W is the radius of W, r is sensors' maximum perception radius, and $\rho > 0$ is a compensator for errors.
- A gate is applied to every landmark in W. Only measurements that are close enough to the prediction position of a landmark are considered as possible candidates to be associated with the landmark. The gate is given by [9]:

$$\tau_{jk} = v_{jk}^T s_j^{-1} v_{jk} < \varepsilon, \ j = 1...N; k = 1...K, \tag{21}$$

where $v_{jk} = z_k - \hat{z}_j$ is the innovation and s_j is it's covariance. τ_{jk} is a squared Mahalanobis distance between z_k and $\hat{z}_j$, and follows the χ^2 distribution. If a measurement is out of the gate of all landmarks, it matches a new landmark.

- For a new landmark, a punishment ε i.e. the gate in (21) is added to the cost of its corresponding node in CT, since new landmarks will give zero residual.

During the pruning procedure, covariance matrices of variables are needed. In our approach, the efficient method presented in [14] is applied.

6 Experiment Results

The algorithm presented in this paper was tested with simulation experiments and a real robot experiment dataset. In these experiments, we have compared our algorithm with EKF-based SLAM [15] using the *nearest neighbor* and the *joint compatibility branch and bound* (JCBB)[8] data association. The experiment results have shown that our algorithm is more accurate.

6.1 Simulation

In order to evaluate our approach, a simulation environment with about 200 land-marks has been created as shown in Fig. 5(a), and significant motion and measurement noise have been added.

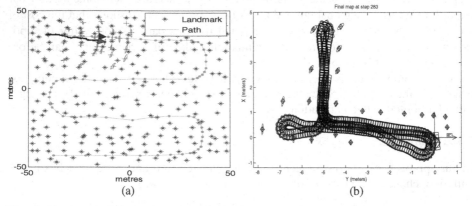

(a)   (b)

Fig. 5. (a). A typical simulation experiment scene. (b). The estimated robot trajectory and 2-sigma ellipses of landmark locations for the 'sjursflat' dataset with our algorithm.

For 20 times we applied EKF-based SLAM with NN and JCBB data association and our algorithm to perform the SLAM process respectively. Every time, we changed the landmarks positions randomly and some landmarks were significantly close to each other. In the experiments, we saved a 5-layered path of the CT tree, and the gate for pruning was $\varepsilon = 6.63$ with a probability of 0.99 for a measurement to match a close landmark. The correct data association rates of every algorithm are shown in Table. 1. Our algorithm had finished all the experiments, and the correct data association rate was 96.8%. Backtracking search had occurred for 157 times. The wrong data associations appeared to landmarks far from the robots, and only observed for one or two times with high measurement noise. The correct data association rate for the EKF-based SLAM with NN was 79.6%. When landmarks were too closed, the NN algorithm failed. The correct data association rate for the EKF-based SLAM with JCBB was 85.3%, and when landmarks were too closed, it worked better than the NN algorithm, but when there is large motion error and landmarks have similar deployments it may accept wrong data association. But our algorithm will accept such wrong data association temporally, and the revise it after several time steps.

Table 1. Comparison of correct data association rate (CDAT) of different algorithm

	Our algorithm	NN	JCBB
CDAT(%)	96.8	79.6	85.3

6.2 Sjursflat Dataset

The 'sjursflat' dataset shows the ground floor of Sjur Vestli's house, recording using a SmartRob 2 with a forward looking SICK LMS200, which has been distributed with 'the CAS Robot Navigation Toolbox' [16].

In our experiment, the same parameters were used as simulations, and a beacon-based slam was done. There were 283 time steps and 20 landmarks were estimated. Our algorithm had gotten similar results as the EKF-based method in [16], and Fig. 6 shows the resulting trajectory and map. In this experiment, our data association algorithm had degenerated as an incremental maximum likelihood method, and no backtracking had appeared, because landmarks were not close to each other, but our algorithm had matched all the 615 measurements, while the NN algorithm matched 609 measurements, for ambiguous associations were ignored.

7 Conclusion

The SLAM problem includes two parts: data association and state estimation. Former algorithms deal with them separately. This paper presents a new on-line algorithm for the SLAM problem. The characteristic of our approach is that there is direct interaction between data association and state estimation. The state estimation process evaluates every data association hypothesis, and provides direct feedback to the data association model. At the same time the data association model can revise incorrect former data association by backtracking the correspondence tree when a potential better state estimation is proposed. We have evaluated our algorithm by simulations and a real robot dataset, and compared it with the classical methods. Experiment results have validated the high accuracy of our algorithm.

We have just essentially invalidated our approach. In the future work, we will try to use this method in SLAM for the RoboCup Rescue Competition.

References

1. Frese, U.: A Discussion of Simultaneous Localization and Mapping. Autonomous Robots 20(1), 25–42 (2006)
2. Leonard, J., Durrant-Whyte, H.: Dynamic Map Building for an Autonomous Mobile Robot. The International Journal on Robotics Research 11(4), 286–298 (1992)
3. Cheeseman, P., Smith, R., Self, M.: A Stochastic Map for Uncertain Spatial Relationships. In: 4th International Symposium on Robotic Research, pp. 467–474. MIT Press, Cambridge (1987)
4. Häehnel, D., Burgard, W., Wegbreit, B., Thrun, S.: Toward Lazy Data Association in SLAM. Intl. J. of Robotics Research, 421–431 (2003)

5. Bar-Shalomand, Y., Fortmann, T.E.: Tracking and Data Association. Academic Press, San Diego (1988)
6. Cox, I.: A Review of Statistical Data Association Techniques for Motion Correspondence. Int'l. J. Computer Vision. 10, 53–65 (1993)
7. Guivant, J., Nebot, E.M.: Optimization of the Simultaneous Localization and Map Building Algorithm for Real Time Implementation. IEEE Trans. Robot. Automat. 17, 242–257 (2001)
8. Neira, J., Tardos, J.D.: Data Association in Atochastic Mapping Using the Joint Compatibility Test. IEEE Trans. Robot. Automat. 17, 890–897 (2001)
9. Zhang, S., Xie, L., Adams, M.D.: An Efficient Data Association Approach to Simultaneous Localization and Map Building. Int'l. J. Robotics Research. 24, 49–60 (2005)
10. Bailey, T., Nebot, E.M., Rosenblatt, J.K., Durrant-Whyte, H.F.: Data association for Mobile Robot Navigation: a Graph Theoretic Approach. In: IEEE Int. Conf. Robot. Automat, pp. 2512–2517. IEEE Press, San Francisco (2000)
11. Montemerlo, M., Thrun, S.: Simultaneous Localization and Mapping with Unknown Data Association Using FastSLAM. In: Proc. IEEE Int. Conf. Robotics Automation, pp. 1985–1991. IEEE Press, Los Alamitos (2003)
12. Dellaert, F., Kaess, M.: Square Root SAM: Simultaneous Localization and Mapping via Square Root Information Smoothing. Technical Report, Georgia Tech. (2005)
13. Kaess, M., Ranganathan, A., Dellaert, F.: Fast Incremental Square Root Information Smoothing. In: Intl. Joint Conf. on Artificial Intelligence, pp. 2129–2134, Hyderabad (2007)
14. Kaess, M., Ranganathan, A., Dellaert, F.: iSAM: Fast Incremental Smoothing and Mapping with Efficient Data Association. In: IEEE Intl. Conf. on Robotics and Automation, pp. 1670–1677. IEEE Press, Rome (2007)
15. Leonard, J., Cox, I., Durrant-Whyte, H.F.: Dynamic Map Building for an Autonomous Mobile Robot. Intl. J. of Robotics Research. 11, 286–289 (1992)
16. Kai, O.: Arras.: The CAS Robot Navigation Toolbox,
   http://www.cas.kth.se/toolbox

Development of an Augmented Environment and Autonomous Learning for Quadruped Robots

Hayato Kobayashi, Tsugutoyo Osaki, Tetsuro Okuyama,
Akira Ishino, and Ayumi Shinohara

Graduate School of Information Science, Tohoku University, Japan
{kobayashi,osaki,okuyama}@shino.ecei.tohoku.ac.jp
{ishino,ayumi}@ecei.tohoku.ac.jp

Abstract. This paper describes an interactive experimental environment for autonomous soccer robots, which is a soccer field augmented by utilizing camera input and projector output. This environment, in a sense, plays an intermediate role between simulated environments and real environments. We can simulate some parts of real environments, e.g., real objects such as robots or a ball, and reflect simulated data into the real environments, e.g., to visualize the positions on the field, so as to create a situation that allows easy debugging of robot programs. As an application in the augmented environment, we address the learning of goalie strategies on real quadruped robots in penalty kicks. Our robots learn and acquire sophisticated strategies in a fully simulated environment, and then they autonomously adapt to real environments in the augmented environment.

1 Introduction

The experiments on real robots, especially quadruped robots, take much more time and cost than those on PCs. It is also an enormous difference since we often need to consume a lot of energy for treating physical objects such as robots. The use of virtual robots is one of the efficient methods to avoid those difficulties, and so there are many studies on dynamic simulated environments [1,2,3,4,5]. However, since simulated environments cannot produce complete, real environments, we finally need to conduct experiments in the real environments where basic skills heavily depend on complex physical interactions.

In this paper, we propose an interactive *augmented environment*, which serves as a bridge between simulated environments and real environments. Augmented environments, also known as *Augmented Reality* (AR), involve the real environments combined with some data generated by computers. In our augmented environment, robots can obtain both the precise positions of physical objects and the imaginary positions of virtual objects despite being in a real environment. Moreover, human programmers can get debugging information overlaid on a soccer field instead of that in a console terminal, allowing them to concentrate on observing the behavior of the robots in the field. This environment, in a sense, plays an intermediate role between simulated environments and real

L. Iocchi et al. (Eds.): RoboCup 2008, LNAI 5399, pp. 109–120, 2009.
© Springer-Verlag Berlin Heidelberg 2009

environments; considering the fact that we can simulate some parts of real environments.

There are several related studies of augmented environments utilizing real robots [6,7]. The study of Guerra et al. [7] is most related to our study. They proposed the CITIZEN Eco-Be! league (recently, also known as the physical visualization sub-league) where CITIZEN's mini robot Eco-Be! plays soccer in an augmented environment created by a display and a camera. This league gave new interesting and challenging research issues in RoboCup. Our augmented environment can be regarded as a generalization of the technique in the Eco-Be! league to the other real robot leagues. We can easily apply our techniques to different types of robots that are used in other leagues, although we focus only on Sony's quadruped robot AIBO in this paper.

As an application in the augmented environment, we address the learning of goalie strategies in penalty kicks. The goalie strategies involve the skills to save an incoming ball kicked by an opponent player, which is critical to not lose the game. The learning can be also regarded as the two dimensional extension of the study of Kobayashi et al. [8]. They studied the *autonomous learning* to trap a moving ball by utilizing reinforcement learning. The goal of the learning was to acquire a good timing in initiating its trapping motion autonomously, depending on the distance and the speed of the ball, whose movement was restricted to one dimension. The two dimensional extension would need the troublesome programs to find a rebounded ball and return it to the initial position, as well as an accurate localization system. In contrast, our robot in augmented environments can avoid these difficulties by adopting a virtual ball that can be positioned and controlled arbitrarily, where the precise positions of the robot and the ball are measured and calculated by the system. In this way, we can perform much more effective experiments, by using real robots.

The remainder of this paper is organized as follows. In Section 2, we describe our system to create an augmented environment for autonomous soccer robots. In Section 3, we propose an autonomous learning method of goalie strategies in penalty kicks and make our robot learn and acquire goalie strategies on their own. Finally, Section 4 presents our conclusions.

2   Augmented Soccer Field System

2.1   Overview of Our System

In this section, we describe the *Augmented Soccer Field System* for our augmented environment. This system consists of a pair of cameras and a pair of projectors in the ceiling with one computer to control them. The process in the computer is separated into two programs: a *recognition program* and a *virtual application*. The recognition program is a program that recognizes objects, such as robots and balls, in a soccer field based on the images from the ceiling cameras. The virtual application is a program that calculates real coordinates in the field based on the results of the recognition program and then sends the positions of the objects to the robots. It also detects collisions between the objects.

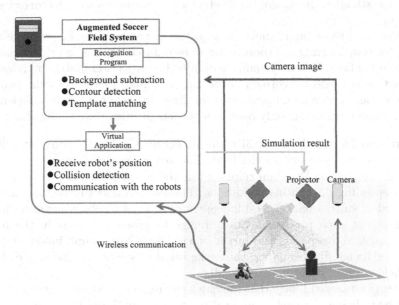

Fig. 1. Overview of our system, which is separated into two programs: A recognition program that recognizes objects with a camera and a virtual application that projects generated data with a projector

Fig. 1 shows the overview of our whole system. Each pair of cameras and projectors was set up on the ceiling located in the center of each half of the field so as to cover the whole field, since one pair can take images with a camera and projects generated data with a projector for a half of the field. We use Dragonfly2 of Point Grey Research Inc. as a camera and LP-XL45 of Sanyo Electric Co., Ltd. as a projector. Only the projector gives our room's lighting when the system is running, so that we can sufficiently recognize projected images on the field. Each projector projects the farther half of the field, since the distance from the ceiling to the field, which is about 2.5 meter height, is too small to cover the whole field.

2.2 Recognition Program

Capturing of Camera Images. Camera images are captured into the computer by utilizing the API for Dragonfly2, and then passed to an object recognition process with the Open Source Computer Vision Library (OpenCV). The processes for two cameras are multithreaded, since they are independent of each other.

Estimation of Camera Parameters. Captured camera images are quite distorted since our cameras have wide-angle lenses. The distortion of images can easily lead to some errors in the object recognition process for the images. Thus, we need to estimate adjustment parameters for each camera so as to correct such distortion.

We also need to estimate different adjustment parameters (i.e., a projection matrix) for transforming the coordinates of recognized objects in camera images to those in the field. In our system, we obtain the projection matrix by referring 8 points that consist of 4 corners of the half field and 4 corners of the penalty area. Since our cameras and projectors are fixed in our room, it is sufficient to estimate those parameters only once unless their positions are changed.

Recognition Method. We utilize the *background subtraction* method [9] in order to recognize objects on the field. The reason comes from the fact that the object recognition based on colors is usually difficult because the colors can change depending on lighting conditions. The background subtraction method is designed to store a background image without any objects in advance, judge whether or not each pixel in a target image in process belongs in the background image, and separate the target image into a foreground image and the background image. In this paper, the background image means the image of our soccer field without any robots and balls.

We create a masked image that indicates the regions of the objects to be recognized by utilizing the background subtraction method and then identify each object by extracting the contour of the object. Since our robot has its direction (or angle), we identify the orientation of each robot by utilizing template matching.

Calculation of Background Subtraction. The most naive subtraction method is to calculate color subtraction between a current target image and the background image for every pixel and then judge whether the pixel belongs to the background image based on a certain threshold. In general, however, the criterion for each pixel should be changed depending on its location, since brightness varies with location. Thus, we suppose that the degree of color change for each pixel follows a normal distribution $N(\mu, \sigma^2)$. We calculate the mean μ and variance σ^2 of each pixel by capturing 100 background images and regard the pixel as background if the difference between μ and the color value of the pixel is greater than $c\sigma$ with some constant c. Each image has 3 color channels (RGB), and so we separately treat each channel, in order that the constant c of a channel is independent of those of the others. We define that the pixels regarded as background must satisfy the above condition in every channel.

Extraction of Contour. We extract the contour of each object from a masked image calculated by the background subtraction method in order to identify the object. Since there can be unknown objects other than robots and balls in the masked image, we utilize the length of a contour line and the size of a contour area for eliminating unknown objects. Then we can obtain the minimum bounding box covering each identified contour as shown in Fig. 2 (a).

Identification of the Orientation of Robots. Although each robot is specified by a bounding box shaped like a non-regular rectangle, the orientation of

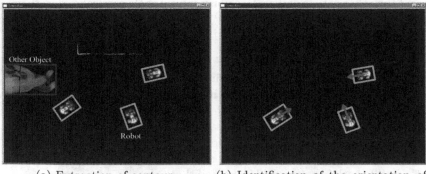

(a) Extraction of contour   (b) Identification of the orientation of
                                        robots

Fig. 2. (a) Results extracting robots as small rectangles and other objects as large rectangles. (b) Results of template matching. An arrow shows the orientation of each robot.

the robot can face either forward or backward. We capture the overhead image of the robot as a template and find a better matching orientation based on the template image. This make it so that we can recognize multiple robots simultaneously without any markers as shown in Fig. 2 (b). Even if we must use different kinds of robots, we only have to exchange the template.

2.3   Virtual Application

The virtual application carries out physical simulations, such as collision detection based on the results of the recognition program and then projects the simulation results on the field. Thus, the real robots can play a soccer game even with a virtual ball. In addition, the virtual application can communicate with real robots, and so the robots can receive the ideal positions of real objects without an accurate localization system.

Simulator. Our simulator is based on Haribote [5], the dynamic simulator for AIBO, developed by utilizing the Open Dynamic Engine (ODE) and the Open Graphical Library (OpenGL). Although Haribote has a virtual robot represented as the exact model of AIBO, in order to represent the robots recognized by the ceiling cameras, we utilize the minimum rectangular solid for covering the robot. The reason is that the rectangular model can more easily synchronize real robots so that the real robot can manipulate a virtual ball in our simulator. Of course, we can utilize the exact model in order to verify some physical motions in the simulator as well.

Sending of the Positions of Objects. The virtual application always sends the precise positions of the physical objects recognized by the recognition program to real robots through UDP. The robots can receive the positions of the ball and the other robots and replace noisy inputs with precise ones in their programs. This indicates that we can verify developed strategies in ideal situations in spite of using real robots.

Projection of the Simulator on the Field.
The virtual application projects the simulator on the field in order to visualize the positions of objects in the simulator. In addition, if we display some debugging information in the simulator, we can see the debugging information (e.g., the robot's position, the ball's position, and the robot's state) overlaid on the field instead of that in a console terminal. Although our projector projects data from an oblique perspective, this fact does not cause problems at all by utilizing the automatic trapezoidal correction in the pro-

Fig. 3. Our simulator that is projected on the field

jector. Fig. 3 shows our simulator projected on the field by the projector. The upper robot is real, and the lower robot and the ball are virtual. In our system, both of the real robot and the virtual robot can manipulate the virtual ball.

2.4 Contributions

The main merit of our system is that we can easily find which program exerts a bad influence to the whole program. For example, if we simulate the robot's position and ball's position, we can verify the validity of strategy programs by utilizing the precise output of localization programs. In the same way, if we simulate the beacons' positions, we can verify the validity of localization programs by utilizing the precise output of vision programs.

Another merit of our system is the visualization of debugging information by the projector. For example, we can visualize the particles in localization programs using a particle filter algorithm as well as the objects recognized by vision programs. We can also visualize the current state of strategy programs and even the error messages of programs as strings.

3   Autonomous Learning of Goalie Strategies

In this section, we address the learning of goalie strategies in a penalty kick as an application of our system. We first describe the rule of penalty kicks and the learning method of goalie strategies. Next we experiment in our simulator so as to ensure the validity of the method, and finally experiment in our augmented environment so as to verify the effectiveness of our augmented soccer field system.

3.1   Penalty Kick

In the rules for the RoboCup standard platform league, the penalty kick is carried out with one attacker and one goalie. If the ball goes into the goal within the time limit (1 minute), the penalty kick is deemed successful. If the time limit expires or if the ball leaves out the field, the penalty kick is deemed unsuccessful. In addition, if the attacker enters the penalty area then the penalty kick is deemed unsuccessful; if the goalie leaves the penalty area then a goal will be awarded to the attacker. The aim of our goalie is to acquire the best strategy to save the goal.

3.2   Learning Method

In this paper, we apply reinforcement learning algorithms [10]. Since reinforcement learning requires no background knowledge, all we need to do is to give the robots some appropriate rewards so that they can successfully learn goalie strategies.

The reinforcement learning process is described as a sequence of states, actions, and rewards,

$$s_0, a_0, r_1, s_1, a_1, r_2, \ldots, s_i, a_i, r_{i+1}, s_{i+1}, a_{i+1}, r_{i+2}, \ldots,$$

which is a reflection of the interaction between a learner and an environment. Here, $s_t \in S$ is a state given from the environment to the learner at time t $(t \geq 0)$, and $a_t \in \mathcal{A}(s_t)$ is an action taken by the learner for the state s_t, where $\mathcal{A}(s_t)$ is the set of actions available in the state s_t. One time step later, the learner receives a numerical reward $r_{t+1} \in \mathcal{R}$, in part as a consequence of its action, and finds itself in a new state s_{t+1}.

- **State.** The state of our learner is represented by a list of 6 scalar variables $(r, \phi, V_v, V_h, X, Y)$. The variables r and ϕ represent the position of the ball on a polar coordinate and refer to the distance and angle from the robot to the ball estimated by our vision system. The variables V_v and V_h represent the velocity of the ball and refer to the vertical and horizontal velocity from the robot. The variables X and Y represent the relative position of the robot from its own side's goal and refer to the values of x-axis and y-axis on a right handed Cartesian coordinate system with its origin at the center of the goal, i.e., an absolute position (2700 mm , 0 mm), and a positive x-axis toward the center of the field, i.e., an absolute position (0 mm, 0 mm). In order to allocate the required minimum state space for our problem, we empirically limited the range of those state variables such that r, ϕ, V_v, V_h, X, and Y are in [0 mm, 1000 mm], [-$\pi/2$ rad, $\pi/2$ rad], [-50 mm, 0 mm], [-50 mm, 50 mm], [-500 mm, 1000 mm], and [-700 mm, 700 mm], respectively.
- **Action.** The action of our learner takes one of the following 10 exclusive actions. One is *save* to stop an incoming ball, which performs the motion to spread out robot's front legs. The action *save* cannot be interrupted for 350 ms until the motion finishes. Another is *stay* to prepare for an opponent's shooting, which moves its head to watch the ball without walking. The others are 8 directional walking actions (i.e., vertically, horizontally, or diagonally) to intercept an opponent's shooting.

- **Reward.** Our learner receives one of the following 5 kinds of rewards. A negative reward is also called a *punishment*.
  - *save_punishment* The punishment is -0.02, if the robot performs *save* action, since the time during *save* motion has the risk losing the ball.
  - *passive_punishment* The punishment is -0.0000001, if the robot performs an action other than *save* action, since the robot should take various actions in the initial phase of learning.
  - *lost_punishment* The punishment is -10, if the ball goes into the goal.
  - *save_reward* The reward is 0.5, if the robot stops the ball by *save* action, since it may be safest to save the goal.
  - *dist_reward* The reward is $1 - |ydist| / 112.5$, if the robot saves the goal during a penalty kick, since it may be the safest saving. The variable *ydist* is the *y*-axis distance of the ball, which is limited in [-225 mm, 225 mm] based on the covering range 450 mm of *save* action. That is, *dist_reward* takes a value in [-1, 1] based on *ydist*.

Algorithm 1 shows the autonomous learning algorithm of goalie strategies used in our study. It is a combination of the episodic SMDP Sarsa(λ) with the linear tile-coding function approximation (also known as CMAC). This is one of the most popular reinforcement learning algorithms, as seen by its use in Kobayashi et al. [8]. In our experiments, we define the period from the starting to the ending of a penalty kick as one *episode*.

Here, F_a is a *feature set* specified by tile coding with each action a. In this paper, we use 6-dimensional tiling, where we set the number of tilings to 32 and the number of tiles to about 400000. We also set the tile widths of r, ϕ, V_v, V_h, X, and Y to a quarter of the ranges, i.e., 250 mm, $\pi/4$ rad, 12.5 mm, 25 mm, 375 mm, and 350 mm, respectively. The vector $\overrightarrow{\theta}$ is a *primary memory vector*, also known as a *learning weight vector*, and Q_a is a *Q-value*, which is represented by the sum of $\overrightarrow{\theta}$ for each value of F_a. The policy ϵ-*greedy* selects a random action with probability ϵ, and otherwise, it selects the action with the maximum Q-value. We set $\epsilon = 0.001$. Moreover, $\overrightarrow{e}$ is an *eligibility trace*, which stores the credit that past action choices should receive for current rewards. λ is a *trace-decay parameter* for the eligibility trace, and we set $\lambda = 0.9$. We set the *learning rate parameter* $\alpha = 0.18$ and the *discount rate parameter* $\gamma = 1.0$.

3.3   Experiments in Simulated Environments

We experiment in the simulated environment (or simulator) to ensure the validity of the algorithm in the previous section. In our simulator we do not treat the dynamics of robots, while the scales of the field, ball, and robots are set to the same values in real environments. Our virtual robot is represented as a rectangular solid, and if *save* action is performed then the width of the rectangular solid stretches out. State inputs have their ideal values without any noise, and action outputs are set to almost the same values as real robots. In experiments from now on, we assume that the attacker dribbles the ball to any point and then shoots it to the goal so as to confuse the goalie. Thus, we set the ball in the simulator to perform such behavior in the beginning of penalty kicks.

Algorithm 1. Algorithm of the autonomous learning of goalie strategies

---

while *still not acquiring goalie strategies* **do**

    initialize penalty shootout settings.;

    $s \leftarrow$ a state observed in the environment;

    forall $a \in \mathcal{A}(s)$ **do**

        $F_a \leftarrow$ set of tiles for a, s;

        $Q_a \leftarrow \sum_{i \in F_a} \theta(i)$;

    end

    $lastAction \leftarrow$ an optimal action selected by ϵ-*greedy*;

    $\vec{e} \leftarrow 0$;

    forall $i \in F_{lastAction}$ **do** $e(i) \leftarrow 1$;

    $reward \leftarrow 0$;

    while *during a penalty kick* **do**

        do *lastAction*;

        if *lastAction = guard* **then**

            $reward \leftarrow save\_punishment$;

        else

            $reward \leftarrow passive\_punishment$;

        end

        $\delta \leftarrow reward - Q_{lastAction}$;

        $s \leftarrow$ a state observed in the environment;

        forall $a \in \mathcal{A}(s)$ **do**

            $F_a \leftarrow$ set of tiles for a, s;

            $Q_a \leftarrow \sum_{i \in F_a} \theta(i)$;

        end

        $lastAction \leftarrow$ an optimal action selected by ϵ-*greedy*;

        $\delta \leftarrow \delta + Q_{lastAction}$;

        $\vec{\theta} \leftarrow \vec{\theta} + \alpha\delta\vec{e}$;

        $Q_{lastAction} \leftarrow \sum_{i \in F_{lastAction}} \theta(i)$;

        $\vec{e} \leftarrow \gamma\lambda\vec{e}$;

        forall $a \in \mathcal{A}(s)$ *s.t.* $a \neq lastAction$ **do**

            forall $i \in F_a$ **do** $e(i) \leftarrow 0$;

        end

        forall $i \in F_{lastAction}$ **do** $e(i) \leftarrow 1$;

    end

    if *lost the goal* **then**

        $reward \leftarrow lost\_punishment$;

    else

        $reward \leftarrow dist\_reward$;

        if *lastAction = guard* **then**

            $reward \leftarrow reward + save\_reward$;

        end

    end

    $\delta \leftarrow reward - Q_{lastAction}$;

    $\vec{\theta} \leftarrow \vec{\theta} + \alpha\delta\vec{e}$;

end

We made our robot on the simulator learn goalie strategies in 2000 episodes, and the experiment took about 20 minutes. Fig. 4 (a) shows the learning process

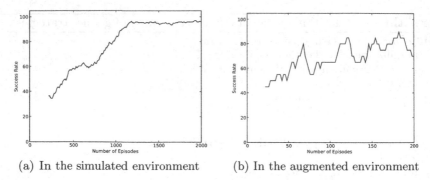

(a) In the simulated environment   (b) In the augmented environment

Fig. 4. Experimental results in our simulated environment and augmented environment, whose success rates indicate how many times the robot saved the goal in the past 200 episodes and 20 episodes, respectively

in the simulated environment in terms of the success rate of saving. The success rate of saving means how many times the robot saved the goal in the past 200 episodes. The figure indicates that the success rate reached more than 90% at 2000 episode. Actually, we repeated the same experiment 5 times and got the average of the final success rates, 93.32%. This result indicates that our virtual robot can successfully acquire a better goalie strategy in penalty kicks.

Fig. 5 shows an acquired goalie strategy (i.e., a state-action mapping) after 2000 episodes. This intuitively visualizes all of the better actions that the learner will select at each state in the environment. The figure indicates that *save* actions shown by squares are mostly on the way to the destination of a kicked ball so as to intercept the incoming ball. Although walking actions shown by arrows have seemingly no regularity, looking at the arrows only around the squares, we find that the directions of the walking actions mostly face to the states with *save* actions, so that the robot can move to intercept the ball if possible. The states with the other walking actions (far from the squares) seem not to be learned yet. The reason is that our virtual robot never loses the ball in the simulator, and so it cannot go through the experience of those states. In the case of the penalty kick, this may not be a big problem, since the robot knows the fixed initial positions of the attacker and the ball in advance.

3.4   Experiments in Augmented Environments

In this section we experiment in our augmented environment, where we set the ball as simulated and project to the field. Since a virtual ball is easily manipulated by our system, our robot does not need to perform troublesome tasks such as restoring the ball, and this advantage easily enables the autonomous learning of goalie strategies. Moreover, our robot can receive the ideal positions of the ball and the robot itself from our system. This also means that we can experiment using real robots in real environments, even if some basic programs such as vision and localization systems are premature or unfinished.

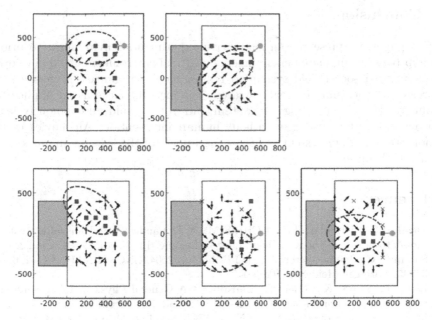

Fig. 5. State-action mapping that was acquired by the learning in the simulator. Each figure shows the left half field, and its coordinate shows the relative position of the robot from the goal, (i.e., X and Y in the state of our learner). Each pair of a circle and arrow shows the velocity of a ball, and each figure indicates which action our robot selects in the situation. A square, cross, and arrow show *save*, *stay*, and walking actions, respectively. In the area enclosed with a dashed circle in each figure, better actions are selected so as to intercept the incoming ball.

We made our robot learn goalie strategies in 200 episodes, starting from the strategy acquired in the previous section, since it takes a lot of time and cost for real robots to perform a lot of actions. The experiment took about 45 minutes with 1 battery. Fig. 4 (b) shows the learning process in the augmented environment in terms of the success rate of saving. In the initial phase of the process, the success rate was only about 50% despite the initial strategy whose success rate was more than 90% in the simulator. This suggests that there is a certain gap between simulated environments and real environments, especially in terms of quadrupedal locomotion. In the latter phase of the process, the success rate increased to about 80%, which may be the maximum value since real robots cannot walk as smoothly as virtual robots. This also suggests that our system can fill in such a gap. We repeated the same experiment 3 times and got the average success rates in the first 50 episodes and the last 50 episodes, which were 56.7% and 70.7%, respectively. The strategies that were acquired in the augmented environment should work well even in real environments as long as our robot can accurately estimate the positions of the ball and the robot itself.

4 Conclusion

In this paper, we presented an interactive experimental environment to bridge the gap between simulated environments and real environments, and developed an augmented soccer field system to realize the environment. Moreover, we addressed the learning of goalie strategies in penalty kicks as an application of our system. By utilizing our system, our robot could autonomously learn and acquire better strategies without human intervention. All movies of the demonstration of our system and the experiments of the learning in our system are available on-line[1].

References

1. Zagal, J.C., del Solar, J.R.: UCHILSIM: A Dynamically and Visually Realistic Simulator for the RoboCup Four Legged League. In: Nardi, D., Riedmiller, M., Sammut, C., Santos-Victor, J. (eds.) RoboCup 2004. LNCS (LNAI), vol. 3276, pp. 34–45. Springer, Heidelberg (2005)
2. Laue, T., Spiess, K., Röfer, T.: SimRobot - A General Physical Robot Simulator and Its Application in RoboCup. In: Bredenfeld, A., Jacoff, A., Noda, I., Takahashi, Y. (eds.) RoboCup 2005. LNCS (LNAI), vol. 4020, pp. 173–183. Springer, Heidelberg (2006)
3. Zaratti, M., Fratarcangeli, M., Iocchi, L.: A 3D Simulator of Multiple Legged Robots based on USARSim. In: Lakemeyer, G., Sklar, E., Sorrenti, D.G., Takahashi, T. (eds.) RoboCup 2006: Robot Soccer World Cup X. LNCS (LNAI), vol. 4434, pp. 13–24. Springer, Heidelberg (2007)
4. Microsoft: Microsoft Robotics Studio (2007), http://msdn.microsoft.com/robotics/
5. Takeshita, K., Okuzumi, T., Kase, S., Hasegawa, Y., Mitsumoto, H., Ueda, R., Umeda, K., Osumi, H., Arai, T.: Technical Report of Team ARAIBO. Technical report, ARAIBO (2007), http://araibo.is-a-geek.com/
6. Sugimoto, M., Kagotani, G., Kojima, M., Nii, H., Nakamura, A., Inami, M.: Augmented coliseum: display-based computing for augmented reality inspiration computing robot. In: SIGGRAPH 2005: ACM SIGGRAPH 2005 Emerging technologies, p. 1. ACM, New York (2005)
7. da Silva Guerra, R., Boedecker, J., Mayer, N., Yanagimachi, S., Hirosawa, Y., Yoshikawa, K., Namekawa, M., Asada, M.: CITIZEN Eco-Be! League: bringing new flexibility for research and education to RoboCup. In: Proceedings of the Meeting of Special Interest Group on AI Challenges, vol. 23, pp. 13–18 (2006)
8. Kobayashi, H., Osaki, T., Williams, E., Ishino, A., Shinohara, A.: Autonomous Learning of Ball Trapping in the Four-legged Robot League. In: Lakemeyer, G., Sklar, E., Sorrenti, D.G., Takahashi, T. (eds.) RoboCup 2006: Robot Soccer World Cup X. LNCS (LNAI), vol. 4434, pp. 86–97. Springer, Heidelberg (2007)
9. Piccardi, M.: Background subtraction techniques: a review. In: IEEE International Conference on Systems, Man and Cybernetics 2004, vol. 4, pp. 3099–3104 (2004)
10. Sutton, R.S., Barto, A.G.: Reinforcement Learning: An Introduction. MIT Press, Cambridge (1998)

---

[1] URL: <http://www.shino.ecei.tohoku.ac.jp/~kobayashi/movies.html#goalie>

Automatic Parameter Optimization for a Dynamic Robot Simulation

Tim Laue[1] and Matthias Hebbel[2]

[1] Deutsches Forschungszentrum für Künstliche Intelligenz GmbH,
Sichere Kognitive Systeme, Enrique-Schmidt-Str. 5, 28359 Bremen, Germany
tim.laue@dfki.de
[2] Robotics Research Institute, Section Information Technology,
Dortmund University of Technology, Otto-Hahn-Str. 8, 44221 Dortmund, Germany
matthias.hebbel@uni-dortmund.de

Abstract. One common problem of dynamic robot simulations is the accuracy of the actuators' behavior and their interaction with the environment. Especially when simulating legged robots which have optimized gaits resulting from machine learning, manually finding a proper configuration within the high-dimensional parameter space of the simulation environment becomes a demanding task. In this paper, we describe a multi-staged approach for automatically optimizing a large set of different simulation parameters. The optimization is carried out offline through an evolutionary algorithm which uses the difference between the recorded data of a real robot and the behavior of the simulation as fitness function. A model of an AIBO robot performing a variety of different walking gaits serves as an example of the approach.

1 Introduction

When working with robots, using a simulation is often of significant importance. On the one hand, it enables the evaluation of different alternatives during the design phase of robot systems and may therefore lead to better decisions and cost savings. On the other hand, it supports the process of software development by providing a replacement for robots that are currently not on-hand (e.g. broken or used by another person) or not able to endure long running experiments. Furthermore, the execution of robot programs inside a simulator offers the possibility of directly debugging and testing them.

One characteristic trait of all simulations is that they can only approximate the real world, this inherent deficit is called *Reality Gap*. This affects all aspects of simulations: the level of detail as well as the characteristics of sensors and actuators. For most of the beforehand mentioned applications, the gap is of minor relevance as long as the simulated robot system performs in a reasonable way.

Within the RoboCup domain, actuator performance is a crucial aspect. Through applying optimization algorithms, impressive results regarding robot velocities have been achieved during the last years, e.g. by Röfer [1] and Hebbel et al. [2] using the AIBO robot, or by Hemker et al. [3] using a humanoid robot.

L. Iocchi et al. (Eds.): RoboCup 2008, LNAI 5399, pp. 121–132, 2009.
© Springer-Verlag Berlin Heidelberg 2009

(a) Semi-elliptical locus   (b) Rectangular locus

Fig. 1. Comparison of controlled (dashed) and real (solid) walk trajectories of an AIBO robot's joint during walking [2]

One common attribute of these algorithms is the strong exploitation of the environment's features, i.e. certain characteristics of the motors or the properties of the ground in this case. This leads to control trajectories that strongly differ from the resulting trajectories of the real robot joints, as shown in Fig. 1. For robot simulations, especially when working with legged robots which have a high number of degrees of freedom, this requires a proper parametrization, i.e. to simulate actuators that behave close to real ones. Otherwise, the simulated robot might not only behave unrealistic but could fail completely.

Currently, there exists a broad range of dynamic robot simulators which are used – not only – by RoboCup teams, e.g. Webots [4], the Übersim [5], SimRobot [6], or Microsoft Robotics Studio [7]. Also the RoboCup Simulation League introduced a fully dynamic robot simulation , additionally aiming towards a closer cooperation with real robots [8]. All these simulations allow a detailed specification of the environment, but demand the user to do this manually what might become an exhausting task given the high number of environmental parameters. Additionally, a once working parameter set is not guaranteed to be compatible with a different walking gait learned at a later point of time.

The contribution of this paper is to present a general multi-staged process which minimizes the reality gap between real and simulated robots regarding the behavior of actuators and their interaction with the environment. This optimization is carried out by using an evolutionary algorithm. The approach is kept general and transferable to different kinds of legged robots, its application is shown using a model of an AIBO robot as example.

A similar concept named *Back to Reality* by Zagal et al. [9], implemented within the UChilsim application [10], also automatically determines environmental parameters of a robot simulation. Nevertheless, their focus is the co-evolution of software parameters and the simulation, rather than the detailed optimization of a general purpose model.

This paper is organized as follows: Sect. 2 gives an overview of relevant parameters within a dynamic robot simulation, Sect. 3 describes the algorithm used for optimization. The multi-staged optimization setup is described in Sect. 4, its results are presented in Sect. 5.

2   Relevant Parameters

Within every robot simulation, there exists a vast number of parameters which influence realism and performance. For our work, we used the SimRobot simulator [6] based on the free *Open Dynamics Engine (ODE)* by Russell Smith [11]. This engine provides a simulation of rigid body dynamics at industrial quality and has already been used in many projects, e. g. in most of the previously mentioned simulators.

In this section, we describe all parameters relevant for a reasonable simulation of actuators and their interaction with the environment. Most of these parameters are generic for all kinds of simulations, some of them are specific to ODE but have similar equivalents in other simulations. We differentiate between parameters that need to be optimized and those that may be kept static.

2.1   Parameters for Optimization

The parametrization of the motors has probably the highest impact on the correctness of the simulation. In our application (as well as in most others), only motor-driven hinge joints have been used to simulate the servo motors of a real robot. For every motor, there are five parameters: the *maximum velocity*, the *maximum torque*, and the three control factors of an – possibly given – *PID* controller. Vendor specifications (if available) of the motors regarding velocity and torque provide a good starting point for the optimization, but are definitely not directly transferable without any further optimization. The control parameters of servo motors are generally not accessible at all, not to mention their implementation.

When considering not only the motion of single joints but the motion of a whole robot, the interaction with the ground needs to be modeled, i.e. the friction between the robot surface and the floor. ODE provides an efficient approximation of the Coulomb fiction model. It is realized by temporarily adding so-called contact joints between colliding objects. For every pair of different surfaces, a set of six parameters is needed to configure these contact joints.

Two additional, ODE-specific parameters are the *Error Reduction Parameter (ERP)* and the *Constraint Force Mixing (CFM)*. They do not have any counterparts in the real world, their purpose is to keep the simulation mathematically stable by adding forces to preserve a correct joint alignment (ERP) and by allowing a certain amount (CFM) of constraint violation. Both can be used as global parameters as well as per joint. To keep the state space for the later optimization small, we assume that these parameters are the same for all joints and do not need to be adapted locally.

2.2   Fixed Parameters

Additionally to all afore-mentioned parameters, there is a huge set of parameters that does not need to be optimized and might be set to fixed values. Among these are the sizes, shapes, and positions of all body parts. Their manual measurement already provides a sufficient estimate. Even when using a dynamically simplified robot model as we do (see Fig. 3c), additionally optimizing these parameters would lead to an extreme growth of the state space.

The masses of all body parts can also be measured manually (after a disassembly of the robot). In case of complex or irreversibly connected parts, it might be useful to optimize the exact masses and their centers. But this was not necessary for the robot used in this work.

One global force is the gravity. We kept it fixed to its standard value.

3   Optimization Algorithm

The goal of the optimization is to find parameters for the physical simulation engine which result in a behavior of the simulated robot as close as possible to the behavior of the real robot. Notice that rather than finding the real parameters like e. g. the real friction between the robot's legs and the carpet, the aim in this work is to find parameters that overall result in a similar behavior of the simulated robot. Thus, the problem itself cannot be specified mathematically which means that the only way to get the quality of the simulation (from now on called *fitness*) for a certain parameter setting is to try them out in the simulation and to measure how close they match the real robot's movements. This section shortly describes the optimization algorithm which has been used to learn the parameter setting. A standard Evolution Strategy with self-adaptation has been used which is well-established for problems with an unknown structure.

The Evolution Strategy has been developed by Schwefel and Rechenberg for the optimization of technical problems [12]. It is inspired by the biological evolution and is working with a population of individuals. Each individual represents a point x in the n–dimensional real-valued search space; the point x is encoded in the object parameters $x_1, \dots, x_n$ of the individual and represents its genes. In terms of Evolution Strategies each individual can be assigned a function value $f(x)$ representing the quality of the solution, the so called fitness which has to be optimized. The process of the evolution is shown in Fig. 2. An amount of μ individuals constitute the parent population and create a new offspring population of the next generation with λ individuals. An offspring individual is created by recombining (mixing) the real-valued genes of ρ randomly selected parents followed by a random mutation of each object parameter such that the

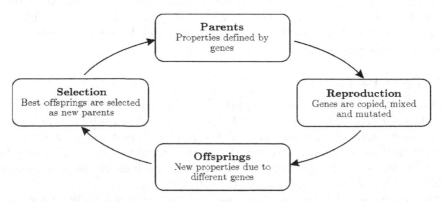

Fig. 2. Optimization cycle of the Evolution Strategy

offsprings differ in their genes from their parents. The fitness of each individual in the offspring generation will be measured and the selection operation chooses the μ best individuals to establish the new parent generation.

The chosen Evolution Strategy uses *self-adaptation* to control the strength of the mutation operation. Each object parameter has its own mutation strength. The values of the mutation strengths have a big effect on the progress speed of the evolution process. With a mutation strength chosen too big, the chance to overshoot the optimum rises, while small mutation strengths increase the probability to get stuck in a local optimum and furthermore the speed to approach the optimum is unnecessary small. Hence it is very important to adapt the mutations strengths to the current situation. Here the mutation strengths are are also included in each individual and are selected and inherited together with the individual's assignments for the object parameters. Thus, they have a higher probability to survive when they encode object parameter variations that produce fitter individuals [13].

After the fitness values of the offspring individuals have been assigned, the selection operator selects the parents for the next generation. Here the next parent generation was only generated from the offsprings, parents of the previous generation always died out, even if their fitness was been better than the best fitness of the best offsprings. Apparently this seems to slow down the convergence speed because previously found better solutions can be forgotten but this trade-off guarantees a good performance of the self-adaptation and also the risk to get stuck in a local optimum is reduced.

4 Experimental Setup

As pointed out earlier, a simulation model consists of a large number of parameters that need to be optimized. For our concrete application, we chose an AIBO robot. This robot has 12 degrees of freedom for locomotion (the additional degrees of freedom for head, ears, tail, and mouth are not relevant for our experiments). By assuming that the motor configuration is symmetrical, i.e. the left legs use the same motors than the right ones, and that within every leg different motors are used, parameters for six motors need to be considered. Furthermore, we have modeled only one kind of surface for the robot's feet, thus only one kind of friction – consisting of six parameters – will be optimized. Together with the two global parameters CFM and ERP, a total of 38 parameters needs to be optimized.

Learning all these parameters simultaneously would be a complex problem, thus the learning is done in three stages. The problem has been split up and the parameters have been clustered into groups which can be evaluated independently from each other. For each stage, a different setup and fitness function are used.

4.1 Motor Velocity and Initial Controller Parameters

In the first step, the maximum velocities and the settings of the PID controllers have to be learned. For this learning step, a special experiment has been set

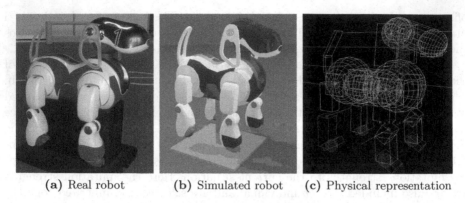

(a) Real robot   (b) Simulated robot   (c) Physical representation

Fig. 3. Setup for learning the maximum velocities

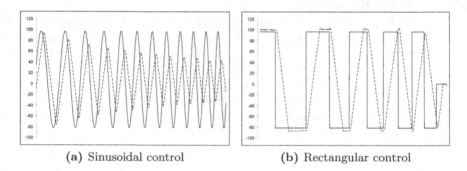

(a) Sinusoidal control   (b) Rectangular control

Fig. 4. An extract from the controlled (solid) and the measured joint movement (dashed) for evaluating the maximum velocities

up in order to minimize the effect of other physical parameters. As shown in Fig. 3, the robot has been placed on a platform with its legs hanging in the air. Two movement sequences have been performed on the real robot and the controlled and the sensor values of each joint have been recorded. The first sequence consisted of a sine movement with increasing frequency. The amplitude of the sine was set to 80% of the maximum angle of the according joint to be able to get the characteristics of the PID controller trying to position the joint at the requested position. Taking the maximum angle of the joint as the target position would give little feedback about the controller because the movement would be stopped abruptly by the hinge stop. In the second joint sequence, a rectangular oscillation has been used instead of the sinusoidial oscillation. Fig 4 shows this control and the resulting sensor curve exemplarily for the front right shoulder joint. The whole movement sequence for this learning step was approximately 60 seconds long.

In this learning step, the simulated robot's legs are controlled exactly like the real robot's legs. The goal is to match the movement of the real robot with the movement of the simulated robot. The fitness which is a measure for the

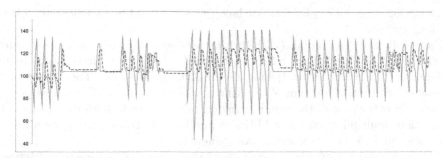

Fig. 5. An extract from the controlled (solid) and the measured joint angle (dashed) of the front right knee joint for evaluating the maximum forces

quality of the matching, is evaluated by summing up the squared differences between the sensor curve of the real robot's movement and the sensor curve of the simulated robot. Consequently, smaller values of the fitness express a better matching of the movements. The time offset between the measured sensor curve and the simulated sensor curve is of no interest and is excluded from the fitness measurement by shifting the simulated sensor curve in a small window in time back and forward; the smallest measured fitness is assigned.

4.2 Motor Torques

The second step serves an intermediate step to get a first estimate of the torques of the joint motors. For this purpose, the robot is walking on the ground as well as doing typical soccer movements, e.g. kicking, getting up, or pushing. Again, the controlled and measured (real) joint values of the physical robot are recorded. The recorded joint controls are used to move the joints of the simulated robot. The resulting joint movements are compared with the real robot's movement. The fitness of the simulation is – as in step 1 – calculated from the summed up squared differences between the real robot's sensor curve and the simulated robot's sensor curve. Figure 5 shows the movement of the front right knee joint of the robot together with the resulting sensor curve. The difference of resulting walk speeds of the simulated and the real robot is ignored yet. Since the maximum forces of the motors certainly depend to some extend on the friction between the robot and the floor (which will be optimized in the following step), the estimated forces in this step are not final.

4.3 Overall Robot Motion

We aim for universally valid parameters for the physical simulation, i.e. a realistic simulation for all walk requests with different walk patterns, instead of just one walk type with different walk requests. For this purpose, a set of 60 different walks of a real robot has been recorded. Among approved walking patterns used for soccer competitions, the set also includes suboptimal walks that contain much stumbling and sliding. In addition to the joint data, the translational and rotational velocities of the robot have been recorded by an external motion tracking system.

Each walk lasts five seconds (with the first second not being taken into account for the fitness). The highest velocities reached by the robot have been about 480mm/s (translation forward) and about 220°/s (rotation). The fitness of a parameter set is the total sum of the fitness of all walks used. The fitness of one single walk is the sum of the squared velocity differences in both directions and in rotation (multiplied with an additional weighting to transform it into a range of values equal to the translational components).

The main purpose of this step is to optimize all global parameters: friction, ERP, and CFM. Additionally, a further modification of the motors' torques and PID control parameters has been allowed, starting from the results of the previous step.

5 Experimental Results

This section presents the results of the three previously explained learning stages. Since each individual of the learning strategy represents a different parameter variation, its fitness can only be evaluated by running a simulation with this parameter setting. Since a lot of simulations are needed to find an optimal parameter set, we distributed the simulations on a compute cluster with 60 nodes. Each node consisted of a PC with Intel Pentium 4 CPU clocked at 2.4 GHz which is able to simulate the robot in real-time.

For each learning stage, the quality of the so far best found solution of each stage is also measured with respect to the resulting robot movement. This was not the primary optimization goal for the first two learning stages, but it is a benchmark to show that the selected learning stages are beneficial regarding the final goal. Table 1 shows the differences between the resulting walk speeds of the real robot and the simulated robot. For each walk direction (x is forward, y is to the robot's left and r is the rotation) the absolute maximum and the standard deviations are calculated, accordingly smaller values represent a better and more realistic simulation. The set of 60 walks was divided into two groups, named A and B. Group A consisted of 40 walk movements and group B of the remaining 20 walks. The entries in the column group A+B are evaluated based on the set union.

In the first stage, the maximum velocities and a preliminary setting for the PID parameters for each joint have been learned. An amount of 5 parents and 25 offsprings have been set as the population size. After 80 generations, the optimization of stage 1 has been aborted because it apparently converged. Fig. 6 shows in comparison an extract of the learning movement of the real robot and the simulated robot with the best learned parameters. The simulated movement gets very close to the real movement. Also the effect of the walk speeds is considerable: all standard deviations and absolute deviations have been decreased significantly. The greatest difference in x direction e. g. was reduced to 386.70 mm/s from 1227.47 mm/s. The maximum velocities in this stage are final

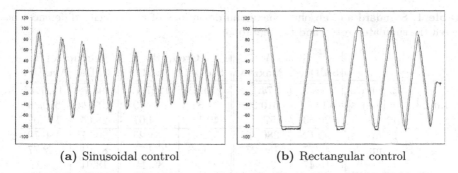

(a) Sinusoidal control   (b) Rectangular control

Fig. 6. An extract from the real (solid) and the simulated joint movement (dashed) after learning the maximum velocities

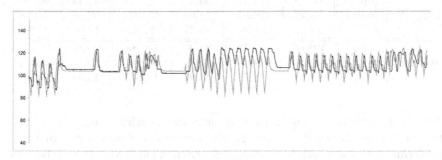

Fig. 7. An extract from the real (solid) and the simulated joint movement (dashed) of the front right knee joint after learning the (preliminary) maximum forces

because the movement of the free swinging legs in this setup is mostly dependent on the maximum available speed. The learned PID parameters instead are just preliminary because until now, there are almost no forces affecting the joints and they certainly also depend on the maximum torque of the motors which will be learned later. Thus the finally learned PID parameters in this stage are taken as starting values for the next learning stage.

In the next stage, the robot learned preliminary values for the maximum torques and the PID controllers. Preliminary for the reason that they are not independent from global physical parameters like e. g. friction between the robot's legs and the ground but will serve as an estimate of the real values for starting torques of the next learning stage. Fig. 7 shows the results of this intermediate stage. The simulated leg movement of the right knee joint gets in some parts close to the real movements, while in other parts the difference is still quite big. However, according to the Table 1 the deviations between the resulting walk speeds are again reduced. In this learning stage, the convergence took longer than in the first stage. After 400 generations the optimization of this stage was aborted, because after approximately 200 generations no significant improvement was observable.

Table 1. Standard and absolute maximum deviations of the velocity differences between the simulated robot and the real robot

		Group A StdDev	Group A MaxDev	Group B StdDev	Group B MaxDev	Group A+B StdDev	Group A+B MaxDev
Initial	x [mm/s]	144.51	574.16	273.33	1227.47	198.78	1227.47
	y [mm/s]	114.67	610.39	91.81	258.69	107.23	610.39
	r [°/s]	20.63	57.87	40.11	134.07	28.65	134.07
Stage 1	x [mm/s]	92.12	386.70	53.81	131.25	80.57	386.70
	y [mm/s]	48.77	180.27	119.93	372.77	82.01	372.77
	r [°/s]	15.47	37.24	24.06	88.24	18.91	88.24
Stage 2	x [mm/s]	89.78	368.90	72.65	264.94	84.49	368.90
	y [mm/s]	50.10	185.76	82.31	258.55	62.44	258.55
	r [°/s]	17.19	42.97	17.76	57.87	17.30	57.87
Stage 3	x [mm/s]	40.00	113.47	55.55	183.40	45.48	183.40
	y [mm/s]	35.49	150.34	75.69	255.98	52.26	255.98
	r [°/s]	10.31	24.06	15.47	41.25	12.61	41.25
Webots	x [mm/s]	123.65	442.58	91.90	199.86	113.01	442.58
	y [mm/s]	147.80	421.33	218.06	502.51	173.35	502.51
	r [°/s]	44.12	178.76	42.97	129.49	43.54	178.76

For the last learning stage, an Evolution Strategy with 9 parents and 50 off-springs has been chosen. In this stage, 32 parameters (8 global parameters and the maximum torques with PID settings for each of the 6 joints) have been optimized at once. The maximum torques and PID parameters for the joint motors were initialized with relatively small mutation strengths because they have already been estimated in the previous learning stage. In this stage, only the 40 different walks in group A have been considered for the parameter optimization. The optimization was not executed on the 20 walks in group B, which serves in this stage as a test case to show that the found parameters for group A are generally valid and not specific for this group. The optimization in this case was stopped after 200 generations. The maximum and standard deviations clearly decreased for the walks of group A, but also the simulated walks in group B got significantly closer to the real walks.

An analysis of the remaining most problematic walks disclosed a shortcoming of our simulation model: only one type of friction is realized between all robot parts and the ground. This is a problem especially for the AIBO robot because it mostly walks on the plastic covers of its elbows which allow some degree of slipping on the ground. Nevertheless, some (especially backward) walks are problematic for our simulation because they make use of the rubber caps at the end of each leg exploiting the good grip between the rubber and the ground.

Finally, to get a rating of the quality of our simulation using the optimized parameters, we compared the walks in group A and B with their performance in the commercially available robot simulator Webots which also includes a dynamic model of the AIBO robot. The results are presented in the last row in Table 1. The simulation with the Webots simulator is mostly better than with our initial setting, but after the first learning stage, the physical simulation with the optimized parameters can significantly outperform the Webots model with the walks in both groups A and B.

6   Conclusion and Future Works

In this paper, we presented an approach for optimizing parameters of a dynamic robot simulation. By dividing the optimization process into different stages, it was possible to deal with the high-dimensionality of this problem. The application of this approach to a model of an AIBO robot whose joint and friction parameters were improved continuously performed well. This is especially revealed by comparing the results with the accuracy of another, well-established robot simulator. The inaccuracies resulting from modeling only one type of friction verify the assumption that a high level of detail is needed for accurate models in dynamic robot simulations.

Our results lead to several starting points for further investigations. Although the final parameters close the reality gap to a large extend, improvements seem still to be possible. These could eventually be reached by taking more parameters – for instance in a fourth learning stage – into account, e.g. the center of mass of certain major body parts or the anchor points of the joints. Since all experiments could be made offline, computing time was not in the focus of our research. The applied Evolution Strategy demonstrated to be able to cope with the optimization problem but it would be interesting to evaluate, if other algorithms are able to provide comparable results in less time.

To further evaluate the presented approach, optimizing parameters for other robot models is a necessary proceeding. This will be done for a humanoid robot in the near future.

Acknowledgments

This work has been partially funded by the Deutsche Forschungsgemeinschaft in the context of the Schwerpunktprogramm 1125 (*Kooperierende Teams mobiler Roboter in dynamischen Umgebungen*).

References

1. Röfer, T.: Evolutionary Gait-Optimization Using a Fitness Function Based on Proprioception. In: Nardi, D., Riedmiller, M., Sammut, C., Santos-Victor, J. (eds.) RoboCup 2004. LNCS (LNAI), vol. 3276, pp. 310–322. Springer, Heidelberg (2005)
2. Hebbel, M., Nistico, W., Fisseler, D.: Learning in a high dimensional space: Fast omnidirectional quadrupedal locomotion. In: Lakemeyer, G., Sklar, E., Sorrenti, D.G., Takahashi, T. (eds.) RoboCup 2006: Robot Soccer World Cup X. LNCS (LNAI), vol. 4434, pp. 314–321. Springer, Heidelberg (2007)
3. Hemker, T., Sakamoto, H., Stelzer, M., von Stryk, O.: Hardware-in-the-loop optimization of the walking speed of a humanoid robot. In: CLAWAR 2006: 9th International Conference on Climbing and Walking Robots, Brussels, Belgium, September 11-14, pp. 614–623 (2006)

4. Michel, O.: Cyberbotics Ltd. - WebotsTM: Professional Mobile Robot Simulation. International Journal of Advanced Robotic Systems 1(1), 39–42 (2004)
5. Go, J., Browning, B., Veloso, M.: Accurate and flexible simulation for dynamic, vision-centric robots. In: Proceedings of International Joint Conference on Autonomous Agents and Multi-Agent Systems (AAMAS 2004) (2004)
6. Laue, T., Spiess, K., Röfer, T.: SimRobot - A General Physical Robot Simulator and Its Application in RoboCup. In: Bredenfeld, A., Jacoff, A., Noda, I., Takahashi, Y. (eds.) RoboCup 2005. LNCS (LNAI), vol. 4020, pp. 173–183. Springer, Heidelberg (2006)
7. Jackson, J.: Microsoft robotics studio: A technical introduction. Robotics and Automation Magazine 14(4), 82–87 (2007)
8. Mayer, N.M., Boedecker, J., da Silva Guerra, R., Obst, O., Asada, M.: 3D2Real: Simulation League Finals in Real Robots. In: Lakemeyer, G., Sklar, E., Sorrenti, D.G., Takahashi, T. (eds.) RoboCup 2006: Robot Soccer World Cup X. LNCS (LNAI), vol. 4434, pp. 25–34. Springer, Heidelberg (2007)
9. Zagal, J.C., Ruiz-del-Solar, J., Vallejos, P.: Back to Reality: Crossing the Reality Gap in Evolutionary Robotics. In: Proceedings of the 5th IFAC/EURON Symposium on Intelligent Autonomous Vehicles - IAV 2004 (2004)
10. Zagal, J.C., del Solar, J.R.: UCHILSIM: A Dynamically and Visually Realistic Simulator for the RoboCup Four Legged League. In: Nardi, D., Riedmiller, M., Sammut, C., Santos-Victor, J. (eds.) RoboCup 2004. LNCS (LNAI), vol. 3276, pp. 34–45. Springer, Heidelberg (2005)
11. Smith, R.: Open Dynamics Engine - ODE (2007), www.ode.org
12. Beyer, H.G., Schwefel, H.P.: Evolution strategies – A comprehensive introduction. Natural Computing 1(1), 3–52 (2002)
13. Beyer, H.G.: Toward a theory of evolution strategies: Self-adaptation. Evolutionary Computation 3(3), 311–347 (1996)

Arbitrary Ball Recognition Based on Omni-Directional Vision for Soccer Robots

Huimin Lu, Hui Zhang, Junhao Xiao, Fei Liu, and Zhiqiang Zheng

College of Mechatronics and Automation,
National University of Defense Technology, Changsha, China
{lhmnew,xjh_nudt,liufei,zqzheng}@nudt.edu.cn,
zhanghui_nudt@126.com

Abstract. Recognizing the arbitrary standard FIFA ball is a significant ability for soccer robots to play competition without the constraint of current color-coded environment. This paper describes a novel method to recognize arbitrary FIFA ball based on omni-directional vision system. Firstly the omni-directional vision system and its calibration for distance map are introduced, and the conclusion that the ball on the field can be imaged to be ellipse approximately is derived by analyzing the imaging character. Then the arbitrary FIFA ball is detected by using image processing algorithm to search the ellipse imaged by the ball according to the derivation. In the actual application, a simple but effective ball tracking algorithm is also used to reduce the running time needed after the ball has been recognized globally. The experiment results show that the novel method can recognize the arbitrary FIFA ball effectively and in real-time.

1 Introduction

The RoboCup Middle Size League competition is a standard real-world test bed for autonomous multi-robot control, robot vision and other relative research subjects. It is still a color-coded environment, though some great changes have taken place in the latest competition rules, such as replacing the blue/yellow goals with white goal nets, no color flag post any more. The final goal of RoboCup is that robot soccer team defeats human champion, so robots will have to be able to play competition without the constraint of current color-coded environment. It is a significant step to realize this for soccer robots to be able to recognize any FIFA ball like human being.

Paper [1] [2] [3] proposed a so called Contracting Curve Density (CCD) algorithm to recognize ball without color labeling. This algorithm fits parametric curve models to image data by using local criteria based on local image statistics to separate adjacent regions. This method can extract the contour of ball even in cluttered environments under different illumination, but the vague position of the ball should be known in advance. So the global detection can not be realized by this method. Paper [4] presented a method to detect and track a ball without special color in real-time even in cluttered environments by integrating the Adaboost feature learning algorithm into a condensation tracking framework. Paper [5] presented a novel scheme for fast color invariant ball detection, in which the edged filtered images serve as the input of an

L. Iocchi et al. (Eds.): RoboCup 2008, LNAI 5399, pp. 133–144, 2009.

Adaboost learning procedure that constructs a cascade of classification and regression trees. This method can detect different soccer balls in different environments, but the false positive rate is also high when there are other round objects. So this method is then improved by combining a biologically inspired attention system-VOCUS [6] with the cascade of classifiers. This combination makes the ball recognition highly robust and eliminates the false detection effectively. Paper [7] proposed an edge-based arc fitting algorithm to detect ball for soccer robots. However, above algorithms are all used in perspective camera vision system in which the field of view is far smaller and the image is also far less complex than that of the omni-directional vision system.

This Paper presents an arbitrary FIFA ball recognition algorithm based on omni-directional vision system for our RoboCup Middle Size League robots-NuBot. In the following part, we will firstly introduce our omni-directional vision system and its calibration for distance map in section 2, and then derive that the ball on the field ground can be imaged to be ellipse approximately and calculate the shape information of the ellipse by analyzing the imaging character in section 3. The image processing algorithm for arbitrary ball recognition will be proposed in section 4. The experiment results and the discussions will be presented in section 5 and section 6 respectively. The conclusion will be given in section 7 finally.

2   Our Omni-Directional Vision System and Its Calibration

The performance of omni-directional vision system is determined almost by the shape of panoramic mirror. It is also the most important factor we have to take into account to calibrate the distance map from the pixel distance of the image coordinate to the metric distance of the real world coordinate at the ground plane.

2.1   The New Panoramic Mirror

We design a new panoramic mirror which is made up of the hyperbolic mirror, the horizontally isometric mirror and the vertically isometric mirror from the inner to the outer. This mirror is different from our former mirror which only includes horizontally isometric mirror and vertically isometric mirror [8]. The only deficiency of our former mirror lies in that the imaging of scene very close to robot is bad, such as the robot itself can not be seen in the panoramic image, which is caused by the difficulty of manufacturing the innermost part of the mirror accurately. So we replace the innermost part of the former mirror with a hyperbolic mirror to solve this problem. The designed profile curve of new mirror and the manufactured mirror are demonstrated in figure 1(a) and figure 1(b). The typical panoramic image captured by our new omni-directional vision system is showed in figure 2(b). Our new omni-directional vision system not only maintains the merit of our former system that is making the imaging resolution of the objects near the robot on the field constant and making the imaging distortion of the objects far from the robot small in vertical direction, but also can have clear imaging of scene very close to robot, such as robot itself.

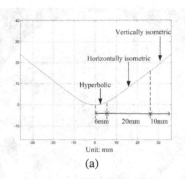
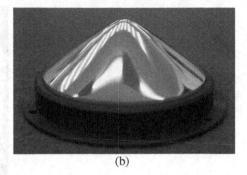

(a)   (b)

Fig. 1. Our new panoramic mirror. (a) The profile curve of the new panoramic mirror. (b) The manufactured panoramic mirror.

2.2 The Calibration for the Distance Map

We can calculate the locations of objects in the robot's Cartesian coordinate only when we have calibrated the distance map from the pixel distance of the image coordinate to the metric distance of the real world coordinate at the ground plane. Furthermore, if this calibration has been done, we can analyze what shape the FIFA ball can be imaged in the panoramic image and derive the shape parameters, which is essential for our arbitrary ball recognition method.

We define the center of the mirror equipped by the robot as the center of robot. For the imaging of omni-directional vision is isotropy, the only thing we have to do in the calibration is to calculate the map $x = f(i)$, in which x is the actual distance from some point to the center of robot on the real world ground, and i is the pixel distance from the pixel imaged by the same point to the center of robot on the image.

The three parts of our panoramic mirror are totally different, so we have to calibrate the distance map respectively in three different forms. Firstly we have known the following parameters according to the mirror designing calculation:

- x_m: the maximal distance to the center of robot the hyperbolic part mirror can observe on the real world ground;
- x_n: the maximal distance to the center of robot the horizontally isometric part mirror can observe on the real world ground;
- h_n: the maximal height to the ground the vertically isometric part mirror can observe in the real world location with distance x_n to the center of robot;
- i_m: the pixel distance to the center of robot on the image mapped by x_m;
- i_n: the pixel distance to the center of robot on the image mapped by x_n;
- i_r: the pixel distance to the center of robot on the image mapped by h_n, and also equal to the radius of the circle surrounding the virtual part of the image.

Above parameters are shown in figure 2. In figure 2(b), the red point on the center of the image is the imaging point of the center of robot.

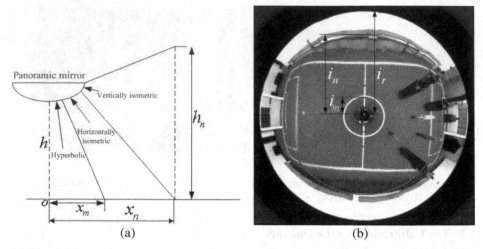

Fig. 2. The parameters of our omni-directional vision system. (a) The sketch of observation range the three part mirrors can observe in real world. (b) The sketch of imaging range of scenes the three part mirrors can observe.

In calibrating the distance map for hyperbolic part mirror, for the hyperbolic mirror is a single viewpoint mirror, we can derive the relationship of the point on the real world ground and the pixel imaged by the point on the panoramic image according the imaging theory of single viewpoint catadioptric vision [9]. We use a lookup table to store the distance map for the hyperbolic part mirror.

In calibrating the distance map for the horizontally isometric part mirror, for the resolution of the imaging of the scenes on the ground is constant, the distance map for this part mirror can be expressed as follows:

$$x = x_m + (i - i_m) * (x_n - x_m) / (i_n - i_m) \qquad (1)$$

In calibrating the distance map for the vertically isometric part mirror, for the resolution of the imaging of the scenes vertical to the ground with distance x_n to the center of robot is constant, and the objects on the same incident ray have the same imaging on the panoramic image, as shown in figure 3, the following 2 equations can be derived:

$$(i - i_n) / (i_r - i_n) = h_c / h_n \qquad (2)$$

$$h_c / h = (x - x_n) / x \qquad (3)$$

So the distance map for the vertically isometric part mirror can be derived with equation (2), (3) as follows:

$$x = (i_r - i_n) * h * x_n / ((i_r - i_n) * h - (i - i_n) * h_n) \qquad (4)$$

The distance map function $x = f(i)$ can be expressed as the following piecewise function:

$$x = f(i) = \begin{cases} f_1(i) & \text{if } i \leq i_m \\ f_2(i) = x_m + (i - i_m) * (x_n - x_m)/(i_n - i_m) & \text{if } i_m < i \leq i_n \\ f_3(i) = (i_r - i_n) * h * x_n /((i_r - i_n) * h - (i - i_n) * h_n) & \text{if } i_n \leq i < i_r \end{cases} \quad (5)$$

In the equation (5), the sub-function $x = f_1(i)$ is replaced with a lookup table in our algorithm.

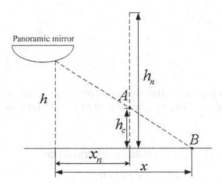

Fig. 3. The sketch of distance map calibration for the vertically isometric part mirror

3 Analysis of the Imaging Character of Ball

In the analysis of the imaging character of ball, we only consider the situation that ball is located on the ground, and we assume that the panoramic mirror is a point with height h to the ground, for the mirror size is far smaller comparing to ball size and the distance from mirror to ball. So the incident rays from ball to mirror can be considered to form a cone tangent to the ball approximately. As we all know, the intersections of a plane and a cone generate conic sections such as circles, ellipses, hyperbolas, parabolas and so on. In this situation, an ellipse is generated by the intersection of the cone and the ground plane, which is shown in figure 4. We define a right hand Cartesian coordinate with the center of robot on the plane as the origin o of the coordinate, with the direction from robot to ball on the plane as x axis. The direction of the major axis of the ellipse coincides with x axis. We assume that the distance between ball and robot is x_b.

The imaging of the ball is the same as the imaging of the ellipse in the omni-directional vision system, so we need to derive the shape parameters of the ellipse on the real world ground and then analyze what shape the ellipse is imaged in the panoramic image. The equation of the ellipse is as follows:

$$\frac{(x - x_0)^2}{a^2} + \frac{y^2}{b^2} = 1 \quad (6)$$

In the equation (6), x_0 determines the location of the ellipse, a and b are the semi-major axis and semi-minor axis that determine the shape of the ellipse.

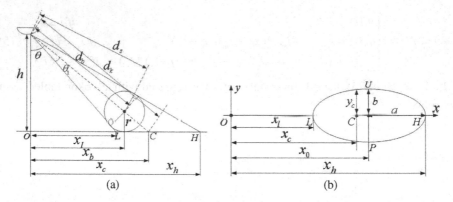

Fig. 4. The sketch of imaging of the ball in omni-directional vision system. (a) The front view of the imaging of the ball. (b) The ellipse generated by the intersection of the cone and the ground plane.

According to figure 4, we get the following equations:

$$x_b = x_c * (h-r)/h \tag{7}$$

$$d_b = \sqrt{(h-r)^2 + x_b^2} \tag{8}$$

$$d_s = \sqrt{d_b^2 - r^2} \tag{9}$$

$$tg\theta_1 = r/d_s \tag{10}$$

$$tg\theta = x_b/(h-r) \tag{11}$$

$$tg(\theta+\theta_1) = (tg\theta+tg\theta_1)/(1-tg\theta tg\theta_1) \tag{12}$$

$$tg(\theta-\theta_1) = (tg\theta-tg\theta_1)/(1+tg\theta tg\theta_1) \tag{13}$$

$$x_l = h*tg(\theta-\theta_1) \tag{14}$$

$$x_h = h*tg(\theta+\theta_1) \tag{15}$$

$$d_k = \sqrt{h^2 + x_c^2} \tag{16}$$

$$y_c = d_k * tg\theta_1 \tag{17}$$

$$a = (x_h - x_l)/2 \tag{18}$$

$$x_0 = (x_h + x_l)/2 \tag{19}$$

The height h of mirror to the ground and the radius r of ball are known in advance. If x_b or x_c is given, we can calculate a and x_0 by substituting the equation (8) ~ (15) into the equation (18) and (19). For the point (x_c, y_c) is located on the ellipse, we can derive the following equation:

$$\frac{(x_c - x_0)^2}{a^2} + \frac{y_c^2}{b^2} = 1 \tag{20}$$

We can derive b by substituting the equation (7), (16) ~ (19) into the equation (20). For processing the panoramic image to detect the ball, we have to derive further what shape this ellipse will be imaged in the panoramic image. We assume that the ellipse still will be imaged as ellipse, and the distance between the center of the ellipse and the center of robot is i on the panoramic image. Then according to the distance map calibration of our omni-directional vision system in the section 2, we can calculate $x_0 = f(i)$. But it is very complex to calculate a and b if only x_0 is given according to the equation (7) ~ (20), so we do a little simplification on this problem by replacing x_0 with x_c, for the point C is very close to the center of ellipse in figure 4(b). The simplification will be verified to be feasible by the experiment in section 5. So we can calculate $x_c = f(i)$, and then derive the real world ellipse parameters x_l, x_h, x_0, a and b from x_c according to the equation (7) ~ (20). For we have calibrated the distance map in section 2, we can use the inverse function of distance map function to derive the semi-major axis a_i and the semi-minor axis b_i of the imaged ellipse on the panoramic image. The calculation functions are also piecewise functions as follows:

$$a_i = \begin{cases} (f_1^{-1}(x_h) - f_1^{-1}(x_l))/2 & \text{if } x_l < x_h < x_m \\ (f_2^{-1}(x_h) - f_1^{-1}(x_l))/2 & \text{if } x_l < x_m \leq x_h < x_n \\ (f_2^{-1}(x_h) - f_2^{-1}(x_l))/2 & \text{if } x_m \leq x_l < x_h < x_n \\ (f_3^{-1}(x_h) - f_2^{-1}(x_l))/2 & \text{if } x_l < x_n \leq x_h \\ (f_3^{-1}(x_h) - f_3^{-1}(x_l))/2 & \text{if } x_n \leq x_l < x_h \end{cases} \qquad (21)$$

$$b_i = \begin{cases} b * f_1^{-1}(x_0)/x_0 & \text{if } x_0 < x_m \\ b * f_2^{-1}(x_0)/x_0 & \text{if } x_m \leq x_0 < x_n \\ b * f_3^{-1}(x_0)/x_0 & \text{if } x_n < x_0 \end{cases} \qquad (22)$$

Up to now, we have finished the derivation on the shape parameters a_i and b_i of the ellipse imaged by the ball on the field ground, given that distance between the center of the ellipse and the center of robot is i on the panoramic image. The process of the calculation can be summarized as follows:

$$i \rightarrow x_c \rightarrow x_l, x_0, x_h, a, b \rightarrow a_i, b_i$$

We store all the values of a_i and b_i varying with i in a lookup table which will be used to detect the arbitrary FIFA ball by image processing in the next section.

4 Image Processing Algorithm for Arbitrary Ball Recognition

We have derived the semi-major axis and the semi-minor axis of the ellipse imaged by the ball on each location of the panoramic image, so we can recognize the arbitrary ball by processing the images to search the possible ellipses according to this character. For the arbitrary FIFA balls have different colors, we can not detect the ball based on color classification as the traditional color objects recognition method. However, the color variations still exist between the pixels belonging to the two sides of the ball contour. So we define two color variation scan methods to detect all of the possible contour points. The first scan method is called rotary scan, in which we define a series of concentric circles with the center of robot on the image as their common centers and we will do the following scan in the concentric circles one by one. In each concentric circle, we search the color variations of every two neighboring pixels, and the color variations are measured by Euclidean distance in YUV color space. If the color variation is higher than a threshold, a possible contour point is found. Then if the distance between every two possible contour points is close to the minor axis value of the ellipse with its center located on this concentric circle, a possible ellipse center point which is the middle point of the two possible contour points is found.

The another scan method is called radial scan, in which we define 360 radial scan rays with the center of robot on the image as their common origins, and we will do the following scan along the radial scan rays one by one. In each radial scan ray, we search the same color variations of every two neighboring pixels as in the rotary scan. If the color variation is higher than a threshold, a possible contour point is found. Then if the distance between every two possible contour points is close to the major axis value of the ellipse with its center located on the middle point of the two possible contour points, a possible ellipse center point which is the middle point is found.

After the two sets of the possible ellipse center points have been acquired by rotary scan and radial scan, we can compare all the points in one set with all the points in the other set one by one. If the two points almost coincide with each other, we can consider that a candidate ellipse exists with the coinciding point as the center of ellipse and also get the equation of this candidate ellipse. Then we match the possible contour points which are detected in the rotary scan and near to the center of the candidate ellipse with the ellipse equation. If enough points match well with the equation, we can verify the candidate ellipse to be real one. Furthermore, for the ball is on the field ground in most situations of the competition, we also combine the traditional image segmentation result based on color classification to reduce the disturbance from outside of the field. After having segmented green field from the image, only those ellipses close to the field are considered to be the final imaging ellipses of the ball.

In the actual application of competition, there will be only one ball needed to recognize on the competition field, so we use the best matching result as the final ball detected. And we also don't need to do above global detection by processing the whole image in every frame. Once the ball is detected globally, we can track the ball by only processing the nearby image area of the ball detected in the last frame with the same image processing algorithm, and the running time needed in the tracking process could be reduced greatly. The nearby image area is dynamically changed with the major-axis and the minor-axis of the ellipse imaged by the ball detected in the last frame. When the ball is lost in the tracking process, the global detection will be restarted.

Up to now, we can recognize a standard FIFA ball without using color classification, so arbitrary standard FIFA balls can be recognized by our method.

5 Experiment Results

Firstly we demonstrate the process and the result of our arbitrary ball recognition algorithm presented in section 4 by processing the typical panoramic image in figure 2(b).

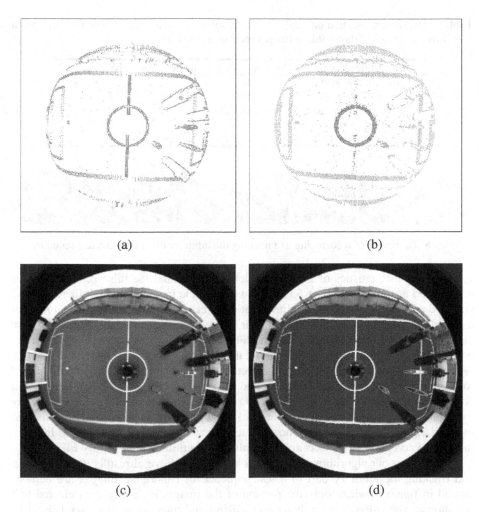

(a)   (b)

(c)   (d)

Fig. 5. The image processing results of our arbitrary ball recognition method. (a) The possible contour points detected by rotary scan. (b) The possible contour points detected by radial scan. (c) The searched candidate ellipse centers. (d) The recognition result of the three arbitrary balls.

The results of the rotary scan and radial scan are shown in figure 5(a), (b) respectively, and the red points in figure 5(a), (b) are the possible contour points. The result of searching the candidate ellipses is demonstrated in figure 5(c) and the center points of the candidate ellipses are denoted as the purple rectangles. The final recognition result is shown in figure 5(d). The recognition result is also verified by combining the image segmentation result based on color classification. The purple rectangles are the center points of the ellipses imaged by the balls on the field ground, the cyan ellipses are the theoretical imaging of the ball on the panoramic image, and the color classification of the green field is also demonstrated in figure 5(d). From figure 5(d) we can find that the three FIFA balls with different colors are all detected correctly.

Table 1. The correct detection rate and the false positives of the arbitrary FIFA ball recognition with different groups of thresholds in image processing algorithm

Different thresholds	The correct detection rate	The false positives
1st group of thresholds	88.34%	7
2nd group of thresholds	89.57%	12
3rd group of thresholds	92.02%	15
4th group of thresholds	94.48%	16
5th group of thresholds	96.32%	17

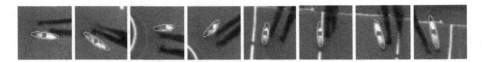

Fig. 6. The results of recognizing and tracking the arbitrary FIFA ball in a test sequence

For doing the statistics of the correct detection rate and the false positives of our recognition method, we collect 55 different panoramic images in which there are 163 standard FIFA balls totally. The adjustment of thresholds in our recognition algorithm especially in matching the possible contour points with the candidate ellipse equation affects the correct detection rate and the false positives. So we process these images with different groups of thresholds and calculate the correct detection rate and the false positives respectively. The correct detection rate and the false positives related to 5 groups of thresholds are demonstrated in table 1. Only global detection is dealt with in our recognition process, and the correct detection rate can be increased greatly by combining our method with object tracking algorithms such as particle filter and other filtering algorithm. So the correct detection rate and the false positives listed in table 1 are acceptable for soccer robots to play competition with arbitrary FIFA ball.

We also test our algorithm in the actual application. Several results of recognizing and tracking the arbitrary ball in a test sequence of panoramic images are demonstrated in figure 6, where only the portion of the images including the detected ball are shown. The video of our robot's recognizing and tracking an arbitrary FIFA ball can be found on our team website: http://www.nubot.com.cn/2008videoen.htm.

RoboCup Middle Size League competition is a highly dynamic environment and robot must process its sensor information as soon as possible. We also test the running time of our recognition algorithm. It takes about 100ms~150ms to realize global detection by processing a whole panoramic image with dimension 444*442. However, once the ball has been recognized globally, the running time can be reduced to several ms~20ms in the tracking process for only the partial image near the ball detected in the last frame is needed to be processed. So our recognition algorithm can meet the real-time requirement of RoboCup Middle Size League competition.

6 Discussion

Comparing to the other existing arbitrary ball recognition methods, our algorithm has following good features:

- Our algorithm doesn't need any learning or training process which is necessary in the recognition algorithm based on Adaboost learning;
- Our algorithm can deal with global detection which is not considered in the CCD algorithm;
- Our algorithm is based on omni-directional vision system, the field of view of which is much wider than those perspective cameras used in other existing methods;
- Our algorithm can incorporate the object tracking algorithms easily to detect the arbitrary ball more efficiently and real-time, and the interim and the final results of our algorithm can also be used as important clues for other recognition methods;
- The idea of our algorithm can also be used in any other omni-directional or perspective vision systems, if the imaging character of ball can be analyzed in advance.

However, there are still some deficiencies in our algorithm, which are shown in figure 7. The first deficiency is that the imaging of ball is occluded partly by robot itself when the ball is very close to robot, so our algorithm fails in recognizing this ball. This deficiency can be solved partially by adding the arbitrary ball recognition algorithm based on perspective camera as we had demonstrated successfully in the second free technical challenge of RoboCup2007 Atlanta. In the algorithm, we applied the Sobel filter to detect all the edge points and their gradient directions on the perspective image firstly, and then recognized the arbitrary ball by using Hough transform based on gradient information to detect the circle imaged by the ball. The second deficiency is that the imaging of ball can not be approximated to be ellipse on the image when the ball is imaged by both of the horizontally isometric part mirror and the vertically isometric part mirror partially, so our algorithm matches the ball with bad accuracy, which is shown in the detection result of the ball with maximal distance to robot in figure 7(b). This deficiency may be solved by incorporating other recognition methods such as Adaboost learning algorithm. The third deficiency is that our algorithm only can deal with the situation that the ball is on the field ground. We have to develop some arbitrary ball recognition method based on stereo-vision system to solve this problem.

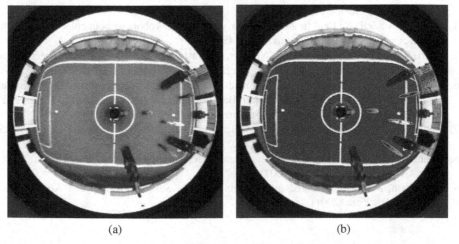

(a)   (b)

Fig. 7. The recognition results of another panoramic image. (a) The original panoramic image. (b) The arbitrary ball recognition result by our algorithm.

7 Conclusion

In this paper, a novel arbitrary FIFA ball recognition algorithm based on our omni-directional vision system is proposed. Firstly we introduce our omni-directional vision system and its calibration for distance map, and then we derive the conclusion that the ball on the field can be imaged to be ellipse approximately by analyzing the imaging character. We also calculate the major axis and the minor axis of the ellipse on different location of the image in advance. In the image processing, we scan the color variation to search the possible major axis and minor axis of the ellipse by radial scan and rotary scan without color classification, and then we can consider that an ellipse imaged from ball may exist if the middle points of a possible major axis and a possible minor axis are very close to each other on the image. Finally we verify the ball by matching the color variation points searched before near the candidate ellipse center with the ellipse equation. Once the ball has been recognized globally, we also use a simple but effective ball tracking algorithm to reduce the running time needed by only processing the nearby image area of the ball detected in last frame. The experiment results show that our novel method can recognize the arbitrary FIFA ball effectively and in real-time.

References

1. Hanek, R., Beetz, M.: The Contracting Curve Density Algorithm: Fitting Parametric Curve Models to Images Using Local Self-Adapting Separation Criteria. International Journal of Computer Vision 59, 233–258 (2004)
2. Hanek, R., Schmitt, T., Buck, S., Beetz, M.: Fast Image-based Object Localization in Natural Scenes. In: Proceedings of the 2002 IEEE/RSJ International Conference on Intelligent Robots and Systems, pp. 116–122 (2002)
3. Hanek, R., Schmitt, T., Buck, S., Beetz, M.: Towards RoboCup without Color Labeling. In: Kaminka, G.A., Lima, P.U., Rojas, R. (eds.) RoboCup 2002. LNCS, vol. 2752, pp. 179–194. Springer, Heidelberg (2003)
4. Treptow, A., Zell, A.: Real-time object tracking for soccer-robots without color information. Robotics and Autonomous Systems 48, 41–48 (2004)
5. Mitri, S., Pervölz, K., Surmann, H., Nüchter, A.: Fast Color-Independent Ball Detection for Mobile Robots. In: Proceedings of IEEE Mechatronics and Robotics, pp. 900–905 (2004)
6. Mitri, S., Frintrop, S., Pervölz, K., Surmann, H., Nüchter, A.: Robust Object Detection at Regions of Interest with an Application in Ball Recognition. In: Proceedings of IEEE International Conference on Robotics and Automation, pp. 125–130 (2005)
7. Coath, G., Musumeci, P.: Adaptive Arc Fitting for Ball Detection in RoboCup. In: APRS Workshop on Digital Image Analysing (2003)
8. Zhang, H., Lu, H., Ji, X., et al.: NuBot Team Description Paper 2007. RoboCup 2007 Atlanta, CD-ROM, Atlanta, USA (2007)
9. Benosman, R., Kang, S.B. (eds.): Panoramic Vision: Sensors, Theory and Applications. Springer, New York (2001)

A Robust Statistical Collision Detection Framework for Quadruped Robots

Tekin Meriçli, Çetin Meriçli, and H. Levent Akın

Department of Computer Engineering
Boğaziçi University, Istanbul, Turkey
{tekin.mericli,cetin.mericli,akin}@boun.edu.tr

Abstract. In order to achieve its tasks in an effective manner, an autonomous mobile robot must be able to detect and quickly recover from collisions. This paper proposes a new solution to the problem of detecting collisions during omnidirectional motion of a quadruped robot equipped with an internal accelerometer. We consider this as an instance of general signal processing and statistical anomaly detection problems. We find that temporal accelerometer readings examined in the frequency domain are good indicators of regularities (normal motion) and novel situations (collisions). In the course of time, the robot builds a probabilistic model that captures its proprioceptive properties while walking without obstruction and uses that model to determine whether there is an abnormality in the case of an unfamiliar pattern. The approach does not depend on walk characteristics and the walking algorithm used, and is insensitive to the surface texture that the robot walks on as long as the surface is flat. The experiments demonstrate quite fast and successful detection of collisions independent of the point of contact with an acceptably low false positive rate.

1 Introduction

One of the most important skills for an autonomous mobile robot is to safely and robustly navigate in both known and unknown environments, such as planetary surfaces [1,2] and disaster areas [3]. Different tasks and operation environments require different mobility configurations. Every mobility configuration has its advantages and disadvantages; however, all have a possibility of suffering from unforeseen failure events, such as collisions.

Detection of and fast recovery from failures are crucial for autonomous systems since improper treatment of such situations may cause loss of control and harm the system or the environment. Inability to detect collisions may lead to a failure of the whole mission. For instance, collision of a mobile service robot with a low doorstep will prevent it from reaching the destination hence accomplishing its goal, or collision of a soccer playing robot with another robot on the field will probably lead to a failure in its main task which is to chase the ball and score goals. Such situations require the mobile robot to be able to detect the unexpected/abnormal event quickly and to infer the cause of the event in order to react appropriately [4].

L. Iocchi et al. (Eds.): RoboCup 2008, LNAI 5399, pp. 145–156, 2009.
© Springer-Verlag Berlin Heidelberg 2009

Obstacle detection and collision avoidance problems are usually solved by processing the data obtained from dedicated hardware such as laser range-finders and touch/bump sensors. However, the task becomes much harder when those sensors are not present. This paper considers a case of a quadruped robot, Sony Aibo ERS-7 [5], with accelerometers and a color camera. It is able to perform vision-based localization. In this case, one possible approach is to use vision to detect collisions; especially in the environments where the structure is known by the robot which can localize itself in that environment [6,7,8]. The difference between vision-based and odometry-based pose (location and orientation) estimates may give an idea about collisions since the pose of the robot will not change as expected if the robot is blocked by an obstacle when trying to move. However, the objects that are used for localization are not always in the field of view of the robot, hence vision-based methods may easily fail. Failure in detecting collisions usually has a severe impact on the pose estimate of the robot when only the odometry feedback is used to update the estimate due to the lack of visual information.

This paper proposes a new approach for detecting collisions during omnidirectional motion of an Aibo robot. There is a limited number of sensors on the robot and those available do not include touch/bump sensors that can be used for detecting collisions directly. Hence, our proposed method utilizes internal accelerometer readings, which are obtained in the time-amplitude domain, and processes this signal in the frequency domain since the robot generates a noisy but regular pattern while walking. Approaching the problem from a statistical novelty detection point of view, our method can discriminate between what is normal and what is not in terms of accelerometer signal patterns. The main advantages of our method are

- being collision-point-independent,
- detecting collisions with an *acceptably low* false positive rate,
- being computationally cheap and fast, which is essential considering limited processing power on Aibo robots that is shared between several processes.

The rest of the paper is organized as follows. In Section 2, previous work on collision detection in literature is discussed. Section 3 restates the problem to be solved and elaborates on our approach. Experiments are described and results are discussed in Section 4. Section 5 summarizes and concludes the paper, providing possible extensions to our work.

2   Related Work

There has been limited research on collision and slippage detection using Aibo as the robotic platform. Vail and Veloso [9] used the accelerometer readings for surface detection and velocity prediction . In their work they used a supervised learning approach, feeding a decision tree with labeled data collected while the robot was marching in place on different types of surfaces. Quinlan *et al* [10,11] showed that it is possible to learn normal motion of the limbs for distinct walking directions and used this to detect slippage of the legs and collisions of the robot with its environment. The main drawbacks of this method are that it is not generalized for omnidirectional walking and that it is not independent of collision point. This method will not be able to detect the collision when the robot touches the obstacle with its chest while the legs are freely moving. Also, they used ERS-210 model in their experiments; however, ERS-7 has much more powerful servo motors and therefore the paws can slip on the surface even though the robot is obstructed, in which case this approach will again fail in detecting collisions. Hoffman and Göhring [12] proposed an approach that compares the desired motion of the actuators; that is the actuator command, with actual motion calculated from servo readings. Their work also considers very limited walking speeds and directions for training; therefore, it has almost the same shortcomings of the work of Quinlan *et al.*

3   Approach

Having a perfect map of the environment with all the static obstacles are marked, makes the life of a mobile robot very easy in terms of avoiding collisions even if it has a very limited number of sensors. In such a case the robot may not even need collision detection capability. However, when there are dynamic objects around, it becomes crucial for the robot to be able to detect collisions since getting stuck may result in the complete failure in the robot's task. Robot soccer environments are perfect examples of such dynamic environments with many robots on a small field [13]. Since 2005, RoboCup 4-legged league games have been played on a $540cm \times 360cm$ field with 4 robots in each team [14]. With that many robots on that small field, it is therefore inevitable to have collisions which may be between two or more robots or between a robot and a static object such as one of the goals or a beacon. Not having the ability of detecting such situations may have significant negative impacts. For instance, consider the case where the robot gets the ball and tries to make a 180° turn to aim the opponent goal. During its motion, if it collides with another robot and is not able to see sufficient number of objects to correctly update its pose estimate, hence relying only on its odometry-based update, it may very well end up with shooting towards an arbitrary direction, even towards its own goal. More generally, while the robot is walking on the field and updating the pose estimate based on odometry feedback, colliding with another object without noticing may cause mislocalization, which results in improper behaviors. Figure 1 shows two robots colliding with each other where neither of them can proceed in the desired direction. This is one example of a set of situations which we want the robot to detect and avoid using our proposed method.

The problem is specified using Aibo details to make it concrete, but the method we propose is general and can be applied to any legged robot with an accelerometer. It is essentially an analysis of the signal obtained by keeping a history of accelerometer readings. The internal accelerometers of Aibo consist of a mass/spring system located in the core of the body and they provide

Fig. 1. One robot colliding with another, which is a very common situation in a robot soccer game. Both are unable to proceed in the desired direction; hence, the result is failure in the task.

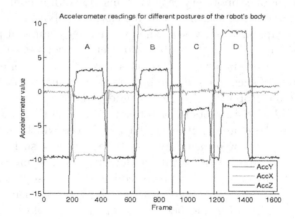

Fig. 2. Accelerometer readings for different body postures. In region A, the robot lays on its left on the ground, in region B, it lays on its right, in region C, it is sitting on its back (i.e. its tail), in region D, it is standing upside down on its head. All the rest correspond to stand still on the paws.

real-valued estimates of the acceleration of the body in x, y, and z axes. These values are read at 30Hz, which is the decision making rate in our implementation, although new values from the sensors can be read at 125Hz. Positive y is defined as the direction the robot is facing, positive x as directly away from the right side of the robot, and positive z as up from the floor. Figure 2 shows the accelerometer readings for different body postures of Aibo.

As it is apparent from Figure 2, the readings can provide useful information when there is a substantial change in the robot's body posture, e.g. when it falls

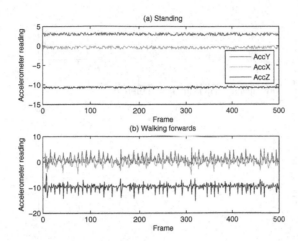

Fig. 3. Accelerometer readings while the robot is (a) standing still and (b) walking forwards with a constant velocity. The positive value on the y axis is a result of the robot's default pose that slightly tilts forward. As it can be seen from the plots, the readings are quite noisy.

over. However, due to the noisy characteristics of the accelerometers, it may be quite challenging to identify minor changes. Figure 3 shows the accelerometer data collected for 500 frames while the robot is (a) standing still and (b) walking forwards with a constant velocity without any obstruction.

Since the default motion system built by Sony for Aibo robots is inadequate in terms of speed and versatility which are necessary to be a good soccer player, most of the RoboCup 4-legged league teams developed inverse kinematics-based motion modules for better locomotion. In our parameterized and flexible motion engine, the loci defined by 3D polygons are traced by the paws in a certain number of intermediate steps and diagonally opposite legs move simultaneously, resulting in a trot gait. Different speeds are obtained by stretching or squeezing the polygons, and walking in different directions is achieved by rotating the polygons around the z axis accordingly. In either case the number of intermediate steps to trace the whole polygon stays the same. This is what makes the motion of the body, hence the accelerometer signals, regular in a certain way independent of speed and direction of the motion.

During the analysis of the accelerometer readings for omnidirectional walking, we observed that the characteristics of those signals change when there is a collision and a different pattern emerges. Those changes, however, are not very obvious or easy to characterize using simple threshold values. Figure 4 illustrates the three signals obtained from the accelerometers while the robot is walking forwards and starts colliding with an obstacle at 300th frame. In order to make the change easier to see, a 4253h-twice smoother [15] was applied (Figure 4b) to the raw signals (Figure 4a).

As it can easily be seen from the plot, the magnitudes of the signals, especially for the x-axis signal, are different during collision compared to regular walk. The

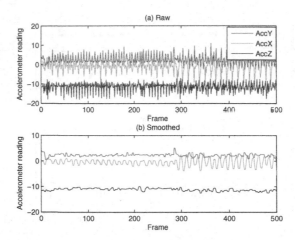

Fig. 4. Accelerometer readings ((a) raw and (b) smoothened using 4253h-twice smoother) during forward walking where the robot collides with an obstacle starting from frame number 300.

reason that x-axis is affected the most is due to the trot gait style, which -by default- makes the robot swing around the y-axis. When the robot collides with an object, it tries to climb over it, and that motion results in a greater magnitude of swinging. Same effect arises when the robot's paws get stuck with the carpet that it walks on. Therefore, we decided to examine the patterns generated on the x-axis.

Buffered temporal readings of x-axis values are transformed to the frequency domain via the application of Fast Fourier Transform (FFT) operation. In order to expose the patterns in the recent readings, FFT operation is applied only to a window keeping the recent history of the readings. Since the buffer is filled in a certain number of frames that is specified by the length of the window that we slide over the buffer, keeping the window size as small as possible is crucial in responding to collisions quickly. Considering our decision making rate of 30 frames/second and necessity to supply an array with a size of a power of 2 to the FFT function, we decided to use a sliding window of size 64, which was enough for FFT to better extract the frequency characteristics of the accelerometer signal.

The change in the frequency/power distribution is obvious as in Figure 5, but since different motion commands produce different signal characteristics and different frequency/power distributions, representing the change in distribution over time is a challenging task. To cope with this problem, we defined Sum of Squared Differences (SSD) of two consecutive frequency/power distributions as the sum of squared differences of each power value and its corresponding previous value. By doing so, we are able to represent the change in frequency/power distribution between two consecutive frames with a single scalar value. Since the frequency/power characteristics differ drastically between normal (unobstructed walking) and novel (collision) situations, we expect the corresponding SSD values to be different as well. The major problem here is that collecting collision data for

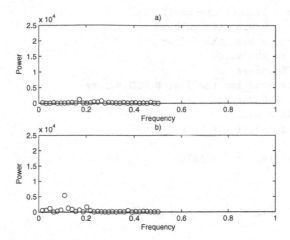

Fig. 5. Frequency/Power representation for a) signal portion recorded under unobstructed walking, b) signal portion recorded under collision situation.

all possible collision situations and for all possible walk speeds and directions is impractical; hence, formulating this problem as a supervised learning problem is neither practical nor flexible enough to capture most of the abnormal occurrings (i.e. collisions). Therefore, we decided to formulate the problem as an instance of statistical anomaly/novelty detection problem.

Novelty detection is the identification of new or unknown data or signal that a machine learning system is not aware of during training [16]. Although it is not possible to train a system on all possible failure situations, it is relatively easy to train a system on known situations and reject the ones that substantially diverge from what is expected. In the collision detection case, it is much easier to train the robot on the normal walking behavior rather than trying to simulate all possible collision cases to provide as negative examples.

We used hypothesis testing which is a well-known statistical technique for answering a *yes*-or-*no* question about a population and assessing the probability that the answer is wrong. In our problem, we are seeking the answer of the question whether the sample of SSDs consisting of SSD values calculated for a certain number of recent consecutive frames are drawn from the population of SSD values calculated under normal circumstances. Formulation of such questioning as a hypothesis testing problem is shown in Equation 1

$$H_0 : \mu_{sample} = \mu_{regular}$$
$$H_1 : \mu_{sample} \neq \mu_{regular}$$

(1)

where, μ_{sample} is the mean of a certain number of recent consecutive SSD values and $\mu_{regular}$ is the learned mean of SSD values calculated under normal walking conditions. Here, the null hypothesis is that the mean of the recent SSD values and the learned mean of SSD values are equal, i.e., the robot is walking without

```
function updateCollisionDetector()
  read recent accelerometer values
  slide the window over the buffer
  perform FFT on the window
  calculate SSD values
  update sample mean for the last N SSD values
  apply one-tailed t-test
  if testResult > tValue
      set collision flag = true
  else
      set collision flag = false
  end
end
```

Fig. 6. Pseudo-code of the collision detector

any obstruction. The alternative hypothesis states that the means are different; that is, the recent SSD values are calculated under a novel situation.

Based on the *Central Limit Theorem* [17], we assumed that the SSD values calculated under regular walking conditions come from a normal distribution. A one-tailed t-test [17] is used to test the hypothesis in order to be able to use small ($N < 30$) samples and hence to be able to respond quickly in case of an unfamiliar pattern. The formula for t-test is shown in Equation 2 where $\sigma_{regular}$ is the standard deviation of the regular SSD values population and N is the sample size. *Type I errors*, or *false positives* are crucial for a mobile robot since such errors fire false collision signals resulting in improper behaviors. Depending on the application domain, there is always a trade-off between false positives and false negatives. In the robot soccer case, Type II errors (false negatives; i.e. the robot not detecting collision while colliding) are preferred over false positives (i.e. the robot detecting collision while walking without obstruction); therefore, we applied a conservative test and rejected the null hypothesis with an expected false rejection probability of $p = 0.001$.

$$T = \frac{\mu_{sample} - \mu_{regular}}{\sigma_{regular}/\sqrt{N}} \tag{2}$$

The pseudo-code summarizing the operations from reading accelerometer values to concluding about collision status is given in Figure 6.

4   Experiments

Experiments are conducted on a regular RoboCup 4-Legged field using an Aibo ERS-7 robot. Experimental study consists of two phases:

- Building a statistical proprioceptive sensory model
- Testing the performance

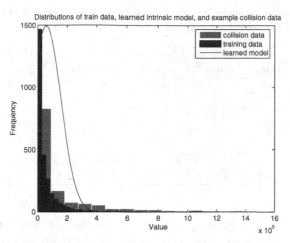

Fig. 7. Distribution of gathered train data, learned model, and the distribution of a test sequence containing collisions

4.1 Building a Statistical Proprioceptive Sensory Model

In order to develop a model of frequency changes under normal walk conditions, we considered two most common situations: self positioning and ball chasing. First, we let the robot go to its predefined initial position on the field omnidirectionally, so that it would walk in various directions with various speeds. After self positioning, we let the robot chase the ball for nearly 1.5 minutes. Ball was manually moved across the field in such a way to make sure that the robot walked omnidirectionally in a symmetric manner. This setup provided us more stratified data. During the execution, a total of 4150 frames were processed and a Gaussian distribution was fit on the data. The distribution of gathered train data, learned model, and the distribution of a test sequence containing collisions are illustrated in Figure 7. As it can be seen in the figure, although the gathered data is not in normal form, learned model is a good estimator of it. The collision data contains normal parts (the left-most bin) and the collision parts. As we mentioned in Section 3, our domain specific preference about Type I and Type II errors is in favor of Type II errors; therefore, we used a very conservative hypothesis testing and considered only extremely far values from the mean as the indicators of novel situations (collisions). We used a sample size of $N = 10$ and an expected false rejection probability $p = 0.001$, and we applied a one-tailed t-test with 9 degrees of freedom yielding a T value limit of 4.297.

4.2 Testing the Performance

Using our code base, a simple behavior for handling the collision situations was implemented for assessing the performance of the collision detection mechanism. This behavior makes the robot walk in the exact opposite of the direction that it was walking when it encounters a collision situation, then makes it to walk

Table 1. Test results comparing the performances of robots featuring collision detection algorithm and not featuring collision detection algorithm

	Collision detection algorithm **ON** / **OFF**	
Trial	**Total Time**	**Obstacle Contact Time**
1	7 / 20.3	0.1 / 17
2	6.5 / 20.7	0.3 / 19
3	6.2 / 22.5	1.8 / 20
4	6.9 / 18.6	0.7 / 17
5	6 / 17	0.3 / 14
Mean	**6.52 / 19.82**	**0.64 / 17.4**
Std. Dev.	**0.43 / 2.1**	**0.24 / 2.3**

sideways to avoid the obstacle, and finally resets the collision detector buffers to the learned sample mean value. The reason behind resetting is to provide the robot empty buffers since the buffers are implemented as sliding windows over a history of accelerometer readings and they still keep the readings obtained during collision even after the collision is detected and handled.

The performance was tested for both collisions during self positioning and ball chasing. For testing collision detection performance during self positioning we placed stationary robots on the field and recorded the average response time during collisions. Experimental setup regarding ball chasing was consisting of a stationary robot placed in front of a ball, 50 cm away from it. Two test robots running the same soccer player code were released from the same point 5 times each and the average time spent in between touching the obstacle for the first time and touching the ball was recorded. Only one of the robots had our proposed collision detection algorithm, all other code being equal.

Compared to the one that does not have the collision detection module, the robot featuring our collision detection algorithm reduced the measured time nearly by a factor of 3, having an average time to reach the ball of 6.52 seconds while it took 19.82 seconds on the average for the other robot to push the obstacle completely and reach the ball. Excluding the time needed to walk from the starting point to the obstacle and from the obstacle to the ball, the average time to detect collisions, which is the time between the first touch to the obstacle and the execution of the retreat behavior, was 0.64 seconds. The detailed results are given in Table 1. Although we cannot claim that the proposed approach results in a zero false positive rate, we did not observe any false positives during our experiments.

5   Conclusions and Future Work

Detecting collisions and recovering from them quickly is crucial for a mobile robot to achieve its tasks effectively. Collision detection on mobile robots that are not equipped with special hardware such as bump sensors is a problem that is challenging yet not studied deeply. In this paper, a robust statistical collision detection framework for a quadruped robot equipped with an accelerometer sensor is presented. The proposed method considers the problem as an instance of general signal processing and statistical anomaly/novelty detection problem.

The main contribution of this paper is proposing a novel collision detection framework for legged robots which is:

- collision-point-independent,
- walk speed and direction independent
- walk surface texture independent as long as the surface is flat
- fully autonomous in building statistical proprioceptive sensory model
- and, computationally cheap and fast

Although it was tested only on four-legged robots, the method can be applied to any kind of legged robot with a periodical walking pattern.

The method is tested on a soccer playing legged robot, specifically on a Sony Aibo ERS-7 four-legged robot. Experimental results showed that the proposed approach provided fast recovery from collisions, hence reducing the time necessary for accomplishing the task. By building a statistical proprioceptive sensory model fully autonomously for normal walk conditions in less than two minutes, the robot was able to detect collisions with a zero false-positive rate during experiments.

Applying this method to bipedal humanoid robots, investigating the underlying properties of the SSD distribution, and developing an online version of the proposed method are among the possible future works.

References

1. Yoshida, K., Hamano, H., Watanabe, T.: Slip-based traction control of a planetary rover. In: 8th International Symposium on Experimental Robotics (ISER). Springer Tracts in Advanced Robotics, vol. 5, pp. 644–653. Springer, Heidelberg (2002)
2. Lacroix, S., Mallet, A., Bonnafous, D., Bauzil, G., Fleury, S., Herrb, M., Chatila, R.: Autonomous rover navigation on unknown terrains, functions and integration. International Journal of Robotics Research 21(10-11), 917–942 (2003)
3. The RoboCup Rescue official website, http://www.robocuprescue.org/
4. Plagemann, C., Fox, D., Burgard, W.: Efficient failure detection on mobile robots using particle filters with gaussian process proposals. In: IJCAI, pp. 2185–2190 (2007)
5. The Sony Aibo robots, http://support.sony-europe.com/aibo/
6. Röfer, T., Jüngel, M.: Vision-based fast and reactive monte-carlo localization. In: Polani, D., Bonarini, A., Browning, B., Yoshida, K. (eds.) Proceedings of the 2003 IEEE International Conference on Robotics and Automation (ICRA), pp. 856–861. IEEE, Los Alamitos (2003)
7. Sridharan, M., Kuhlmann, G., Stone, P.: Practical vision-based monte carlo localization on a legged robot. In: IEEE International Conference on Robotics and Automation (April 2005)
8. Kaplan, K., Çelik, B., Meriçli, T., Meriçli, Ç., Akın, H.L.: Practical extensions to vision-based monte carlo localization methods for robot soccer domain. In: Bredenfeld, A., Jacoff, A., Noda, I., Takahashi, Y. (eds.) RoboCup 2005. LNCS, vol. 4020, pp. 624–631. Springer, Heidelberg (2006)

9. Vail, D., Veloso, M.: Learning from accelerometer data on a legged robot. In: Proceedings of the 5th IFAC/EURON Symposium on Intelligent Autonomous Vehicles (2004)
10. Quinlan, M.J., Murch, C.L., Middleton, R.H., Chalup, S.K.: Traction monitoring for collision detection with legged robots. In: Polani, D., Browning, B., Bonarini, A., Yoshida, K. (eds.) RoboCup 2003. LNCS, vol. 3020, pp. 374–384. Springer, Heidelberg (2004)
11. Quinlan, M., Chalup, S., Middleton, R.: Techniques for improving vision and locomotion on the sony aibo robot. In: Australasian Conference on Robotics and Automation, Brisbane (2003)
12. Hoffmann, J., Göhring, D.: Sensor-actuator-comparison as a basis for collision detection for a quadruped robot. In: Nardi, D., Riedmiller, M., Sammut, C., Santos-Victor, J. (eds.) RoboCup 2004. LNCS, vol. 3276, pp. 150–159. Springer, Heidelberg (2005)
13. RoboCup International Robot Soccer Competition, http://www.robocup.org
14. The RoboCup Standard Platform League, http://www.tzi.de/spl
15. Velleman, P.F.: Definition and comparison of robust nonlinear data smoothing algorithms. Journal of the American Statistical Association 75(371), 609–615 (1980)
16. Markou, M., Singh, S.: Novelty detection: A review - parts 1 and 2. Signal Processing 83, 2481–2521 (2003)
17. Walpole, R.E., Myers, R.H., Myers, S.L., Ye, K.: Probability and Statistics for Engineers and Scientists, 8th edn. Prentice-Hall, Inc., New Jersey (2007)

Increasing Foot Clearance in Biped Walking: Independence of Body Vibration Amplitude from Foot Clearance

Hamid Reza Moballegh, Mojgan Mohajer, and Raul Rojas

Institut Für Informatik, FU-Berlin Takustr. 9 D-14195 Berlin, Germany
moballegh@gmail.com, {mohajer,rojas}@inf.fu-berlin.de

Abstract. A method of foot clearance achievement and frontal plane stability of 3D bipeds is mathematically analyzed. The independency of the robot's body vibration amplitude from the foot clearance is also proven. The analyzed method takes advantage of the sideways vibration of the body generated by periodically shortening each leg to obtain the required foot clearance. A mathematical model of the biped in the frontal plane is suggested and analyzed in two separate phases, resulted in the calculation of the steady state working point. It is demonstrated that in steady state, the amplitude of the body vibration becomes independent from the leg length vibration amplitude. A direct advantage of the proof is the possibility of achieving of high foot-to-ground distances by increasing the leg length vibration amplitude as the robot reaches its steady state. The results have been verified using both simulations and real robot experiments. To guarantee the stability of the robot in the transient phases a method is suggested based on an energy criterion.

1 Introduction

Stability and efficiency are two main aspects in today's walking humanoid research. There are known successful robots with high stability but poor performance [1]. Since they usually try to ignore the natural properties of the biped's body and urge predefined trajectories using powerful and exact servomotors. On the other hand there are several works poorly based on the mechanical properties of the biped which has been categorized under "passive Dynamic Walking" [1,2,3,4,5]. These approaches usually try to minimize or even avoid the existence of actuators. These solutions present almost human efficient robots, however they suffer from very limited stability. As reported in many of the published results, foot clearance is one of the key features in successful walking [7]. In some simulated 2D passive dynamic walking approaches any premature ground contact known as foot scuffing is easily ignored [2,5]. Some others have tried to make special test grounds which allow the flying foot to continue flying even if it goes below the ground surface [3,6]. Adding knees or using telescopic feet can be useful in a great extent [4,7], but synchronization of the foot clearance mechanism with the walking gait needs special solutions. In addition, in 3D form, side stability of the robot becomes a critical problem as the foot clearance increases. Some presented solutions to this problem are [8,10,11].

In servoed walking one of the common solutions is to transfer the COM projection, or more generally the ZMP to the stance foot support area before lifting the flying foot [9]. However the calculation of these parameters and proper reaction times need exact and reliable feedback data and control from the servos which cannot be expected from commercially available low-cost products.

One of the foot clearance achievement methods is to excite the foot lifting mechanism in sequence with a harmonic function which results in the side vibration of the body. This method has been commonly used in RoboCup humanoid league, however a theoretical analysis of its behavior is still missing. Towards this goal, a mathematical model of the biped in the frontal plane is suggested in section 2. The model is analyzed in two phases called "Flying" and "Heel Strike" phases. The steady state working point of the robot is calculated as the result of intersection between two functions extracted from the analysis of the phases. It is then shown that the steady state working point is independent from foot clearance. In section 3 the simulated results of the model are presented and the divergence of the method is demonstrated. In addition a method is suggested to stabilize the transient state of the robot and finally, experimental results are discussed in section 4.

2   Modeling the Biped in Frontal Plane

Fig. 1a shows the mathematical model used in this paper. The Model is similar to the 2D point foot walker described in [2] however it describes the behavior of the biped in the frontal plane instead of the sagittal plane. The whole mass of the model m is assumed to be placed at the hip joint. Leg opening angle α stands for the distance between the feet contact points and remains constant. The flying leg is shortened to l_f for the period of T seconds, while the stance leg has the maximum length l_s. The role of the legs will be exchanged as the period is elapsed.

The model is analyzed in two phases: The flying phase and the heel strike phase. In the flying phase, the model can be considered as an inverted pendulum hinged to the contact point of the stance foot to the ground. The flying phase starts with an initial angle β_0 equal to $\frac{\pi-\alpha}{2}$ (see Fig 1a). The initial speed is normally positive, so that the stance angle β starts to increase. The center of mass decelerates and stops after a while if (and it is normally the case) it does not have enough energy to reach the vertical position. The center of mass flies back again until the period is elapsed. The stance angle at this moment is called β_{hs}.

In the heel strike phase, the flying foot is re-extended to its original length and the stance foot is shortened. This usually leads to an impact between the flying foot and the ground surface, which can be simplified and modeled in two components of gaining and loosing energy. This simplification is possible since it can be shown that in steady state β_{hs} is very near to $\frac{\pi-\alpha}{2}$. Considering the inelastic impact between the heel and the ground surface, there is a loss of kinetic

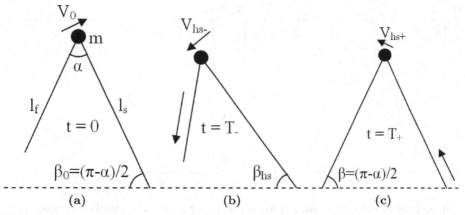

Fig. 1. Biped model (a) in initial conditions of the flying phase. (b) before heel strike and (c) after heel strike.

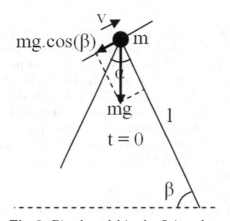

Fig. 2. Biped model in the flying phase

energy caused by the sudden change of velocity vector of the center of mass. On the other hand it gains potential energy due to the lifting of its center of mass.

2.1 Analysis of the Flying Phase

Fig. 2 shows the biped in the flying phase. As mentioned, the biped acts as an inverted pendulum in this phase. The tangent component of the weight is the only accelerating force applied to the model. Equation 1 describes the motion in this phase.

$$\ddot{\beta} = \frac{g \cos \beta}{l} \qquad (1)$$

where:

$$\beta_0 = \frac{\pi - \alpha}{2}, \quad v_0 = v$$

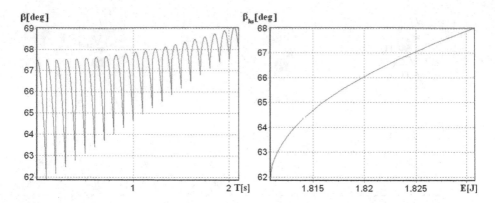

Fig. 3. Left: Simulation of several flying phases with increasing initial energies. Right: β_{hs} as a function of the energy of the model. Angles are presented in degrees to increase the readability.

The goal of analyzing the flying phase is to determine the value of β_{hs} as a function of the energy of the robot. The energy of the biped is a proper parameter showing the initial condition of the robot. As it remains constant during the flight, it can also be used in the analysis of the heel strike. The time of the flight is constant and equal to the period of the leg length vibrations. To solve the differential equation the normal numerical integration method over the period T is used. Fig. 3a shows the simulation of several flying phases with different initial energies. In Fig. 3b β_{hs} is presented as a function of the energy of the model. Following values has been given to the fixed variables used in the calculations through out the paper:

$$m = 1 \; Kg, \quad l = 0.2 \; m, \quad T = 0.1 \; s, \quad \alpha = \frac{\pi}{4} \; rad$$

2.2 Analysis of the Heel Strike Phase

In Fig. 1b and Fig. 1c the model is shown just before and after the heel strike. The changes in energy level of the model in the heel strike phase as discussed above can be considered separately over the kinetic and potential components of the energy of the biped. Equation 2 presents the energy of the model in terms of kinetic and potential energies in initial state.

$$E = mgl \cos \frac{\alpha}{2} + \frac{1}{2}mv_0^2 \tag{2}$$

As the period time T is elapsed the height of the center of mass is reduced to $l \sin(\beta_{hs})$. This difference in the potential energy is converted into kinetic energy. The whole energy of the biped can be presented with equation 3:

$$E = E_{hs-} = mgl \sin \beta_{hs} + mgl(\cos \frac{\alpha}{2} - \sin \beta_{hs}) + \frac{1}{2}mv_0^2 \tag{3}$$

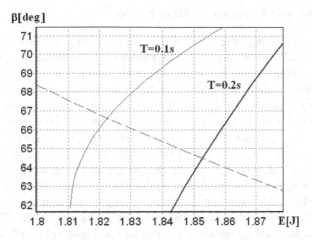

Fig. 4. Steady state condition relating E to β_{hs} (dashed) intersected with the function extracted in section 2.1 (solid) for two different values of T

The first term indicates the potential energy and the second and the third one show the kinetic energy of the system. Upon the occurrence of the heel strike, the potential energy of the COM will be increased to $mgl\cos(\frac{\alpha}{2})$. On the other hand, due to substitution of the stand point only the tangent component of the velocity vector remains unchanged, and the centrifugal component will be eliminated due to the inelastic impact. The new energy can be shown as in equation 4:

$$E_{hs+} = mgl\cos\frac{\alpha}{2} + (mgl(\cos\frac{\alpha}{2} - \sin\beta_{hs}) + \frac{1}{2}mv_0^2)\cos^2\alpha \qquad (4)$$

Equation 4 can be simplified and rephrased in terms of E:

$$E_{hs+} = mgl\cos\frac{\alpha}{2} + (E - mgl\sin\beta_{hs})\cos^2\alpha \qquad (5)$$

The next step is to apply the steady state conditions to the model. In steady state, the energy of the biped should remain unchanged. This can be shown as:

$$E_{hs+} = E_{hs-} = E \qquad (6)$$

By substituting equation 7 in equation 6, the steady state energy of the biped can be presented as the following function of the heel strike stance angle.

$$E(\beta_{hs}) = \frac{mgl\cos\frac{\alpha}{2} - (mgl\sin\beta_{hs})\cos^2\alpha}{\sin^2\alpha} \qquad (7)$$

The function is shown in Fig. 4 together with the numerical solution of the equation 1. The intersection point between these two functions determines the steady state working point. As it is observed, the length of the flying leg has no effect on the steady state working point. However, it will be further shown that this value is subject to a certain maximum. It means that the body vibration amplitude becomes independent from the foot clearance as the foot clearance exceeds a certain minimum.

2.3   Delayed and Premature Heel Strike

When the initial velocity of the center of mass exceeds a certain value, the stance angle remains still higher than $\frac{\pi-\alpha}{2}$ at the end of the flying phase. The result is that the re-extended leg does not reach the ground level at the time of T. The heel strike is called to be delayed in this case. A delayed heel strike has only the energy loss component and therefore cannot occur in steady state. It is however a useful event in transient state, as it reduces the energy of the biped and stabilizes the process. An early heel strike can also happen in which the biped lands on its shortened flying foot due to the lack of energy. A premature heel strike is the only event in which the length of the flying foot plays a role. Actually, the length of the flying foot limits the minimum stance foot angle. As observed in Fig. 4, the less the stance foot angle becomes the higher amount of energy will be pumped in the system by the next heel strike. Therefore this parameter can be used to limit the energy of the biped to avoid instabilities as the biped goes through its transient condition. This method will be discussed further in the paper.

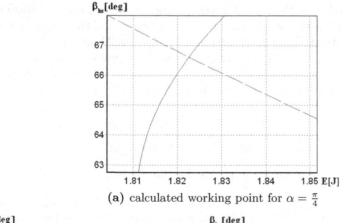

(a) calculated working point for $\alpha = \frac{\pi}{4}$

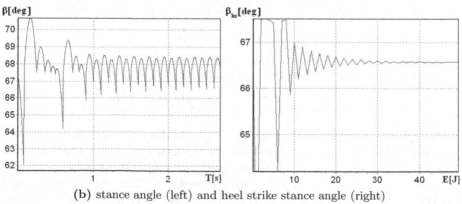

(b) stance angle (left) and heel strike stance angle (right)

Fig. 5. Simulation results: the model converges to the calculated steady state working point

3 Simulation of the Biped Model

In order to test the convergence of the method and to compare the steady state working point with the one calculated in the previous analysis, the flying and the heel strike phases are simulated. The results are shown in Fig. 5 and Fig. 6 using two different leg opening angles. Other parameters remain unchanged. The model reaches its steady state after a few cycles of vibration and the steady state working point in both cases matches the values calculated in analysis.

3.1 Stabilization of the Robot in Transient State

As observed so far, the biped model seems to have an internal feedback loop which regulates the energy. If the biped has too much energy it takes longer for it to fly back and therefore the stance angle of the heel strike becomes larger which means fewer energy for the next flight and vice versa. However this internal

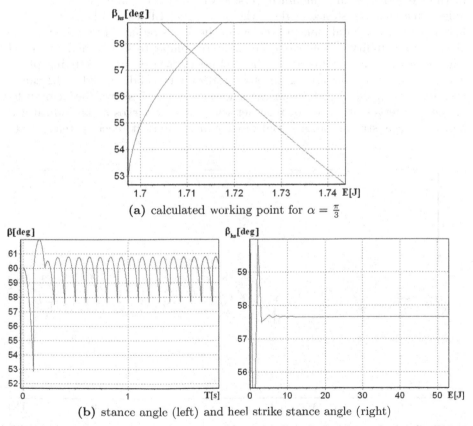

(a) calculated working point for $\alpha = \frac{\pi}{3}$

(b) stance angle (left) and heel strike stance angle (right)

Fig. 6. Simulation results: the model converges to the calculated steady state working point

feedback loop can also make the system diverge under certain circumstances. To avoid this, the amount of energy pumped in the system should be limited before the system reaches its steady state and also when it leaves the steady state under any disturbances. The peak-to-peak value of the stance angle can be used as a suitable indicator of the energy of the robot. It is then enough to calculate this value in each period and assign a proportion of it to the foot clearance. The value can also be low pass filtered to avoid rush changes of it. Applying this method in simulations shows a remarkable improvement in the convergence time and stability of the biped.

4 Experimental Results

The idea of increasing the foot clearance after the biped reaches the steady state has been successfully applied to the humanoid robot "LOLOS" which is going to attend RoboCup 2008 competitions. LOLOS is a 60 cm humanoid with 18 degrees of freedom. Our robot uses an implementation of the introduced method together with a modified version of passive dynamic walking. To be able to measure the stance angle of the robot, the stiffness of the servo motors belonging to the frontal plane movement of the feet is reduced to minimum. This gives the possibility to the robot to act more similar to the point foot model. The stance foot area remains on the ground surface during the flying phase. The peak-to-peak value used in the stabilization method discussed in 3.1 can be therefore extracted directly from the side servos. Fig. 7 shows the angle value of the side servos of the robot together with its knee servos as an indicator of foot clearance, starting from initial rest condition until it reaches steady state.

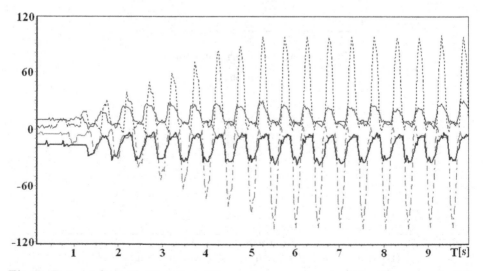

Fig. 7. Test results with the robot: The angle measurement of the side servos (solid) together with knee servos as an indicator of foot clearance (dashed)

The independency of the side vibration amplitude from the foot clearance is observed clearly in the figure. Vibration amplitude remains almost constant as foot clearance increase during the first five seconds.

5 Conclusion

We have theoretically analyzed the toddle like foot clearance achievement method and demonstrated that side stability in this method is independent from foot clearance. Hence, it is possible to achieve high foot-to-ground distances by increasing the leg length vibration amplitude as the robot reaches it steady state. We have examined this idea using simulations and real robot experiments. The experiment platform has been able to reach a maximum speed of 50 cm/s while no foot scuffing has been observed. The results show the success of the method.

References

1. Collins, S., Ruina, A., Tedrake, R., Wisse, M.: Efficient bipedal robots based on passive-dynamic walkers. Science 307, 1082–1085 (2005)
2. Garcia, M., Chatterjee, A., Ruina, A., Coleman, M.: The simplest walking model: Stability, complexity, and scaling. Journal of Biomechanical Engineering - Transactions of the ASME 120(2), 281–288 (1998)
3. McGeer, T.: Passive dynamic walking. International Journal of Robotics Research 9(2), 62–82 (1990)
4. McGeer, T.: Passive walking with knees. In: IEEE International Conference on Robotics and Automation (ICRA), vol. 33, pp. 1640–1645 (1990)
5. Goswami, A., Thuilot, B., Espiau, B.: Compass-Like Biped Robot Part I: Stability and Bifurcation of Passive Gaits. Technical Report, INRIA, RR-2996 (October 1996)
6. Asano, F., Yamakita, M., Furuta, K.: Virtual passive dynamic walking and energy-based control laws. In: Proceedings of 2000 IEEE/RSJ International Conference on Intelligent Robots and Systems (IROS 2000), vol. 2, pp. 1149–1154 (2000)
7. Collins, S.H., Wisse, M., Ruina, A.: A Three-Dimensional Passive-Dynamic Walking Robot with Two Legs and Knees. International Journal of Robotics Research 20(2), 607–615 (2001)
8. Wisse, M., Schwab, A.L.: Skateboards, Bicycles, and Three-dimensional Biped Walking Machines: Velocity-dependent Stability by Means of Lean-to-yaw Coupling. International Journal of Robotics Research 24(6), 417–429 (2005)
9. Kajita, S., Kanehiro, F., Kaneko, K., Fujiwara, K., Harada, K., Yokoi, K., Hirukawa, H.: Biped walking pattern generation by using preview control of zero-moment point. In: Proceedings of IEEE International Conference on Robotics and Automation (ICRA), apos; 2003, vol. 14-19(2), pp. 1620–1626 (September 2003)
10. Tedrake, R., Zhang, T.W., Fong, M.-f., Seung, H.S.: Actuating a simple 3D passive dynamic walker. In: Proceedings of IEEE International Conference on Robotics and Automation (ICRA), apos; 2004, vol. 26(5), pp. 4656–4661 (2004)
11. Donelan, J., Shipman, D., Kram, R., Kuo, A.: Mechanical and metabolic requirements for active lateral stabilization in human walking. Journal of Biomechanics 6(37), 827–835 (2004)

Adapting ADDIE Model for Human Robot Interaction in Soccer Robotics Domain

Rajesh Elara Mohan, Carlos A. Acosta Calderon, Changjiu Zhou, Tianwu Yang, Liandong Zhang, and Yongming Yang

Advanced Robotics and Intelligent Control Centre,
School of Electrical and Electronics Engineering,
Singapore Polytechnic, 500 Dover Road, Singapore
{ZhouCJ,MohanRajesh}@sp.edu.sg
www.robo-erectus.org

Abstract. Though human robot interaction has attracted lots of attention in the recent years due to increasing presence of robots in the marketplace, very little work has been done with systematic human robot interaction development using process models. Analysis, Design, Develop, Implement and Evaluate (ADDIE) model is one of the well established systematic process models for developing instructions in instructional design (ID) community. This paper focuses on two issues: 1) guidance in adapting ADDIE model for human robot interaction, and 2) performance improvement in adopting ADDIE model for human robot interaction in soccer robotics domain. Evaluations were performed and adequate results were obtained. ADDIE modelled human robot interaction was used by our Robo-Erectus Junior humanoid robots in the 2 versus 2 humanoid leagues of RoboCup 2007.

1 Introduction

Growing popularity and increasing viable application domains has contributed to greater presence of robots in the commercial marketplace. Among various robotic platforms, especially humanoid robotic research has seen a rapid growth in the recent years due to the ability of humanoid robots to behave and interact like humans. Humanoid robots might provide day-to-day support in the home and the workplace, doing laundry or dishes, assisting in the care of the elderly, or acting as a caretaker for individuals within a home or institution [1]. Many of these tasks will involve a close interaction between the robot and the people it serves. Researchers have studied the humanoid robot development, including control, emotional expressiveness, humanoid-humanoid collaboration, human robot interaction and perception [2][3][4]. However, there have been only fewer studies with systematic human robot interaction development using process models.

The Analysis, Design, Develop, Implement and Evaluate (ADDIE) model is one of the most widely used systematic process model in instructional design community. ADDIE model provides a generic and systematic framework to the instructional

L. Iocchi et al. (Eds.): RoboCup 2008, LNAI 5399, pp. 166–176, 2009.
© Springer-Verlag Berlin Heidelberg 2009

design process that can be applied to any learning solutions. Most of the currently used instructional design models are simple variants of the ADDIE model. ADDIE model comprises of five elements namely, analysis, design, develop, implement and evaluate. The five elements are ongoing activities that continue throughout the life of an instructional program. After building an instructional program, the other phases do not end once the instructional program is implemented. The five elements work like a loop. They are continually repeated on a regular basis to see if further improvements can be made [5]. ADDIE model being a mature systematic process model can be modified and applied to the human robot interaction development for performance improvement.

Soccer robotics serves as an excellent platform for evaluating performance of an ADDIE modelled human robot interaction. RoboCup is an international joint research initiative to foster artificial intelligence, robotics, and related research. RoboCup uses soccer game as a domain of research, aiming at innovations to be applied for socially significant problems. The ultimate goal of the RoboCup initiative is to develop a team of fully autonomous humanoid robots that can win against the human world champion team in soccer. Our Robo-Erectus has participated in the humanoid league of RoboCup since 2002, collecting several awards since then. Various technologies are incorporated into a soccer robot including, multi robot cooperation, human robot interaction, sensor fusion, artificial intelligence, real time processing, navigation and localization.

Every league has its own requirements on the robot hardware and software systems. Humanoid 2 versus 2 soccer match is played with two teams of two robots in autonomous mode on a 4.5m x 3m field. Each robot must have a human-like body plan with two legs, two arms, and one head, which are attached to a trunk and fit into a cylinder of diameter 0.55*H, where H is the height of the humanoid robot. During the soccer match, system failures occur due to the hardware and software complexity of the humanoid robot operating in autonomous mode. The corresponding team request for stoppage of game with a timeout period of 120s to find fault and service their robot and the game starts after the elapse of 120s [6]. Therefore, humanoid 2 versus 2 soccer match requires human robot interaction design to be able to detect the system fault with the robot as early as possible. Experimental results have shown that the interaction design must be able to find the fault with the robot in atleast 60s or within half the timeout period so as to enable the user to fix the fault within the timeout period. Delay in finding faults during timeout period might provide an edge for the opponent team as they can start the playing game.

This paper focuses on adapting ADDIE model for human robot interaction development in soccer domain and evaluates the performance of the developed interaction. The ADDIE modelled human robot interaction design of Robo-Erectus Junior was evaluated and the results of the experiments showed that, the interaction design was able to find faults in an average time of 22.91s as compared to 57.47s with the unmodelled interaction design. Also, the interaction design was able to detect the fault within 60s in 100% of the cases. Our Robo-Erectus Junior version of humanoid robot was equipped with the evaluated interaction design in RoboCup 2007 humanoid soccer league.

Section 2 discusses ADDIE model and its application in human robot interaction, while section 3 provides a brief introduction of our humanoid platform, Robo-Erectus Junior and its ADDIE modelled human robot interaction. Section 4 summarizes the experiments performed and the results. Finally, section 5 presents the conclusion of the paper.

2   ADDIE Model

ADDIE model is considered to be the most universal instructional designer model. It is widely used by instructional designers and training developers. The ADDIE model encompasses five elements of the instructional design as shows in Figure 1. Each element of the ADDIE model is a critical step, where the instructional designer makes crucial decisions to deliver effective instructions. Table 1 defines each of the five elements of ADDIE model [7]. This guideline is very flexible for building a road map for the entire training project. The ADDIE model is the most commonly used model and can be applied to almost every learning application [8] [9]. However, these elements of ADDIE model cannot be directly applied for human robot interaction system due to its architecture, autonomous nature and dynamically changing environment.

2.1   ADDIE Model for HRI

In this subsection, we modify the elements of ADDIE model so as to make them more applicable to human robot interaction in soccer robotics domain. We modified the elements thru independent review based on the humanoid soccer league objectives by a panel of four researchers working in interaction design and humanoid soccer robotics areas at our lab.

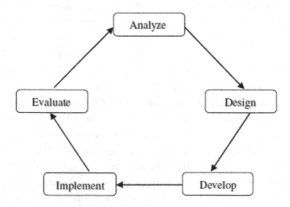

Fig. 1. ADDIE Instructional Design Model

Table 1. Elements of ADDIE model

No.	ADDIE Model
1.	**Analysis** In the analysis element, the instructional problem is validated, the instructional goals and objectives are specified and the learning environment, learner's behaviour and skills are studied. The designer develops a perfect understanding of the gaps between the desired outcomes and the present condition. Some common questions analyzed in this element include, 1. Who is the audience? 2. What do they learn? 3. What are the delivery options? 4. What is the timeline for project completion? 5. What constraints exist?
2.	**Design** In the design element, the designer selects and documents the specific learning objectives, evaluation tools, lesson plan and delivery tools. The design process must be both systematic and specific. Some common steps in design element includes, 11. Documentation of instructional design strategies 12. Design user interface and experience 13. Create prototype design materials
3.	**Develop** In the develop element, the designer completes the actual creation of the learning materials designed to meet the learning objectives. A detailed action plan specifying step by step procedure for implementation of the instructional strategy is prepared. Designers work to develop technologies, perform debugging, review and revise strategies.
4.	**Implement** In the implement element, the designer delivers the instructions to the target group. In this element, training is provided to the facilitator on the course curriculum, learning outcomes, method of delivery and evaluation procedures. The delivery environment should be prepared for implementation of the course.
5.	**Evaluate** In the evaluate element, the effectiveness of the instructional content and the achievement of learning objectives are evaluated. Evaluation is administered in two phases: formative and summative. Formative evaluation is performed within each element of the ADDIE process whereas the summative evaluation is performed after the implementation to determine its effectiveness in satisfying instructional objectives.

Table 2 defines the elements of modified ADDIE model for human robot interaction system. All the five elements of ADDIE remains with only modifications introduced to the definition of the elements. The modified ADDIE model was adopted for the development of human robot interaction for our soccer playing Robo-Erectus Junior humanoid robot.

Table 2. Modified ADDIE Model

No.	Modified ADDIE Model
1.	**Analysis** In the analysis element, the human robot interaction problem is validated, the human robot interaction goals and objectives are specified, the application environment, and robot operator's requirements are studied. The designer develops a perfect understanding of the gaps between the desired outcomes and the present condition. Some common questions analyzed in this element include, 1. What is the application domain? 2. What are the requirements on human robot interaction? 3. What are the control modalities (joystick/computer, etc)? 4. What is the timeline for project completion? 5. What constraints exist?
2.	**Design** In the design element, the designer selects and documents the specific human robot interaction objectives, evaluation tools, and human robot interaction plan and control modalities. The design process must be both systematic and specific. Some common steps in design element includes, 11. Documentation of human robot interaction design strategies 12. Design user interface and experience 13. Create prototype human robot interaction designs
3.	**Develop** In the develop element, the designer completes the actual creation of the human robot interaction to meet the interaction objectives. A detailed action plan specifying step by step procedure for implementation of the human robot interaction strategy is prepared. Designers work to develop technologies, perform debugging, review and revise strategies.
4.	**Implement** In the implement element, the designer delivers the human robot interaction to the robot operator. In this element, training is provided to the robot operator on the usage of the human robot interface, shortcuts, and help details. The robot operator adopts the ADDIE modelled human robot interaction to operate the robot.
5.	**Evaluate** In the evaluate element, the effectiveness of the human robot interaction and the achievement of human robot interaction objectives are evaluated. Evaluation is administered in two phases: formative and summative. Formative evaluation is performed within each element of the ADDIE process whereas the summative evaluation is performed after the implementation to determine its effectiveness in satisfying human robot interaction objectives.

3   Robo-Erectus Junior – A Humanoid

This section describes the Robo-Erectus Junior humanoid robot and its ADDIE modelled human robot interaction systems. The Robo-Erectus project is developed in the Advanced Robotics and Intelligent Control Centre of Singapore Polytechnic. Robo-Erectus Junior is one of the foremost leading soccer playing humanoid robots in the

Fig. 2. Robo-Erectus Junior, the Latest Generation of the Family Robo-Erectus

RoboCup Humanoid League. Robo-Erectus Junior won the 2nd place in the Humanoid Walk competition at the RoboCup 2002 and got 1st place in the Humanoid Free Performance competition at the RoboCup 2003. In 2004, Robo-Erectus Junior won the 2nd place in Humanoid Walk, Penalty Kick, and Free Performance. The aim of the Robo-Erectus Junior development team is to develop a low-cost humanoid platform. The development of Robo-Erectus Junior has gone through many stages either in the design of its mechanical structure, electronic control system and gait movement control. Figure 2 shows the physical design of Robo-Erectus Junior. Robo-Erectus Junior has been designed to cope with the complexity of a 2 versus 2 soccer game. It has three processors each for vision, artificial intelligence and control. Table 3 shows the specification of the processors used in Robo-Erectus Junior. The platform is equipped with three sensors: an USB camera to capture images, a tilt sensor to detect a fall, and a compass to detect their direction [10]. To communicate with its teammates,

Table 3. Processor Specification of Robo-Erectus Junior

Features	Artificial Intelligence Processor	Vision Processor	Control Processor
Processor	Intel ARM Xscale	Intel ARM Xscale	ATMEL ATmega-128
Speed	400Mhz	400Mhz	16Mhz
Memory	16MB	32MB	4KB
Storage	16MB	16MB	132KB
Interface	RS232, WIFI	RS232, USB	RS232, RS485

Table 4. Physical Specification of Robo-Erectus Junior

Weight	Dimension			Speed
	Height	Width	Depth	Walking
3.2 Kg	480 mm	270 mm	150 mm	2 m/min

Robo-Erectus Junior uses a wireless network connected to the artificial intelligence processor. The vision processor performs recognition and tracking of objects of interest including ball, goal, field lines, goal post teammate and the opponents based on a blob finder based algorithm. The further processing of detected blobs, wireless communications and decision making are performed by the artificial intelligence processor which selects and implements the soccer skills (like walk to the ball, pass ball, kick ball, dive....) the robot is to perform. Finally, the control processor handles the low level control of motor based on the soccer skill selected by the artificial intelligence processor. Robo-Erectus Junior was fabricated to participate in RoboCup 2007 in the KidSize category. Table 4 shows the physical specifications of Robo- Erectus Junior. It is powered by two high-current Lithiumpolymer rechargeable batteries, which are located in each foot. Each battery cell has a weight of only 110g providing 12v which means about 15 minutes of operation [11].

In the RoboCup 2007 competitions, Robo-Erectus Junior participated in the 2 versus 2 Soccer Games and the Technical Challenges.

3.1 Human Robot Interaction System

Figure 3 shows the human robot interaction system of Robo-Erectus Junior, containing wealth of information. A separate window shows the video streaming from the robot. On the lower right of the interface is the score board showing the current status of the soccer match to the left of which is the simulated actual match with the field localized Robo-Erectus Junior and its team mates. On the top of the interface is the control buttons for team configuration and game status signals.

The human robot interaction design was developed to be homogeneous so as to adopt the same interaction design for multiple robot platforms. The user can view and operate both the team configuration controls and game status signal controls at the same time. The team configuration controls include options for assigning robot identification numbers to identify each and individual robots in own team as well as the opponent team, goal colours to identify own and opponent goals, and individual roles of robots like striker, goalie or hybrid. In striker mode, the robot aims to score goals by kicking the ball to the opponent goal. In goalie mode, the robot defends its goal area by stopping the ball from the opponent. In hybrid mode, the robot plays the role of striker as well as goalie depending on the circumstance. The team configuration controls also include options for viewing the sensor and actuator status which include display of sensor data and actuator position data.

The game status signal controls provide robot external information on the status of the match including kick off, time out, time off, resume, ball out, indirect kick and catch. Kick-off is a way of starting or restarting play at the start of the match, after a goal has been scored, at the start of the second half of the match, or at the start of each

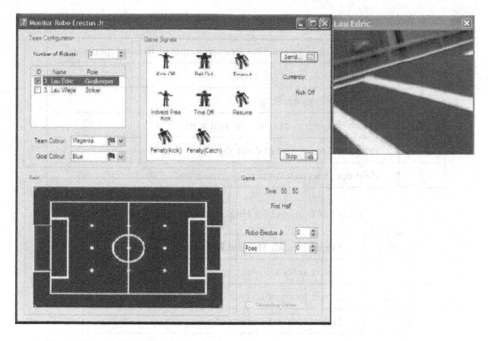

Fig. 3. Human Robot Interaction System of Robo-Erectus Junior Humanoid Robot

period of extra time. Depressing the kick off icon would start the robot in the autonomous mode for a soccer game. A team may request for timeout period of 120s to service their robots for software/hardware faults during a soccer match causing a stoppage of the game. Each team may take atmost one timeout per half period of the match. Depressing the timeout icon would perform the test routine for the checking software faults with the artificial intelligence algorithms of the humanoid robot.

Time off indicates the half time period break in the middle of the soccer game. Depressing the timeoff icon pause the robot's status and depressing the resume icon would start the robot from pause state into autonomous soccer playing mode. An indirect free kick is awarded to the opposing team if a player commits any offenses in a manner considered by the referee to be careless, reckless or using excessive force. Depressing the indirect kick icon would enable the robot to start in autonomous soccer playing mode with an indirect free kick. Depressing the catch icon would perform a routine check to test the goalie robot's ability to dive for catching an incoming ball. The controls under the game control signal can also be used to test the functionality of the robot hardware (sensors, onboard controllers, joints and actuators) and software (implemented algorithms).

4   Experimental Results

The modified ADDIE model was adopted for the development of human robot interaction for our soccer playing Robo-Erectus Junior humanoid robot. In the analysis stage, the human robot interaction goal was determined for 2 versus 2 humanoid

Table 5. Human Robot Interaction Requirements of Robot Operators

No.	Human Robot Interaction Requirements of Robot Operators
1	Simple to use
2	Visibility of system status
3	Match between system and the real world
4	Prioritized placement of controls
5	Help operators to recognize, diagnose, and recover from errors.

Table 6. Objectives for Human Robot Interaction

No.	Objectives for Human Robot Interaction
1	Detect software fault
2	Detect locomotion fault
3	Detect sensor & actuator fault

soccer leagues by expert researchers as, *Human robot interaction design must be able to detect the system fault in the robot within 60s of the timeout period.* A survey was conducted with the robot operators on their requirements for the human robot interaction. Table 5 shows the key requirements specified by the robot operators. The human robot interaction objectives were derived from the interaction goal as shown in Table 6. In the design stage, a blueprint detailing interaction objectives, evaluation tools, and human robot interaction plan and control modalities was created. Interaction strategy was then developed for localization of video displays, sensor displays, and control buttons. Computer keyboard control was chosen as the only control modality. One to one and survey based formative evaluations were conducted at the end of each of the five stage of ADDIE model. RoboCup 2 versus 2 humanoid soccer league mimicked soccer matches were selected for summative evaluation of the human robot interaction design. In the develop stage, the actual human robot interaction for soccer playing humanoid robot was created. Three experienced research staffs from our lab selected the appropriate interaction components and developed the human robot interaction design for our soccer playing humanoid robot, Robo-Erectus Junior. In the implementation stage, the human robot interaction was passed to the robot operator. The robot operator was also trained on the effective use of the developed human robot interaction.

For summative evaluations, we conducted 20 RoboCup 2 versus 2 humanoid soccer league mimicked soccer matches with other teams in our lab. Each match had two equal periods of 10 minutes. Players were entitled to an interval at half-time. Results of the experiment showed that 41 faults occurred during 20 soccer matches. There were 7 faults due to artificial intelligence algorithm and 34 faults due to locomotion related issues. In all the 41 cases, the fault was detected without having to make a replacement of the robot within 60s of the timeout period. Figure 4 shows the 41 faults and their corresponding fault detection time period with the ADDIE modelled human robot interaction.

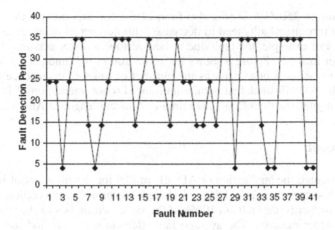

Fig. 4. Fault Detection Period Versus Fault Number

Table 7. Software and Hardware Faults & Their Average Fault Detection Time Periods

Type of Fault	Total Number of Soccer Matches	Number of faults	Average Fault Detection Periods (s)
Software	20	13	8.78s
Hardware	20	28	29.48s

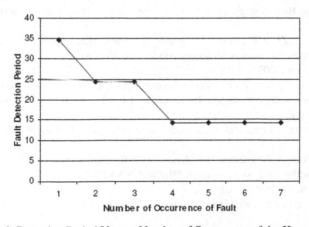

Fig. 5. Fault Detection Period Versus Number of Occurrence of the Knee Joint Fault

The average fault detection time was found to be 22.91s. Table 7 shows the number of software and hardware faults and their average fault detection time periods.

Experiments with the unmodelled human robot interaction design showed that the average fault detection time was found to be 57.47s. Therefore, the ADDIE model helped in improving the fault detection performance of human robot interaction in soccer domain by 250.85%. It was also found from the experiments that time period for detecting repeating faults tend to decrease with number of occurrence of that particular fault. For example, the fault due to heated foot actuator occurred 6 times during the soccer matches. Figure 5 shows the reduction in the time period for detection of fault due to heated foot actuator as the number of times of occurrence of that fault increases. The ADDIE modelled human-humanoid robot interaction of Robo-Erectus Junior was adopted for RoboCup 2007 humanoid soccer league at Atlanta, USA

5 Conclusion

We have presented the application of ADDIE model for development of human robot interaction in soccer domain. ADDIE modelled human robot interaction was able to detect software/hardware faults in 100% of the cases within 60s of the timeout period during the soccer matches. The average fault detection time period for the ADDIE modelled interaction design was found to be 22.91s as compared to 57.47s with the unmodelled interaction design improving the performance. ADDIE model provides a systematic approach for human robot interaction development that can be readily adopted by soccer robotics community with ease.

References

1. Metev, S.M., Veiko, V.P.: Task Structure and User Attributes as Elements of Human-Robot Interaction Design. In: Proc. IEEE RO-MAN 2006 Conference, Hatfield, United Kingdom (2006)
2. Weiss, N., Hildebrand, L.: An exemplary robot soccer vision. In: CLAWAR/EURON Workshop on Robots in Entertainment, Leisure and Hobby, Vienna, Austria (December 2004)
3. Calderon, C.A., Zhou, C., Yue, P.K., Wong, M., Elara, M.R.: A distributed embedded control architecture for humanoid soccer robots. In: Proceeding of CLAWAR Conference 2007, Singapore (2007)
4. Hu, L., Zhou, C., Sun, Z.: Estimating probability distribution with Q-learning for biped gait generation and optimization. In: Proc. Of IEEE Int. Conf. on Intelligent Robots and Systems (2006)
5. [Online] http://itsinfo.tamu.edu/workshops/handouts/pdf_handouts/addie
6. RoboCup, Compiled by Emanuele Menegatti: RoboCup Soccer Humanoid League Rules and Setup for the 2007 competition in Atlanta, USA (2007), http://www.robocup.org
7. [Online] http://en.wikipedia.org/wiki/ADDIE_Model
8. [Online] http://www.syl.com/hb/instructionaldesignersmodelstheaddiemodel.html
9. [Online] http://www.e-learningguru.com/articles/art2_1.html
10. Zhou, C., Yue, P.K.: Robo-Erectus: A low cost autonomous humanoid soccer robot. Advanced Robotics 18(7), 717–720 (2004)
11. Team Robo-Erectus website (2007), http://www.robo-erectus.org

A Proposal of Bridging Activities between RoboCupJunior and Senior Leagues

Yasunori Nagasaka[1], Tatsumasa Kitahara[2], and
Tomoichi Takahashi[3]

[1] Chubu University, Kasugai, Aichi 4878501, Japan
any@isc.chubu.ac.jp
[2] Kyoto University, Kyoto 6068501, Japan
kitahara@e-kagaku.com
[3] Meijo University, Nagoya, Aichi 4688502, Japan
ttaka@ccmfs.meijo-u.ac.jp

Abstract. We propose new bridging activities between RoboCupJunior (RCJ) and RoboCup senior leagues. Many of RCJ graduates cannot find a suitable way to progress to the senior leagues. Therefore, they often retire from the RoboCup at this stage. A major problem is the technical gap between RCJ and senior leagues. However, if some top-level RCJ teams are provided with sufficient technical advice, they can join senior leagues in a short time. Our proposal aims to achieve this. Since authers have managed RCJ Soccer Secondary class (RCJSS) and RoboCup Soccer Small-Size League (SSL) in RoboCup national competition and know them well, the bridging activities for them are proposed. The first-level activities include the following. (1) RCJSS members visit an SSL competition, watch the game, and understand its rules, and vice versa. (2) We have realized an inter-league game using only the current field and robots in RoboCup National Open 2007. (3) A greater level of interaction is required. When the RCJSS teams require technical assistance or advice, if they are friendly with the members of any SSL team, they can freely seek assistance. The second-level is to form a bridge league between RCJSS and SSL. (1) Rules are introduced from SSL since this league aims to be a stepping-stone to full SSL. The field size is reduced to half the current size of an SSL field. The number of robots is reduced from five to three in order to reduce the parts cost. (2) We prepare development kits for transition. RCJSS teams can develop their robot systems rapidly by utilizing these kits for transition.

1 Introduction

We would like to ask the following question to RoboCup senior league researchers. Do you know any of the teams that participate in the RoboCupJunior (RCJ)? In fact, we actually asked many senior league teams this question during the RoboCup JapanOpen 2007. We also asked many RCJ teams whether they were aware of any senior league team. From our questions, we found that 8% (3/39) of the teams in the senior league were aware of RCJ teams, while 18%

L. Iocchi et al. (Eds.): RoboCup 2008, LNAI 5399, pp. 177–188, 2009.

(2/11) of RCJ teams knew senior league teams. Overall, only 10% (5/50) of the teams were aware of teams belonging to another league. This clearly shows that the interaction between the teams belonging to the RCJ and senior leagues is insufficient.

The objective of RCJ is described in its official website [1]. According to the league representatives, the RCJ is a project-oriented educational initiative, which is designed to introduce the RoboCup competition to primary and secondary school children, and its focus is on education. From the viewpoint of such an ongoing educational aim, the lack of interaction between the RCJ and senior leagues is not desirable. We believe that interaction and cooperation between the two leagues should be meaningful and effective to improve the current situation. Some activities that will bridge the gap between the two leagues are necessary for this.

We propose new bridging activities between the RCJ and senior leagues. The present authors have been managing RCJ (particularly RCJ soccer) and RoboCup Soccer Small-Size League (SSL) in RoboCup national competitions, and therefore, we are aware of the potential difficulties as well as the possibilities. Hence, we propose an intermediate league between RCJ Soccer Secondary class (RCJSS) and SSL as a specific application of the proposed bridging activities.

Guided by similar desires and motivations, Anderson et al. have proposed the "ULeague", which is meant for undergraduate students [2]. ULeague is designed as an entry-level league for SSL. Since both ULeague and our proposed league have identical objectives – that is, the formation of an entry-level senior league for RCJ graduates or undergraduate students and focus on SSL – they have identical features in many respects. However, they also have several differences. This is discussed in a later section in greater detail.

Our proposal consists of activities divided into three levels. The first level refers to the initiation of immediate actions, which will be effective in making the interactions between the RCJ and senior leagues more active. The second is the establishment of a bridge league between RCJSS and SSL and providing software- and hardware-development kits for RCJSS teams in order to ease their transition from RCJ to our bridge league or even to the full SSL. The third-level activities would consist of other kinds of assistance.

In Sect. 2, we survey the current status of science education by referring to the PISA report. In Sect. 3, we describe the problems associated with RCJ and senior leagues. In Sect. 4, the details of our proposal are presented. In Sect. 5, we describe the difference between our proposal and previous studies, for example, that on ULeague by Anderson et al. [2]. In Sect. 6, the conclusions from this paper are given.

2   Science Education and RoboCupJunior

The OECD Programme for International Student Assessment (PISA) [3] periodically assesses young students' knowledge and skills. In the latest report titled "PISA 2006 Science Competencies for Tomorrow's World" [3], the relationship between students and science is investigated. The chapter titled "Attitudes to science : A profile of student engagement in science" provide an executive summary as well as a more comprehensive analysis of knowledge and skills. In this chapter, for the question, "Do students support scientific enquiry?" the answer is as follows. "In general, students showed a strong support for scientific enquiry. For example, 93% said that science was important for understanding the natural world." However, for the question "Are students interested in science?" the majority of students reported that they were motivated to learn science, but only a minority reported taking a close interest.

From the viewpoint of science education, the current RCJ activities have achieved success and have made major contributions. Today, there are many branches for managing local RCJ competitions in many countries. Only the winner from these competitions can advance to the next stage, that is, the national championships. Subsequently, only the winner from the national championships can advance to the next stage – the world championships. Thus, the RCJ is established in many countries, and they have become the largest league in terms of the number of participants [1], [4]–[6].

RCJ focuses on introducing students to science and technology and their education, while the senior leagues focus on research and development of robotics and AI. The proposed bridge league will play the role of closing the gap between them. It focuses on higher education in science and technology, and training in conducting scientific experiments, manufacturing devices, and technological advances.

3   Problems with RoboCupJunior and Senior Leagues

3.1   RoboCupJunior

The RCJ mainly focuses on education. In addition, the league plays a role in finding potential future robotic researchers and providing them with encouragement and motivation. However, with respect to encouraging young students, the current activities are not sufficient. Hence, many of the RCJ graduates cannot find the right path to the senior leagues. Moreover, these students often retire from the RoboCup. Some go to a university, but it is unclear whether they return to RoboCup. This implies that after we discover the next generation of RoboCup researchers, we lose them at the end of RCJ secondary class activities. Hence, it is important that we maintain the interest of the students in RoboCup (and in the world of science and technology) from the viewpoint of continuing their education as well as keeping the RoboCup active.

This tendency of students was indicated by Anderson [2]. One of the authors of the present paper, who is currently managing an RCJ, also strongly agrees with Anderson's thoughts on the subject. Why do not or cannot many of RCJ graduates advance to the senior leagues?

The major problem is the considerable technical gap between the RCJ and senior leagues. During the early years of the RoboCup, a new team was able to join one of the senior leagues without too much difficulty, since the technical level was not so high then. However, nowadays, the rules have become more difficult and the required robot specifications have increased. These changes are made in order to keep with the technical advancements, but this has made joining each league to become more and more difficult.

On the other hand, with respect to the technical skills of the RCJSS teams, some of the top-level teams already have good skills to the best of our knowledge. Those teams can build original robots for an RCJSS competition on their own. This robot building procedure includes mechanical processing, assembling parts, and making control circuits. Moreover, certain advanced students even begin to write a simple computer program after several months of training. If such students are given adequate technical guidance, they can join a senior league within a short time. This is the objective of our bridging activities.

3.2 Senior Leagues, Soccer Small-Size League

The technology of all RoboCup leagues is advancing every year. Hence, the level of competition increases to a great extent. This is true for SSL as well. The movement of the robots has become very smooth, fast, and accurate. Most robots have a powerful kicking device. The field size has been increasing every two or three years. These technical advancements are keeping in line with the objectives of the RoboCup. However, as described above, such advancements become a technical barrier for new teams.

For example, if a certain team wants to compete in SSL, they must have an understanding of a wide range of technical fields such as mechanical engineering, electronic engineering, control, wireless communication, computer programming, image processing, and AI. Moreover, they have to combine these techniques into the building of an actual working robot system. The total performance of the system depends on the technical field in which they are weakest. Even if only one element of the abovementioned technical fields is weak, the robots will not work correctly. Therefore if a laboratory belonging to faculty of engineering in a university attempts to develop the SSL robot system, it is not easy to cover all of above technical fields with higher level. Another drawback is the total cost of the parts. Expensive parts are required to build competitive robots. This problem of cost was also pointed out by Anderson [2].

Therefore, there must be some support for new teams. However, at present, there is no such effective support system. Hence, we believe that our proposed bridging activities are effective for not only RCJSS graduates but also the new teams in SSL, and it is effective to expand the teams and make SSL more active.

4   Bridging Activities between RoboCupJunior and Senior Leagues

As mentioned above, both RCJ and senior leagues have their own drawbacks. We can solve them by preparing and showing a next step or a method for transitioning from RCJ to senior leagues by mutual cooperation.

As an example, we use the gaps between RCJSS and SSL and propose the following three levels of activities for bridging it.

4.1 First-Level Activities: Immediate Actions

First-level activities are the immediate actions we can perform. They consist of the following three activities.

(1) Understanding Partner League. We must understand the partner league first of all. RCJSS members must visit the SSL field and watch SSL games during the competition, and vice versa; We must understand the rules and regulations of RCJSS and SSL each other.

(2) Realizing Inter-League Games. We can realize an inter-league game by using the current field and robots. RCJSS robots require a special ball that radiates infrared rays. Further, RCJSS needs a field with white to black gradations. Therefore, SSL teams will have to adjust their ball, field surface, and vision settings to those of the RCJSS. On the other hand, SSL teams require a camera above the field. Further, the SSL vision system requires an orange-colored ball and a color marker on the robot to indicate blue or yellow in order to distinguish the two teams. Therefore, the ball should be painted orange. The team's respective color marker is placed on the RCJSS robot. Thus, an inter-league game between RCJSS and SSL teams can be realized.

The advantage of this style is that there is no need for a serious modification of both the RCJSS and SSL robot systems, except a minor change in the vision system setting of the SSL team. The objectives of this game are (a) realizing interaction, and (b) motivating the RCJSS members by showing them how the SSL robots run fast and precisely, and sophisticated strategy of SSL system.

(3) Communicating as Much as Possible. We need to make an effort to know and talk to each other. When the RCJSS teams develop new robots for the SSL competition, they may sometimes require technical assistance or advice. In such cases, if they have a knowledgeable and approachable SSL team working with them, they can easily sort out their problems.

Achievement for Year 2007. As one of the activities, the authors participated in a RoboCupJunior Japan Soccer Summer Camp 2007, and gave a lecture on the SSL robot design, its specifications, and the difference between RCJ soccer and SSL. Figure 1 shows a scene from the lecture. Subsequently, the participating students enquired about the mechanical construction and electronic circuit design displaying considerable enthusiasm. Although the allotted time was limited, we had a lively and thorough discussion.

We prepared a ball painted with orange color for the inter-league game. Figure 2 (left-hand side) shows the ball. A ball used in the RCJ soccer is originally covered with transparent plastic. We painted the inner part of the ball with the same color as that of the SSL ball, barring the front part of the radiator,

Fig. 1. Lecture at RoboCupJunior Japan Soccer Summer Camp 2007

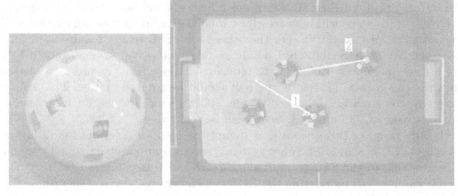

Fig. 2. Painted RCJ soccer ball (lefthand side), and checking of SSL vision system using RCJ soccer field (righthand side)

which is meant for infrared rays. This ball can be detected on the field by both RCJSS and SSL robots.

Figure 2 (right-hand side) shows a screen image of our vision system for SSL. We tested whether the vision system functions correctly on the RCJ soccer field. It can be seen that the ball position and the blue and yellow markers (indicating robots' positions) including the direction are correctly detected even on the RCJ soccer field, which has black to white gradation. Thus, we verified that the gradation of the RCJ soccer field does not affect the SSL vision system.

The inter-league game between the RCJSS and SSL teams was actually conducted during the RoboCup JapanOpen 2007. Figure 3 shows the field for the inter-league game when it was in progress. The RCJ soccer field was placed at the corner of the SSL field. The goal of the SSL field can be seen at the left-hand side of the RCJ soccer field. The SSL teams were able to use their cameras directly with a minor change in the settings. The RCJ soccer robots were covered with black paper because the SSL vision system is sensitive to various colors

Fig. 3. Inter-league game during RoboCup JapanOpen 2007

present in the body of the RCJ soccer robots. The RCJ soccer robots also have their blue team markers at the top.

Three games were conducted and successfully completed. We have confirmed that the proposed inter-league game is possible. The participants from RCJSS must understand that they can actually compete with the SSL teams.

4.2   Second-Level Activities: Forming a Bridge League

Second-level activities include the formation of a bridge league between RCJSS and SSL and various related supporting activities. This forms the main activities of this proposal. The league is positioned as a freshmen league or a novice league of SSL.

(1) Forming a Bridge League. The basic rules are derived from SSL since this league aims to be a path toward promotion to a complete SSL league. The main difference is the size of the field and the number of robots in one team.

The size of the current (Year 2008) SSL field is 6100 × 4200 mm. A field of this size results in many advantages to top-level SSL teams; however, new teams will have several problems. For example, two or more cameras will be required to cover the entire field. Moreover, it will be difficult to prepare such a large space, which will also have to include 4000-mm high frames for cameras. Hence, the size of the fields for the bridge league should be reduced to a smaller size. We determined the field size to be 3200 × 1800 mm, but the actual dimensions is not that important. Since the field size is not large, new teams will be able to have the field in their working space.

Another change is the decrease in the number of robots from five to three. This is to reduce the cost of the parts of the robots and to reduce overcrowding on the field.

Fig. 4. Main board used in the engineering freshman course (lefthand side) and an example of robots (righthand side)

(2) Development Kit for Transition. We are preparing a software-development kit for image processing and wireless communications. These two fields are not required for RCJSS. However, they are indispensable for the SSL robot control system. Hence, RCJSS teams who intend to progress to SSL must need these techniques. By preparing and offering these kits, the RCJSS teams can rapidly develop a working robot system for the bridge league.

We also have a plan to develop kits for other technical areas such as mechanical design, manufacturing control boards, and microcomputer programming. For such kits, the authors already have a resource, which is currently used as the course for an engineering freshman major [7]; we will utilize this resource for this purpose. Figure 4 (lefthand side) shows the main board used in abovementioned course. Figure 4 (righthand side) shows an example of robots built by freshman.

(3) Technical Subjects for Transition. For the abovementioned development kits, we defined certain technical subjects, which must be understood first and cleared to build robots. If RCJSS graduates have already become familiar with SSL teams through the first-level activities, they can consult such SSL teams for determining solutions to their problems. The following lists the seven main technical fields for building the SSL robots.

1. Design of the robot, Mechanical processing
2. Microcomputer programming
3. Electronic circuit, Control board
4. Assembling the robot
5. Wireless communication
6. Image processing
7. AI, Team strategy

Each technical field has several subjects. Table 1 shows those subjects.

For the advanced RCJSS graduates who are able to complete all subjects, we defined next advanced subjects. Table 2 shows those subjects.

Completing the subjects of the technical field one to four, beginner teams can build a moving robot following the direction described as the program of

Table 1. Technical subjects to be cleared for building the SSL robots

Technical field	Subjects
1. Design, Mechanical processing	Designing the robot body. Preparing material. Preparing tools. Processing. Assembling. Checking functions.
2. Microcomputer programming	Preparing microcomputer. Preparing software development tools. Programming practice. Writing an easy program. Executing it. Understanding hardware functions, I/O etc.
3. Electronic circuit, Control board	Preparing tools. Designing the power supply circuit. Designing the motor driver circuit. Designing the wiring diagram. Preparing the electronic parts. Developing the circuit board by etching. Soldering. Testing the functions.
4. Assembling robot body	Assembling robot body. Writing the motor speed control program. Writing the robot motion control program. Testing the functions.
5. Wireless communication	Preparing the device. Making the interface circuit. Writing a program that sends and receives signals. Designing the protocol. Sending the robot motion commands. Checking the transmission speed.
6. Image processing	Preparing a camera and an image capture board. Writing a program which captures color images. Writing a program which discriminate colors. Writing a program which estimate the robot and ball positions. Checking the image processing speed.
7. AI, Team strategy	Recognizing the teammates and enemy positions. Calculating the moving path. Robots move following the path. Robots kick a ball after moving. Planning the next action of each robot. Realizing the combination of robots. Making a team strategy.

microcomputer. Further, completing the one to five, they can direct a robot by wireless communication. Moreover, completing the one to seven, they can compete in the bridge league. If they have sufficient vision system to cover an entire SSL field, they can compete even in full SSL. Finally, completing all subjects, they can compete in full SSL against top teams of the world.

Table 2. Advanced technical subjects

Technical field	Subjects
8. Advanced subjects	Realizing robust image processing against the change of illumination or shades on the field. Planning the moving path using smooth curves. Using rotary encoder to detect the rotation precisely. Realizing the feedback control such as PID method for acculate control. Writing a robot simulator. Realizing pass of ball between teammate robots.

4.3 Third Level Activities: Other Supports

Third level activities include several supports for RCJSS students from outside of RoboCup competition. They includes;

(1) Internship in Senior League Team. RCJ students temporally belong to the senior league team as an intern. There, they can take charge of some development under the senior league members coaching. They can experience the work of senior leagues.

(2) Evaluating the Career of RCJ in AO Entrance Examination. We request universities which have the AO (Admissions Office) entrance examination to consider the career and the result of competition of RCJ and RoboCup as one of the standard of selection. This aims to make a certain path to a university to continue RoboCup activities there.

5 Comparison with Previous Work, ULeague

ULeague is the entry level league of SSL proposed by Anderson [2] by similar motivation as ours. We compared our proposal and ULeague. In this section, we show the similar points and differences.

5.1 Similarity

- Motivation. That is to give the RCJSS graduates a next step or adequate path to SSL.
- Focusing on SSL. Proposed league is positioned as an entry level league of SSL.
- Equipments. Field size, material of surface, orange colored golf ball.
- Main object of the proposal. From high school students to undergraduates.

5.2 Differences

- In our proposal, it is mainly taken into account that already existing SSL team can participate in the new league with as little change as possible. In other words, we want to realize the game by both RCJSS and SSL teams as many as possible. For RCJSS graduates teams, they can play with existing SSL team in the same field, even the world champion team, by actual play. We expect RCJSS graduates teams to have confidence and to be motivated. We then strongly recommend the participation of existing SSL teams.
- Regulation of robots is identical to SSL. Because of using same regulation, existing SSL team can participate easily. Further, for RCJSS graduates teams, if they could develop the image processing system for full SSL field size, they can participate in full SSL immediately.
- We have a plan to provide sample implementation of image processing system (including ULeague video server, if possible) and image processing server computer in competition. However, we also recommend to develop their own image processing system to RCJSS graduates teams. This is for their future entry to full SSL, and for advance of their developing skills.
- Our proposal includes not only the proposal of new entry level league, but also many side activities. As immediate actions, there are inter-league game, encouragement of interchange and communication, and encouragement of consultation to familiar SSL teams. As second level activities, there are proposal of bridge league, providing hardware- and software-development kits, and technical subjects to build fully functional robot system.

5.3 Others

These are the specific points to our proposal. We cannot confirm whether these points are same or not between ULeague and our proposal from the paper [2].

- If others than undergraduates participate in the new league, they are welcome for us. Even a laboratory of an university, master course students, or a company, they are welcome.
- If the participants became strong, we recommend to progress to full SSL.
- In our proposal, number of robots is three. This aims for reduction of total parts cost and to dissolve the overcrowding of robots in the field.

6 Conclusion

The bridging activities between RCJ and the senior leagues, particularly between RCJSS and SSL, were described in this paper.

We believe most participants of RoboCup can understand the necessity of such bridging activities between RCJ and senior leagues. However, actual activities are not fixed in past. Though the ULeague was once demonstrated in RoboCup world championships, it was not fixed after demonstration. Our proposal is the second challenge of bridging activities between RCJ and the senior leagues. We hope that our proposal become one trial to spread RoboCup senior league activities to RCJSS graduates and beginners.

We have conducted first level activities this year. The results of those activities were reported. On the other hand, our second and third level activities need much time. Actually we cannot release the development kits yet. However, we continue to prepare those kits with patience, because they are indispensable to realize the bridge league.

Our proposal is for only RCJSS and SSL now, but similar activities must be needed to other leagues. After investigating the effectiveness of our trial, we want to apply our proposal to other leagues.

References

1. http://rcj.sci.brooklyn.cuny.edu/
2. Anderson, J., et al.: Toward an Undergraduate League for RoboCup. In: Polani, D., Browning, B., Bonarini, A., Yoshida, K. (eds.) RoboCup 2003. LNCS (LNAI), vol. 3020, pp. 670–677. Springer, Heidelberg (2004)
3. http://www.oecd.org/pages/0,3417,en_32252351_32235731_1_1_1_1_1,00.html
4. http://www.robocup.org/
5. Sklar, E., Eguchi, A., Johnson, J.: RoboCupJunior: Laerning with Educational Robotics. In: Kaminka, G.A., Lima, P.U., Rojas, R. (eds.) RoboCup 2002. LNCS (LNAI), vol. 2752, pp. 238–253. Springer, Heidelberg (2003)
6. Sklar, E., Eguchi, A.: RoboCupJunior – Four Years Later. In: Nardi, D., Riedmiller, M., Sammut, C., Santos-Victor, J. (eds.) RoboCup 2004. LNCS (LNAI), vol. 3276, pp. 172–183. Springer, Heidelberg (2005)
7. Nagasaka, Y., et al.: A New Practice Course for Freshmen Using RoboCup Based Small Robots. In: Bredenfeld, A., Jacoff, A., Noda, I., Takahashi, Y. (eds.) RoboCup 2005. LNCS (LNAI), vol. 4020, pp. 428–435. Springer, Heidelberg (2006)

A Collaborative Multi-robot Localization Method without Robot Identification

Nezih Ergin Özkucur, Barış Kurt, and H. Levent Akın

Boğaziçi University, Department of Computer Engineering, Artificial Intelligence
Laboratory, 34342 Istanbul, Turkey
{nezih.ozkucur,baris.kurt,akin}@boun.edu.tr

Abstract. This paper introduces a method for multi-robot localization
which can be applied to more than two robots without identifying each of
them individually. The Monte Carlo localization for the single robot case
is extended using negative landmark information and shared belief state
in addition to perception. A robot perceives a teammate and broadcasts
its observation without identifying the teammate, and whenever a robot
receives an observation, the observation is processed only if the robot
decides that the observation concerns itself. The experiments are based
on scenarios where it is impossible for a single robot to localize precisely
and the shared information is ambiguous. We demonstrate successful
robot identification and localization results in different scenarios.

1 Introduction

Map based localization is an important and challenging problem for autonomous
mobile robots. A precise solution to this problem is essential especially in the
robot soccer domain and very difficult to obtain due to partial observability and
high sensory noise.

In the robot soccer domain, the map is landmark based and position estimation can be classified as kidnapped robot problem, where the robot can initially
be anywhere on the map and can be kidnapped at any time. Particle filter based
Monte Carlo Localization (MCL) [10] methods are the most popular ones in
dealing with this particular problem.

In the RoboCup Standard Platform League, the solution of the localization
problem continuously gets harder with the decreasing number of unique landmarks on the soccer field, and it becomes impossible to estimate the robot's
position using visual information on some parts of the soccer field. In [3], a
solution to this problem is proposed. The perception model is improved with
negative landmark information, where the unseen landmarks lead to a decrease
in the importance of a particle if the detection of those landmarks is expected
according to the pose of that particle.

Beyond the single-robot case, collaborative multi-robot localization seems
promising for improving the accuracy of pose estimation. In the collaborative
multi-agent case, the robots act in the same environment and can share information among themselves. Fox *et al.* [2] demonstrated a probabilistic collaborative multi-robot localization approach that outperforms single robot localization

L. Iocchi et al. (Eds.): RoboCup 2008, LNAI 5399, pp. 189–199, 2009.

methods. They modified both Markov and Monte Carlo localization algorithms for the multi-robot case successfully. It was also shown that collaboration makes it possible for a team of robots to get localized in cases where a single robot localization is impossible. Köse *et al.* [5] extended the Reverse Monte Carlo Algorithm [6] for the multi robot case, where a mobile robot is tracked and supervised by two other robots and can get localized using the shared information. In [7], another multi-robot localization method based on Extended Kalman Filter (EKF) is proposed. The agents share and merge their belief states that are represented in terms of relative bearing, relative distance, and relative orientation with Gaussian uncertainty.

While improving pose estimation, multi-robot localization introduces new challenges. One of the challenges is the uncertainty representation and merging of belief states. Another difficulty arises when the robots cannot identify each other. In the robot soccer domain, a sophisticated visual robot perception can give good position and orientation estimates for a teammate, but it is practically impossible to identify the teammate by visual perception since there is no distinguishing mark except a small player number on the robot's jersey. This identification is important since the information coming from a teammate robot can only be used if a data association method exists that relates the information with the robot itself. In the previous studies, these problem were assumed to be solved.

In this paper, we introduce a collaborative distributed method which does not rely on *a priori* information about robot identification. In our approach, the information about a robot's position is processed by the relevant robot, only if the information is decided to be of relevance to that robot after a probabilistic decision making process based on MCL particles.

The rest of this paper is organized as follows. The proposed approach is elaborated in Section 2. Section 3 explains the experiments and discusses the results on different scenarios. Section 4, summarizes and concludes the paper, discussing the advantages and shortcomings of the proposed method and points out some possible future work.

2   Proposed Approach

2.1   Single Robot Case

Our approach is based on Monte Carlo localization for a single robot. We additionally employed the method of using negative information which is described in [3,4]. This method is necessary especially for the cases where a robot can see either no landmark or just a single landmark. In such cases, without using negative information, a robot's belief on its position will be distributed over the entire field; however, using negative information can give a rough estimate on the robot's position.

The Monte Carlo algorithm is based on Particle Filtering. The belief state for state $\mathbf{x}_t$ at time t is $bel(\mathbf{x}_t)$ and represented as a set of particles S where $S_t = \{\mathbf{x}_t^i : i = 1...N\}$ and N is the number of particles. We represent perceptions

procedure $MCL\_Update(S_t, o_t, u_t)$
1: **for all** $\mathbf{x}_t^i$ *in* S_t **do**
2:     $\mathbf{x}_{t+1}^{\hat{i}} = motionUpdate(\mathbf{x}_t^i, u_t)$
3:     $w_{t+1}^i = p(\mathbf{x}_{t+1}^{\hat{i}}|o_t)$
4: **end for**
5: $S_{t+1} = resample(\mathbf{x}_{t+1}^{\hat{i}}, w_{t+1}^i)$

Fig. 1. MCL Algorithm

(observations) at time t as o_t and control inputs as u_t. Initially all the particles are randomly generated. The outline of the MCL algorithm can be seen in Fig. 1.

In our case, the state of the robot is $\mathbf{x}_t = [x_t, y_t, \theta_t]$, and the control input is the odometry signal as $u_t = [\Delta x_t, \Delta y_t, \Delta \theta_t]$. In the $motionUpdate$ step (line 2 in Fig. 1), the translated value of a particle is calculated as $\mathbf{x}_{t+1}^{\hat{i}} = [x_t + \Delta x_t, y_t + \Delta y_t, \theta_t + \Delta \theta_t]$.

In the third step of MCL algorithm (in Fig. 1), w_t^i is the importance of the particle and calculated as the likelihood of a particle given the observations. In our landmark based environment, observation o_t at time t is a set of polar coordinates for landmarks in the environment with the origin being the robot itself. We denote $o_t = \{[p_t^i, \theta_t^i] : i = 1...M\}$ where M is the number of landmarks.

To calculate likelihood (similarity) of a particle, we used the method described in [3]. The method, unlike the traditional ones, calculates the likelihood using both positive and negative evidence on the landmarks. For a particle, $w_t^i = \prod_{j=1}^{M} sm(k_t^j, p_t^j, \theta_t^j)$ where k_t^j is a boolean variable that takes 1 if the landmark j at time t is known or 0 otherwise, and the similarity function sm is defined as

$$sm(k_t^j, p_t^j, \theta_t^j) = \begin{cases} gl(\theta_t^j - \theta_{exp,t}^j) & \text{if } k_t^j = 1 \text{ and } j \text{ is a goal} \\ gl(p_t^j - p_{exp,t}^j) * gl(\theta_t^j - \theta_{exp,t}^j) & \text{if } k_t^j = 1 \text{ and } j \text{ is a beacon} \\ gl(\pi - abs(p_{exp,t}^j)) & \text{if } k_t^j = 0 \text{ and } j \text{ is in FOV} \\ 1 & \text{otherwise} \end{cases}$$

$$(1)$$

The gl function stands for Gaussian likelihood function with mean 0. The goals are the bearing only landmarks and beacons are distance and bearing landmarks. With the similarity function in Eq. 1, the accuracy of the likelihood function for unseen landmarks is increased. When a landmark is expected straight ahead for a particle but was not seen, it ends up getting assigned the minimum likelihood value. The actual horizontal field of view (FOV) of the robots is 22.5 degrees to both sides, but we provide a short landmark history for perceptions, which is described in Section 2.2. Consequently, we took the FOV of the robot as 115 degrees wide to both sizes, which is the maximum value when the robot is panning its head. We did not state any probability model if the landmark is not expected to be seen and indeed is not seen. In the resampling step, we copied the particles with small distortions using the roulette-wheel [8] method and inject some random particles to the environment in each step.

2.2 Perception Filtering

In the soccer environment, the instant landmark perceptions are very noisy and the FOV of the robot is limited. To increase the accuracy of perception and provide a short history for the localization module, we track the landmarks with Kalman Filters.

To model the problem for Kalman Filtering, the state to estimate is $o_t^j = [p_t^j, \theta_t^j]$ for landmark j at time t, and the measurement vector has the same form as the state. The control input is calculated from the odometry readings as $[sqrt(x'^2 + y'^2) - p_t^i, atan(y'/x') - \Delta\theta - \theta_t^i]$ where $x' = p_t^i * cos(\theta_t^i) - \Delta\theta$ and $y' = p_t^i * sin(\theta_t^i) - \Delta\theta$. With these vectors, our state transition, control input, and measurement correction steps are all linear, and corresponding matrices for the Kalman Model are all identity matrices.

To use a Kalman Model for landmarks, we have to assume both measurement and transition noises to be Gaussian. The noise parameters are tuned manually with real world data, taken from sample runs.

2.3 Multi-agent Case

In the robot soccer environment, the robots can only estimate the orientation of a teammate reliably. Estimating the distance of a teammate is not trivial and not accurate. Our approach is basically as follows. When a robot perceives another robot, it broadcasts a message containing a ray in the global coordinate system and the belief uncertainty. The rays are formed with particles with the highest importance weights. The starting point of a ray is the position of the particle and the direction is the orientation of the observed teammate. In the particle filter approach, uncertainty is represented with particles. If we form the ray set with only the best particle, it would cause a loss of uncertainty knowledge. On the other hand, forming the ray set with the entire particle set would be impractical due to communication overheads. We decided to form the ray set with K particles, and K is selected empirically. In our experiments, while N is 100, K is chosen to be 20. Formally, the belief state representation of a teammate k at time t is $R_k^t = [x_i, y_i, \theta_i] : i = 1...K$, where each particle represents the start point and direction of a ray. In Fig. 2, a sample particle set and formed ray set is given.

The data association problem in our case corresponds to associating the message to a robot other than the sender. When the other robots receive the message, they calculate ownership likelihood for the message and broadcast it. This negotiation process requires a second communication operation. The calculation of ownership likelihood for a particle set $p(S_n|R)$ is given in as

$$p(S_n|R) = \prod_{i=1}^{N}\prod_{j=1}^{K} gl\left(atan\left(\frac{y_{particle,i} - y_{ray,j}}{x_{particle,i} - x_{ray,j}}\right) - \theta_{ray,j}\right) \qquad (2)$$

In Eq. 2, S_n represents the particle set of robot n and R represent a ray set generated by an observer robot. In the implementation, the product operation causes the value to converge to zero, so the values are normalized after calculation. In Fig. 2, likelihoods for three different particles are given.

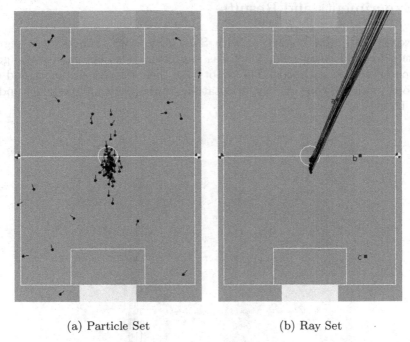

(a) Particle Set   (b) Ray Set

Fig. 2. A sample belief representation for a teammate observation. A robot stands at
(0,0), faces the blue goal and observes a teammate on its right. (a) The particle set
of the robot. (b) Generated ray set for the observed teammate. The likelihood of blue
points for ray set are 0.923425 for (a), 0.142321 for (b) and 6.79625×10^{-5} for (c).

procedure $MultiAgent\_MCL\_Update(S_t, o_t, u_t, R_t)$
1: **for all** $\mathbf{x}_t^i$ in S_t **do**
2:     $\mathbf{x}_{t+1}^{\hat{i}} = motionUpdate(\mathbf{x}_t^i, u_t)$
3:     $w_{t+1}^i = p(\mathbf{x}_{t+1}^{\hat{i}}|o_t)$
4: **end for**
5: **if** R_t not NULL and $p(S_t|R_t) = max_n p(S_t^n|R_t)$ **then**
6:     **for** $i = 1$ to N **do**
7:         $w_{t+1}^i = w_{t+1}^i * \prod_{j=1}^{K} gl(atan(\frac{y_{particle,i} - y_{ray,j}}{x_{particle,i} - x_{ray,j}}) - \theta_{ray,j})$
8:     **end for**
9: **end if**
10: $S_{t+1} = resample(\mathbf{x}_{t+1}^{\hat{i}}, w_{t+1}^i)$

Fig. 3. Multi-Agent MCL Algorithm

When the teammates receive likelihoods of other teammates for a message
from observer, the robot with the maximum likelihood value assumes ownership
of the message and incorporates it to the MCL update step. The updated MCL
Update algorithm is given in Fig. 3. The main change in the algorithm from the
single case is that if a message is owned, importance of each particle is multiplied
with likelihoods with rays in the message.

3   Experiments and Results

Our experiments were conducted with Sony Aibo ERS7 robots on a regular Robocup 4-legged soccer field, which has four color coded landmarks: two goals and one beacon on each side. The robots use their relative distances and orientations to the beacons and the orientations to the goals for the belief update in MCL.

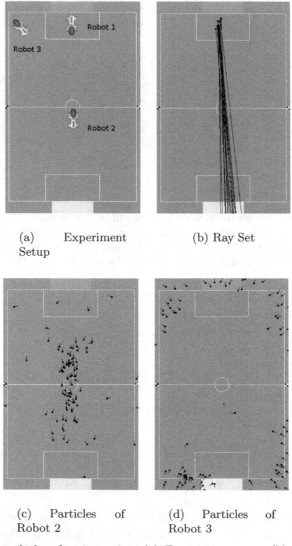

(a)   Experiment
Setup

(b) Ray Set

(c)   Particles   of
Robot 2

(d)   Particles   of
Robot 3

Fig. 4. Data association demonstration. (a) Experiment setup. (b) $Robot_1$ observes $Robot_2$ and generates ray set. (c) Particle set of $Robot_2$. Best estimate is near center of field. (d) Particle set of $Robot_3$. Best estimation is near blue goal. The owner of message is $Robot_2$.

3.1   Experiment 1

In the first experiment, we demonstrate a data association scenario. As we noted in the previous section, the ownership likelihoods for an incoming message are not calculated with only the best position estimate but also with the whole particle set. The benefit of using all of the particle set can be seen in the scenario where the robots have different particle sets but the best position estimations are very close. The experimental setup is given in Fig. 4-a. $Robot_3$ cannot observe any landmark so the particles are spread across the corners (Fig. 4-d). On the other hand, $Robot_2$ have converged to the actual position with some variance (Fig. 4-c). When $Robot_1$ observes $Robot_2$ and generates a ray set (Fig. 4-b), the best position estimations of the $Robot_1$ and $Robot_2$ are both on the ray set. This would be an ambiguity, however by calculating ownership likelihoods using all the particle set, $Robot_2$ owns the message.

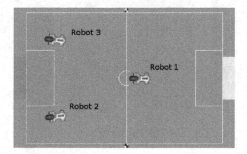

Fig. 5. Experiment 2 Setup: $Robot_1$ can perceive the goal, two beacons and the other two robots. $Robot_2$ and $Robot_3$ can only perceive the goal.

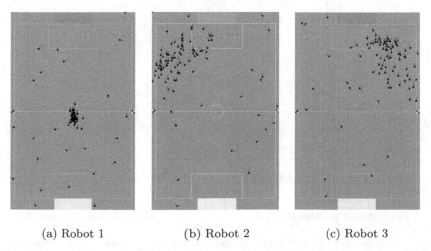

(a) Robot 1	(b) Robot 2	(c) Robot 3

Fig. 6. MCL particles when there is no collaboration. (a)$Robot_1$ localizes confidently. (b) and (c) $Robot_2$ and $Robot_3$ have high uncertainties about their positions.

3.2   Experiment 2

In the second experiment, we placed three robots on the field, as shown in
Fig. 5. $Robot_1$ can perceive three landmarks: two beacons and the goal, as well
as the other two robots. On the other hand, $Robot_2$ and $Robot_3$ can only perceive the goal. Fig. 6 shows the distribution of MCL particles when there is no
collaboration among the robots. $Robot_1$ localizes itself precisely with the help of
landmarks it perceives. On the other hand the particles of $Robot_2$ and $Robot_3$

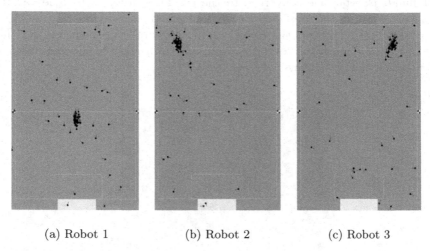

(a) Robot 1   (b) Robot 2   (c) Robot 3

Fig. 7. MCL particles when there is collaboration. (a)$Robot_1$ localizes confidently. (b)
and (c) $Robot_2$ and $Robot_3$ improve their localizations by using the information that
comes from the $Robot_1$.

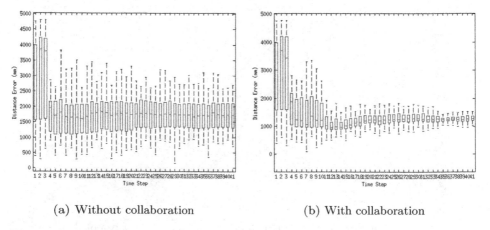

(a) Without collaboration   (b) With collaboration

Fig. 8. Distance errors of $Robot_2$. (a) Robot does not get any team message. The
variance is high. (b) The robot gets team message after time step 10. Note that the
variance of particles decreases.

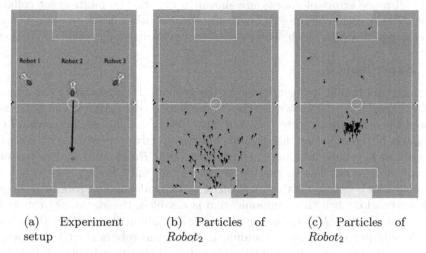

(a) Experiment
setup

(b) Particles of
$Robot_2$

(c) Particles of
$Robot_2$

Fig. 9. Experiment 3 (a) The experiment setup. $Robot_1$ and $Robot_3$ are stationary, $Robot_2$ goes to ball. (b) Particles of $Robot_2$ when there is no collaboration. (c) Particles of $Robot_2$ when there is collaboration.

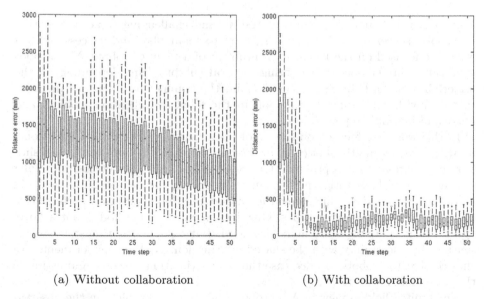

(a) Without collaboration

(b) With collaboration

Fig. 10. Distance errors of $Robot_2$. (a) Robot does not get any team message. The variance is high. (b) The robot gets team message after time step 10. Note that the variance of particles decreases.

are distributed on a large area, showing high uncertainty about their positions, since only a single landmark is perceived. When the robots collaborate as shown in Fig. 7, the uncertainties of $Robot_2$ and $Robot_3$ decrease and their particles converge along the rays that are sent from $Robot_1$.

The distance errors of $Robot_2$ are shown in Fig. 8. The multi-agent collaboration decreased the variance dramatically. On the other hand, the bias did not decrease significantly, which shows the effect of bearing-only robot perceptions.

3.3 Experiment 3

Our third experiment was conducted in the Webots [1] simulation environment. In this experiment, as shown in Fig. 9 we placed two stationary robots, and allowed the other robot go to ball. Initially, $Robot_1$ and $Robot_3$ perceive all the landmarks by panning their heads, and $Robot_2$ can only perceives a goal. The scenario of this experiment is more complicated, since $Robot_1$ and $Robot_3$ perceive not only $Robot_2$, but also each other. Furthermore, since $Robot_2$ is perceived by two robots, its belief uncertainty is reduced by two independent robots messages. In this scenario, when the communication is disabled, the stationary robots localize perfectly, but $Robot_2$ has high uncertainty about its position (Fig. 9-b), since it only perceives a single landmark. When the robots start to communicate, all team messages are correctly shared, and the uncertainty of $Robot_2$ is decreased (Fig. 9-c). The distance errors and the variances are given in Fig. 10.

4 Conclusion

Multi-Agent collaboration is an interesting and challenging topic in robotics. In robotic soccer, collaboration is the key to team play and success. In this work, we focused on the localization problem of a team of robots. Multi-robot localization in the soccer field brings several sub-problems depending on the underlying method. In the soccer domain, the common problems of methods that deal with multi-robot localization are the robot identification problem, and noisy and incomplete perceptions.

In this work, we focused on the robot identification problem and have come up with a robust multi-robot collaboration mechanism. Previous works had similar contributions to this problem, but most of them did not aim for improving the teammates' localization performance and mostly focused on building world models. A few studies which aimed for improving the localization performance had some unrealistic assumptions. Our data association method is not a novel approach; it is simply maximum likelihood, but we made it robust by representing the uncertainty with the shared information. With our experiments, we showed that the robots can decrease the pose estimation variance and estimate the pose more accurately.

The limited field of view in Aibo robots was another problem in the observation models with negative information. We have overcome this with the egocentric world model module and also provided short history for the localization module.

Acknowledgements

This project is supported by Boğaziçi University Research Fund Project 06HA102.

References

1. http://www.cyberbotics.com/products/webots/index.html
2. Fox, D., Burgard, W., Kruppa, H., Thrun, S.: A probabilistic approach to collaborative multi-robot localization. Auton. Robots 8(3), 325–344 (2000)
3. Hoffman, J., Spranger, M., Gohring, D., Jungel, M.: Exploiting the unexpected: Negative evidence modeling and proprioceptive motion modeling for improved markov localization. In: Bredenfeld, A., Jacoff, A., Noda, I., Takahashi, Y. (eds.) RoboCup 2005. LNCS, vol. 4020, pp. 24–35. Springer, Heidelberg (2006)
4. Hoffman, J., Spranger, M., Gohring, D., Jungel, M.: Making use of what you don't see: negative information in markov localization. pp. 2947–2952 (2005)
5. Köse, H., Akın, H.L.: A collaborative multi-robot localization method for a team of autonomous mobile robots. In: ICAR 2007,13th International Conference on Advanced Robotics, Jeju, Korea, pp. 641–646 (2007)
6. Köse, H., Akın, H.L.: The reverse monte carlo localization algorithm. Robot. Auton. Syst. 55(6), 480–489 (2007)
7. Martinelli, A., Pont, F., Siegwart, R.: Multi-Robot Localization Using Relative Observations. In: Proceedings of ICRA 2005 (2005)
8. Rekleitis, I.M.: A particle filter tutorial for mobile robot localization. Technical Report TR-CIM-04-02, Centre for Intelligent Machines, McGill University, 3480 University St., Montreal, Québec, CANADA H3A 2A7 (2004)
9. Roumeliotis, S.I., Rekleitis, I.M.: Propagation of uncertainty in cooperative multi-robot localization: Analysis and experimental results. Auton. Robots 17(1), 41–54 (2004)
10. Thrun, S., Fox, D., Burgard, W., Dellaert, F.: Robust monte carlo localization for mobile robots. Artificial Intelligence 128(1-2), 99–141 (2001)
11. Welch, G., Bishop, G.: An introduction to the kalman filter. Technical report, Chapel Hill, NC, USA (1995)

Teamwork Design Based on Petri Net Plans

Pier Francesco Palamara[1], Vittorio A. Ziparo[1], Luca Iocchi[1], Daniele Nardi[1],
and Pedro Lima[2],*

[1] Dipartimento di Informatica e Sistemistica – "Sapienza" University of Rome, Italy
{ziparo,iocchi,nardi}@dis.uniroma1.it
[2] Institute for Systems and Robotics - ISR/IST, Lisbon, Portugal
pal@isr.ist.utl.pt

Abstract. This paper presents a design of cooperative behaviors through Petri
Net Plans, based on the principles provided by Cohen and Levesque's Joint Commitments Theory. Petri Net Plans are a formal tool that has proved very effective
for the representation of multi-robot plans, providing all the means necessary for
the design of cooperation. The Joint Commitment theory is used as a guideline
to present a general multi-robot Petri Net Plan for teamwork, that can be used to
model a wide range of cooperative behaviors. As an example we describe the implementation of a robotic-soccer passing task, performed by Sony AIBO robots.

1 Introduction

The design of complex robotic behaviors in dynamic, partially observable and unpredictable environments is a crucial task for the development of effective robotic applications. The annual RoboCup soccer competitions provide an ideal testbed for the
development of robotic behavior control techniques, as the design of behaviors in the
robotic-soccer environment requires the definition of expressive plans for the performance of complex tasks.

Petri Nets [5] are an appealing modeling tool for Discrete Events Systems, that has
been used in several works for the modeling of robotic behaviors. [2] provides an interesting formal approach for the modeling and analysis of single-robot tasks, and in [7]
it is shown how Petri Nets can be used to model a multi-robot coordination algorithm
for environment exploration. In [11] Petri Nets' semantics is used for the definition of
a formal language for the representation of generic behaviors: Petri Net Plans (PNPs),
which are adopted in this work. The graphical approach of this representation framework allows for an intuitive design of complex plans, and multi-robot PNPs are able to
represent multi-robot interactions.

Cooperation in multi-robot systems plays an important role, as teamwork can lead
to consistent performance improvements. Several RoboCup teams achieve cooperation
facilitating interaction through the assignment of individual behaviors, as for instance
through the tactical placements of the team members in the soccer field. Some works
have studied the possibility of a structured approach to the design of cooperation, for
which coordination and synchronization is required. In [9], synchronization through

---

* This work was partially supported by Fundação para a Ciência e a Tecnologia (ISR/IST pluri-
annual funding) through the POS_Conhecimento Program that includes FEDER funds.

L. Iocchi et al. (Eds.): RoboCup 2008, LNAI 5399, pp. 200–211, 2009.

explicit communication is used to attain cooperation on real robots. Implicit communication is used in [6] for the performance of a pass behavior among the members of a team in the RoboCup Simulation League, while in [4] (also in the RoboCup Simulation League) a neural network is employed to learn the conditions associated to the performance of a pass. The RoboCup Virtual RescueRobots League represents another testbed for teamwork, and some interesting applications can be found. In these works, the engagement of cooperation is not usually explicitly modeled, and it is difficult to handle situations, such as action failures, in which the robots have to withdraw the cooperative execution. In [8], the Joint Commitment theory [1] has been used to guide the implementation of cooperative passes through finite state automata. In our work the principles for cooperation outlined by the theory are modeled through Petri Nets, which are provably more expressive than finite state automata.

The Joint Commitment theory provides a detailed formal specification for the design of generic cooperative behaviors. Its prescriptive approach is easily expressed using PNPs, which provide the required level of expressiveness and intuitiveness, maintaining the desired generality to allow the design of a wide range of cooperative tasks. Building upon these characteristics of Petri Net Plans and the Joint Commitment theory, in this paper we present a general model for the representation of multi-robot cooperative behaviors. A robotic-soccer passing task is detailed, as an example in the RoboCup Four Legged League.

not related to the RoboCup competitions.

explicit communication. The guidelines for the design of cooperation are provided by the Joint Commitment theory, which is easily integrated in Petri Net Plans for the implementation of effective and consistent cooperation.

The paper is organized as follows.

Section 2 briefly describes the key elements and operators of Petri Net Plans. Section 3 introduces the Join Commitment theory, showing how it is used as a guideline for the design of multi-robot Petri Net Plans for teamwork. Section 4 describes how this approach has been used in the domain of the technical challenges in the RoboCup Four Legged League.

2   Petri Net Plans

Petri Net Plans [11] are a behavior representation framework that allows the design of highly expressive plans in dynamic, partially observable and unpredictable environments. Note that PNPs do not follow a generative approach, but are a tool for graphical representation of plans. PNPs are based on Petri Nets ([5]), a graphical modeling language for dynamic systems, which is used to represent the many features that are required for behavior modeling, such as non-instantaneous actions, sensing, loops, concurrency, action failures, and action synchronization in a multi-agent context.

The basic structures of a PNP are non-instantaneous ordinary and non-instantaneous sensing actions, shown in Figures 1 and 2.

In an ordinary action two transitions and three places are employed: p_i, p_e and p_o are, respectively, the initial, execution and termination places. A token in p_e represents the execution phase of the non-instantaneous action. The firing of the transitions t_s

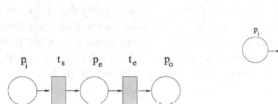

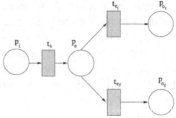

Fig. 1. A non-instantaneous ordinary action   **Fig. 2.** A non-instantaneous sensing action

and t_e represents, respectively, the starting and the ending of the action. Transitions may be labelled with conditions (typically expressed through a propositional formula) that control their firing. In a sensing action (Figure 2), the ordinary action structure is enriched through an additional transition and a place. Depending on the value of the sensed condition, the corresponding transition is fired (t_{et} is fired if the sensed condition is *true*, t_{ef} is fired otherwise). Ordinary and sensing actions can also be modeled as instantaneous. In this case a single transition is used to represent the start, execution and ending of actions (in the case of a sending action two transitions are used according to the value of the sensed condition). An additional structure, called *no-action*, can be used to connect the structures during the design of a plan. This structure is represented by a single place with no transitions.

In a PNP these elementary structures are combined, through a set of operators, to achieve action sequences, loops, interruption, conditional and parallel execution. These operators are detailed in [11]. t_{interr} becomes *true* during the execution of the high-lighted ordinary action, the token is removed from the execution place p_e, causing the action to be interrupted.

structure consents to concurrently start the execution of multiple actions, while the join operator allows to wait for the termination of multiple actions.

2.1 Sub-plans

In the design of a PNP, sub-plans can be used for a higher modularity and readability. A sub-plan is represented as an ordinary action but refers to a PNP rather than to a primitive behavior.

A plan execution module, running on the robot, takes care of dynamically loading sub-plans in case a super-plan invokes its execution. In particular, whenever a start transition of a subplan is fired, the marking of the subplan is set to the initial one. The subplan will then be executed, possibly concurrently with other primitive behaviors or subplans, until it reaches its goal marking or a condition labeling its ending transition is met. Moreover, subplans allow a more powerful use of interrupts which can be used to inhibit an entire behavior at once. This is a very important feature which will be used, as described in the following, to provide a generic implementation of teamwork through PNPs.

Defined over the *end* transition in the super-plan are met), and can be interrupted as an ordinary action.

2.2 Multi Robot Plans

Petri Net Plans are also able to represent multi-robot plans [10], through the union of n single robot PNPs enriched with synchronization constraints among the action of different robots. The model we present allows for the design of plans for small teams of robots, such as the ones used in Robocup, and may also be scaled up to medium size teams with an appropriate use of sub-plans. Multi-robot Petri Net Plans are produced in a centralized manner, and then automatically divided, implementing the *centralized planning for distributed plans approach* [3].

Each action of a multi-robot PNP is labeled with the unique ID of the robot that performs it. At execution time each robot divides the multi-robot plan into a single agent plan, for its individual execution. Two operators are used to attain synchronization: the $softsync$ operator and the $hardsync$ operator. Figure 3 shows the structure of the hard sync operator, used to synchronize the execution of two actions.

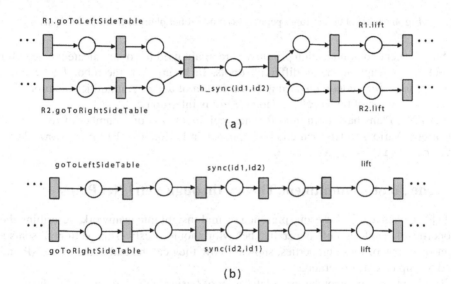

Fig. 3. Hard synchronization operator: (a) multi-robot plan (b) single-robot plans

The hard sync operator relies on the single-robot $sync$ primitive, used to establish a communication link between the two robots to exchange information and synchronize the execution. In the example shown in Figure 3, one robot moves to a side of a table to lift it, while the other robot reaches the other side. The hard sync operator ensures that the table will be lifted only after the robots have successfully terminated their preparation phase.

The soft sync operator provides the possibility to establish a precedence relation among the actions of the individual robots in the multi-robot plan. This structure can be used to allow a robot to asynchronously notify the ending of an action to its partners, starting the execution of another action without waiting for a synchronization with other robots (see [10] for further details).

The case of a multi-robot action interruption is shown in Figure 4.

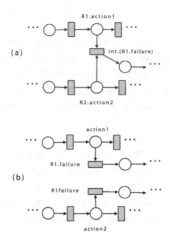

Fig. 4. Multi-robot interrupt operator: (a) multi-robot plan (b) single-robot plans

Single agent communication primitives are again used to communicate the need for an action interruption among different robots. In Figure 4, if the robot $R1$ becomes aware of a *failure* condition during the execution of *action1*, it notifies the robot $R2$, and the execution of both *action1* and *action2* is interrupted.

Petri Net Plans have been used for the implementation of a number of robotic applications. Various videos and complete plans can be found at http://www.dis.uniroma1.it/~ziparo/pnp.

3 The Joint Commitment Theory through Petri Net Plans

In [1] P. Cohen and H. Levesque present a formal insight into teamwork, describing the properties that a design of cooperative behavior should satisfy. This section presents a brief overview of these properties, showing how they can be embodied in a PNP and used to implement cooperation.

The Joint Commitment theory isolates a set of basic characteristics that all the cooperating members of a team should share. Too strong and too weak specification of these characteristics are avoided, in order not to set unnecessary constraints on the design, and at the same time to maintain the possibility of a consistent design of cooperative behaviors, given the potential divergence on the mental states of the team members. The theory is rooted in the concept of *commitment*, that is established among the team members that decide to perform teamwork. To summarize, a set of team members that are committed to the execution of a cooperative behavior will continue their individual action execution until one of the following conditions hold:

1. The behavior was concluded successfully
2. The behavior will never be concluded successfully (it is impossible)
3. The behavior became irrelevant

The prescriptive approach of the Joint Commitment (JC) theory can be used to provide a systematic design of cooperative behaviors in a multi-robot team.

3.1   Petri Net Plans for Teamwork

Given the intuitive and expressive behavior programming approach provided by the Petri Net Plans framework, it is easy to embody the specifications provided by the JC theory in the design of multi-robot plans for cooperative tasks. The multi-robot interrupt operator shown in the previous section is used to consistently interrupt the action execution among the different robots that are engaged in a cooperation (being *committed*), in case the behavior becomes irrelevant or fails. The successful conclusion of the individual actions is implemented in the multi-robot plan through a hard-sync operator.

Figure 5 shows a multi-robot Petri Net Plan for the performance of a cooperative behavior, according to the specifications provided by the JC theory.

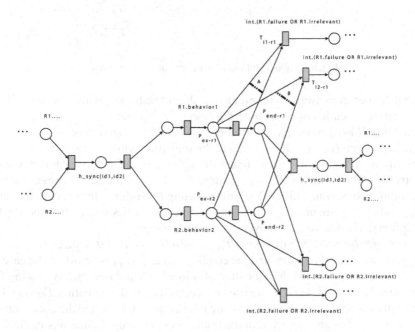

Fig. 5. A Petri Net Plan for a cooperative behavior

After a first synchronization (during which the commitment is established), the two robots start the cooperation, executing their individual behaviors (i.e. *behavior*1 and *behavior*2) which are represented as sub-plans. Following the guideline provided by the JC theory, the commitment is broken if one of the above listed conditions holds. In case one of the engaged robots senses that his behavior became irrelevant or that it has failed, the multi-agent interrupts ensure the event is communicated to the partner, and the execution of the individual actions is interrupted. In the case of successful termination of both *behavior*1 and *behavior*2, a hard sync is performed to successfully end the commitment. It may be possible that one of the two robots successfully terminates the execution of the cooperative behavior while the other is still performing some actions. The possible occurrence of this situation is reflected in the plan shown in Figure 5. If

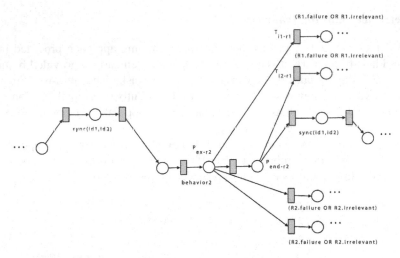

Fig. 6. Single agent plan for the cooperative behavior

the robot $R1$ becomes aware the commitment should be broken while the robot $R2$ has terminated the execution of *behavior2*, the places P_{ex-r1} and P_{end-r2} are marked with a token. Once $Robot1$ has successfully communicated its will to break the commitment, the condition over T_{i1-r1} will be true, and the transition will fire (the edges leading to the transition are highlighted in the figure through the segment A, which is not part of the Petri Net). The transition T_{i2-r1} is instead enabled in the case $R1$ senses the interrupt condition to be true while $R2$ is still executing *behavior2*. In this case the places P_{ex-r1} and P_{ex-r2} are marked with a token, and once $R1$ has successfully notified the interruption to $R2$, the condition over T_{i2-r1} will be true.

Executing *behavior2*, the transition T_{i1-r1} will be *enabled* (the places P_{ex-r1} and P_{ex-r2} are marked with a token and the condition over T_{i1-r1} is *true*). In the case that $Robot2$ has already finished the execution of *behavior2*, and is currently waiting for a synchronization from $Robot1$, to interrupt the commitment, the transition T_{i2-r1} will be *enabled* (the places P_{ex-r1} and P_{end-r2} are marked with a token and the condition over T_{i2-r1} is *true*). The arcs involved in the firing of the interrupt transitions in the figure have been linked through the segment A (for the transition T_{i1-r1}) and B (transition T_{i2-r1}).

The structure of this Petri Net Plan prevents a deadlock to occur in the described situation. In Figure 6 the single agent plan for the robot $R2$ is shown.

Only one of the two interrupt transitions (T_{i1-r1} and T_{i2-r1}) connected to the execution of the behavior *behavior2* can fire during the execution, as the other robot will only handle one of the two possible multi-robot interrupt communications. The communication required to evaluate the interrupt conditions in this Petri Net is implemented through the underlying communication layer.

3.2 Applications

Teamwork is very beneficial, if not unavoidable, in many robotic applications. The structure shown in the PNP of Figure 5 can be used as a model for a wide range of cooperative tasks that require the establishment of an explicit *commitment* among robots.

As an example, in the RoboCup Rescue domain, consider a mini UGV and a mini UAV proceeding in formation during the exploration of a terrain. The two vehicles are *committed* to the cooperative exploration. While committed, the mini UAV and the mini UGV perform coordinated individual behaviors for the exploration. The formation is in this case a necessary condition for the success of the cooperation. In case, for some reason, the formation is broken (e.g. the mini UGV looses visual contact with the mini UAV), the commitment is broken (through a communication action), and the cooperative exploration is interrupted. This interruption leads to the execution of individual behaviors that will allow the reestablishment of the formation (e.g. the mini UAV performs a behavior to facilitate its detection, while the mini UGV seeks its partner). The described behaviors can be easily represented in the PNP framework, making use of the structure of Figure 5 to handle the commitment of the two vehicles.

Explicit cooperation for the execution of complex tasks may be required in the RoboCup Soccer scenario as well. Consider the example of a pass between two robotic-soccer players. If the conditions for a pass hold, a commitment is established. The robots will need to agree on the allocation of the required tasks (pass and receive the ball). Suppose the passer robot looses the ball (e.g., an adversary robot intercepts it before the pass can take place): the failure of the pass needs to be communicated to the receiving robot, which is meantime preparing to receive the pass, and the execution of the individual cooperative behaviors needs to be interrupted. This example has been implemented in the RoboCup Four Legged League scenario, and will be detailed in the next section.

4   An Example in the Robotic-Soccer Environment

In the past editions of the RoboCup competitions the development of cooperative behaviors has been encouraged. The Passing Challenge, proposed in 2006 (Bremen, Germany) and 2007 (Atlanta, USA) in the Four Legged League, directly addresses the problem of cooperation. In this technical challenge the robots are placed in three spots on the soccer field with the task of passing a ball. Passing the ball to a robot which was not engaged in the last pass has a higher score reward, and a pass is considered valid if the robot intercepts the ball within a certain distance from its assigned position. In this paper, we describe a passing task with a more specific focus on cooperation, exogenous events and dynamic assignment of tasks, ignoring the aspects related to the localization of the robots in the field.

The implementation of this task requires (besides the development of basic functions such as vision, locomotion and primitive behviors) synchronization and coordination. The implemented system has been presented at the seventh international conference on Autonomous Agents and Multi-Agent Systems (AAMAS08), and received the "Best Robotic Demo Award". The multi-robot PNPs written for this task, as shown below, reflect the principles of the JC theory.

The assignment of the roles for the pass behavior is performed in the multi-robot PNP at the first stage of the task execution: two of the three robots select the roles of *Passer* and *Receiver*, according to the position of the ball in the field and the previously performed passes (a robot that recently passed the ball has a lower probability of

Fig. 7. Two robots passing the ball during the passing task

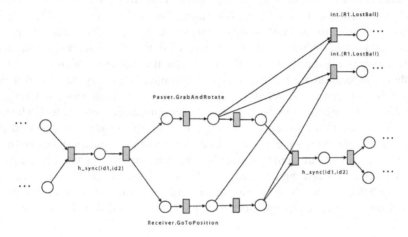

Fig. 8. Preparation phase of the pass behavior

being assigned with the role of *Receiver*). To allow role assignment the robots initially exchange their local information on the distance from the ball, and synchronize for a consistent allocation of roles. For more details on the assignment of the cooperative roles see [10].

The synchronized execution of actions required by the passing task is implemented through Petri Net Plans, which embody, as shown in the previous sections, the guidelines provided by the Joint Commitment Theory. A first synchronization is used to commit the robots to the execution of the pass. The hard synchronization operator is used for this purpose. The robot that has been assigned with the *Passer* role reaches the ball, grabs it and rotates towards its partner. Meanwhile the *Receiver* robot reaches the desired position and prepares to intercept the passed ball, rotating towards the *Passer*. At the end of this phase, the robots renew their commitment through another synchronization. The hard sync operator is again used to ensure both the robots have completed their task before they can proceed with the pass. This preparation phase is prone to action failures, due to the difficulty of implementing reliable grab and rotation primitives with AIBO robots, and due to possible occurrence of exogenous events (e.g. collisions with other robots) that may interfere with the predicted performance of the primitives. Reflecting the principles of the JC theory, the robots break their commitment if and

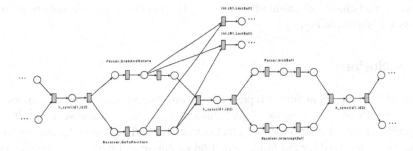

Fig. 9. Multi-robot Petri Net Plan for the pass behavior

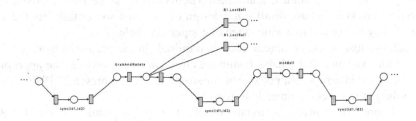

Fig. 10. Single-robot Petri Net Plan for the pass behavior: Passer

when a failure occurs during this phase (in this particular task the cooperative behavior is never considered irrelevant, as the robots have the unique task of passing the ball).

Figure 8 shows the Petri Net Plan for the execution of this first part of the task. The *LostBall* condition becomes *true* in case the *Passer* robot realizes that the ball has been lost during the grab or the rotation phases. The ball may in fact roll away from the robot, causing the need for a new task assignment procedure. If control of the ball is lost by the *Passer* robot, the *Receiver* robot needs to be notified, in order to break its commitment to the current execution of the pass. A multi-robot interrupt operator is used to consistently interrupt the execution of the actions of both the *Passer* and the *Receiver*.

If the commitment is successfully maintained the pass can take place. The *Passer* robot kicks the ball towards the receiver, which in the meantime performs an intercept behavior. This phase does not require particular attention for action interruption, as the kick and the intercept behaviors are atomically performed and the pass behavior is concluded both in the case of success and in the case of failure of the pass. A further synchronization (through a hard sync operator) is performed to exchange information about the outcome of the behavior, and the commitment is broken. The final multi-robot plan for the pass behavior is shown in Figure 9, while Figure 10 shows the single agent plan executed by the *Passer* robot.

Necessary high level programming. Using the modularity that Petri Net Plans offer through the possibility of using sub-plans, it is possible to implement the multi-robot behavior for the pass, that requires synchronization among the robots, and represents the core of the behavior definition for this task. The plan for the two robots performing the pass was shown in Figure 3.

5 Conclusions

The use of Petri Net Plans for the representation and execution of robotic behaviors has proven very effective. Besides the formal characteristics of the framework, and its intuitive graphical interface, an appealing characteristic of PNPs is the systematic approach that has been provided for the implementation of single and multi-robot behaviors. In this work we have introduced a general model for the design of cooperation through PNPs, building upon the multi-robot synchronization operators, aided by the specifications provided by the Joint Commitment Theory. To illustrate the effectiveness of the proposed model we have detailed the design of a robotic-soccer task, but the same approach may be applied to a wide range of cooperative behaviors.

To achieve teamwork, communication is required. In the presented work we have assumed the existence of a reliable communication channel. However, the appropriate use of behavior interruptions, not shown for simplicity in the presented PNPs, allows the handling of noisy communications as well.

As a future work towards a structured definition of cooperation in the RoboCup domain, we are working on the integration of cooperative behaviors in the soccer competitions. To this extent, we are currently developing a system for the establishment of commitments among the team members during the soccer games, using a task assignment algorithm based on utility functions.

References

1. Cohen, P.R., Levesque, H.J.: Teamwork. Special Issue on Cognitive Science and Artificial Intelligence 25, 486–512 (1991)
2. Costelha, H., Lima, P.: Modelling, analysis and execution of robotic tasks using petri nets. In: IEEE/RSJ International Conference on Intelligent Robots and Systems, 2007. IROS 2007, October 29 - November 2, vol. 2, pp. 1449–1454 (2007)
3. Durfee, E.H.: Distributed Problem Solving and Planning. In: Multiagent Systems. A Modern Approach to Distributed Artificial Intelligence. MIT Press, Cambridge (2000)
4. Matsubara, H., Noda, I., Hiraki, K.: Learning of cooperative actions in multiagent systems: A case study of pass play in soccer. In: Sen, S. (ed.) Working Notes for the AAAI Symposium on Adaptation, Co-evolution and Learning in Multiagent Systems, Stanford University, CA, pp. 63–67 (1996)
5. Murata, T.: Petri nets: Properties, analysis and applications. Proceedings of the IEEE 77(4), 541–580 (1989)
6. Pagello, E., D'Angelo, A., Montesello, F., Garelli, F., Ferrari, C.: Cooperative behaviors in multi-robot systems through implicit communication. Robotics and Autonomous Systems 29(1), 65–77 (1999)
7. Sheng, W., Yang, Q.: Peer-to-peer multi-robot coordination algorithms: petri net based analysis and design. In: Proceedings, 2005 IEEE/ASME International Conference on Advanced Intelligent Mechatronics, July 24-28, pp. 1407–1412 (2005)

8. van der Vecht, B., Lima, P.U.: Formulation and implementation of relational behaviours for multi-robot cooperative systems. In: Nardi, D., Riedmiller, M., Sammut, C., Santos-Victor, J. (eds.) RoboCup 2004. LNCS, vol. 3276, pp. 516–523. Springer, Heidelberg (2005)

9. Yokota, K., Ozaki, K., Watanabe, N., Matsumoto, A., Koyama, D., Ishikawa, T., Kawabata, K., Kaetsu, H., Asama, H.: Uttori united: Cooperative team play based on communication. In: Asada, M., Kitano, H. (eds.) RoboCup 1998. LNCS, vol. 1604, pp. 479–484. Springer, Heidelberg (1999)

10. Ziparo, V., Iocchi, L., Nardi, D., Palamara, P., Costelha, H.: Pnp: A formal model for representation and execution of multi-robot plans. In: Padgham, M., Parkes, Parsons (eds.) Proc. of 7th Int. Conf. on Autonomous Agents and Multiagent Systems (AAMAS 2008), Estoril, Portugal, pp. 79–86. IFAAMAS Press (May 2008)

11. Ziparo, V.A., Iocchi, L.: Petri net plans. In: Proceedings of Fourth International Workshop on Modelling of Objects, Components, and Agents (MOCA), Turku, Finland, Bericht 272, FBI-HH-B-272/06, pp. 267–290 (2006)

Bayesian Spatiotemporal Context Integration Sources in Robot Vision Systems*

Rodrigo Palma-Amestoy, Pablo Guerrero, Javier Ruiz-del-Solar, and C. Garretón

Department of Electrical Engineering, Universidad de Chile
{ropalma,pguerrer,jruizd}@ing.uchile.cl

Abstract. Having as a main motivation the development of robust and high performing robot vision systems that can operate in dynamic environments, we propose a bayesian spatiotemporal context-based vision system for a mobile robot with a mobile camera, which uses three different context-coherence instances: current frame coherence, last frame coherence and high level tracking coherence (coherence with tracked objects). We choose as a first application for this vision system, the detection of static objects in the RoboCup Standard Platform League domain. The system has been validated using real video sequences and has presented satisfactory results. A relevant conclusion is that the last frame coherence appears to be not very important in the tested cases, while the coherence with the tracked objects appears to be the most important context level considered.

1 Introduction

Visual perception of objects in complex and dynamical scenes with cluttered backgrounds is a very difficult task which humans can solve satisfactorily. However, computer and robot vision systems perform very badly in this kind of environments. One of the reasons of this large difference in performance is the use of context or contextual information by humans. Several studies in human perception have shown that the human visual system makes extensive use of the strong relationships between objects and their environment for facilitating the object detection and perception [1][3][5][6][12].

Context can play a useful role in visual perception in at least three forms: reducing the perceptual aliasing, increasing the perceptual abilities in hard conditions, speeding up the perceptions. From the visual perception point of view, it is possible to define at least six different types of context: low-level context, physical spatial context, temporal context, objects configuration context, scene context and situation context. More detailed explanation can be found in [17]. Low-level context is frequently used in computer vision. Most of the systems performing color or texture perception use low-level context in some degree (see for example [13]). Scene context have been also addressed in some computer vision [10] and image retrieval [4] systems. However, we believe that not enough attention has been given in robotic and

---

* This research was partially supported by FONDECYT (Chile) under Project Number 1090250.

L. Iocchi et al. (Eds.): RoboCup 2008, LNAI 5399, pp. 212–224, 2009.
© Springer-Verlag Berlin Heidelberg 2009

computer vision to the other relevant context information here mentioned, especially in spatiotemporal context levels.

Having as main motivation the development of a robust and high performing robot vision system that can operate in dynamic environment in real-time, in this work we propose a generic vision system for a mobile robot with a mobile camera, which employs spatiotemporal context. Although other systems, as for example [1][3][5][12], use contextual information, to the best of our knowledge this is one of the first work in which context integration is addressed in an integral and robust fashion. We believe that the use of a bayesian-based context filter is the most innovative contributions of this work.

We choose as a first application for our vision system, the detection of static objects in the RoboCup Standard Platform (SP) League domain. We select this application domain mainly because static objects in the field (beacons, goals and field lines) are part of a fixed and previously known 3D layout, where it is possible to use several relationships between objects to calculate the defined context instances.

This paper is organized as follows. The proposed spatiotemporal context based vision system is described in detail in section 2. In section 3, the proposed system is validated using real video sequences. Finally, conclusions of this work are given in section 4.

2 Proposed Context Based Vision System

The proposed vision system is summarized in the block diagram shown in figure 1. The first input used is the sensor information given by the camera and encoders (odometry). Odometry is used in several stages to estimate the horizon position and to correct the images between the different frames (see [18] for more details). The image of the camera is given to the preprocessor module, where color segmentation is performed and blobs of each color of interest are generated. These blobs are the first object candidates. We will call $\{C^k\}$ to the object candidates at time step k.

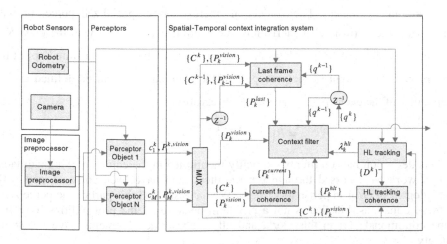

Fig. 1. Block diagram of the proposed general vision system

Each perceptors module evaluates the blob candidates with a model of the real objects. This module selects the best candidates c_i^k, and calculates an a priori probability that the candidate is correctly detected. These probabilities in the time k are called $\{P_k^{vision}\}$.

The spatiotemporal context integration stage has five modules. Current frame coherence, last frame coherence and high level tracking coherence modules give a measurement of the coherence of each current candidate with the respective context instance: with all other current detections, with last detections, and with high-level tracking estimations. The output of these modules are the probabilities $\{P_k^{current}\}$, $\{P_k^{last}\}$, and $\{P_k^{hlt}\}$. The HLT (High-Level Tracking) module maintains an estimation of the objects' pose based on the information given by all detected objects along the time. This module calculates a confidence of these estimations, which is called λ_k^{hlt}. The context filter module uses the information $\{P_k^{current}\}$, $\{P_k^{last}\}$, $\{P_k^{hlt}\}$, and λ_k^{hlt} to calculate an a posteriori probability for each current candidate given all the context instances mentioned before. The context filter module is the more relevant contribution of this work. It implements a bayesian filter to integrate all context information given by each module exposed above. This module can be represented by a function of all context instances whose result is the a posteriori probability that an object is correctly detected given all past detections, which is called ($\{q^k\}$).

2.1 Perceptors

Let c_i^k be the observation of the object i at time step k defined by $c_i^k = \left(\mathbf{x}_i^k, \mathbf{y}_i^k, \eta_i^k, \sigma_{y_i^k}, \sigma_{\eta_i^k} \right)^T$, where $\mathbf{x}_i^k$ is the relative pose of the object with respect to the robot, and ($\mathbf{y}_i^k, \eta_i^k$) and ($\sigma_{y_i^k}, \sigma_{\eta_i^k}$) are the horizon position and angle with their corresponding tolerances. Each object of interest has a specialized perceptor that evaluates some intrinsic characteristic of the candidate c_i^k related with the class K^i. We define $[c_i^k]^{OK}$ and $[c_i^k]^{NO.OK}$ as the events where c_i^k has been generated or not by the object i. The output of the preceptor of the candidate c_i^k can be defined as the probability of the event $[c_i^k]^{OK}$ given the observation c_i^k:

$$P^{vision} = P\left([c_i^k]^{OK} \mid c_i^k \right) \tag{1}$$

This definition has a term not explicitly mentioned in the equation. All candidates in this work have passed through binary filters, and have been characterized with some degree of error in perceptors stages. We have shelved this part of the perceptors in these equations, but that is not a problem, because all the probabilities have the same conditional part in this work, and all algebraic developments have the same validity.

2.2   HLT Module

The HLT module is intended to maintain information about all the objects detected in the past, although they are currently not observed (for instance, in any moment you have an estimation of the relative position of the objects that are behind you). This tracking stage is basically a state estimator for each object of interest; where the state to be estimated, for fixed objects, is the relative pose $\mathbf{x}_k^i$ of the object with respect to the robot and not in the camera space. For this reasons it is possible to say that the HLT module needs a transformation of the coordinated system. We define $F^k = T(C^k)$ and $f_j^k = t(c_j^k)$, where $T()$ and $t()$ correspond to the transformation functions from the camera point of view to the field point of view. The relative pose of the objects respect to the robot, is less dynamic and more traceable than the parameters in the camera point of view.

2.3   Context instances Calculation in the RoboCup SP League

We will consider three different context instances separately. The first one is the coherence filtering between all detected objects in the current frame. The second one is the coherence filtering between current and last frame´s detected object, and the third one is the coherence filtering with high level tracking estimator.

We have preferred consider last frame coherence and HLT coherence separately, because last detections may have very relevant information about objects in the current frame. Due that the HLT has an estimation of the object´s pose, which is given by a bayesian filter that integrates the information of the all detected objects in the time; the information of the last frame has a low importance in HLT. In the other hand, we think that to considerate more than one past frame is too noisy and it is better to have an estimation with HLT in these cases.

In this approach we have used two kinds of relationships that can be checked between physical detected objects. The first one, *Horizon Orientation Alignment*, must be checked between candidates belonging to the same image, or at most between candidates of very close images, when the camera's pose change is bounded. The second one, *Relative Position or Distance Limits*, may be checked between candidates or objects of different images, considering the movements of the camera between images:

- Horizon Orientation Alignment. In the RoboCup´s environment, several objects have almost fixed orientation with respect to a vertical axis. Using this quality, it is possible to find a horizon angle that is coherent with the orientation of the object in the image. Horizontal angles of correct candidates must have similar values, and furthermore, they are expected to be similar to the angle of the visual horizontal obtained from the horizontal points.
- Relative Position or Distance Limits. In some specific situations, objects are part of a fixed layout. The robot may know this layout a priori from two different sources: previous information about it, or a map learned from observations. In both cases, the robot can check if the relative position between two objects, or at least their distances (when objects has radial symmetry), is maintained.

2.3.1 Current Frame Coherence

We can define the current frame context coherence as the probability of the event $[c_i^k]^{OK}$ given all other detection in the current frame. If $C^k = \{c_i^k\}_{i=0}^{M}$ is the vector of observations in time step k, then the current frame context coherence may be defined like $P^{curr} = P\left([c_i^k]^{OK} \mid C^k\right)$.

However, this probability must be calculated with comparisons between pairs of objects given that they are correctly detected $P\left([c_i^k]^{OK} \mid [c_j^k]^{OK}\right)$. In section 2.4 we will show the relation established between these probabilities.

In a RoboCup SP League soccer field, there are many objects that have spatial relationships between them. These objects are goals, beacons and field lines. This static objects in the field are part of a fixed and previously known 3D layout, thus it is possible to use several of the proposed relationships between objects to calculate a candidate's coherence (for more details about object configuration in RoboCup Four Legged League, see description in [14]).

We consider three terms to calculate the coherence between two objects in the same frame:

$$P\left([c_j^k]^{OK} \mid [c_i^k]^{OK}\right) = \\ P_{hor}\left([c_i^k]^{OK} \mid [c_j^k]^{OK}\right) \cdot P_{dist}\left([c_i^k]^{OK} \mid [c_j^k]^{OK}\right) P_{lat}\left([c_i^k]^{OK} \mid [c_j^k]^{OK}\right) \tag{2}$$

In this equation, horizontal coherence is related with horizontal position and orientation alignment. In the sense of the relative position and distance limits, we are able to use distances between the objects and laterality. Laterality and distances information comes from the fact that the robot is always moving in an area that is surrounded by the fixed objects. For that reason, it is always possible to determine, for any pair of candidates, which of them should be to the right of the other and their approximated distances.

We define the horizontal coherence term using a triangular function:

$$P_{hor}\left([c_i^k]^{OK} \mid [c_j^k]^{OK}\right) = \\ tri\left(\Delta\eta_k^{i,j}, \sigma_{\Delta\eta_k^{i,j}}\right) \cdot tri\left(\Delta\eta_k^{j,i}, \sigma_{\Delta\eta_k^{j,i}}\right); tri(\Delta x, \sigma) = \begin{cases} 1 - \dfrac{\Delta x}{\sigma} & \Delta x < \sigma \\ 0 & otherwise \end{cases}$$

$$\Delta\eta_k^{i,j} = \left|\eta_k^i - \eta_k^{i,j}\right|; \eta_k^{i,j} = \eta_k^{j,i} = \sphericalangle\left(y_k^i - y_k^j\right)$$

$$\sigma_{\Delta\eta_k^{i,j}} = \sigma_{\Delta\eta_k^{j,i}} = \sigma_{\eta_k^i} + \sigma_{\eta_k^j} + \tan^{-1}\left(\frac{\sigma_{y_k^i} + \sigma_{y_k^j}}{\left|y_k^i - y_k^j\right|}\right) \tag{3}$$

The distance coherence $P_{dist}\left([c_i^k]^{OK} \mid c_j^k\right)$ is also approximated using a triangular function:

$$P_{dist}\left([c_i^k]^{OK} \mid [c_j^k]^{OK}\right) = tri\left(\Delta \mathbf{x}_k^{i,j}, \sigma_{\Delta \mathbf{x}_k^{i,j}}\right); \Delta \mathbf{x}_k^{i,j} = \left|\mathbf{x}_k^i - \mathbf{x}_k^j\right| \tag{4}$$

where $\mathbf{x}_k^i$, $\mathbf{x}_k^j$ are the relative detected positions of c_i^k and c_j^k respectively.

The lateral coherence $P_{lat}\left([c_i^k]^{OK} \mid [c_j^k]^{OK}\right)$ is defined as binary function, which is equal to 1 if the lateral relation between c_i^k and c_j^k is the expected one, and 0 otherwise.

2.3.2 Last Frame Coherence

Analogously to the previous subsection, we can define the coherence between the candidate and the objects in the past frame as $P^{last} = P\left([c_i^k]^{OK} \mid C^{k-1}\right)$. However as well as the previous subsection, we just can calculate the relationship between a pair of objects given that they are correctly detected. We assume the same model that in the current frame:

$$
\begin{aligned}
&P\left([c_i^k]^{OK} \mid [c_j^{k-1}]^{OK}\right) \\
&= P_{hor}\left([c_i^k]^{OK} \mid [c_j^{k-1}]^{OK}\right) \cdot P_{dist}\left([c_i^k]^{OK} \mid [c_j^{k-1}]^{OK}\right) \cdot P_{lat}\left([c_i^k]^{OK} \mid [c_j^{k-1}]^{OK}\right)
\end{aligned} \tag{5}
$$

The calculation of these terms is totally analogous with the current frame coherence, with only two differences: $\mathbf{y}_k^j$ and η_k^j are modified using the encoder´s information and the tolerances $\sigma_{\eta_k^j}$ and $\sigma_{\mathbf{y}_k^j}$ are increased to meet the uncertainty generated by the possible camera and robot movements.

2.3.3 High Level Tracking Coherence

The HLT module maintains an estimation of the objects with the information given by all time steps from zero until $k-1$. Let $\left\{F^n\right\}_{n=0}^{k-1}$ be the information of all frames from zero to $k-1$, we call $\{D^k\}^- = \{d_i^k\}_{n=0}^M$ the estimation calculated by the HLT using $\left\{F^n\right\}_{n=0}^{k-1}$. The HLT coherence will be defined as $P^{hlt} = P\left([f_k^i]^{OK} \mid \{D^k\}^-\right)$.

Again, the relation between two objects needs to be calculated.

$$P\left([f_i^k]^{OK} \mid [d_j^k]^{OK}\right) = P_{lat}\left([f_i^k]^{OK} \mid [d_j^k]^{OK}\right) P_{dist}\left([f_i^k]^{OK} \mid [d_j^k]^{OK}\right) \tag{6}$$

In this case we can not consider the terms related with horizon alignment but just the term related with relative position and distances limits. The calculus of P_{lat} and P_{dist} are the same that in the current coherence, but the observations c_i^k must be converted to the field point of view as was written on the equation.

When an object is detected and it is not being tracked, the HLT module creates a new state estimator for it and initializes it with all the values coming from the detection process. In particular, the coherence is initialized with the a posteriori probability obtained by the candidate that has generated the detection. However, as the robot moves, odometry errors accumulate and high-level estimations become unreliable. If a set of high-level estimations is self-coherent, but moves too far from real poses of tracked objects, then all the new observations may become incoherent and will be rejected. To avoid this situation, high-level estimations are also evaluated in the coherence filter. In order to inhibit the self-confirmation of an obsolete set of estimations, the confidence HLT_k^{conf} is only checked with respect to the current observations, but it is smoothed to avoid a single outlier observation discarding all the objects being tracked. Thus, the confidence of a tracked object is updated using:

$$\{\lambda_k^{conf}\}_i = \beta \cdot \{\lambda_{k-1}^{conf}\}_i + (1-\beta) \cdot \frac{\sum_{j=1}^{N} P\left([d_i^k]^{OK} \mid [f_j^k]^{OK}\right) \cdot P\left([f_j^k]^{OK} \mid f_j^k\right)}{\sum_{j=1}^{N} P\left([f_j^k]^{OK} \mid f_j^k\right)} \tag{7}$$

where β is a smoothing factor.

2.4 Context Filter

Let us define the probability a posteriori that we are interested. The most general spatiotemporal context that we can define is the probability that an object is correct, given all other detections from init frame to current frame k. Then we define q_i^k as:

$$q_i^k = P\left([c_i^k]^{OK} \mid \left\{C^n\right\}_{n=0}^k\right) \tag{8}$$

We can assume independence between detections in different times as is shown in [19]. Then we have $P\left(\left\{C^n\right\}_{n=0}^k \mid [c_i^k]^{OK}\right) = P\left(C^k \mid [c_i^k]^{OK}\right) \cdot P\left(\left\{C^n\right\}_{n=0}^{k-1} \mid [c_i^k]^{OK}\right)$. We apply Bayes theorem in a convenient way:

$$q_k^i = \frac{P\left(C^k \mid [c_i^k]^{OK}\right) \cdot P\left(\left\{C^n\right\}_{n=0}^{k-1} \mid [c_i^k]^{OK}\right) \cdot P\left([c_i^k]^{OK}\right)}{P\left(\left\{C^n\right\}_{n=0}^k\right)}$$

$$q_k^i = \frac{P\left([c_i^k]^{OK} \mid C^k\right) \cdot P\left([c_k^i]^{OK} \mid \left\{C^n\right\}_{n=0}^{k-1}\right)}{P\left([c_i^k]^{OK}\right)} \tag{9}$$

Here, $P\left([c_i^k]^{OK} \mid C^k\right)$ is the coherence between objects in the current frame P^{curr} and $P\left([c_k^i]^{OK} \mid \left\{C^n\right\}_{n=0}^{k-1}\right)$ have the information about all other detections in the past. In our case we will separate it into the last frame coherence and HLT coherence.

2.4.1 Current Frame Coherence Integration

To calculate the current frame coherence, we decompose $P\left([c_i^k]^{OK} \mid C^k\right)$ in:

$$P\left([c_i^k]^{OK} \mid C^k\right) = \frac{P\left(C^k \mid [c_i^k]^{OK}\right) \cdot P\left([c_i^k]^{OK}\right)}{P\left(C^k\right)}$$

(10)

$$P\left(C^k \mid [c_i^k]^{OK}\right) = \prod_{j=1}^{M} P\left(c_j^k \mid [c_i^k]^{OK}\right)$$

$$P\left(c_j^k \mid [c_i^k]^{OK}\right) =$$
$$\left[P\left(c_j^k \mid [c_j^k]^{OK}\right) \cdot P\left([c_j^k]^{OK} \mid [c_i^k]^{OK}\right)\right] +$$
$$\left[P\left(c_j^k \mid [c_j^k]^{NO.OK}\right) \cdot P\left([c_j^k]^{NO.OK} \mid [c_i^k]^{OK}\right)\right]$$

(11)

Note that we have applied total probabilities theorem to obtain the probability that we need as a function of $P\left([c_j^k]^{OK} \mid [c_i^k]^{OK}\right)$ and $P\left(c_j^k \mid [c_j^k]^{OK}\right)$. Note that $P\left([c_j^k]^{OK} \mid [c_i^k]^{OK}\right)$ is symmetric, then $P\left([c_j^k]^{OK} \mid [c_i^k]^{OK}\right) = P\left([c_i^k]^{OK} \mid [c_j^k]^{OK}\right)$ is the output of the calculus of current context coherence defined in (2). $P\left(c_j^k \mid [c_j^k]^{OK}\right)$ is the a posteriori probability of perceptor modules, so we can apply Bayes and obtain the a priori probability of perceptor modules:

$$P\left(c_j^k \mid [c_j^k]^{OK}\right) = \frac{P\left([c_j^k]^{OK} \mid c_j^k\right) \cdot P\left(c_j^k\right)}{P\left([c_j^k]^{OK}\right)}$$

(12)

where $P\left([c_j^k]^{OK} \mid c_j^k\right)$ is directly the output of perceptor module defined in (1). Clearly, $\Pr\left([c_j^k]^{NO.OK} \mid [c_i^k]^{OK}\right) = 1 - \Pr\left([c_j^k]^{OK} \mid [c_i^k]^{OK}\right)$ and applying Bayes and complementary probabilities, the term $\Pr\left(c_j^k \mid [c_j^k]^{NO.OK}\right)$ can be calculated as

$$P\left(c_j^k \mid [c_j^k]^{NO.OK}\right) = \frac{P\left(c_j^k\right) \cdot \left(1 - P\left([c_j^k]^{OK} \mid c_j^k\right)\right)}{P\left([c_j^k]^{NO.OK}\right)}.$$

All other probabilities no explicitly calculated here, can be estimated statistically.

2.4.2 Past Frames Coherence Integration

The term $P\left([c_k^i]^{OK} \mid \{C^n\}_{n=0}^{k-1}\right)$ considers the information of all detected objects along the time. Each candidate can be represented into the camera coordinate system, or into the field coordinate system. Assuming independence between the probabilities calculated in both coordinate systems, the problem was decomposed considering both coordinate systems separately. In the camera coordinate system, just the last frame detections are considered, because more than one past frame would introduce too much noise to the problem, due to the highly dynamical nature of the objects. Hence, we just need to calculate the term $P\left([c_k^i]^{OK} \mid C^{k-1}\right)$. In future works, it is possible to face the problem with more details, considering an estimation of the objects in the camera coordinate system to take into account more than one past frame. On the other hand, the HLT module gives an estimation of the objects in the field coordinate system, considering all detections along the time. The HLT module performs a bayesian estimation of the objects; therefore, we can assume the Markov principle, which say that the probability $P\left([f_k^i]^{OK} \mid \{F^n\}_{n=0}^{k-1}\right)$ can be substituted by $P\left([f_k^i]^{OK} \mid \{D^k\}\right)$ (see subsection 2.3.3). Applying Bayes and assuming F^k, C^k statistically independent, we obtain:

$$P\left([c_i^k]^{OK} \mid F^{k-1}, C^{k-1}\right) = \frac{P\left(\{D^k\} \mid [f_i^k]^{OK}\right) \cdot P\left(C^{k-1} \mid [c_i^k]^{OK}\right) \cdot P\left([c_i^k]^{OK}\right)}{P\left(D^k\right) \cdot P\left(C^{k-1}\right)} \tag{13}$$

where, $P\left(C^{k-1} \mid [c_i^k]^{OK}\right) = \prod_{j=1}^{M} P\left(c_j^{k-1} \mid [c_i^k]^{OK}\right)$ and as in (11):

$$P\left(c_j^{k-1} \mid [c_i^k]^{OK}\right) =$$
$$\left[P\left(c_j^{k-1} \mid [c_j^{k-1}]^{OK}\right) \cdot P\left([c_j^{k-1}]^{OK} \mid [c_i^k]^{OK}\right)\right] + \tag{14}$$
$$\left[P\left(c_j^{k-1} \mid [c_j^{k-1}]^{NO.OK}\right) \cdot P\left([c_j^{k-1}]^{NO.OK} \mid [c_i^k]^{OK}\right)\right]$$

and $P\left(c_j^{k-1} \mid [c_j^{k-1}]^{OK}\right)$ is the a posteriori probability q_{k-1}^j, calculated in the past frame. $P\left([c_j^{k-1}]^{OK} \mid [c_i^k]^{OK}\right) = \Pr\left([c_i^k]^{OK} \mid [c_j^{k-1}]^{OK}\right)$ is the last frame coherence defined in (5). All other terms, can be calculated analogously to the current frame case. On the other hand, $P\left(D^k \mid [f_i^k]^{OK}\right) = \prod_{j=1}^{M} P\left(d_j^k \mid [f_i^k]^{OK}\right)$, then, applying total probabilities theorem we obtain:

$$P\left(d_j^k \mid [f_i^k]^{OK}\right) =$$
$$\left[P\left(d_j^k \mid [d_j^k]^{OK}\right) \cdot P\left([d_j^k]^{OK} \mid [f_i^k]^{OK}\right)\right] + \tag{15}$$
$$\left[P\left(d_j^k \mid [d_j^k]^{NO.OK}\right) \cdot P\left([d_j^k]^{NO.OK} \mid [f_i^k]^{OK}\right)\right]$$

where $\Pr\left(d_j^k \mid [d_j^k]^{OK}\right)$ is the confidence $\{\lambda_k^{conf}\}_j$ defined in (7) by HLT module, and $P\left([d_j^k]^{OK} \mid [f_i^k]^{OK}\right) = P\left([f_i^k]^{OK} \mid [d_j^k]^{OK}\right)$ is the coherence with the HLT module's estimation defined in (6). All other terms can be calculated in the same way already explained.

3 Experimental Results

Our vision system was tested using real data sequences obtained by an AIBO Robot inside a RoboCup Four Legged Soccer field. The detection rates were measured in

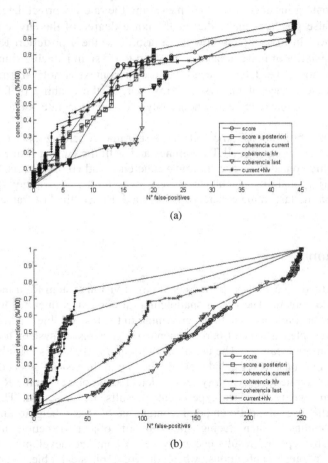

(a)

(b)

Fig. 2. ROC curves using different context instances. Score: it is the a priori probability given by perceptors modules. Score a posteriori: it is the a posteriori probability calculated by the proposed context integration system. Coherence instances: are the a posteriori probability given by each context instance.

two different situations: a low noise situation with few false objects, and a noisier situation, with much more false objects. In the first situation, false object presents were "natural" objects, like the cyan blinds and some other real, colored objects of our laboratory, which are naturally placed around the field. These objects appear in approximately 20% of the frames. In the second situation, additional false objects were added: one false goal and one false beacon over the ground plane, and one false goal and one false beacon in the border of the field. Both situations can be observed in real games of the RoboCup due to the non-controlled conditions of the environment. The public can wears with the same colors of the interesting objects and several other objects of different colors can be founded around the field.

In this work, ROC curves with the number of false-positives in the x-axis have been used to evaluate the system. These ROC curves permit to compare the utility of the different context instances proposed, measuring the rate of correct detection given a number of false positives that indicates the noise degree of the environment. The results are shown in Fig. 2. Note how the a priori and the a posteriori ROC curves evolve as the quantity of noise is increased. When the system is facing situations with low amount of noise (i.e. false objects), the use of context is not very important to improve the performance of the system. However, as the quantity of false objects grows, the use of context increases noticeably the detection rate for a given false positive rate.

An important observation is the fact that last frame coherence appears not to be very important compared with HLT coherence and with the current frame coherence. In fact, if we only consider the current frame coherence and HLT coherence instances, the a posteriori probability calculated is very near to the a posteriori probability calculated when the last frame coherence is included. Hence, the last frame coherence is irrelevant.

4 Conclusions

We have presented a general-propose context based vision system for a mobile robot having a mobile camera. The use of spatiotemporal context is intended to make the vision system robust to noise and high performing in the task of object detection.

We have presented a general-purpose context based vision system for a mobile robot having a mobile camera. The use of spatiotemporal context is intended to make the vision system robust to noise and high performing in the task of object detection.

We have first applied our vision system to detect static objects in the RoboCup SP League domain, and preliminary experimental results are presented. These results confirm that the use of spatiotemporal context is of great help to improve the performance obtained when facing the task of object detection in a noisy environment. The reported results encourage us to continue developing our system and to test it in other applications, where different physical objects and lighting conditions may exist.

As future work, we propose to include some other context instances, and integrate these to the bayesian context filter. In the other hand, it is possible to research about:

what is the best way to calculate the different context instances and how to extend the bayesian approach to the HLT estimation.

Although we have satisfactory results, we believe that the system may be improved considerably by facing these issues.

References

1. Torralba, A., Sinha, P.: On Statistical Context Priming for Object Detection. In: International Conference on Computer Vision (2001)
2. Torralba, A.: Modeling global scene factors in attention. JOSA - A 20, 7 (2003)
3. Cameron, D., Barnes, N.: Knowledge-based autonomous dynamic color calibration. In: Polani, D., Browning, B., Bonarini, A., Yoshida, K. (eds.) RoboCup 2003. LNCS, vol. 3020, pp. 226–237. Springer, Heidelberg (2004)
4. Oliva, A., Torralba, A., Guerin-Dugue, A., Herault, J.: Global semantic classification of scenes using power spectrum templates. In: Proceedings of The Challenge of Image Retrieval (CIR 1999), Newcastle, UK. BCS Electronic Workshops in Computing series. Springer, Heidelberg (1999)
5. Jüngel, M., Hoffmann, J., Lötzsch, M.: A real time auto adjusting vision system for robotic soccer. In: Polani, D., Browning, B., Bonarini, A., Yoshida, K. (eds.) RoboCup 2003. LNCS, vol. 3020, pp. 214–225. Springer, Heidelberg (2004)
6. Oliva, A.: Gist of the Scene. Neurobiology of Attention, pp. 251–256. Elsevier, San Diego (2003)
7. Foucher, S., Gouaillier, V., Gagnon, L.: Global semantic classification of scenes using ridgelet transform. In: Human Vision and Electronic Imaging IX. Proceedings of the SPIE, vol. 5292, pp. 402–413 (2004)
8. Torralba, A., Oliva, A.: Statistics of Natural Image Categories. In: Network: Computation in Neural Systems, (14), pp. 391–412 (August 2003)
9. Spillman, L., Werner, J. (eds.): Visual Perception: The Neurophysiological Foundations. Academic Press, London (1990)
10. Oliva, A., Torralba, A.: Modeling the Shape of the Scene: A Holistic Representation of the Spatial Envelope. International Journal of Computer Vision 42(3), 145–175 (2001)
11. Potter, M.C., Staub, A., Rado, J., O'Connor, D.H.: Recognition memory for briefly presented pictures: The time course of rapid forgetting. Journal of Experimental Psychology. Human Perception and Performance 28, 1163–1175 (2002)
12. Strat, T.: Employing contextual information in computer vision. In: Proceedings of DARPA Image Understanding Workshop (1993)
13. Ruiz-del-Solar, J., Verschae, R.: Skin Detection using Neighborhood Information. In: Proc. 6th Int. Conf. on Face and Gesture Recognition – FG 2004, 463 – 468, Seoul, Korea (May 2004)
14. RoboCup Technical Comitee, RoboCup Four-Legged League Rule Book (2006), http://www.tzi.de/4legged/bin/view/Website/WebHome
15. Stehling, R., Nascimento, M., Falcao, A.: On 'Shapes' of Colors for Content-Based Image Retrieval. In: Proceedings of the International Workshop on Multimedia Information Retrieval, pp. 171–174 (2000)
16. Zagal, J.C., Ruiz-del-Solar, J., Guerrero, P., Palma, R.: Evolving Visual Object Recognition for Legged Robots. In: Polani, D., Browning, B., Bonarini, A., Yoshida, K. (eds.) RoboCup 2003. LNCS, vol. 3020, pp. 181–191. Springer, Heidelberg (2004)

17. Guerrero, P., Ruiz-del-Solar, J., Palma-Amestoy, R.: Spatiotemporal Context in Robot Vision: Detection of Static Objects in the RoboCup Four Legged League. In: Proc. 1st Int. Workshop on Robot Vision, in 2nd Int. Conf. on Computer Vision Theory and Appl. – VISAPP 2007, pp. 136 – 148, March 8 – 11, Barcelona, Spain (2007)
18. Ruiz-del-Solar, J., Guerrero, P., Vallejos, P., Loncomilla, P., Palma-Amestoy, R., Astudillo, P., Dodds, R., Testart, J., Monasterio, D., Marinkovic, A.: UChile1 Strikes Back. In: Team Description Paper, 3rd IEEE Latin American Robotics Symposium – LARS 2006, October 26 27, Santiago, Chile (CD Proceedings) (2006)
19. Torralba, A., Murphy, K., Freeman, W., Rubin, M.: Context-based vision system for place and object recognition. In: Proc. Intl. Conf. on Computer Vision - ICCV 2003, October 13-18, Nice, France (2003)

Towards Cooperative and Decentralized Mapping in the Jacobs Virtual Rescue Team

Max Pfingsthorn, Yashodhan Nevatia, Todor Stoyanov, Ravi Rathnam,
Stefan Markov, and Andreas Birk

Jacobs University Bremen, Campus Ring 1, 28759 Bremen, Germany

Abstract. The task of mapping and exploring an unknown environment remains one of the fundamental problems of mobile robotics. It is a task that can intuitively benefit significantly from a multi-robot approach. In this paper, we describe the design of the multi-robot mapping system used in the Jacobs Virtual Rescue team. The team competed in the World Cup 2007 and won the second place. It is shown how the recently proposed pose graph map representation facilitates not only map merging but also allows transmitting map updates efficiently.

1 Introduction

The task of mapping and exploring an unknown environment remains one of the fundamental problems of mobile robotics. While many advances have been made in mechatronics, and the development of remotely controlled and autonomous single robot systems, the development of teams containing multiple autonomous robots is still an open research question.

Using cooperative robot teams for urban search and rescue (exploration and mapping) [1], space robotics [2], construction [3], or other tasks seems intuitively better than using uncoordinated teams or single robots. Using multiple robots has the obvious advantage of allowing more goals to be pursued simultaneously. Additionally, receiving data from multiple robots improves our confidence in the data, and allows the use of heterogeneous sensors (beyond the capacity of a single robot) for confirmation as well as heterogeneous manipulators.

The main challenge for a robot team is to be able to fuse their individual observations into one map. Extending robotic mapping to robot teams is the fundamental step for higher team functions, such as exploration or path planning. Distributed or cooperative mapping has been studied extensively [4, 5, 6, 7, 8, 9, 10, 11, 12, 13], but often requires a significant change in the system architecture or core algorithms.

This paper presents a simple, but effective, approach to realize multi-robot cooperative mapping with a significantly reduced required bandwidth compared to the previously mentioned alternatives. The design presented here was implemented by the Jacobs University team that participated in RoboCup 2007 and won the second place in the Rescue Virtual Robots Competition.

The design was tested on simulated robots in the Urban Search And Rescue Simulation (USARSim) Environment. USARSim is a high fidelity robotics

L. Iocchi et al. (Eds.): RoboCup 2008, LNAI 5399, pp. 225–234, 2009.
© Springer-Verlag Berlin Heidelberg 2009

Fig. 1. Left: A Jacobs land robot cooperating with an aerial robot at the European Land Robotics Trials (ELROB) 2007 in Monte Ceneri, Switzerland. In this technology evaluation event, the aerial robot has to search for hazard areas like sites of fire in a forest, which the land robot then has to reach. Right: Two Jacobs robots support first responders at ELROB 2007 in the situation assessment after a simulated terrorist attack with NBC substances at a large public event. The robots can operate fully autonomously and perform a coordinated exploration of the area.

simulator [14], designed primarily to aid the development of controllers for individual robots as well as robot teams. Implemented on top of Unreal Tournament 2004, it uses the high-end Unreal Engine 2 and Karma Physics Engine to accurately model the robots, the environment and interactions between the two. In addition to the simulated environment, the group also has experience with using cooperative real robot teams in USAR scenarios (figure 1).

In the following, the proposed approach is described and evaluated in terms of generated maps and required communication bandwidth in comparison to other occupancy grid based mapping algorithms.

2   Cooperative Mapping

The presented cooperative mapping approach combines two very successful techniques, a rao-blackwellized particle filter mapping algorithm [15], and the recently proposed pose graph map representation [16, 17].

The pose graph map consists of all laser range scans (plus additional information such as victim locations in USAR) and their relation to another. The nodes in the graph denote poses these observations were made from, and edges contain transformations between two such poses. These transformations also contain uncertainty information and might be generated by a motion model or a scan matching operation [17].

When paired with a particle filter, these transformations are not necessary anymore. Each observation simply corresponds to a specific pose in each particle. This simplified version of the pose graph map then only consists of a list of observations, e.g. laser range scans, and a number of sets of poses for these observations.

Intuitively, this map representation can be viewed as stopping a step short of actually rendering the corresponding standard occupancy grid map. All information is retained, but instead of actually drawing the specific laser range scan into the occupancy grid map from the specific pose, pose and laser range scan are stored. This makes it very straight forward to render an occupancy grid when needed, e.g. for display or processing, but also facilitates efficient transmission via a network or the merging of maps from multiple robots.

In the presented approach, each robot r_i in the team runs a separate mapping process M_i. Each robot always transmits its new observations $o_{i,t}$, along with the current best pose estimate $p^*_{i,t}$, to all other team members. In each particle P_n of M_i consists of a possible pose for each observation $\{p_{i,t|n}|t = 1...T\}$. Each robot r_i also merges all received information from all other robots into its own pose graph map m_i. Therefore, each robot's map m_i is equivalent to the global map m_G.

It is important to note that the individual mapping process M_i is not required to process the input from other robots to achieve cooperative mapping. If a mapping process of some robot M_j updates its estimates for the previously made observations, e.g. in the event of a loop closure, the robot r_j would simply broadcast new pose estimates $p^*_{j,t}$. Robot r_i replaces the old estimates in its local map m_i with the newly transmitted ones in order to reflect this update.

In fact, the implementation in the RoboCup Rescue team only uses the particle filter mapping algorithm for localization. Mapping is exclusively done in the pose graph map, no fused information is used in the particle filter mapper. While this might negate a small advantage of a truly multi-robot SLAM approach, it is effective, efficient, and very easy to implement. A fused map is readily available to important other tasks that benefit tremendously from it, e.g. exploration and planning.

In some cases, it is also not necessary to process the combined pose graph in order to fuse maps from multiple robots. Often it is sufficient to know their relative starting poses. Rotating and translating the individual subsets of poses to match those starting locations and subsequently rendering the combined pose graph map already provides very satisfying results. In fact, the maps shown in section 4.1 were fused in this manner.

For cases where explicit alignment and correction of partial maps is necessary, it is possible to use algorithms presented by Olson et al. [16, 18, 19] or Grisetti et al. [20].

3   Incremental Map Updates

One specific important property of the presented approach is that map updates are especially efficient to share amongst team members. In this section, we compare the efficiency of the update messages for pose graphs and the standard occupancy grid representation.

There are various ways to transmit updates to both the occupancy grid and pose graph. Several message types are defined below and their sizes are briefly discussed in order to give an indication of the factors involved.

The most efficient way to update a grid map is to send partial updates, either through a list of changed grid cells with their new values, or through sending a complete subset of the map defined by a bounding box. In the following, these are named *CellList* and *BoundingBox*, respectively.

To be as efficient as possible, *BestGrid* dynamically chooses the best of *CellList* and *BoundingBox* for each update since each message excels in different scenarios. If many cells close to each other change, it might be best to use *BoundingBox*. Otherwise, *CellList* will probably perform better.

Specifically, the two message types require the following amount of memory:

1. *CellList*: Transmits $(x_c, y_c, v_c)^{N_c}$ which requires $3 \cdot N_c \cdot 4$ bytes. Let $CellList(N_c)$ denote this size.
2. *BoundingBox*: Transmits $(x_1, y_1, x_2, y_2, v_c^{(x_2-x_1)(y_2-y_1)})$ which requires $(4 + N_b) \cdot 4$ bytes. Let $BoundingBox(N_b)$ denote this size.
3. *BestGrid*: Let $BestGrid(N_c, N_b) = min(CellList(N_c), BoundingBox(N_b))$.

with the number of changed cells N_c and the number of cells in a bounding box N_b. The above also assumes that 4-byte *floats* or *ints* are used to represent the data.

For the pose graph, the plain laser range scan is sent, along with the pose of the sensor when the scan was taken. If the current particle changes, the updated poses of all previously sent laser range scans are transmitted. Since both messages are necessary to transmit the simplified pose graph completely, they will be jointly named *PoseGraph*.

It is important to notice that the regular update message for *PoseGraph* has constant size since it depends only on the physical resolution of the sensor. Again assuming 4-byte *floats* or *ints*, the message will usually consist of $(3 + N_r) \cdot 4$ bytes. Usually N_r, the number of beams of the laser range sensor, is either 181 or 361 resulting in a message size of 736 or 1456 bytes, respectively.

In case the poses of previous observations are included, the message requires $(3 + N_r + 3 \cdot N_s) \cdot 4$ bytes, with N_s being the number of previous scans.

In order to compare the occupancy grid updates to pose graph updates, it is necessary to look at corresponding corner cases. We briefly describe two cases where one representation is considerably more efficient than the other.

The first case is best for the standard occupancy grid. Here, the map update only changes a single cell. This situation can occur when the map is already well known and new laser range scans only add very small bits of new information. While a complete scan is integrated in the map, it only contributes new information to this one single grid cell. It is important that it does add new information at least to one cell, otherwise the scan would be discarded completely. In contrast to the occupancy grid, the pose graph structure will send the entire scan. Concretely, the *CellList* message would transmit 3 numbers while the *PoseGraph* message transmits 184 or 364. In the worst case for the pose graph, it needs to transmit only 361 numbers more.

The second case is best for the pose graph. In this case, we assume that the newest laser range scan reveals the maximum amount of free space, that is, it only contains maximum range readings in the direction of previously unknown space.

Again, the *PoseGraph* message transmits 184 or 364 numbers, as it only needs to transmit the laser range scan. The occupancy grid map, however, changes dramatically. Assuming a maximum range of $20m$ for the laser scanner, a cell size of $0.2m$, and that the scan was taken perpendicular to a major axis (i.e. it fits into a $20m \times 40m$ axis aligned box), the resulting *BoundingBox* message would transmit 20004 numbers. Here, the *BoundingBox* message needs to transmit 19640 numbers more. That is 54.4 times more than in the worst case for the pose graph. This estimate is rather conservative, as most laser scanners have a maximum range of $80m$ and the maps shown below use a map cell size of $0.1m$. Also, sensor readings are never perfectly aligned with a major axis on the map.

These results already indicate that on average, the pose graph representation should be much more efficient than the standard occupancy grid representation. Section 4.2 compares the two representations given data from the USARSim simulator environment.

4   Results

4.1   Combined Maps

Figures 2 and 3 show maps generated by the presented multi-robot mapping approach.

The maps shown in figure 2 shows the two delivered maps from the Rescue Virtual Robots competition final round at the 2007 World Cup in Atlanta. A team of four robots was used for both runs. Their paths are included in the maps as dark red lines.

Figure 3 shows two maps generated from the "compWorldDay2" environment distributed with the USARSim simulator. This simulated office building was used

Fig. 2. Maps generated by a robot team of four robots in the RoboCup World Cup final round in 2007

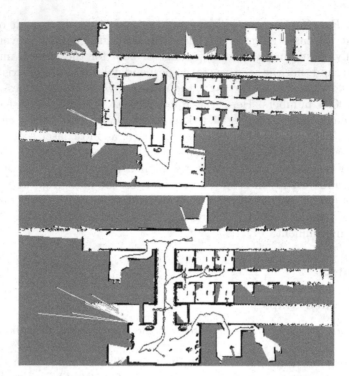

Fig. 3. Top: Map generated by single remote-controlled robot. Bottom: Map generated by an autonomous team of three robots.

in the 2006 World Cup competition. As a comparison, both a single robot map as well as a multi-robot map is shown. Interestingly, the quality is very similar. There are no visible artifacts from map merging, such as slightly misaligned walls. This is especially notable since only the starting poses of the robots were used for fusing the maps.

Up to four simulated ActivRobot P2AT platforms with SICK-LMS200 laser scanners were used in the competitions as well as for the following experiments.

4.2   Bandwidth Requirements

The theoretical discussion from section 3 show that the benefits of using the pose graph are significant in the extreme cases. While these cases rarely occur in practice, if ever, they do give a good indication of the potential savings the pose graph can provide. For instance, the bigger the influenced area of a single new scan, the better the pose graph should fare. This implies the difference between the two representations should be even more pronounced in an outdoor setting. We limit our experiments to indoor environments as it is a more common scenario.

Given the enormous difference between the two presented corner cases, it is reasonable to assume that the break-even-point, i.e. the amount of change in the map for which both representations generate update messages with the same size, should fall well below the average case. This means that, usually, the map changes much more than required by the break-even-point. Such an observation would imply that the pose graph representation would also significantly outperform the standard occupancy grid representation in the average case.

Figure 4 shows experimental results from a simulated robot.

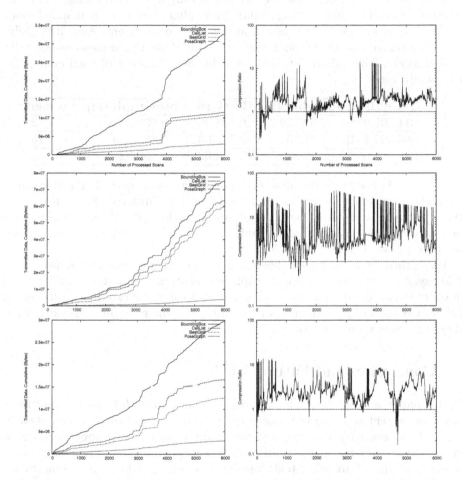

Fig. 4. Cumulative bandwidth required and comparison of individual update messages for different runs in the environment shown in figure 3

The first row of plots in figure 4 corresponds the iterative transmission of the map shown on the top in figure 3. The other plots represent two other runs in the same environment. Maps are very similar to figure 3 and thus not reproduced.

Most importantly, the left column shows a comparison of the cumulative used bandwidth for each update message. Naturally, *BestGrid* is the most efficient for the occupancy grid. However, the bandwidth required by the pose graph consistently stays below the one for the occupancy grid. This is especially visible in the middle left plot.

The right column in figure 4 shows the size ratio of pose graph versus occupancy grid update messages. Here, only the *BestGrid* values were used for clarity. In most instances, the pose graph messages are much smaller than the occupancy grid ones. All three sets contain some outliers that are in favor of the occupancy grid. As discussed earlier, this may happen if only a few cells change value, for example when changing position slightly in an almost completely mapped room Such small rooms are fairly common in the used environment. Also, it is visible that the average case in this environment is well above the break-even-point discussed before. The following table shows the exact amount of total cumulative bandwidth used.

	CellList	BoundingBox	BestGrid	PoseGraph	Improvement
Set 1	11,501,076	32,729,252	10,703,968	2,932,252	3.7
Set 2	63,354,144	75,161,548	59,541,032	3,823,200	15.6
Set 3	16,678,608	29,844,104	12,534,620	2,956,928	4.2

The rows in the previous table correspond to rows in figure 4. These numbers show the exact amount of bytes that needed to be transferred for each message type. In the most extreme case, the occupancy grid updates require 15.6 times more bandwidth than the pose graph ones. In the most modest one, it still requires 3.7 times more bandwidth.

While these numbers depend heavily on the environment and the sensors used, it is reasonable to expect the pose graph representation to perform similarly well in other indoor environments. With different sensors, for example with a larger field of view or range, or in outdoor scenarios, the difference in bandwidth usage should be even more pronounced.

5   Conclusion and Future Work

In this paper, a simple, yet effective and efficient approach to multi-robot mapping was presented. Using the pose graph map representation, it is very straight forward to iteratively combine multiple maps from different robots. However, most importantly for many multi-robot scenarios, the pose graph is very efficient to transmit to other team members. Even in indoor environments the standard occupancy grid uses 3 to 15 times more bandwidth than the pose graph representation. This number should increase in outdoor environments.

As a logical next step, the fused pose graph map should be used in the particle filter mapping algorithm in order to reuse other robot's observations. Currently, the presented system only uses the particle filter for localization given the robot's own sensor observations.

In order to further improve the bandwidth requirements for the pose graph representation, the group is already working on a compression method for transmitting raw laser scans.

Acknowledgments

Please note the name-change of our institution. The Swiss Jacobs Foundation invests 200 Million Euro in **International University Bremen (IUB)** over a five-year period starting from 2007. To date this is the largest donation ever given in Europe by a private foundation to a science institution. In appreciation of the benefactors and to further promote the university's unique profile in higher education and research, the boards of IUB have decided to change the university's name to **Jacobs University Bremen (Jacobs)**. Hence the two different names and abbreviations for the same institution may be found in this paper, especially in the references to previously published material.

References

1. Wang, J., Lewis, M.: Human control for cooperating robot teams. In: HRI 2007: Proceeding of the ACM/IEEE international conference on Human-robot interaction, pp. 9–16. ACM Press, New York (2007)
2. Elfes, A., Dolan, J., Podnar, G., Mau, S., Bergerman, M.: Safe and efficient robotic space exploration with tele-supervised autonomous robots. In: Proceedings of the AAAI Spring Symposium, pp. 104–113 (March 2006) (to appear)
3. Brookshire, J., Singh, S., Simmons, R.: Preliminary results in sliding autonomy for coordinated teams. In: Proceedings of The 2004 Spring Symposium Series (March 2004)
4. Fox, D., Ko, J., Konolige, K., Limketkai, B., Schulz, D., Stewart, B.: Distributed multirobot exploration and mapping. Proceedings of the IEEE 94(7), 1325–1339 (2006)
5. Thrun, S., Burgard, W., Fox, D.: A real-time algorithm for mobile robot mapping with applications to multi-robot and 3d mapping. In: Proceedings of the IEEE International Conference on Robotics and Automation, pp. 321–328 (2000)
6. Ko, J., Stewart, B., Fox, D., Konolige, K., Limketkai, B.: A practical, decision-theoretic approach to multi-robot mapping and exploration. In: Proc. of the IEEE/RSJ International Conference on Intelligent Robots and Systems (IROS) (2003)
7. Williams, S.B., Dissanayake, G., Durrant-Whyte, H.: Towards multi-vehicle simultaneous localisation and mapping. In: Proceedings of the 2002 IEEE International Conference on Robotics and Automation, ICRA. IEEE Computer Society Press, Los Alamitos (2002)
8. Fenwick, J.W., Newman, P.M., Leonard, J.J.: Cooperative concurrent mapping and localization. In: Proceedings of the 2002 IEEE International Conference on Robotics and Automation, ICRA. IEEE Computer Society Press, Los Alamitos (2002)
9. Roy, N., Dudek, G.: Collaborative exploration and rendezvous: Algorithms, performance bounds and observations. Autonomous Robots 11 (2001)
10. Thrun, S.: A probabilistic online mapping algorithm for teams of mobile robots. International Journal of Robotics Research 20(5), 335–363 (2001)
11. Burgard, W., Fox, D., Moors, M., Simmons, R., Thrun, S.: Collaborative multi-robot exploration. In: Proceedings of the IEEE International Conference on Robotics and Automation. IEEE Press, Los Alamitos (2000)

12. Simmons, R.G., Apfelbaum, D., Burgard, W., Fox, D., Moors, M., Thrun, S., Younes, H.L.S.: Coordination for multi-robot exploration and mapping. In: Proceedings of the Seventeenth National Conference on Artificial Intelligence, pp. 852–858 (2000)

13. Fox, D., Burgard, W., Kruppa, H., Thrun, S.: A probabilistic approach to collaborative multi-robot localization. Automous Robots, Special Issue on Heterogeneous Multi-Robot Systems 8(3), 325–344 (2000)

14 Carpin, S , Lewis, M , Wang, J., Balakirsky, S., Scrapper, C.: Bridging the gap between simulation and reality in urban search and rescue. In: Lakemeyer, G., Sklar, E., Sorrenti, D.G., Takahashi, T. (eds.) RoboCup 2006: Robot Soccer World Cup X. LNCS (LNAI), vol. 4434, pp. 1–12. Springer, Heidelberg (2007)

15. Grisetti, G., Stachniss, C., Burgard, W.: Improving grid-based slam with rao-blackwellized particle filters by adaptive proposals and selective resampling. In: Proceedings of the IEEE International Conference on Robotics and Automation, ICRA (2005)

16. Olson, E., Leonard, J., Teller, S.: Fast iterative alignment of pose graphs with poor initial estimates. In: Leonard, J. (ed.) Proceedings 2006 IEEE International Conference on Robotics and Automation, 2006. ICRA 2006, pp. 2262–2269 (2006)

17. Pfingsthorn, M., Slamet, B., Visser, A.: A scalable hybrid multi-robot slam method for highly detailed maps. In: Visser, U., Ribeiro, F., Ohashi, T., Dellaert, F. (eds.) RoboCup 2007: Robot Soccer World Cup XI. LNCS (LNAI), vol. 5001, pp. 457–464. Springer, Heidelberg (2008)

18. Olson, E., Leonard, J., Teller, S.: Spatially-adaptive learning rates for online incremental slam. In: Proceedings of Robotics: Science and Systems, Atlanta, GA, USA (June 2007)

19. Olson, E., Walter, M., Teller, S., Leonard, J.: Single-cluster spectral graph partitioning for robotics applications. In: Proceedings of Robotics: Science and Systems, Cambridge, USA (June 2005)

20. Grisetti, G., Stachniss, C., Grzonka, S., Burgard, W.: A tree parameterization for efficiently computing maximum likelihood maps using gradient descent. In: Proceedings of Robotics: Science and Systems, Atlanta, GA, USA (June 2007)

Robust Supporting Role in Coordinated Two-Robot Soccer Attack

Mike Phillips and Manuela Veloso

Computer Science Department, Carnegie Mellon University
mlphilli@andrew.cmu.edu, veloso@cmu.edu

Abstract. In spite of the great success of the fully autonomous distributed AIBO robot soccer league, a standing challenge is the creation of an effective planned, rather than emergent, coordinated two-robot attack, where one robot is the main attacker and goes to the ball and the other robot "supports the attack." While the main attacker has its navigation conceptually driven by following the ball and aiming at scoring, the supporter objectives are not as clear. In this work, we investigate this distributed, limited perception, two-robot soccer attack with emphasis on the overlooked supporting robot role. We contribute a region-based positioning of the supporting robot for a possible pass or for the recovery of a lost ball. The algorithm includes a safe path navigation that does not endanger the possible scoring of the teammate attacker by crossing in between it and the goal. We then further present how the supporter enables pass evaluation, under the concept that it is in a better position to visually assess a pass than the attacker, which is focused on the ball and surrounded by the opponent defense. We show extensive statistically significant lab experiments, using our AIBO robots, which show the effectiveness of the positioning algorithm compared both to a previous supporter algorithm and to a single attacker. Additional experimental results provide solid evidence of the effectiveness of our passing evaluation algorithm. The algorithms are ready to incorporate in different RoboCup standard platform robot teams.

1 Introduction

We conduct research on creating teams of robots that work cooperatively on tasks in dynamic environments. In this paper, we focus on our work within the domain of the RoboCup [4] 4-Legged robot soccer using the Sony AIBO robots. The robots are fully autonomous with onboard control, actuators, and in particular, limited visual perception. They can wirelessly communicate with each other. These features are general to other robot platforms and are the basis for our work for on the needed real-time coordination in robot soccer. While the RoboCup 4-Legged league has been very successful, we have not seen coordinated passes, for example, such as in the RoboCup small-size robot league. In this paper, we investigate a two-robot attack with distributed, limited perception equipped AIBO robots. We present cooperation algorithms for positioning and coordination and show successful passing lab results.

L. Iocchi et al. (Eds.): RoboCup 2008, LNAI 5399, pp. 235–246, 2009.
© Springer-Verlag Berlin Heidelberg 2009

More specifically, one robot has the role of *attacker* whose job is to go to the ball, move it upfield, and score goals. This robot's role is very clear since it is ball and goal oriented. This paper is concerned with the interesting problem that arises when a second attacker is added to the team, known as the *supporter*. The supporting role, whose task is not quite as clear, when used effectively should shorten the time it takes to score goals. This goal can be achieved through proper positioning, to recover a lost ball more quickly, and to receive passes. At the same time, it is critical for the supporter to not hinder the attacker's performance by crowding the ball, blocking the attacker's shots on the goal, or causing hesitation. With these objectives in mind, we present sets of algorithms. We first provide an approach for selecting a supporting position while dealing with supporting constraints, such as not crossing the attacker's shot on goal with domain constraints, like a bounded field. We present details as to how a supporter gets to its support point quickly while accounting for opponent movement.

Secondly, we give insight into how a supporter can be used to evaluate the option of passing. The attacker is defined to be the robot in possession of the ball, making all decisions regarding how the ball is moved. We argue that this generally accepted concept for deciding how the ball should be moved is not ideal in the presence of limited perception and a highly dynamic environment, such as robot soccer. We introduce an approach *to outsource* some of the decision making to the supporting robot, which we note has more information about the game state than the attacker does. The attacker has less information about the ball's surroundings because the attacker's perception is completely focused on the very close ball. Furthermore the ball's attractive nature draws opponents which obstruct the camera's already limited view of the attacker.

Teamwork and the positioning of a supporter have been addressed with a few approaches. The supporter robot has been positioned using potential [1]. However, this potential-field based algorithm does not take into consideration the constraint of not crossing over the attacker's shot on goal, does not take opponents into account, and does not include planned passing. Another related algorithm tries to position the supporter to be in the place where it has the highest probability of being of use, applied in simulation and in the centralized controlled small-size robot soccer [2] with centralized perception. Our problem is inspired by these efforts, but made more difficult by the robots' distributed control and local view from a small and directional onboard camera. Furthermore we address the dynamic positioning for passes in the presence of teammates and opponents, in addition to static passing [7].

First, we briefly overview the RoboCup AIBO league, and our team's role framework. We then discuss region-based positioning and the pass determination of the supporter robot. Each part includes experimental results.

The RoboCup 4-Legged league has teams of four Sony AIBO robots competing against each other. The Sony AIBO (Figure 1) is a 4-legged robot whose primary sensor is a low resolution camera, with a small field of view (less than 60 degrees). This camera is mounted in the robot's head (at the tip of its nose) and can pan 90 degrees in both directions as well as having two joints for tilting.

Fig. 1. A typical attacker situation: an AIBO robot surrounded by other robots

Our team uses a team control approach with *roles* being used to divide the tasks of robot soccer amongst the four players. The four roles that we use are goalie, defender, attacker, and offensive supporter. In CMDash'07, a specific robot took the defender role and kept it throughout the game unless the other two team robots would get penalized and out of the game. The goalie and defender are used for defense with the former being in the goal box while the latter defends from outside. The other two roles, which are the ones of interest for this paper, are the attacker and supporter. The attacker's role is well defined, being the robot that acquires the ball, moves it up the field, and finally scores goals. The supporter's role is to assist the attacker. Our play-based framework allows the attacker and supporter to swap roles when the supporter is evaluated to be more likely to attain the ball [8]. Therefore, the supporter's objectives involve trying to prevent the loss of possession of the ball to opponents, and to recover lost balls as quickly as possible, while not interfering with the attacker's objectives.

Each robot in the distributed CMDash'07 team uses a world model that maps perceptual data into positioning of all the objects in the domain, namely ball and robots. Robots communicate to improve their assessment of the complete world. Our robots follow an individual and shared world model approach [5]. The individual version contains Gaussian locations of recently seen objects, and therefore smoothes the movement of objects, which removes spurious readings. It also allows objects to continue to be modeled based on previous perceptions and models of one robot's actions, even if they have not been seen for a few frames. The shared world model includes the shared ball and teammate locations communicated between robots on the team. The world model is updated based on the confidence of the information perceived, communicated and its recency [6]. Due to compounded localization and vision errors, teammates do not yet pass position information about the other robots they have seen.

2   Region-Based Supporter

Under our framework, the supporter only exists when the team has an attacker that is either: in possession of the ball or has claimed the ball as its own, based on the role selection algorithm (the attacker/supporter swap, mentioned earlier). The supporter's purpose is to help the team keep possession of the ball. It can do this in two key ways: strategic positioning and receiving passes.

2.1 Positioning

Proper positioning of a supporter increases the number of balls that are recovered by our team. We develop an algorithm that selects a good *support point*, while also meeting several other objectives to maximize performance:

- The supporter should be in a location where the ball is likely to roll if our attacker loses possession of the ball.
- When near the goal, the supporter should be placed to recover balls that are deflected by the goalie without getting in the way of the attacker's shots.
- In the offensive side of the field, it is important that the supporter does not obstruct a shot on goal at any time. This objective can cause the ideal supporting point to be unreachable due to the field bounds. The supporter is then *pinched* and moving getting to the desired support point would require it to go outside the boundaries. Therefore, the supporter needs to have a "next best" point to support that is reachable.
- If the ball is in the defensive half of the field, the supporter should stay on the offensive side to allow defenders to *clear* the ball upfield to it.

Some of these objectives only apply to certain parts of the field. Thus, our positioning algorithm uses three regions (offense, defense, goal), as seen in Figure 2, to better handle the cases. The region is chosen based on the position of the ball. Each region slighly overlaps with its neighbors (overlap not shown in Figure 2 in order to prevent oscillation. When the ball is in an overlapping area the robot maintains the region that the ball was in previously.

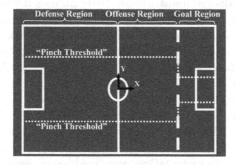

Fig. 2. Field with regions. Neighboring regions slightly overlap to prevent oscillation. Note that we assume that the positive x axis points toward the opponent's goal.

Before we present our region-based supporter positioning algorithm, we define a few functions that the algorithms uses.

The function *AligntoCloseSide* takes the ball's coordinates (b_x, b_y) and the robot's y (r_y), and a minimum value x_{min}. It returns a tuple representing a support point with coordinates (p_x, p_y), respectively set to b_x, with a lower bound, x_{min} and a constant offset from the ball, $C_{yoffset}$. p_y is on the side of b_y that the robot is closer to; p_y is bounded at $\pm C_{ymax}$ just inside the field bounds.

A similar function, $AlignToFarSide$, selects p_y by picking the side of the ball the robot is farthest from.

- $(p_x, p_y) \leftarrow$ AlignToCloseSide(b_x, b_y, r_y, x_{min})
  $p_x \leftarrow Max(b_x, x_{min})$
  $p_y \leftarrow Max(Min((b_y + C_{yoffset} * Sign(r_y - b_y)), C_{ymax}), -C_{ymax})$
  return (p_x, p_y)

The function $AlignUpField$ takes the ball's coordinates (b_x, b_y) and robot's r_y and returns a support point that is upfield of the ball, keeping its p_x bounded at the top of the field, C_{xmax}. The p_y value is set to the edge of the field that the robot is closer to than the ball.

- $(p_x, p_y) \leftarrow$ AlignUpField(b_x, b_y, r_y)
  return $(Min(b_x + C_{xoffset}, C_{xmax}), C_{ymax} * Sign(r_y - b_y))$

The function $Pinched$ returns a boolean that tells if the robot's r_y is "pinched" between the ball's b_y and the nearest field boundary. The ball's b_x is important since we introduce different pinch constants for different regions.

- $bool \leftarrow$ Pinched(b_x, b_y, r_y)
  return $((b_x > C_{goal, offense \ bound}$ **and** $abs(b_y) > C_{goal \ pinch})$ **or**
  $(b_x \leq C_{goal, offense \ bound}$ **and** $abs(b_y) > C_{offense \ pinch}))$ **and**
  $(abs(r_y) > abs(b_y)$ **and** $Sign(r_y) == Sign(b_y))$

The function $PathCollideWithAttack$ checks if the path from the robot (r_x, r_y) to the support point (p_x, p_y) intersects with the attacker. The attacker is estimated to be on the line segment straight back from the ball (b_x, b_y) at a distance $C_{robot \ length}$.

- $bool \leftarrow$ PathCollideWithAttack$(b_x, b_y, r_x, r_y, p_x, p_y)$
  return Intersects$((((r_x, r_y), (p_x, p_y)), ((b_x, b_y), (b_x - C_{robot \ length}, b_y)))$

The function $GoBehindAttacker$ returns a support point that is behind the attacker. If the robot is not already behind the attacker, then it moves backward. Otherwise the supporter moves horizontally toward the desired support point's y coordinate, p_y.

- $(p_x, p_y) \leftarrow$ GoBehindAttacker$(b_x, b_y, r_x, r_y, p_y)$
  if $r_x > b_x - C_{robot \ length}$ **and** $abs(b_y - r_y) < C_{robot \ length}$
  return $(r_x - 2 * C_{robot \ length}, r_y)$
  else if $r_x \leq b_x - C_{robot \ length}$
  return (r_x, p_y)

The function $CloseGoalCorner$ returns a support point with fixed x and y coordinates that line up on the corner of the goal box. It chooses from the two corners by selecting the corner that is on the same side of the ball as the robot. A similar function, $FarGoalCorner$ (not shown) chooses the corner on the opposite side of the ball.

- $(p_x, p_y) \leftarrow$ CloseGoalCorner(b_y, r_y)
  return $(C_{goal \ box \ bottom}, C_{goal \ box \ side} * Sign(r_y - b_y))$

(a)

GETOFFENSEPOINT(b, r)
$(p_x, p_y) \leftarrow ALIGNTOCLOSESIDE(b_x, b_y, r_y, b_x)$
if PINCHED(b_x, b_y, r_y)
   if $r_x > b_x$
      $(p_x, p_y) \leftarrow ALIGNUPFIELD(b_x, b_y, r_y)$
   else
      $(p_x, p_y) \leftarrow ALIGNTOFARSIDE(b_x, b_y, r_y, b_x)$
if PATHCOLLIDEWITHATTACK($b_x, b_y, r_x, r_y, p_x, p_y$)
   $(p_x, p_y) \leftarrow GOBEHINDATTACKER(b_x, b_y, r_x, r_y, p_y)$
return p

(b)

GETGOALPOINT(b, r)
$(p_x, p_y) \leftarrow CLOSEGOALCORNER(b_y, r_y)$
if PINCHED(b_x, b_y, r_y)
   if $r_x > b_x$
      $(p_x, p_y) \leftarrow ALIGNUPFIELD(b_x, b_y, r_y)$
   else
      $(p_x, p_y) \leftarrow FARGOALCORNER(b_y, r_y)$
if PATHCOLLIDEWITHATTACK($b_x, b_y, r_x, r_y, p_x, p_y$)
   $(p_x, p_y) \leftarrow GOBEHINDATTACKER(b_x, b_y, r_x, r_y, p_y)$
return p

(c)

GETDEFENSEPOINT(b, r)
$(p_x, p_y) \leftarrow ALIGNTOCLOSESIDE(b_x, b_y, r_y, C_{midline})$
if PINCHED(b_x, b_y, r_y)
   $(p_x, p_y) \leftarrow ALIGNTOFARSIDE(b_x, b_y, r_y, C_{midline})$
return p

Fig. 3. Region-based support point for (a) offense, (b) goal, (c) defense, regions

Supporter Positioning in the Offense Region. The supporter in the offense region (Figure 3 (a)) keeps the same x coordinate as the ball while maintaining a constant distance offset from the ball's y position. If the supporter becomes *pinched* and is behind the ball (in terms of x), it safely and effectively goes around behind the ball and the attacker, resuming then its usual positioning. A more difficult case arises when the supporter is ahead of the ball. Going around the ball now is not only too time consuming, but it may actually be impossible to do without walking out of bounds. Therefore, to remain in bounds and to not block a shot on goal, the supporter moves upfield to allow for an upfield pass. This position is not as desirable and therefore, as soon as the supporter becomes un-pinched or gets the chance to go behind the attacker, it goes back to its normal behavior. All motion is capped by the field boundaries.

Supporter Positioning in the Goal Region. In the goal region the focus is on recovering the ball after the opposing goalie deflects it. Therefore, the supporter chooses to sit on the goal box corner (Figure 3 (b)). The supporter still handles being "pinched" the same way, except that the thresholds are moved in more, so the ball never gets too close to the supporter. If the attacker switches sides, the supporter moves to the other goal box corner.

Supporter Positioning in the Defense Region. Finally, the supporter's behavior (Figure 3 (c)) when the ball is in the defense region is very similar to that of the offensive region. However the supporter never goes below the midline, so that when a defensive robot clears the ball, there is already a robot in the offensive region. There is also no concern with interrupting a shot on goal since the defense region is far away from the opponent's goal. As a result, when the supporter gets "pinched," it always freely crosses in front of the ball's shot on goal to get the better supporting position.

2.2   Navigation to the Support Point

The region-based positioning algorithm shows how the optimal support point is selected. However, this does not address how the robot actually moves to the selected point quickly and safely while getting enough information from the field to not hinder the support point selection algorithm. Therefore, we have some key objectives that need to be addressed.

- The robot should move to the point quickly by prioritizing its walking directions. The AIBO walks fastest forward, then backward, then sideways.
- When the supporter is going around opponents, it should keep itself between them and the ball in order to get to the support point safely.
- The robot must maintain a good view of the ball in order to choose good supporting points. This is difficult because the supporter is not moving to the ball like the attacker.
- In the event that the supporter loses sight of the ball for more than a moment, it needs some way to continue to make acceptable positioning decisions.

Our algorithm NavToPointSeeBall (pseudo code not shown due to limited space) first determines whether it has seen the ball recently. If it hasn't, then it chooses the support point based on the attacker's estimated ball location.

The algorithm prioritizes the primary walking directions so that the robot can choose the fastest walking direction that still allows it to see the ball at all times. With the proper constants, our algorithm makes minimal use of the slower sideways walk, while choosing a direction that allows the ball to remain centralized in the AIBO's field of view. In our actual implementation, we overlap the FOV regions slightly to prevent oscillation between walking directions.

Our algorithm extends the NavToPoint behavior, developed previously to be used with object detection [3]. The algorithm in that paper describes how to detect opponent robots. This behavior was written to make use of this opponent detection when walking forward to a target point. We extended this algorithm to use all four primary walking directions. This is still most effective when walking forward as the robot is most likely to see opponents. Another key change was to always go towards the ball when going around an opponent, putting the supporter between the opponent and the ball.

3   Positioning Experimental Results

In order to test the effectiveness of our algorithms, we performed three sets of experiments. All three sets involve a defensive team of a goalie and defender protecting the goal. The goalie always begins in the goal box while the defender is started a short distance in front of it. The ball is started just past the midline on the defensive side. All three sets have the same attacker which starts a small distance behind the ball. Figure 4 shows the initial field state for each trial.

The three sets of experiments only differ in the supporter of the offensive team. The first set uses our new positioning algorithms. The second set uses a

Fig. 4. The starting configuration for our experiments

Table 1. Supporter positioning results

	Region Supporter	Previous Supporter	No Supporter
Mean time to score (seconds)	75.2	133.5	164.8
Standard Deviation	67.2	113.5	133.0
Goals with supporter assistance	26/30	25/30	0/30

previously developed algorithm, where both offensive robots are attackers until one gets close to the ball. When this occurs, the farther robot cannot get closer than a constant distance to the ball, to prevent crowding. The third set is a control group with no supporter and only one robot as attacker. All robots on the field were given time to localize before each test began. We timed the time to go from the starting configuration to scoring a goal. We ran the trials following standard RoboCup rules for the 4-Legged League. After a goal was scored, the time was recorded and the field was reset, to make each of the goals independent of each other. We ran 30 trials of each approach. Table 1 shows the resulting analysis of the data.

We performed unpaired t-tests using the data we collected. When comparing our supporter positioning algorithms with the previous supporter algorithm, the t-test gives us a two-tailed P value of 0.0188, which is statistically significant at the standard 0.05 level. The comparison between our algorithms and the control group with no supporter achieved a two-tailed P value of 0.0017, even more significant. These results provide solid evidence of the value of our algorithms. The algorithms described in this paper were used at RoboCup 2007 and were instrumental in our team's achievement of 3rd place in our division.

4   Pass Determination

This section focuses on the problem of the specific coordination between the attacker and supporter robots towards effective passing.

4.1 Distributed Decision Making

As stated previously, in robot soccer, the ball acts as an attractor for robots on both teams. The field around the ball is therefore dense with robots, leading us to two important realizations:

- The attacker with the ball is more likely to be in an opponent-rich environment than any teammate.
- The increased number of opponents trying to get close to the ball can easily cause the attacker's limited camera to become obstructed.

We consider that the attacker only has a few ways to move the ball: shoot on goal (if close enough), pass, dodge, or dribble (if too far to shoot on goal). A pass to a teammate seems like an obvious choice to escape the situation where the attacker has opponents closing in on it, since it quickly moves the ball from a point of high opponent density to one that is much lower. Also, the attacker has very limited information about whether a pass is possible or not. This is due to many factors. First, the general view of the attacker may be obstructed by opponents which could leave other opponents unnoticed or leave the attacker mislocalized. Also, due to the attacker's closeness to the ball, the attacker may know very little about the ball's surroundings. Most importantly, the attacker cannot afford to let go of the ball and look around to determine if passes are possible. Our idea is to have the decision making be distributed to players who are better informed to make a particular decision. This allows the attacker to make better decisions since they are being sent to it from more informed teammates. This *outsourcing* of decision-making also has the added benefit of being efficient since it reduces the amount of computation the attacker must perform. The attacker now merely has to look up the most recent message. Although in our situation we deal with the decision of whether to pass or not, our concept of outsourcing decisions away from the focal player to teammates can be generalized. Specifically, we show how outsourcing can greatly improve accuracy when determining whether a pass to another player (in this case an offensive supporter) is possible.

We use an offensive supporter as the pass receiver and as the robot to which the pass evaluation algorithm is outsourced. The supporter robot is a good outsourcing choice as it can be well positioned to maintain an offset distance from the attacker robot and near the goal as described in the previous section.

4.2 Pass Evaluation Algorithm

The algorithm we present assumes that opponent and teammate robots are detected and represented in the world model, as in [3]. Our pass evaluation algorithm, assumes that it runs on a supporter that is well positioned to keep the ball in view (we used our earlier described positioning algorithm). The following algorithm uses the positions of robots in the world model and the location of the ball to determine whether a pass is likely to be successful, meaning it is not likely to be intercepted by another robot. Figure 5 shows the pseudo code for our algorithm, EvaluatePass, running on the supporter robot for distributed decision

```
EvaluatePass(posn, ball, robots)
open ← IsOpen(posn, ball, robots)
open ← ApplySmoothing(open)
if open ≠ +1
    ResetOpenTimer()
if OpenTime() > OPEN_THRESH
    SendMessage(MSG_OPEN)
else if open = −1
    SendMessage(MSG_NOT_OPEN)
```

Fig. 5. Complete pass evaluation algorithm

making. It makes use of IsOpen and ApplySmoothing, two main functions of the algorithm. They evaluate the pass on a single vision frame and apply smoothing across a sequence of noisy vision frames, respectively.

EvaluatePass sends a MSG_OPEN message to its teammate (SendMessage) if a pass is possible from the ball to the supporter. A pass is considered possible if open has been +1 for an amount of time OPEN_THRESH. If open is -1, the message MSG_NOT_OPEN is sent to its teammate, stating that a pass is not possible. Lastly, if open is 0, no messages are sent since the supporter is currently unsure, which means the attacker uses the last sent +1/-1 message. The function IsOpen determines whether the supporter is open to receive a pass based only on the current vision frame.

The IsOpen function determines if the supporter can receive a pass based on the current vision frame. If the robot has not seen the ball recently, then it certainly cannot receive a pass. If the ball was seen recently, then the function uses vector math to determine if a robot is in the passing lane. It does this by "drawing" a rectangle from the supporter to the ball, with a predefined width and checks if a robot is contained within that rectangle. It allows for up to one teammate to be in the rectangle since that is likely the attacker. However, if there are opponents, or more than one teammate in the rectangle, then a -1 is returned. If the passing lane is clear though, then +1 is returned.

The returned decisions of IsOpen are fed into ApplySmoothing, a simple smoothing function. ApplySmoothing returns the decision +1/-1 if it has been seen for a predefined number of consecutive frames, otherwise it returns 0.

5   Passing Experimental Results

In order to test our algorithm, we show that it performs well in assorted scenarios that make use of all of the attacker's ball movement options. The attacker is in the shooting region so it can choose to shoot on goal, pass, or dodge. In all cases we have the attacker, supporter, and ball start at the same locations. Figure 6-A shows the set up for when the attacker should simply shoot on goal since there is no opponent stopping it. Figure 6-B now has a static opponent placed in such a way that the shot on goal is blocked and it is advantageous to pass to the open teammate. Finally, Figure 6-C displays the setup for the case when the shot on

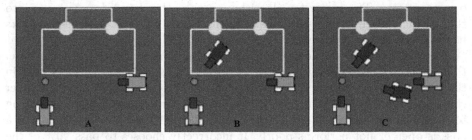

Fig. 6. Several scenarios for the attacker-supporter distributed decision making A - open shot, open pass; B - blocked shot, open pass; C - blocked shot, blocked pass

Table 2. Pass success results

Scenario	Pass Determination	
(expected action)	Attacker	Supporter
Goal open (shoot on goal)	100%	100%
Goal blocked, Pass open (pass to teammate)	80%	100%
Goal blocked, Pass blocked (dodge opponent)	10%	100%

goal is blocked as well as the pass to our teammate meaning that the attacker should resort to a dodge to attempt to not lose the ball.

We ran 10 trials for each case for both algorithms. Successes were based on whether the attacker performed the best ball movement action, given the scenario. The two algorithms we are using are identical except one of them has the attacker make all decisions from its world model, while the other lets the supporter determine if a pass is possible.

Both algorithms perform well on an open goal which is a good check since the two are identical up to that point. If there is not an opponent in front, the priority goes to shooting on goal. The next case is when the goal is blocked but the pass is open, making a pass ideal because the ball would be movef from a place of high opponent density to a lower one. Here our algorithm performs well again. The supporter saw an open passing lane from itself to the ball and told the attacker that it was open. When the attacker grabbed the ball and saw opponent in front of it, it then had to decide whether to pass or not. It retrieved the answer that the supporter had sent and then executed the pass.

The other algorithm is based only on what the attacker has in its world model. The data shows that 80% of the time, the attacker's world model had no opponent in the passing lane and therefore it passed, which is good since there was not an opponent in the passing lane. The 20% in which the attacker instead chose to dodge is interesting because as the attacker went to the ball, it must have seen an opponent on the edge of its field of view and added it to its world model. This makes sense because the edges of the camera's field of view are not as accurate and head movement tends to blur the image further. It is then possible that the attacker got this false reading randomly or by seeing its own teammate. The supporter algorithm on the other hand does not run into this problem since the passing lane is in the center of its field of view and its head is relatively stationary.

The last case is when the goal and the passing lane are blocked. In this case, the supporter algorithm once again performs well as it detects an opponent in the passing lane. It then sends messages to the attacker so that if the attacker chooses not to shoot on goal, then it also chooses not to pass, instead defaulting to dodging. Not surprisingly, the algorithm where the attacker determines if a pass is open or not performs poorly. It incorrectly chooses to pass 90% of the time, giving the ball to the opponent, as it does not detect the opponent in the passing lane.

6   Conclusion

In this paper, we contributed a positioning and pass evaluation algorithms for the challenging supporter role in distributed robot soccer. We specifically introduced a region-based positioning to capture multiple supporting strategies as a function of the game challenges. We discussed the problems of having a single focal robot making all the decisions, especially when the hardware limits the amount of information the robot has. We introduced the outsourcing of passing decision-making to the more informed supporter teammate. We have fully implemented and experimentally analyzed the supporter algorithms. We showed statistically significant lab results of their effectiveness.

References

1. Vail, D., Veloso, M.: Dynamic Multi-Robot Coordination. In: Multi-Robot Systems. Kluwer, Dordrecht (2003)
2. Veloso, M., Stone, P., Bowling, M.: Anticipation as a Key for Collaboration in a Team of Agents: A Case Study in Robotic Soccer. In: Proceedings of SPIE Sensor Fusion and Decentralized Control II (1999)
3. Fasola, J., Veloso, M.: Real-Time Object Detection using Segmented and Grayscale Images. In: Proceedings of ICRA 2006 (2006)
4. Kitano, H., Kuyinoshi, Y., Noda, I., Asada, M., Matsubara, H., Osawa, E.: RoboCup: A challenge problem for AI. AI Magazine 18(1) (1997)
5. Marling, C., Tomko, M., Gillen, M., Alexander, D., Chelber, D.: Case-based reasoning for planning and world modeling in the RoboCup small-size league. In: Proceedings of the IJCAI Workshop on Issues in Designing Physical Agents for Dynamic Real-Time Environments (2003)
6. Rybski, P., Veloso, M.: Handling diverse information sources: Prioritized multi-hypothesis world modeling. Technical Report CMU-CS-06-182 (2006)
7. Palamara, P.F., et al.: A Robotic Soccer Passing Task Using Petri Net Plans. In: Proceedings of AAMAS (2008)
8. McMillen, C., Veloso, M.: Distributed, Play-Based Role Assignment for Robot Teams in Dynamic Environments. In: Proceedings of DARS (2006)

A Novel Approach to Efficient Error Correction for the SwissRanger Time-of-Flight 3D Camera*

Jann Poppinga and Andreas Birk

Jacobs University Bremen**
Campus Ring 1
28759 Bremen, Germany
{j.poppinga,a.birk}@jacobs-university.de

Abstract. 3D data acquisition gets increasingly important for mobile robotics in general and for Safety, Security and Rescue Robotics (SSRR) in particular. 3D data allows for example to estimate the size of gaps, to construct realistic maps of unstructured disaster environments, or to detect human victims from shape. The SwissRanger SR-3000 time-of-flight 3D camera is a popular device for acquiring 3D range data as it offers the fast update rates of a camera with a resolution of 176×144 pixel at a reasonable cost. But the SR-3000 suffers - like any device using phase differences of modulated light - from the fundamental problem of wrap around error, i.e., distances that are a multiple of the wavelength of the modulated ranging signal can not be distinguished. The standard solution to this problem is to use an amplitude threshold, i.e., to discard pixels which are relatively dark and hence assumed to be far away. Here, a significant improvement to the standard method is presented that relates measured brightness and distance and also takes the geometry of the modulated light source into account. It is shown that a significantly higher amount of valid range data can be acquired with this new method.

1 Introduction

The acquisition of 3D range data is of increasing interest for mobile robots. This holds especially for domains like Safety, Security and Rescue Robotics (SSRR) where robots have to perform in complex, unstructured environments, which require 3D data for perception and world modeling. Examples of the usage of according range data include 3D map building [1][2][3][4][5], semantic environment classification [6] or the detection of drivable terrain [7].

A popular choice for according sensors are 3D laser range finders [8][9][10]. There main drawback is that they are typically based on 2D devices, which are supplemented with an actuator for an additional degree of freedom. The overall time to acquisition the data is hence slow. It is typically in the order of several

---

* This work was supported by the German Research Foundation (DFG).
** Formerly International University Bremen.

L. Iocchi et al. (Eds.): RoboCup 2008, LNAI 5399, pp. 247–258, 2009.

seconds per snapshot. Since a few years, the alternative technology of time-of-flight cameras is emerging [11]. The cameras use a near infrared modulated light source. They measure at the pixel level the phase shift - and hence time of flight - of the light returned by the environment location, which lies on the end of the optical beam corresponding to this pixel. The main advantage of this technology is that it allows high acquisition rates of 20 to 30 Hz and more. Examples of time-of-flight cameras are the SwissRanger SR-3000 [12][13] and its successor, the SR-3100, which are also popular in the RoboCup Rescue League. The SwissRanger SR-3100 is produced by MESA Imaging AG, Zurich, Switzerland. They provide a 176x144 pixel range image at 25 Hz update frequency.

Like any time-of-flight camera, the SwissRangers suffer from the fundamental problem of wrap around error. This means that distances that are a multiple of the wavelength of the modulated ranging signal can not be distinguished. The standard solution to this problem is to use an amplitude threshold, i.e., to discard pixels which are relatively dark and hence assumed to be far away. A significant improvement to this state of the art method is presented. The main idea is to take the geometry of the modulated light source into account. It is shown through extensive experiments that a significantly higher amount of valid range data can be acquired with this new method.

2   The Wrap around Error and Its Standard Solution

The SwissRanger's principle problem is that it cannot detect whether a measurement is erroneous due to being beyond the maximum range. This is because it not only relies on active sensing (illumination), but it also measures distance via phase-shift. Since phase-shift is assumed to be in $[0°, 360°[$, a phase-shift of $370°$ is measured as $10°$. Obviously, this also corrupts the distance measurement. An object beyond the maximum distance of 7.5 m, which is e.g. at 8 m, is reported as being at 0.5 m. This is called the *wrap-around error*, a fundamental problem of any time-of-flight camera.

The standard solution to this problem is to set an *amplitude threshold*, i.e., to drop pixels which are too dark. As will be shown in the experiment section, this solution is very crude and produces large portions of both false negatives and false positives.

3   A Novel Solution: Adaptive Amplitude Threshold

In the following section we first explain how we identify pixels with a wrapped around measurement, then we use this knowledge and the wrong measurement to extend the SwissRanger's range.

3.1   Identification of Erroneous Pixels

The proposed approach is to sanity-check the reported distance by relating it to the brightness. The basic idea is that a wrapped around pixel p will be reported with a low distance value d_p (measured from the image plane), but its brightness b_p is significantly darker than it were in case of a correct small distance. Since the illumination is not uniform, the pixel's position in the image array x_p, y_p is also taken into account. These two factors are used to calculate a minimum expected brightness for each pixel (equation 1).

$$b_p > \frac{\text{bw}_p \text{AAT}}{d_p^2}, \tag{1}$$

where AAT is the *advanced amplitude threshold* and bw_p is the brightness of pixel p when viewing a white wall at roughly one meter, approximated by

$$\text{bw}_p := B - ((x_p - offset_x)^2 + (y_p - offset_y)^2) \tag{2}$$

(B being the *brightness constant*). The approximation is reasonably close, except for the very edges of the image (see figure 1).

The reasoning for this is that the measurement is based solely on light emitted by the camera's illumination unit, hence the perceived brightness of an object decreases quadratically with the distance. The second observation is that the brightness produced by the imager decreases towards the edges of the images (see figure 1(a)). This is an effect, which is common to time-of-flight cameras due to the usage of multiple near infrared LEDs as light sources. They are grouped in a rectangular pattern around the cameras' lens. Hence, the center of the image is better illuminated than the border regions. To compensate for this effect, the brightness is also normalized. We approximate the brightness distribution by the above function 2. For the accuracy see figure 1(b).

In the SR-3000, according to our measurements, the center of the brightness pattern is not in the center of the image. Instead, it is shifted upward. Consequently, the offset in y direction is shifted, namely $offset_y = 61$. In x direction it is at half the image height, i.e., $offset_x = 88$, as one would expect. In the SR-3100, the image is brightest in the center. In experiments calibrating the parameters B (brightness constant) and the AAT in various scenes, we determined the values $12,000$ and 0.2 respectively as working very well.

3.2   Additional Error Sources

Material that reflects near-IR light poorly naturally appears very dark to the SwissRanger. Objects of such material can be filtered out alongside with the wrapped-around pixels. As the measurement of these objects tend to be noisy and sometimes also offset, filtering the resulting pixels out is desirable.

Furthermore, bright sun light can cause areas of random noise in the SwissRanger. This also applies to indirect perception through glass bricks or when seen on other objects. Most of these pixels are filtered out by the proposed method, except for those which by coincident have a roughly correct distance.

As a third case, pixels that appear to close due to light scattering are excluded. As mentioned in [12], bright light from close obstacles can fail to get completely absorbed by the imager. The remaining light can get reflected back to the imager

(a) Grayscale image of a white wall as captured by the Swiss-Ranger

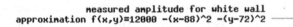

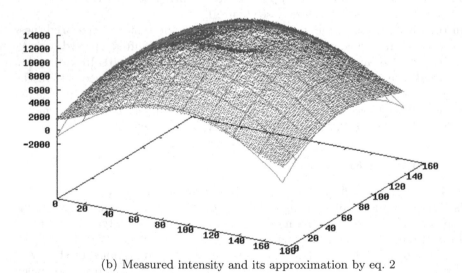

(b) Measured intensity and its approximation by eq. 2

Fig. 1. Irregular distribution of near-IR light from the camera's illumination unit

e.g. by the lens. This way, rather many pixels can get tainted, showing too close values. A large portion of these will be excluded by the presented criterion.

3.3   Correction of Wrong Values

Once the incorrect pixels due to wrap around errors are identified, they do not have to be dropped. Since the principles of the error source are regular and well-defined, they can be undone. We call this process *unwrapping*.

For this, first the perceived distance of a given wrapped-around point is calculated as the Euclidean distance d from its coordinates to the origin. Then, the maximum range (7.5 m) is added to d. Finally, the new coordinates are calculated from the distance and the known angles of the beam corresponding to the pixel in question.

This method assumes that all wrong pixels have wrapped around once. Under certain conditions, pixels can also wrap around more than once. However, in typical indoor scenes obstacles further away than 15 m are only rarely encountered. Other filtered out pixels (e.g. too dark, see previous section) are misinterpreted as being far away. In practice, rather few pixels are affected. Still, it has to be considered for each application if a dense and partially incorrect image with a doubled range or a sparse, short range but almost perfectly correct one is preferable.

    (a) Rack and door   (b) Corner window   (c) Gates

Fig. 2. Three test scenes comparing the proposed method to the standard method; here normal images of the scenes from a web-cam next to the SwissRanger are shown

Table 1. Results on all six scenes

Scene	Discarded Pixels [%]		
	Proposed method	AT 160	AT 240
Rack and door	16.7	46.3	60.3
Corner window	21.7	43.3	54.7
Gates	56.3	92.8	100
Close boxes	38.1	35.0	–
Dark material	13.1	28.0	43.1
Sun	44.9	73.1	84.6
mean	31.8	53.1	68.5
median	44.8	29.9	60.3

(a) Unfiltered distance image: wrap-around in the corner and behind the door

(b) Unfiltered distance image: wrap-around in the corner, noise on the window

(c) Unfiltered distance image: mostly wrapped around, noise on windows

(d) Proposed method: few wrong pixels remain – 16.7 % d. p.

(e) Proposed method: no wrapped-around pixels remain – 21.7 % d. p.

(f) Proposed method: no wrap-around, very little noise – 56.3 % d. p.

(g) AT 160: some wrong pixels remain, correct ones are discarded – 46.3 % d. p.

(h) AT 160: some wrong pixels remain, correct ones are discarded – 43.3 % d. p.

(i) AT 160: almost all valid pixels removed while some wrap-around is kept – 92.8 % d. p.

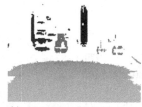 

(j) AT 240 – 60.3 % d. p.   (k) AT 240 – 54.7 % d. p.   (l) AT 240 – 2 valid pixels

Fig. 3. Three test scenes comparing the proposed method to the standard method of the SwissRanger: setting an amplitude threshold (AT). Here, two AT values presented: 240, high enough to exclude every wrapped around pixel and 160, which only almost high enough, but not as over-restrictive. For each filtered image, the fraction of discarded pixels (d. p.) is given.

4   Experiments and Results

4.1   Only Exclusion

We applied the approach presented in the previous section to snapshots taken around our lab with the SwissRanger SR-3100. To compare it to the standard abilities of the device, we also used its default error exclusion technique, the *amplitude threshold* (AT). This term designates merely the exclusion of pixels which are too dark, regardless of their distance. Depending on the situations, different AT work best. These, however, have to be manually chosen every time. Unfortunately, when using a widely applicable AT value, this standard technique excludes far too many pixels (60.3 % on average). As an alternative, we use a lower AT value which preserves more pixels (it only discards an average of 44.8 %). As can be expected, quite some of these are wrong. In figure 3, three example images are presented together with a complete distance image of the scene and a color image that was taken by a web-cam which is mounted on the robot right next to the SwissRanger.

As mentioned earlier, not only wrapped around pixels can be filtered out. This is illustrated in figure 5. Here, the improvement in performance towards the conventional method is not as big as with the wrap-around, but still apparent. For a summary of all results, refer to table 1.

4.2   Analysis

The exclusion criterion (eq. 1) consists of two factors: Wrap around is detected by comparing the reported brightness to inverted squared reported distance to detect wrap-around. In order to be more accurate, brightness is normalized with a function modeling the illumination unit's geometry. Figure 6 shows the contributions of the two components.

4.3   Wrap Around Correction

The information, which pixels are invalid due to wrap-around can be used to correct their value. This is demonstrated in figure 7. One problem here is that

(a) Close boxes   (b) Dark material   (c) Sun

Fig. 4. Three more test scenes showing other sources of error that can be filtered out; here normal images of the scenes from a webcam next to the SwissRanger are shown

(a) Unfiltered distance image: Dark obstacles in the back appear too close

(b) No wrap-around, but noisy and too close trouser due to dark material

(c) Sun causes error and random noise

(d) Unfiltered distance image with objects removed, for comparison

(e) Proposed method: dark pixels removed – 13.1 % d. p.

(f) Proposed method: very little noise remains – 44.9 % d. p.

(g) Proposed method: correct pixels of both back- and foreground kept – 38.1 % d. p.

(h) AT 160: dark objects discarded, on edges more than in center – 28.0 % d. p.

(i) AT 160: no noise, but more parts of floor and locker removed – 73.1 % d. p.

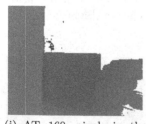  

(j) AT 160: pixels in the front preserved, those in the back gone – 35.0 % d. p.

(k) AT 240 – 43.1 % d. p.

(l) AT 240 – 84.6 % d. p.

Fig. 5. Three more test scenes showing other sources of error that can be filtered out. From left column to right: light scattering, too low reflectivity, and sun light. Here, an AT of 160 is sufficient, most AT 240 images are still given for completeness.

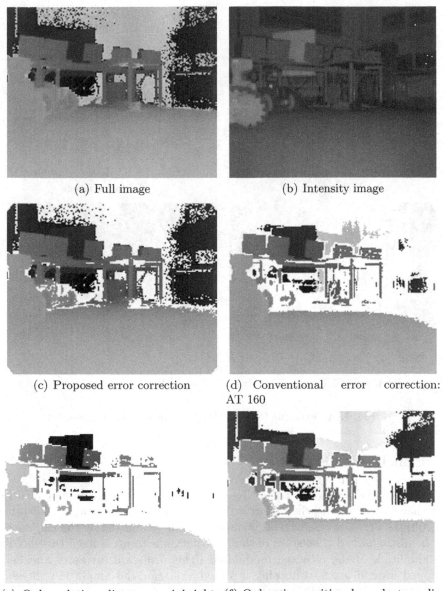

(a) Full image

(b) Intensity image

(c) Proposed error correction

(d) Conventional error correction: AT 160

(e) Only relating distance and brightness: Wrap-around removed, but too many excluded pixels, esp. on the edges

(f) Only using position dependent amplitude threshold: Not increasing strictness towards the edges (as in (d) and (e)), but keeping wrap-around

Fig. 6. Using different error exclusion criteria on the scene in figure 2(a)

some of the pixels classified as invalid are not wrapped around but excluded due to other reasons, e.g. because they are too dark. These very few pixels hence get a wrong value in the corrected image.

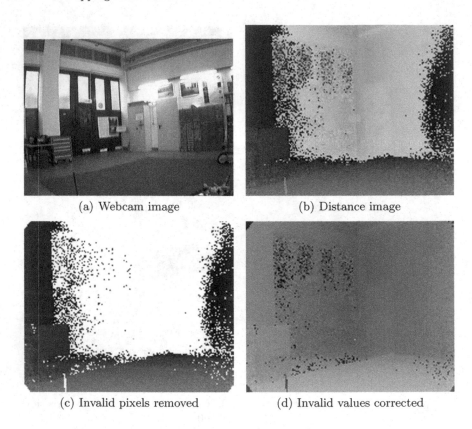

(a) Webcam image   (b) Distance image

(c) Invalid pixels removed   (d) Invalid values corrected

Fig. 7. An example for correcting wrapped-around pixels

5   Conclusion

Time-of-flight cameras are a promising technology for 3D range data acquisition. But they suffer from a principle drawback of wrap around error, i.e., distances at multiples of the wavelength of the modulated light emitted by the camera can not be distinguished. The state of the art solution is to use a fixed threshold on the amplitude of the returned modulated light to discard far away and hence wrapped around pixels. Choosing a suited value for the amplitude threshold (AT) is very difficult as it is strongly environment dependent. Furthermore, the approach tends to discard a significant number of valuable information. Here, a novel approach is presented, which takes the geometry of the light source into account. This allows a physically more plausible estimate of the to be expected amplitudes per distance than a fixed threshold. This leads to much higher amounts of valid data in time-of-flight camera images as illustrated with a Swiss-Ranger SR-3000. Furthermore, the quality of the correction even allows to detect and compensate for wrap around and hence to increase the nominal range of this time-of-flight camera.

Acknowledgments

Please note the name-change of our institution. The Swiss Jacobs Foundation invests 200 Million Euro in **International University Bremen (IUB)** over a five-year period starting from 2007. To date this is the largest donation ever given in Europe by a private foundation to a science institution. In appreciation of the benefactors and to further promote the university's unique profile in higher education and research, the boards of IUB have decided to change the university's name to **Jacobs University Bremen (Jacobs)**. Hence the two different names and abbreviations for the same institution may be found in this paper, especially in the references to previously published material.

Furthermore, the authors gratefully acknowledge the financial support of the *Deutsche Forschungsgemeinschaft* (DFG) for their research.

References

1. Howard, A., Wolf, D.F., Sukhatme, G.S.: Towards 3D mapping in large urban environments. In: Proceedings of the IEEE/RSJ International Conference on Intelligent Robots and Systems (IROS), Sendai, Japan (2004)
2. Thrun, S., Haehnel, D., Montemerlo, M., Triebel, R., Burgard, W., Baker, C., Omohundro, Z., Thayer, S., Whittaker, W.: A system for volumetric robotic mapping of abandoned mines. In: Proc. IEEE International Conference on Robotics and Automation (ICRA), Taipei, Taiwan (2003)
3. Hähnel, D., Burgard, W., Thrun, S.: Learning compact 3D models of indoor and outdoor environments with a mobile robot. Robotics and Autonomous Systems 44(1), 15–27 (2003)
4. Davison, J., Kita, N.: 3D simultaneous localisation and map-building using active vision for a robot moving on undulating terrain. In: IEEE Conference on Computer Vision and Pattern Recognition, Hawaii, December 8-14 (2001)
5. Liu, Y., Emery, R., Chakrabarti, D., Burgard, W., Thrun, S.: Using em to learn 3d models of indoor environments with mobile robots. In: 18th Conf. on Machine Learning, Williams College (2001)
6. Nüchter, A., Wulf, O., Lingemann, K., Hertzberg, J., Wagner, B., Surmann, H.: 3D mapping with semantic knowledge. In: Bredenfeld, A., Jacoff, A., Noda, I., Takahashi, Y. (eds.) RoboCup 2005. LNCS (LNAI), vol. 4020, pp. 335–346. Springer, Heidelberg (2006)
7. Poppinga, J., Birk, A., Pathak, K.: Hough based terrain classification for realtime detection of drivable ground. Journal of Field Robotics 25(1-2), 67–88 (2008)
8. Wulf, O., Wagner, B.: Fast 3D-scanning methods for laser measurement systems. In: International Conference on Control Systems and Computer Science (CSCS 14) (2003)
9. Surmann, H., Nuechter, A., Hertzberg, J.: An autonomous mobile robot with a 3D laser range finder for 3D exploration and digitalization of indoor environments. Robotics and Autonomous Systems 45(3-4), 181–198 (2003)
10. Wulf, O., Brenneke, C., Wagner, B.: Colored 2D maps for robot navigation with 3D sensor data. In: IEEE/RSJ International Conference on Intelligent Robots and Systems (IROS), vol. 3, pp. 2991–2996. IEEE Press, Los Alamitos (2004)

11. Weingarten, J., Gruener, G., Siegwart, R.: A state-of-the-art 3D sensor for robot navigation. In: IEEE/RSJ International Conference on Intelligent Robots and Systems (IROS), vol. 3, pp. 2155–2160. IEEE Press, Los Alamitos (2004)
12. CSEM: The SwissRanger, Manual V1.02. CSEM SA, 8048 Zurich, Switzerland (2006)
13. Lange, R., Seitz, P.: Solid-state time-of-flight range camera. IEEE Journal of Quantum Electronics 37(3), 390–397 (2001)

Autonomous Evolution of High-Speed Quadruped Gaits Using Particle Swarm Optimization

Chunxia Rong, Qining Wang, Yan Huang, Guangming Xie, and Long Wang

Intelligent Control Laboratory, College of Engineering,
Peking University, Beijing 100871, China
xiegming@mech.pku.edu.cn
http://www.mech.pku.edu.cn/robot/fourleg/

Abstract. This paper presents a novel evolutionary computation approach to optimize fast forward gaits for a quadruped robot with three motor-driven joints on each limb. Our learning approach uses Particle Swarm Optimization to search for a set of parameters automatically aiming to develop the fastest gait that an actual quadruped robot can possibly achieve, based on the concept of parameterized representation for quadruped gaits. In addition, we analyze the computational cost of Particle Swarm Optimization taking the memory requirements and processing limitation into consideration. Real robot experiments show that the evolutionary approach is effective in developing quadruped gaits. Satisfactory results are obtained in less than an hour by the autonomous learning process, which starts with randomly generated parameters instead of any hand-tuned parameters.

1 Introduction

Over the past years, plenty of publications have been presented in the biomechanics literature which explained and compared the dynamics of different high-speed gaits including gallop, canter, bound, and fast trot (eg. [15], [16]). To study and implement legged locomotion, various robot systems have been created (eg. [3], [4], [5]). However, most of the high-speed machines have passive mechanisms which may be not easy to perform different gaits. To understand and apply high-speed dynamic gaits, researchers have implemented different algorithms or hand-tune methods in the simulation [6] and real robot applications [2], [7], [8]. Much published research in learning gaits for different quadruped robot platforms used genetic algorithm based methods. Different from genetic algorithms, Particle Swarm Optimization (PSO) described in [1], [13], [14] eliminated the crossover and mutation operations. Instead, the concept of velocity was incorporated in the searching procedure for each solution to follow the best solutions found so far. PSO can be implemented in a few lines of computer code and requires only primitive mathematical operators. Taking the memory and processing limitation onboard into account, PSO is more appropriate in gaits

L. Iocchi et al. (Eds.): RoboCup 2008, LNAI 5399, pp. 259–270, 2009.

learning comparing with the genetic algorithm based methods for quadruped robots, especially those commercial robots with kinds of motors.

Our research focused on the gait optimization of legged robot with motor-driven joints. The commercial available quadruped robot, namely the Sony Aibo robot, which is the standard hardware platform for RoboCup four-legged league, is the main platform that we analyze and implement algorithms. Aibo is a quadruped robot with three degrees of freedom in each of its legs. The locomotion is determined by a series of joint positions for the three joints in each of its legs. Early research in gait learning for this robot employed joint positions directly as parameters to define a gait, which was the case in the first attempt to generate learned gait for Aibo. However, being lack of consistency in representing the gaits, these parameters failed to exhibit the gait in a clear way. Most of the recent research used higher lever parameters to symbolize the gait which focus on the stance of the body and the trajectories of paw. An inverse kinematics algorithm was then implemented to convert these higher lever parameters into joint angles. The general high-lever parameters used to describe the gait for Aibo can be divided into three groups. One group is for determining the gait patter by the relative phase for each leg. [11] mentioned that there exist eight types of gait patters for quadruped animals in nature. [7] described three of the most effective gaits for quadruped robot especially for Aibo, which are the crawl, trot and pace. Another group of the parameters is associated with the stance of robot. The last group of parameters describes the locus of the gait. Most of the gaits developed for Aibo based on this high lever parameter represent method differ in the shaped of the locus of paws or the representation of the locus, that is the actual parameters used to trace out the locus, eg. [9] [10].

In this paper, we present the implementation of Particle Swarm Optimization in generating high-speed gaits for the quadruped robot, specifically the Aibo. First, an overview of the basic PSO and Adaptive PSO (APSO) are introduced. Our gait learning method is based on APSO. With the knowledge of using higher lever parameters to represent the gait which focus on the stance of the body and the trajectories of the paw, the inverse kinematics model is explained. Moreover, the control parameters and optimization problem are proposed. In addition, how to implement PSO in the quadruped gaits learning is introduced in detail. The whole learning process is running automatically by the robot.

In the following section, we introduce the basic PSO and APSO specifically. In section III, we present the optimization problem in gaits learning. In section IV, the process of implementing PSO in quadruped gaits learning is introduced in detail. At last, gaits optimization results are shown in experiments with the Sony Aibo ERS-7 robot.

2   Particle Swarm Optimization

2.1   Overview of the Basic PSO

Particle Swarm Optimization (PSO) is a stochastic optimization technique, inspired by social behavior of bird flocking or fish schooling [12]. It is created by

Dr. Eberhart and Dr. Kennedy in 1995 [1]. Similar with Genetic Algorithms, PSO method searches for optimal solutions through iterations of a population of individuals, which are called a swarm of particles in PSO. However, the crossover and mutation operation are replaced with moving inside the solution space decided by the so-called velocity of each particle. PSO has proved to be effective in solving many global optimization problems and in some areas outperform many other optimization approaches including Genetic Algorithms.

PSO theory derives from imitation of social behavior of bird flocking or fish schooling. It is discovered that each bird, when hunting for food in a bird flock, changes its flying direction based on two aspects: one is the information of food found by itself; the other is information of flying directions of other birds. When one of the birds gets food, the whole flock has food. It is similar to social behavior of human being. People's decision making is not only influenced by their own experience but also affected by other people's behavior.

For an optimization procedure, hunting food by bird flock becomes searching for an optimal solution to this problem. One solution of the problem corresponds to the position of one bird (called particle) in the searching space. Each particle remembers the best position which was found by itself so far, and this information together with its current position makes up the personal experience of that particle. Besides, every particle is informed of the best value obtained so far by particles in its neighborhood. When a particle takes the whole flock as its topological neighbors, the best value is a global one. Each particle then changes its position in according to its velocity relied on this information: the personal best position, current position and the global best position.

In the realization of the PSO algorithm, a swarm of N particles is constructed inside a D-dimensional real valued solution space, where each position can be a potential solution for the optimization problem. The position of each particle is denoted X_i ($0 < i < N$), a D-dimensional vector. Each particle has a velocity parameter V_i ($0 < i < N$), which is also a D-dimensional vector. It specifies that the length and the direction of X_i should be modified during iteration. A fitness value attached to each location represents how well the location suits the optimization problem. The fitness value can be calculated by the objective function of the optimization problem.

At each iteration, the personal best position $pbest_i$ ($0 < i < N$) and the global best position $gbest$ are updated according to fitness values of the swarm. The following equation is employed to adjust the velocity of each particle:

$$v_{id}^{k+1} = v_{id}^k + c_1 r_{1d}{}^k (pbest_{id}^k - x_{id}^k) + c_2 r_{2d}^k (gbest_d^k - x_{id}^k) \tag{1}$$

Where v_{id}^k is one component of V_i (d donates the component number) at iteration k. Similarly, x_{id}^k is one component of X_i at iteration k. The velocity in equation (1) consists of three parts. One is its current velocity value, which can be thought as its momentum. The second part is the influence of the personal best. It tries to direct the particle back to the best place it has found. The last part associated with the global best attempts to move the particle toward the $gbest$. c_1 and c_2 are acceleration factors. They are used to tune the maximum length of flying in each direction. r_1 and r_2 are random numbers uniformly distributed between 0 and 1. They contribute to the stochastic vibration of the algorithm. It should be noted that each component of the velocity has new random numbers, not that all the components share the same one. In order to prevent particles from flying outside the searching space, the amplitude of the velocity is constrained inside a spectrum $[-v_d^{max}, +v_d^{max}]$. If v_d^{max} is too big, the particle may fly beyond the optimal solution. If v_d^{max} is too small, the particle will easily step into the local optimum. Usually, v_d^{max} is decided by the following equation:

$$v_d^{max} = kx_d^{max} \tag{2}$$

where $0.1 \leq k \leq 1$. Now the current position of particle i can be updated by the following equation:

$$x_{id}^{k+1} = x_{id}^k + v_{id}^{k+1} \tag{3}$$

PSO algorithm is considerably easy to realize in computer coding and only a few primitive mathematical operators are involved. Furthermore, it has the advantage of multiple points searching at the same time. Most importantly, the speed of converging is remarkably high in many learning processes. It is a critical virtue when it comes to learning gaits in a physical robot, because it minimizes damage to the robot.

The basic PSO is an algorithm base on stochastic searching, so it has strong ability in global searching. However, in the final stage of searching procedure, it is difficult to converge to a local optimum because the velocity still has much momentum. To improve the local searching ability in the final stage of optimization process, the influence of previous velocity on the current velocity needs to decrease. Thus, we proposed the using of adaptive PSO with changing inertia weight in this study.

2.2   Adaptive PSO with Changing Inertia Weight

In equation (1), by multiplying inertia weight to the momentum part of the velocity vibration can control the impact of previous velocity on the current velocity. The update equation for velocity with inertial weight is as follows:

$$v_{id}^{k+1} = wv_{id}^k + c_1 r_{1d}^k(pbest_{id}^k - x_{id}^k) + c_2 r_{2d}^k(gbest_d^k - x_{id}^k) \tag{4}$$

where w is the inertia weight. PSO with larger inertial weight results in better global searching ability for the reason that the search area is expanded with more momentum. Small inertial weight limits the search area thus improving local searching ability. Empirical results show that PSO has faster convergent rate when w falls in the range from 0.8 to 1.2. With the intention of realizing both fast global search at the beginning and intensive searching in the final stage of iteration, the value of w should vary gradually from high to low. It is similar to the annealing temperature of Simulated Annealing Algorithm. In this way, both global searching in a broaden area at the beginning and intensive search in a currently effective area at the end can be realized.

3   Optimization Problem

3.1   Inverse Kinematics Model

The high-lever parameters that we adopt to represent the gait need to be transferred to joint angles of legs before they can be implemented by the robot. An inverse kinematics model can be used to solve this problem. For a linked structure with several straight parts connecting with each other, the position of the end of this structure relative to the starting point can be decided by all angles of linked parts and only one position results from the same angle values. The definition of the kinematics model is the process of calculating the position of the end of a linked structure when given the angles and length of all linked parts.

In this robot Aibo case, given the angles of all the joints of the leg, the paw positions relative to the shoulder or the hip will be decided. Inverse kinematics does the reverse. Given the position of the end of the structure, inverse kinematics calculates out what angles the joints need to be in to reach that end point. In this study, the inverse kinematics is used to calculate necessary joint angles to reach the paw position determined by gait parameters. Fig.1 shows the inverse kinematics model and coordinates for Aibo. The shoulder or hip joint is the origin of the coordinate system. l_1 is the length of the upper limb, while l_2 is the length of the lower limb. Paw position is represented by point (x, y, z). The figures and equations below only give the view and algorithm to get the solution for left fore leg of robot. In according to the symmetrical characteristic of legs, all other legs can use the same equations with some signs changing.

The following equations shows the inverse kinematics model:

$$x = l_2 \cos\theta_1 \sin\theta_3 + l_2 \sin\theta_1 \cos\theta_2 \cos\theta_3 + l_1 \sin\theta_1 \cos\theta_2$$
$$y = l_1 \sin\theta_2 + l_2 \sin\theta_2 \cos\theta_3 \tag{5}$$
$$z = l_2 \sin\theta_1 \sin\theta_3 - l_2 \cos\theta_1 \cos\theta_2 \cos\theta_3 - l_1 \cos\theta_1 \cos\theta_2$$

The inverse kinematics equation to get θ_1, θ_2, θ_3 by the already known paw position (x, y, z) is as follows:

$$\theta_3 = \cos^{-1}\frac{x^2 + y^2 + z^2 - l_1^2 - l_2^2}{2l_1l_2}$$
$$\theta_2 = \sin^{-1}\frac{y}{l_2\cos\theta_3 + l_1}$$
$$\theta_1 = -\tan^{-1}\frac{a}{b} \pm \cos^{-1}\frac{x}{a^2 + b^2} \tag{6}$$

where $a = l_2 \sin\theta_3$, $b = -l_2 \cos\theta_2 \cos\theta_3 - l_1 \cos\theta_2$.

One problem with the inverse kinematics is that it always has more than one solution for the same end point position. However, as to Aibo, only one solution is feasible due to the restriction on the joint structure. As a result, when using inverse kinematics to calculate joint angles, it is necessary to take joint structure limitation into consideration to get the right solution. Otherwise, it will possibly cause some physical damage to the robot platform.

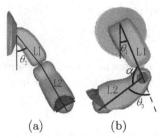

(a)   (b)

Fig. 1. The inverse kinematics model and coordinates for Aibo. (a) is the front view of left fore leg. (b) is the side view of left fore leg.

3.2   Control Parameters

Before we run the learning gait procedure, the control parameters representing a gait need to be decided. There are two rules based on which we choose our parameters: One is the sufficient representation of the gait that makes it possible to get a high-performance gait in an expanded area. The other one is the attempt to limit the number of control parameters in order to reduce the training time. These two rules are to some extent contradicted with each other. We have to find a better way to compromise these two policies manually. We have done some work on the robot's gait patters and found out that trot gait is almost always the most effective pattern in terms of both stability and speediness, thus we limit the gait pattern to mere trot gait.

For stance parameters, based on our observation and analyze of the motion for Aibo, we conclude that forward-leaning posture can speed up the walking, thus

Table 1. Control parameters in gaits evolution

fore height	vertical height from paw to chest
hind height	vertical height from paw to hip
fore width	transverse distance between paw and chest
hind width	transverse distance between paw and hip
fore length	forward distance between paw and chest
hind length	forward distance between paw and hip
step length	time for one complete step in 0.008 second units
fore step height	fore height of the locus
hind step height	hind height of the locus
fore step width	fore width of the locus
hind step width	hind width of the locus
fore ground time	fore paw fraction of time spent on ground
hind ground time	hind paw fraction of time spent on ground
fore lift time	fore paw fraction of time spent on lifting
hind lift time	hind paw fraction of time spent on lifting
fore lowing time	time spent on fore paw lowing around locus
hind lowing time	time spent on hind paw lowing around locus

we constrain the range of stance parameters to keep robot in forward-leaning posture, that is the height of hip higher than that of chest. As to locus, we choose rectangle shape because it has proved to be effective in quadruped gaits and it is simple to be represented. And because of the symmetry of right and left side when moving straight forward, we use the same locus for right legs and left legs. In all, we choose our parameters of gait as shown in Table 1.

4   Implementation of PSO

Given the parametrization of the walking defined above, we formulate the problem as an optimization problem in a continuous multi-dimensional real-value space. The goal of the optimization procedure is to find a possibly fastest forward gait for the robot, therefore the objective function of the optimization problem is simply the forward speed of the walking parameters. Particle Swarm Optimization is then employed to solve this problem with a particle corresponding to a set of parameters. A predetermined number of sets of parameters construct a particle swarm which will expose to learning by PSO, with the forward speed of each parameter being the fitness.

4.1   Initialization

Initially, a swarm of particles are generated in the solution space, which is a set of feasible gait parameters. These particles can be represented by $\{p_1, \varLambda, p_N\}$ (where $N = 10$ in this case).These sets of parameters are acquired by random generation within the parameter limits decided by the robot mechanism. A lot of previous work done on learning gaits start from a hand-tune set of parameters. Comparing with previous work, random generation of initial values has the advantage of less human intervention, and more importantly, has more possibility to lead to different optimal values among different experiments. Initial velocities for all particles are also generated randomly in the same solution space within given ranges. The width of the range is chosen to be half of that of the corresponding parameters. Velocity calculated later is also constrained inside the spectrum. The spectrum is denoted by $(-V^{max}, +V^{max})$, where $V^{max} = \frac{1}{4}(x^{max} - x^{min})$, with (x^{min}, x^{max}) as the changing range of particle P_i. The ranges we chosen turn out to be appropriate to avoid the two problems mentioned in Section II.A. To expedite the search process, c_1 and c_2 are set to 2. The initial *pbests* are equal to the current particle locations. There is no need to keep track of *gbest* while it can be acquired from *pbests*, that is the *pbest* with the best fitness is *gbest*.

4.2   Evaluation

The evaluation of parameters is performed using sole speed. Since the relation between gait parameters and speed is impossible to acquired, we do not know the true objective function. There is no sufficiently accurate simulator for Aibo due to the dynamics complexity. As a result, we have to perform the learning procedure on real robots.

In order to automatically acquiring speed for each parameter set, the robot has to be able to localize itself. We use black and while bar for Aibo to localize, because given the low resolution of Aibo's camera, it is faster and more accurate to detect black-while edge than other things. We put two pieces of boards with the same black and while bars in parallel so the robot can walk between them.

During evaluation procedure, the robot walks to a fixed initial position relative to one of the boards, then load the parameter set needed to be evaluated, walk for a fixed time, 5s, stop and determine the current position. It should be noted that both before and after the walk, robot is in static posture, so the localization is better compare to localizing while running. Now the starting and ending location have been acquired from detecting the bars, speed can be calculated out. After that, robot turns around by 90 degree, and localizes according to the other board, if the position is far from the fixed position, adjust it or else go to the next step, loading another set of parameters and then begin another trial. The total time for testing one set of parameter including turning, localizing, and walking time. Because of the ease of localizing, usually it takes less than 3s to turn and get to the right position. As a result, the test time of one particle is less than 8s.

4.3   Modification

After all particles of the swarm are evaluated, *pbests* are updated by comparing them with corresponding particles. If the performance of P_i is better than *pbest$_i$*, which means the fitness value of P_i is higher than that of *pbest$_i$*, *pbest$_i$* will be replaced by the new position of P_i. In addition, the fitness value of P_i is recorded as fitness value of *pbest$_i$* for future comparing. Subsequently, the new *gbest*, the best among *pbests* can be acquired. It should be noted that the update of *gbest* is not done anytime a particle is evaluated but after the whole swarm is evaluated. The difference does not change the principle of the algorithm or empirically influence the converge rate.

As mentioned in section II, in order to realize global search in a broaden area at the beginning of the learning procedure and intensive search in a currently effective area at the end, we employ adaptive PSO with piecewise linearity declining inertial weight to perform the learning procedure. When inertial weight value w is around 1, it presents global search characteristics and results in fast converge rate. when w is a lot less than 1, intensive search is realized.

5   Experimental Results

Using the method described above, we take two separate experiences and achieve favorable results. In the first experience, since large inertial weight will extend the searching area, resulting in a long time of training, we take a conservative move and reduce inertial weight quickly from the start with initial value being 1. The inertial weight is determined by equation (7). Fig. 2(a) shows the vibration of w through iterations. By iteration 15, w has decreased to 0.1. The global search is diminished, while the intensive search is enhanced. Fig. 3(a) shows the result through iterations. We can see that the learning process is converging quite fast from 1 to 10 iteration. After that, the result improve slowly but firmly until around 25 iteration. Although we get a high-performance gait in a short time in this experiment, we think it is possible that we can have a better result when extending the search area a little by not reducing w so fast. So we tried to use another equation (8) to update w, Fig. 2(b) shows the vibration of w, and Fig. 3(b) shows the learning result.

$$w = 1 - 0.06 \times iter(iteration \leq 15)$$
$$w = 0.1 - 0.01 \times (iter - 15)(15 < iteration \leq 25)$$
$$w = 0(iteration > 25) \tag{7}$$

$$w = 1.2 - 0.02 \times iter(iteration \leq 10)$$
$$w = 1 - 0.085 \times (iter - 10)(10 < iteration \leq 20)$$
$$w = 0.15 - 0.03 \times (iter - 20)(20 < iteration \leq 25)$$
$$w = 0(iteration > 25) \tag{8}$$

We can see that the second experiment achiever better result than the first one. It is interesting that they both reach their peak in the 25 iteration, that's the time when w become zero. It's possible that PSO has little local optimization when current velocity is no longer influenced by previous velocity which is contradicted to what we assumed.

We can also note that there are both advantage and disadvantage comparing these experiments with each other. For one thing, the learning curve of the first experiment is a lot smoother than that of the second one. It means that the second learning process has more undulation. In fact, during the second experiment, there are still new sets of parameters that perform very poor after the 10

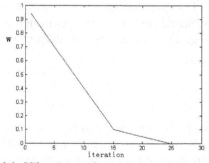

(a) Vibration of inertial weight w through itcrations in the first experiment.

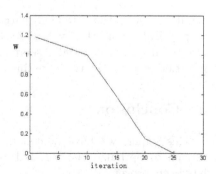

(b) Vibration of inertial weight w through iteration in the second experiment.

Fig. 2. Vibration of inertial weight w through iteration in real robot experiments

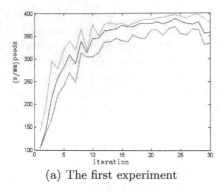
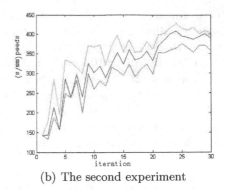

(a) The first experiment   (b) The second experiment

Fig. 3. Optimization results. (a) is the best(in the green line), average of the whole swarm(in the red line)and average of the best half part of the swarm(in the blue line) in real robot experiments. (b) the best result of every iteration in both the two experiments. The green is the first one, and the blue is the second one.

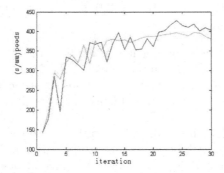

Fig. 4. The best result of every iteration in both the two experiments. The green line is the first one, while the blue line is the second one.

iteration due to the extended searching area. This problem cause more damage to physical robot. However, the second experiment acquire better parameters also because of the extended searching area. Fig. 4 shows the best result of every iteration in both the two experiments.

6   Conclusion

In this paper, we have demonstrated a novel evolutionary computation approach to optimize fast forward gaits using Particle Swarm Optimism. PSO has been proven to be remarkable effective in generating optimal gaits in the robot platform Aibo. Our method was easily coded and computationally inexpensive. Moreover, by using PSO, the evolution converged extremely fast and the training time was largely reduced. That is an essential advantage for physical robot learning, minimizing possible damage to the robot. Another contribution of our method was its initial sets of parameters are randomly generated inside the value range instead of mutation from a hand-tune set of parameters. It reduced the human work as well as generating evolutional results varied a lot in different experiences. Through experiments which took about 40 minutes each, we achieved several high-performance sets of gait parameters which differ a lot from each other. These gait parameter sets were among the fastest forward gaits ever developed for the same robot platform.

In the future, we will compare different high-performance gait parameters and analyze the dynamics model of the robot and in an attempt to get a deeper sight into the relation between parameter and its performance. After that, we will be able to generate more effective gaits in less learning time. Through analysis, we find that the gait actually executed by robot differ significantly from the one we design. There are several reasons accounting for that. The most important one is the interaction with environment prevents the implement of some strokes of robot legs. Although with learning approach, factors that cause the difference between actual gait and planned gait do not have to be taken into consideration. However, we assume that if the planned gait and actual gait can conform with each other, Aibo will walk more stable and fast. In order to solve the problem, the analysis of dynamics between the robot and the environment is necessary. In this gait learning procedure, we only evolve fast forward gait and choose forward speed as the fitness. Later on, we will try to learn effective gaits in other directions, for example, gaits for walking backward, sideward and turning. We also consider exploring optimal omnidirectional gaits. With gaits working well at all directions, robots will be able to perform more flexibly and reliably.

Acknowledgements

The authors gratefully acknowledge the contribution of the team members of the sharPKUngfu Legged Robot Team. This work was supported in part by the National Science Foundation of China (NSFC) under Contracts 60674050, 60528007, and 60635010, by the National 973 Program (2002CB312200), by the National 863 Program (2006AA04Z258) and 11-5 Project (A2120061303).

References

1. Eberhart, R., Kennedy, J.: Particle Swarm Optimization. In: Proceedings of the IEEE Conference on Neural Network, Pelfh., Australia, pp. 1942–1948 (1995)
2. Papadopoulos, D., Buehler, M.: Stable running in a quadruped robot with compliant legs. In: Proceedings of the IEEE International Conference on Robotics and Automation, San Francisco, CA, pp. 444–449 (2000)
3. Holmes, P., Full, R.J., Koditschek, D., Guckenheimer, J.: The dynamics of legged locomotion: Models, analyses, and challenges. SIAM Review 48(2), 207–304 (2006)
4. Raibert, M.H.: Legged robots that balance. MIT Press, Cambridge (1986)
5. Collins, S., Ruina, A., Tedrake, R., Wisse, M.: Efficient bipedal robots based on passive dynamic walkers. Science 307, 1082–1085 (2005)

6. Krasny, D.P., Orin, D.E.: Generating high-speed dynamic running gaits in a quadruped robot using an evolutionary search. IEEE Transactions on Systems, Man, and Cybernatics - Part B: Cybernetics 34(4), 1685–1696 (2004)
7. Hornby, G.S., Fujita, M., Takamura, S., Yamamoto, T., Hanagata, O.: Autonomous evolution of gaits with the Sony quadruped robot. In: Banzhaf, W., et al. (eds.) Proceedings of the Genetic and Evolutionary Computation Conference, Orlando, Florida, USA, vol. 2, pp. 1297–1304. Morgan Kaufmann, San Francisco (1999)
8. Kim, M.S., Uther, W.: Automatic gait optimisation for quadruped robots. In: Australasian Conference on Robotics and Automation (2003)
9. Röfer, T., Burkhard, H.D., Düffert, U., Hoffmann, J., Göhring, D., Jüngel, M., Lötzsch, M., Stryk, O.v., Brunn, R., Kallnik, M., Kunz, M., Petters, S., Risler, M., Stelzer, M., Dahm, I., Wachter, M., Engel, K., Osterhues, A., Schumann, C., Ziegler, J.: GermanTeam RoboCup 2004. Technical report (2004)
10. Röfer, T., Burkhard, H.D., Düffert, U., Hoffmann, J., Göhring, D., Jüngel, M., Lötzsch, M., Stryk, O.v., Brunn, R., Kallnik, M., Kunz, M., Petters, S., Risler, M., Stelzer, M., Dahm, I., Wachter, M., Engel, K., Osterhues, A., Schumann, C., Ziegler, J.: GermanTeam RoboCup 2005. Technical report (2005)
11. Ian Stewart. Nature's Numbers (1996)
12. Reynolds, C.: Flocks, herds, and schools:Adistributed behavioral model. Comp. Graph. 21(4), 25–34 (1987)
13. Angeline, P.: Using selection to improve particle swarm optimization. In: Proc. IEEE Int. Conf. Evolutionary Computation, pp. 84–89 (1998)
14. Naka, S., Genji, T., Yura, T., Fukuyama, Y.: Hybrid particle swarm optimization based distribution state estimation using constriction factor approach. In: Proc. Int. Conf. SCIS & ISIS, vol. 2, pp. 1083–1088 (2002)
15. Alexander, R.M., Jayes, A.S.: A dynamic similarity hypothesis for the gaits of quadrupedal mammals. J. Zoology Lond. 201, 135–152 (1983)
16. Alexander, R.M., Jayes, A.S., Ker, R.F.: Estimates of energy cost for quadrupedal running gaits. J. Zoolog. 190, 155–192 (1980)

Designing Fall Sequences That Minimize Robot Damage in Robot Soccer*

Javier Ruiz-del-Solar[1], Rodrigo Palma-Amestoy[1], Paul Vallejos[1], R. Marchant[1], and P. Zegers[2]

[1] Department of Electrical Engineering, Universidad de Chile
[2] College of Engineering, Universidad de Los Andes, Chile
jruizd@ing.uchile.cl

Abstract. In this paper is proposed a methodology for the analysis and design of fall sequences that minimize robot damage. This methodology minimizes joint/articulation injuries, as well as damage of valuable body parts (cameras and processing units). The methodology is validated using humanoid Nao robots and a realistic simulator. The obtained results show that fall sequences designed using the proposed methodology produce less damage than standard, uncontrolled falls.

1 Introduction

In soccer, as in many other sports that allow contact among players, it is usual that players fall down, as consequence of fouls, collisions with other players or objects, or extreme body actions, such as fast movements or ball kickings from unstable body positions. In addition, soccer players can intentionally fall down to block the ball trajectory (defense player) or to intercept the ball (goalkeeper). Therefore, we can affirm that the management of falls – e.g. how to avoid an unintentional fall, how to fall without damaging the body, how to achieve fast recovering of the standing position after a fall - is an essential ability of good soccer players. In general, this is also true for any physical human activity.

Given the fact that one of the RoboCup main goals is allowing robots play soccer as humans do, the correct management of falls in legged robots, especially in biped humanoid robots, which are highly unstable systems, is a very relevant matter. However, to the best of our knowledge this issue has almost not been addressed in the RoboCup and other mobile robotics communities. The current situation in the RoboCup is that: (i) In case of an unintentional fall, the standard situation is that robots do not realize they are falling down. Therefore, they do not perform any action for diminishing the fall damage. After the fall, they recognize they are on the ground using their internal sensors, and they start the standing up sequence of movements. There are some few examples of systems in the literature that detect unstable situations and avoid the fall [1][4][5][6]; (ii) In most of the cases, unintentional or intentional falls, robots fall as deadweight, without using a fall sequence that can

---

* This research was partially supported by FONDECYT (Chile) under Project Number 1061158.

L. Iocchi et al. (Eds.): RoboCup 2008, LNAI 5399, pp. 271–283, 2009.

allow them to dissipate some of the kinetic energy of the fall or to protect some valuable body parts, as humans do. Two of the few works that address this issue are [2][3], although none of them was developed in the context of RoboCup; and (iii) The damage of robot components or environments parts after a fall is a real problem. This is one of the reasons for limiting the size of robots in some RoboCup leagues (e.g. RoboCup TeenSize league).

In this context, the aim of this paper is to address the management of falls in robot soccer. In concrete, we propose a methodology for designing fall sequences that minimize joint/articulation injuries, as well as the damage of valuable body parts (cameras and processing units). This fall sequences can be activated in case of an intentional fall or in case of a detected unintentional fall. It is not our intention to cover exhaustively this topic in this article, but to focus the attention of the community into this important problem. This paper is organized as follows. In section 2, some related work about human's fall studies is presented. The analysis of these studies suggests some guidelines for the management of falls in robots. The here-proposed methodology for designing fall sequences that minimize the robot damage is described in section 3. In section 4 some experimental results of the application of this procedure in Nao robots are presented. Finally, in section 5 some conclusions of this work are given.

2   Related Work

Uncontrolled falls in biped robots have been largely ignored except for some very interesting works such as [2][3][7]. In general, falls are characterized by violent impacts that quickly dissipate and transfer important amounts of kinetic energy through joints, bones, and tissues. Falling looks like a dual event of walking. Avoiding falls gives a complementary view to walking, which is a controlled fall. In this section, we will analyses the literature about human's falls, which can give us some insights on how to manage falls in robots. There are several approaches to study falls in humans: what has been reported in medical literature, the studies done by people in biomechanics, the techniques developed in martial arts, and the efforts of the animation industry. These will be here reviewed.

Medical Studies. In [8] a fall is defined as "an unintentional event that results in a person's coming to rest on the ground or on another lower level." Falls are in general the result of the convergence of several intrinsic (muscle weakness, visual deficit, poor balance, gait defects, etc.), pharmacological (walking under the influence of alcohol, being under strong medication, etc.), environmental (uneven terrain, poor lighting conditions), and behavioral related (daily tasks, sports, violence, etc.) factors [10][11][12]. The rapid transmission of forces through the body that follow an impact against a surface cause injuries according to the energy-absorbing characteristics of the surfaces that receive the impact, the magnitude and direction of the forces, and the capacity of the tissues to damage [11]. Fall prevention in human has focused on balance and gait impairments, which are mainly affected by the interaction of the sensory (ability to determine whether the center of gravity of the body is within the support of the body or not), neuromuscular (transmission speed of the nervous impulses), and musculoskeletal systems (the available muscular force determine the range of possible movements), and their integration by the central nervous system (Parkinson's disease).

In general, falls from a standing height produce forces that are one order of magnitude greater than those necessary to break any bone of an elderly woman [11]. However, approximately no more than 10% of falls in older people cause fractures [10][11]. The most common examples of serious injuries, besides tissue and organ damage, are hip and wrist fractures [11]. This is a clear indicator that people constantly use fall-handling strategies that manage to reduce injury-producing falls. Wrist injuries are also interesting because they indicate an active intent of people to stop their falls using limbs to shift the impact away to less important organs or bones.

Biomechanical Approach. Even though medical literature has been studying falls for several decades, the needs posed by high efficiency sports and the possibilities created by technological advances such as motion capture equipments has spawned new approaches to the science of human movement, also called biomechanics. This has made possible to understand human dynamics with greater detail and to generate more precise mathematical models of this biological machine: (i) The musculoskeletal system is now modeled as a combination of something that exerts the force, a spring, and a damping system [13][14]; (ii) Machine learning approaches have been used to classify movements in order to understand their relationships, and to prove the existence of clusters of movement patterns [15]; (iii) It has been possible to determine the role of the center of mass of the body in all types of movements, i.e. rock climbing [16]; (iv) Another important aspect are interactions with external objects, where, for example, it has been possible to determine that people require several minutes to adjust their movements to changing asymmetrical loads [17][18], or how vision and limbs coordinate to follow and manipulate balls while juggling [19]; and (v) Control of synchronization between many people exhibiting rhythmic movements under some conditions has been proven to be independent from force control [20].

Of special interest, from the point of view of studying falls, are studies of people displaying fast interactions with the ground or objects. Of special interest is the study of Gittoes *et al* [21], which proves that soft tissue strongly contributed to reduce loading when landing on the ground. Other studies of voluntary fast transitions show that it is possible to generate complex pattern activation patterns that allow to control sudden movements with an amazing degree of control [22][23][24][25].

Martial Arts. In the previous paragraphs we have pointed out the biological aspects of falling. But, Is it possible to control a fall in order to minimize damage? Is it possible to modify a fall in order to achieve some dynamic objective? These questions have been answered long ago by martial arts (check [26]). Of these, Judo and Taekwondo, are perfect examples. Both disciplines teach how to fall from different positions: forward, backward, and sideways. All these techniques are extremely effective in the sense that produce a sequence of movements that vary the geometry of the human body in order to lower the force of the impacts, and spread the kinetic energy transfer through a wider contact area, a longer lapse of time, and limb movement. Moreover, some of these techniques are designed to allow the fighter to move away from the attacker and prepare himself to continue the combat by quickly recovering an upright stance.

Human Dynamics Simulation. The constant pressure of the computer animation market for more realistic special effects has spurred a lot of research in this topic in the last decades [27][28][29]. This research has even tackled problems not studied by other disciplines, such as the reproduction of destructive movements that are impossible to study in humans due to their nature [30][31]. Given that the tools created by this industry are completely located in simulated environments, many of them naturally produce very realistic falls with physically plausible kinematics and dynamical interactions.

Discussion. In general, human falls are characterized by unexpected impacts that affect the whole body. Overall, a fall should not be analyzed as a very local impact. On the contrary, its forces propagate along the entire body and help to distribute and dissipate its effects. In general falls can be classified into three cases according to the degree of awareness of the person falling: (i) The person is not aware of the fall and only passive elements of the body help to absorb the impacts. Given that every fall has a great potential for causing important body damage, there is strong evidence that the very nature of human joints, modeled with springs and dampers, and soft tissue passively contribute to ameliorate the effects of falls. In addition, medical literature reports that external padding may also help to diminish the effects of the falls; (ii) The person detects when a fall is initiated and responds accordingly. This explains, for example, why wrist fractures are common: limbs are commonly used to change the impact points in order to redistribute the impact zones along the different surfaces of the body and time; and (iii) As showed in martial arts, if the person can predict a fall, then he can take control of the fall and change it into a fluid movement that helps to quickly recover the desired behavior. This is one step further into distributing the fall into different and wider surfaces of the body and time, aiming towards diminishing or even eliminating fall damage.

When fall are considered from the point of view of robots, things change. For example, medical literature does not talk about joint damage but of bone fractures. In robots it is more plausible that the opposite is more important: it is always possible to build very strong limbs, while the motors in the joints are the ones that suffer damage during impacts. In this sense there has been work done that points out the problems that need to be solved in order to map human movement into robot movement [9].

In general, the study of human falls suggests that for robots it is important to: (i) Design a body that passively helps as much as possible to walk and to fall. A body that uses joint models with springs and dampers as movable parts that act like soft tissue, and padding specially designed to protect important, frail, and/or expensive parts; (ii) Detect a fall as soon as possible to trigger fall-related movements that allows reducing the fall damage; and (iii) If possible, predict falls and redesign normal moves in order to lower the probability of a fall, or to simply control a fall in order to eliminate it as much as possible or to reduce the damage. This work is directly related with the second and third point.

3   Design of Fall Sequences That Minimize Robot Damage

Let us consider a humanoid robot with n rotational articulations/joints q_i. Each articulation is composed by a DC motor, a gearbox and mechanical elements that fix these components (e.g. an articulation built up using a standard servomotor). In this model each articulation can rotate in a given angular operational range:

$$\theta_i^{\min} \leq \theta_i \leq \theta_i^{\max}; i = 1,...,n \qquad (1)$$

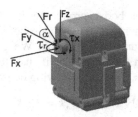

Fig. 1. External forces and torques that can damage an articulation, illustrated for the case of a Dynamixel DX117 motor

The dynamics of each joint i can be characterized in terms of the forces and torques applied in the different axes. Due to the symmetry of the joints, only the axial and radial forces need to be considered. The axial force $F_a = F_x$ and the magnitude of the radial force $F_r = \sqrt{F_y^2 + F_z^2}$ (see fig. 1) are external forces that can damage the articulation due to impacts produced during a fall. The rotational torque $\tau_{rot} = \tau_x$ is an external torque applied in the direction of rotation of the joint, while the radial torque $\tau_{rad} = \tau_r$ is the torque produced by the radial force. These forces and torques can be produced directly by the fall impacts or be transmitted by the robot body to the articulation. In the short period of time after an impact, a joint can be damaged if the linear or angular impulses (i.e. the integral of each external forces or torques over the time period) surpass a given magnitude that depends on the physical properties of the articulation (motor characteristics, gear material, etc.). Let us define J_{fa}, J_{fr}, J_{rot} and J_{rad} as the impulses produced by the axial force, radial force, rotational torque, and radial torque respectively. The damage can be avoided if the following relations hold:

$$J_{i,fa} \leq J_{i,fa}^{\max}; J_{i,fr} \leq J_{i,fr}^{\max}; J_{i,rot} \leq J_{i,rot}^{\max}; J_{i,rad} \leq J_{i,rad}^{\max}; i = 1,...,n \qquad (2)$$

with $J_{i,fa}^{\max}, J_{i,fr}^{\max}, J_{i,rot}^{\max}, J_{i,rad}^{\max}$ threshold values that depends on the physical properties of the joint.

In addition to the joints' damage, the robot body (mainly frames) can be damaged if the intensity of the fall surpasses a given threshold. Therefore, we need a global measure of the fall intensity. In the biomechanics literature and in studies about falls in humans the impact velocity v_{imp} is used as a measure of the fall intensity. From the physics point of view, in rigid body collisions the damage is produced by the change of momentum of the colliding objects. Given that in our case collisions are produced between the robot and the ground, which has a much larger mass than the robot, we can assume that the impact velocity is an adequate measure of the fall impact. Hence, to avoid robot body damage, the following should hold:

$$v_{imp} \leq v_{imp}^{\max} \qquad (3)$$

with $v_{imp}^{\max}$ the maximal impact velocity that do not produce damage in the robot.

Naturally, (2) and (3) are related because the change of the total momentum, which depends on the impact velocity, is equal to the total impulse. This total impulse is then propagated through the robot body, producing local impulses in the joints. An important additional requirement to avoid the robot damage is that valuable body parts (CPU, cameras, etc.) should be protected. We assume that these parts will be protected if they do not touch the ground or if they touch it at a low speed. Let us consider K valuable body parts, then the following constraints should hold:

$$p_z^k > 0 \vee v_z^k \leq v_{k,z}^{\max}; k = 1,..,K \tag{4}$$

with p_z^k and v_z^k the vertical position and speed of each valuable part, respectively.

Let us define the joints' positions during the whole fall period as $\Theta_{T_{Fall}} = \{\Theta(t)\}_{t=0,...,T_{Fall}}$, with $\Theta(t)$ a vector containing the joints' positions at time step t and T_{Fall} the fall period. The process of designing a fall sequence is modeled as a search for the $\Theta_{T_{Fall}}$ that minimizes the damage produced by the fall in robot's joints, frames and valuable parts. From (1)-(4), the expression to be minimized, for a given $\Theta_{T_{Fall}}$, is given by:

$$f(\Theta_{T_{Fall}}) = \sum_{i=1}^{n} \alpha_{i1} J_{i,fa} + \alpha_{i2} J_{i,fr} + \alpha_{i3} J_{i,rot} + \alpha_{i4} J_{i,radial} + \beta v_{imp}$$

subject to

$$J_{i,fa} \leq J_{i,fa}^{\max}; J_{i,fr} \leq J_{i,fr}^{\max}; J_{i,rot} \leq J_{i,rot}^{\max}; J_{i,rad} \leq J_{i,rad}^{\max}; i = 1,...,n \tag{5}$$

$$p_z^k(t) > 0 \vee v_z^k(t) \leq v_{k,z}^{\max}; k = 1,..,K; t = 0,...,T_{Fall}$$

$$\theta_i^{\min} \leq \theta_i(t) \leq \theta_i^{\max}; i = 1,...,n; t = 0,...,T_{Fall}$$

with α_{ij} weight factors that depends on the importance of each joint (e.g. the neck joint is far more important than a finger joint for a human), on its mechanical properties (e.g. the specific motor model used in the joint), and on the importance of the different forces and torque impulses for the specific joint. β is the weighting factor of the global impact measure and n the number of joints. It should be noted that in (5), the force and torque impulses should be measured in the short period of time after an impact.

As already explained, the process of designing a fall sequence consists on searching for a $\Theta_{T_{Fall}}$ that minimizes an expression that quantifies the damage (eq. (5)). However, to implement directly this search process is highly complex because: (i) when working directly with real robots a large amount of experiments is required, which would eventually damage the robots, (ii) it requires measuring in each joint two linear and two rotational impulse values, as well as the impact velocity, in real-time (at a rate of few milliseconds), and (iii) the high-dimensionality of the parameter space; the search process requires the determination of the position of each joint during the whole fall period.

The first two problems can be overcome if a realistic simulator is employed for the analysis and design of the fall sequences. Using this computational tool, robot damage due to extensive experiments is avoided. In addition, if the simulator is realistic enough, all physical quantities that need to be known for evaluating (5) can be easily determined. The high dimensionality of the parameter space is the hardest problem to be tackled. As a suboptimal design strategy, we propose a human-based design procedure consisting on iteratively applying the following consecutive steps: (i) synthesis of fall sequences using a simulation tool, and (ii) quantitative analysis of the obtained sequences using eq. (5). The proposed procedure consists of the following main components: (i) An interactive tool is employed by a human operator for the synthesis of fall sequences (see example in figure 2). For each frame of the sequence, the designer set up all joints' positions; (ii) Each fall sequence is executed in a realistic simulator, and the global damage function given by (5) is used to evaluate the potential damage in the robot body; and (iii) The seeds of the design process, i.e. initial values for the joints' positions during the whole fall period, are examples of human falls, obtained either from standard videos of falls (e.g. martial arts or human sports) or from data acquired using motion capture equipments (e.g. exoskeletons). In the next section we will describe how the proposed strategy has been used in the design of falls sequences for Nao robots.

4 Experimental Results

To validate the proposed strategy it was chosen the problem of designing falls for Nao robots [36]. This robot was selected for the following reasons: (i) the robot correspond to an humanoid robot with 22 degrees of freedom (see table 1 for details about the joints and their motion ranges), which represents a complex problem from the point of view of designing fall sequences, (ii) it exists a realistic simulator available for this robot (Webots [34]), (iii) the simulator and robot controller are URBI-compatible [35], which allows building an interactive interface for designing falls, without modifying the simulator or accessing to its source code, and (iv) we have a team in the SPL league (this league use Nao robots), that already classified for the 2008 RoboCup world-competition.

The first step was to build a user interface that allow designing the falls, i.e. specifying the joints positions for the whole frame period ($\Theta(t); t = 0,...,T_{EndFall}$) and measuring the velocity and impulses values of the falls. This interface was built using URBI. Figure 2 shows its appearance. It can be seen that: (i) the main window has the values of all joint positions for the current frame; these values can be modified by the designer (user), (ii) the bottom window contains the commands for executing the fall, either continuously or in frame-to-frame mode, and (iii) the right window has the values of the radial, axial and angular impulses, the impact velocity, and flags that indicates if any of these values have surpassed their maximal thresholds. In addition, some flags indicate if valuable parts, camera and CPU in this case, touch the floor. The user employs this tool in the designing process, and he can see simultaneously the resulting fall sequence directly in the main simulator window (see examples in fig. 3).

For the purpose of showing the potentiality of this fall designing approach, two "bad" fall sequences and three "good" fall sequences were designed; "good"/"bad" means low/high damage. The designed sequences are: *FrontHead*: frontal fall where the robot impacts the floor with its face; *FrontHand*: frontal fall where the only robot action is to put its arm, in a rigid position, to avoid touching the floor with its face; *FrontLow*: frontal fall where the robot folds its legs in order to lower its center of mass before the impact; *FrontTurn*: frontal fall where the robot turn its body before

Table 1. Nao's joint names and motion ranges

Body Part	Joint Name	Motion	Range (degrees)
Head	HeadYaw	Head joint twist (Z)	-120 to 120
	HeadPitch	Head joint front & back (Y)	-45 to 45
Left arm	LShoulderPitch	Left shoulder joint front & back (Y)	-120 to 120
	LShoulderRoll	Left shoulder joint right & left (Z)	0 to 95
	LElbowRoll	Left shoulder joint twist (X)	-120 to 120
	LElbowYaw	Left elbow joint (Z)	0 to 90
Left leg	LHipYawPitch	Left hip joint twist (Z45°)	-90 to 0
	LHipPitch	Left hip joint front & back (Y)	-100 to 25
	LHipRoll	Left hip joint right and left (X)	-25 to 45
	LKneePitch	Left knee joint (Y)	0 to 130
	LAnklePitch	Left ankle joint front & back (Y)	-75 to 45
	LAnkleRoll	Left ankle joint right & left (X)	-45 to 25
Right leg	RHipYawPitch	Right hip joint twist (Z45°)	-90 to 0
	RHipPitch	Right hip joint front and back (Y)	-100 to 25
	RHipRoll	Right hip joint right & left (X)	-45 to 25
	RKneePitch	Right knee joint (Y)	0 to 130
	RAnklePitch	Right ankle joint front & back (Y)	-75 to 45
	RAnkleRoll	Right ankle right & left (X)	-25 to 45
Right arm	RShoulderPitch	Right shoulder joint front & back (Y)	-120 to 120
	RShoulderRoll	Right shoulder joint right & left (Z)	-95 to 0
	RElbowRoll	Right shoulder joint twist (X)	-120 to 120
	RElbowYaw	Right elbow joint (Z)	-90 to 0

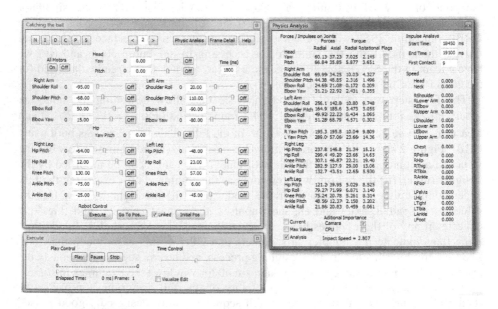

Fig. 2. Image capture of the interactive tool used for the falls design

touching the floor; and *BackLow*: back fall where the robot separates and folds its legs in order to lower its center of mass before the impact. Videos of all sequences are available in [37]. Two of these sequences are exemplified in figure 3.

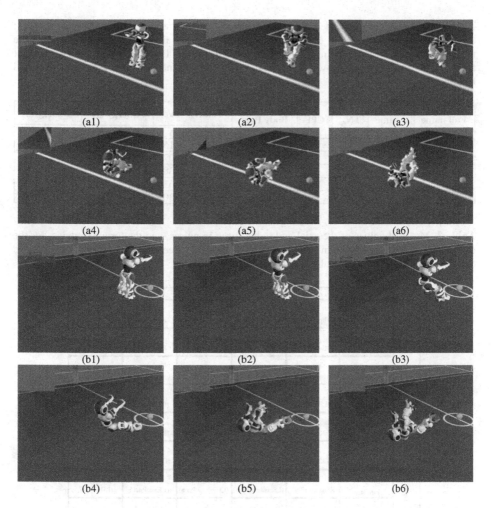

Fig. 3. Simulation sequences showing selected frames of the falls under analysis. *FrontTurn*: (a1)-(a6). *BackLow*: (b1)-(b6).

It is important to stress that the process of finding "good" falls is a very difficult one, due to the huge size of the search space. In our case the *FrontTurn* fall was designed by seeing videos of martial arts, while the *FrontLow* and *BackLow* falls were designed analyzing data generated by a motion capture exoskeleton (Gypsy-5 from Animazoo [33]).

In tables 2 and 3 quantitative measurements of the fall sequences under analysis are shown. Table 2 contains global measures about the intensity of the impact: velocity, kinetic energy and potential energy before (B) and after (A) the impact. In addition, for each case, the body part that first touches the floor (*Body Part*), the fall duration in frames ($\Delta Frame$), the number of body contact points during the whole fall sequence (TN_{cp}), and an index consisting in the sum of frames during which the contact points touched the floor (*CPF*), are shown. Table 3 shows, for each fall

Table 2. Global measures of the falls under comparison. ΔFrame: fall duration in frames. TN$_{cp}$: total number of contact point during the fall. CPF: sum of all the contact points times the number of frames that each point is in contact with the ground. Body Part: Part that first contact the body. B: Before collision. A: After collision. K Energy: Kinetic energy. P Energy. Potential Energy.

Fall name	ΔFrame	TN$_{cp}$	CPF	Body Part	Velocity (m/s)		K Energy (J)		P Energy (J)	
					B	A	B	A	B	A
FrontHead	6	1	6	Head	2.28	0.15	0.91	0.004	0.07	0.25
FrontHand	6	6	26	LlowerArm	2.33	0.07	0.40	0.001	0.21	0.28
FrontLow	8	12	37	RHip	1.16	0.38	0.08	0.009	0.20	0.20
BackLow	8	7	26	RHip	1.11	0.63	0.07	0.02	0.26	0.20
FrontTurn	13	9	19	RHip	1.38	0.30	0.11	0.005	0.01	0.05

Table 3. Maximal values of the axial and radial force impulses, and the torque impulse, for the fall sequences under analysis. In each case, the value of the three largest impulses in kg m/s and the articulation where this impulse was produced, are indicated.

	Max. Values	FrontLow	BackLow	FrontTurn	FrontHand	FrontHead
Axial force impulse	First max	66.9 (RHipYawPitch)	129.4 (RHipYawPitch)	113.6 (RHipYawPitch)	173.9 (RHipYawPitch)	216.8 (RAnkleRoll)
	Second max	63.2 (LHipYawPitch)	127.8 (LHipYawPitch)	73.0 (RHipPitch)	173.2 (LHipYawPitch)	214.9 (LAnkleRoll)
	Third max	55.3 (LHipRoll)	69.6 (HeadYaw)	69.3 (LHipYawPitch)	104.1 (LShoulderRoll)	211.0 (HeadYaw)
Radial force impulse	First max	121.8 (LHipPitch)	273.6 (RHipRoll)	222.0 (RHipRoll)	379.2 (LAnkleRoll)	484.3 (HeadPitch)
	Second max	118.5 (RKneePitch)	267.1 (LHipRoll)	194.6 (RHipPitch)	378.2 (RAnkleRoll)	414.1 (RAnklePitch)
	Third max	115.9 (LKneePitch)	234.2 (RAnkleRoll)	157.2 (LKneePitch)	374.0 (LAnklePitch)	407.0 (LAnklePitch)
Axial torque impulse	First max	5.7 (RShoulderPitch)	8.5 (LShoulderPitch)	10.7 (LHipYawPitch)	19.1 (LKneePitch)	42.5 (LHipPitch)
	Second max	5.4 (RKneePitch)	8.5 (RShoulderPitch)	8.5 (LHipPitch)	19.1 (LHipPitch)	42.5 (LKneePitch)
	Third max	5.3 (LKneePitch)	7.7 (RHipRoll)	7.6 (RKneePitch)	18.2 (RKneePitch)	41.9 (RHipPitch)

sequence, the three maximal impulse's values produced by the radial and axial forces and the external torque, and the articulation where these maxima are produced.

From the results displayed in table 2 we can conclude that, modulating a fall in order to spread the impacts into more contact points and longer periods of time effectively decrease the velocities and energies involved. In addition, a lowering of the center of mass, as in the case of the *FrontLow*, *BackLow* and *FrontTurn* falls, helps to decrease the impact velocity. In table 3 it can be observed that maximal impulses are much lower of the case of longer falls sequences, with several contact points and a lowering in the center of mass. For instance, in the case of the *FrontHead* fall, the maximal rotational impulse is between 4 and 7.5 times larger than the rotational impulse produced in the good sequences (*FrontLow*, *BackLow*, and

FrontTurn). For the same fall, the maximal linear impulses are at least, two times larger than the ones of the good sequences. A similar situation can be observed for the case of the *FrontHand* fall. In addition, given the fact that good falls sequences were inspired by the way in which humans falls, good fall sequences tend to concentrate the largest impulses, in articulation that are robust for humans: hip, knee and shoulder.

5 Conclusions

In this paper, a methodology for designing fall sequences that minimize the robot damage was presented. This methodology includes the use of a realistic simulator for the design process, and an interactive tool that allows the human designer to select the fall sequence parameters (joints values, sequences extension, etc.), and to observe indices that indicate the quality of the obtained sequence.

Simulations using a 22 DOF humanoid robot show that modeling falls after what is observed in humans greatly decreases the robot damage. Thus, longer falls sequences, with several contact points, and a lowering of the center of mass produce less damage in the robot. In addition, fall sequences that protect valuable body parts, as the head, can also be designed using the proposed methodology. As a future work, we would like to: (i) make a more extensive analysis of the fall sequences, in particular to study the incidence of the impulse forces in the articulations, (ii) to use the designed fall sequences in our humanoid robots, in real soccer games, and (iii) to advance in the automation of the fall design process by using motion capture devices.

References

1. Baltes, J., McGrath, S., Anderson, J.: The use of gyroscope feedback in the control of the walking gaits for a small humanoid robot. In: Nardi, D., Riedmiller, M., Sammut, C., Santos-Victor, J. (eds.) RoboCup 2004. LNCS, vol. 3276, pp. 628–635. Springer, Heidelberg (2005)
2. Fujiwara, K., Kanehiro, F., Kajita, S., Kaneko, K., Yokoi, K., Hirukawa, H.: UKEMI: falling motion control to minimize damage to biped humanoid robot. In: Proc. 2002 IEEE/RSJ Int. Conference on Intelligent Robots and System, vol. 3, pp. 2521–2526 (October 2002)
3. Fujiwara, K., Kanehiro, F., Saito, H., Kajita, S., Harada, K., Hirukawa, H.: Falling Motion Control of a Humanoid Robot Trained by Virtual Supplementary Tests. In: Proc. 2004 IEEE Int. Conf. on Robotics and Automation, pp. 1077–1082 (2004)
4. Höhn, O., Gacnik, J., Gerth, W.: Detection and Classification of Posture Instabilities of Bipedal Robots. In: Proc. of the 8th Int. Conf. on Climbing and Walking Robots and the Support Technologies for Mobile Machines - CLAWAR, pp. 409–416 (2005)
5. Morisawa, M., Kajita, S., Harada, K., Fujiwara, K., Fumio Kanehiro, K.K., Hirukawa, H.: Emergency Stop Algorithm for Walking Humanoid Robots. In: Proc. 2005 IEEE/RSJ Int. Conf. on Intelligent Robots and Systems, pp. 2109–2115 (2005)
6. Renner, R., Behnke, S.: Instability Detection and Fall Avoidance for a Humanoid using Attitude Sensors and Reflexes. In: Proc. 2006 IEEE/RSJ Int. Conf. on Intelligent Robots and Systems, pp. 2967–2973 (October 2006)

7. Pagilla, P.R., Yu, B.: An Experimental Study of Planar Impact of a Robot Manipulator. IEEE/ASME Trans. on Mechatronics 9(1), 123–128 (2004)
8. Kellog International Work Group on the Prevention of Falls by the Elderly. The Prevention of Falls in Later Life, Danish Medical Bulletin 34(4), 1–24 (1987)
9. Lopes, M., Santos-Victor, J.: A Developmental Roadmap for Learning by Imitation in Robots. IEEE Trans. on Systems, Man, and Cybernetics-Part B: Cybernetics 37(2), 159–168 (2007)
10. King, M.B.: Evaluating the Older Person Who Falls. In: Masdeu, J.C., Sudarsky, L., Wolfson, L. (eds.) Gait Disorders of Aging: Falls and Therapeutics Strategies. Lippincott-Raven Publishers (1997)
11. Nevitt, M.C.: Falls in the Elderly: Risk Factors and Prevention. In: Masdeu, J.C., Sudarsky, L., Wolfson, L. (eds.) Gait Disorders of Aging: Falls and Therapeutics Strategies. Lippincott-Raven Publishers (1997)
12. Rubenstein, L.Z., Josephson, K.R.: Interventions to Reduce the Multifactorial Risks for Falling. In: Masdeu, J.C., Sudarsky, L., Wolfson, L. (eds.) Gait Disorders of Aging: Falls and Therapeutics Strategies. Lippincott-Raven Publishers (1997)
13. Nikanjam, M., Kursa, K., Lehman, S., Lattanza, L., Diao, E., Rempel, E.D.: Finger Flexor Motor Control Patterns during Active Flexion: An In Vivo Tendon Force Study. Human Movement Science 26, 1–10 (2007)
14. Lussanet, M.H.E., Smeets, J.B.J., Brenner, E.: Relative Damping Improves Linear Mass-Spring Models of Goal-Directed Movements. Human Movement Science 21, 85–100 (2002)
15. Perl, J.: A Neural Network Approach to Movement Pattern Analysis. Human Movement Science 23, 605–620 (2004)
16. Sibella, F., Frosio, I., Schena, F., Borghese, N.A.: 3D Analysis of the Body Center of Mass in Rock Climbing. Human Movement Science 26, 851–852 (2007)
17. Smith, J.D., Martin, P.E.: Walking Patterns Change Rapidly Following Asymmetrical Lower Extremity Loading. Human Movement Science 26, 412–425 (2007)
18. Teulier, C., Delignies: The Nature of the Transition between Novice and Skilled Coordination during Learning to Swing. Human Movement Science 26, 376–392 (2007)
19. Huys, R., Daffertshofer, A., Beek, P.J.: Multiple Time Scales and Subsystem Embedding in The Learning of Juggling. Human Movement Science 23, 315–336 (2004)
20. Rousanoglou, E.N., Boudolod, K.D.: Rhythmic Performance during a Whole Body Movement: Dynamic Analysis of Force-Time Curves. Human Movement Science 25, 393–408 (2006)
21. Gittoes, M.J., Brewin, M.A., Kerwin, D.G.: Soft Tissue Contributions to Impact Forces Simulated Using a Four-Segment Wobbling Mass Model of Forefoot-Heel Landings. Human Movement Science 25, 775–787 (2006)
22. Katsumata, H.: A Functional Modulation for Timing a Movement: A Coordinative Structure in Baseball Hitting. Human Movement Science 26, 27–47 (2007)
23. Caljouw, S.R., van der Kamp, J.: Bi-Phasic Hitting with Constraints on Impact Velocity and temporal Precision. Human Movement Science 24, 206–217 (2005)
24. Ertan, H., Kentel, B., Tumer, S.T., Korkusuz, F.: Activation Patterns in Forearm Muscles during Archery Shooting. Human Movement Science 22, 37–45 (2003)
25. Loseby, P.N., Piek, J.P., Barret, N.C.: The Influence of Speed and Force on Bimanual Finger Tapping Patterns. Human Movement Science 20, 531–547 (2001)
26. Groen, B.E., Weerdesteyn, V., Duysens, J.: Martial Arts Fall Techniques Decrease the Impact Forces at the Hip during Sideways Falling. Journal of Biomechanics 40, 458–462 (2007)

27. Pavey, E.: Ready for Their Close-Ups: Breathing Life into Digitally Animated Faces. IEEE Spectrum Magazine, 42–47 (April 2007)
28. Abe, Y., Popovic, J.: Interactive Animation of Dynamic Manipulation. In: Proc. Eurographics/ACM SIGGRAPH Symposium on Computer Animation (2006)
29. Liu, C.K., Hertzmann, A., Popovic, Z.: Learning Physics-Based Motion Style with Nonlinear Inverse Optimization. ACM Transactions on Graphics 24(3), 1071–1081 (2005)
30. Reil, T., Husbands, P.: Evolution of Central Pattern Generators for Bipedal Walking in a Real-Time Physics Environment. IEEE Trans. on Evolutionary Computation 6(2), 159–168 (2002)
31. Hornby, G.S., Pollack, J.B.: Creating High-Level Components with a Generative Representation for Body-Brain Evolution. Artificial Life 8, 223–246 (2002)
32. Zagal, J., Ruiz-del-Solar, J.: UCHILSIM: A Dynamically and Visually Realistic Simulator for the RoboCup Four Legged League. In: Nardi, D., Riedmiller, M., Sammut, C., Santos-Victor, J. (eds.) RoboCup 2004. LNCS, vol. 3276, pp. 34–45. Springer, Heidelberg (2005)
33. Animazoo official website, http://www.animazoo.com/
34. Webots official website, http://www.cyberbotics.com/
35. URBI official wbesite, http://www.urbiforge.com/
36. Aldebaran robotics official website, http://www.aldebaran-robotics.com/
37. Fall sequences videos, http://www.robocup.cl/fallingvideos

The Use of Scripts Based on Conceptual Dependency Primitives for the Operation of Service Mobile Robots

Jesus Savage[1], Alfredo Weitzenfeld[2], Francisco Ayala[1], and Sergio Cuellar[1]

[1] Bio-Robotics Laboratory
Department of Electrical Engineering
Universidad Nacional Autonoma de Mexico, UNAM
Mexico City, Mexico
savage@servidor.unam.mx
[2] CANNES
Department of Computer Engineering
ITAM
Mexico City, Mexico
alfredo@itam.mx

Abstract. This paper describes a Human-Robot interaction subsystem that is part of a robotics architecture, the ViRbot, used to control the operation of service mobile robots. The Human/Robot Interface subsystem consists of tree modules: Natural Language Understanding, Speech Generation and Robot's Facial Expressions. To demonstrate the utility of this Human-Robot interaction subsystem it is presented a set of applications that allows a user to command a mobile robot through spoken commands. The mobile robot accomplish the required commands using an actions planner and reactive behaviors. In the ViRbot architecture the actions planner module uses Conceptual Dependency (CD) primitives as the base for representing the problem domain. After a command is spoken a CD representation of it is generated, a rule base system takes this CD representation, and using the state of the environment generates other subtasks represented by CDs to accomplish the command. In this paper is also presented how to represent context through scripts. Using scripts it is easy to make inferences about events for which there are incomplete information or are ambiguous. Scripts serve to encode common sense knowledge. Scripts are also used to fill the gaps between seemingly unrelated events.

1 Introduction

There is a need for reliable Human-Robot interaction systems as experienced by the proliferation of new humanoid robots in particular in Japan with advanced interaction capabilities including spoken language. This can also be appreciated by new competitions to exalt the robots social interaction with humans such as a new league in the robots' competition Robocup: RoboCup@Home. The goal of

L. Iocchi et al. (Eds.): RoboCup 2008, LNAI 5399, pp. 284–295, 2009.

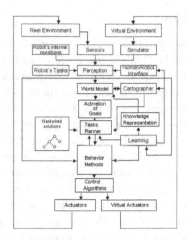

Fig. 1. The ViRbot System consists of several subsystems that control the operation of a mobile robot

this league is to promote the development of real-world applications and human-machine interaction with autonomous robots, or as they put it: "the aim is to foster the development of useful robotic applications that can assist humans in everyday life" [1]. One of the test of this competition is that a service robot helps a person to prepare a cooking recipe, to solve this task in this paper it is proposed the use of scripts that describe a set of possible of actions that a robot and persons can do under certain conditions. Also in this paper is presented the Human/Robot interface subsystem of a mobile robot architecture, the ViRbot system, whose goal is to operate autonomous robots that carry out daily service jobs in houses, offices and factories. The ViRbot system [2] divides the operation of a mobile robot in several subsystems (see figure 1). Each subsystem has a specific function that contributes to the final operation of the robot.

2   Human/Robot Interface

The Human/Robot Interface subsystem in the ViRbot architecture has tree modules: Natural Language Understanding, Speech Generation and Robot's Facial Expressions. In this paper is presented the Natural Language Understanding module.

2.1   Natural Language Understanding

The natural language understanding module finds a symbolic representation of spoken commands given to a robot. It consists of a speech recognition system coupled with Conceptual Dependency techniques [3].

2.2 Speech Recognition

For the speech recognition system it was used the Microsoft Speech SDK engine [4]. One of the advantage of this speech recognition system is that it accepts continuous speech without training, also is freely available and with the C++ source code included, it means that it can be modified as needed. It allows the use of grammars, that are specified using XML notation, which constrains the sentences that can be uttered and with that feature the number of recognition errors it is reduced. Using a grammar the transitions from one word to another is restricted, reducing the perplexity considerable. Perplexity specifies the degree of sophistication in a recognition task [5]. Almost every speech recognition system currently developed has the problem of insertion words, words incorrectly added by the speech recognition system. These worlds may cause a robot to fail to perform the asked command, then it is necessary to find a mechanism that, even if these errors exists, that a robot should be able to perform the required commands. One of the goals of this research is to find an appropriated representation of the spoken commands that can be used by an actions planner.

2.3 Conceptual Dependency

One way to represent a spoken command is by describing the relationships of objects mentioned in the input sentence [6]. During this process the main event described in the sentence and participants are found. In this work the participants are any actors and recipients of the actions. The roles the participants play in the event are determined, as are the conditions under which the event took place. The key verb in the sentence can be used to associate the structure to be filled by the event participants, objects, actions, and the relationship between them.

Dominey [7] describes the Event Perceiver System that could adaptively acquire a limited grammar that extracts the meaning of narrated events that are translated into(action,object, recipient).

Another approach is the use of Conceptual Dependency, this technique represents the meaning contained in a sentence. Conceptual Dependency is a theory developed by Schank for representing meaning. This technique finds the structure and meaning of a sentence in a single step. CDs are especially useful when there is not a strict sentence grammar.

One of the main advantages of CDs is that they allow rule base systems to be built which make inferences from a natural language system in the same way humans beings do. CDs facilitate the use of inference rules because many inferences are already contained in the representation itself. The CD representation uses conceptual primitives and not the actual words contained in the sentence. These primitives represent thoughts, actions, and the relationships between them.

Some of the more commonly used CD primitives are, as defined by Schank [8]:

ATRANS: Transfer of ownership, possession, or control of an object (e.g. give.)
PTRANS: Transfer of the physical location of an object (e.g. go.)
ATTEND: Focus a sense organ (e.g. point.)

MOVE: Movement of a body part by its owner (e.g. kick.)
GRASP: Grasping of an object by an actor (e.g. take.)
PROPEL: The application of a physical force to an object (e.g. push.)
SPEAK: Production of sounds (e.g. say.)

Each action primitive represents several verbs which have similar meaning. For instance give, buy, and take have the same representation, i.e., the transference of an object from one entity to another.

Each primitive is represented by a set of rules and a data structure containing the following categories, in which the sentence components are classified:

An Actor: The entity that performs the ACT.
An ACT: Performed by the actor, done to an object.
An Object: The entity the action is performed on.
A Direction: The location that an ACT is directed towards.

The user's spoken input is converted into a CD representation using a two step process. The CDs are formed first by finding the main verb in the spoken sentence and choosing the CD primitive associated with that verb. Once the CD primitive has been chosen the other components of the sentence are used to fill the CD structure.

For example, in the sentence "Robot, go to the kitchen", when the verb "go" is found a PTRANS structure is issued. PTRANS encodes the transfer of the physical location of an object, has the following representation:

(PTRANS (ACTOR NIL) (OBJECT NIL) (FROM NIL) (TO NIL))

The empty (NIL) slots are filled by finding relevant elements in the sentence. So the actor is the robot, the object is the robot (meaning that the robot is moving itself), and the robot will go from the living room to the kitchen (assuming the robot was initially in the living room). The final PTRANS representation is:

(PTRANS (ACTOR Robot) (OBJECT Robot) (FROM living-room) (TO kitchen))

Another example, the phrase: John gave the book to Mary, can be represented by the following CD:

(ATRANS (ACTOR John) (OBJECT book) (FROM John) (TO Mary))

The initial structure issued to represent a sentence contains NILs of the knowledge representation that need to be filled.

(ATRANS (ACTOR NIL) (OBJECT NIL) (FROM NIL) (TO NIL))

These NILs can be filled either by the previous or following sentences, or by inference using the context surrounding the sentence.

In continuous speech recognition each sentence is represented by a CD. The NILs in the sentence's knowledge representation (i.e. the CD structure) may represent some of the sentence's words that either were wrongly recognized or were not part of the recognition vocabulary. The knowledge NILs will be filled by a mechanism in which an inference engine will use rules and context to look for the needed information.

There are also primitives that represent the state in which an object is, for the previous example the book now belongs to Mary is represented by:

(POSSESS (OBJECT Mary) VALUE Book))

Conceptual dependencies can also be used with multi-modal input [9]. If the user said "Put the newspaper over there", while pointing at the table top, separate CDs will be generated for the speech and gesture input with empty slots for the unknown information (assuming the newspaper was initially on the floor):

Speech:
(PTRANS (ACTOR Robot) (OBJECT Newspaper) (FROM Floor) (TO NIL))
Gesture:
(ATTEND (ACTOR User) (OBJECT Hand) (FROM NIL) (TO Table_top))

Empty slots can be filled by examining CDs generated by other modalities at the same time, and combining then to form a single representation of the desired command:

(PTRANS (ACTOR Robot) (OBJECT Newspaper) (FROM Floor) (TO Table_top))

In this section it was explained how to transform a multi-modal input into a CD structure which can be manipulated more easily. CD structures facilitate the inference process, by reducing the large number of possible inputs into a small number of actions. The final CDs encode the users commands to the robot. To carry out these commands an actions planning module must be used as described in the following section. CDs are suitable for the representation of commands and for asking simple questions to a robot, but they are not suitable for the representation of complex sentences.

3   Planner Subsystem

The ViRbot organization consists of several hierarchical layers. After receiving the CD representation from the Human/Robot interface the Perception subsystem perceives a new situation that needs to be validated by the World Model subsystem. The World Model validates the situation by the information provided by the Cartographer and the Knowledge Representation subsystem. The Planner subsystem takes as an input the output of the World Model subsystem and tries to take care of the situation presented. Planning is defined as the process of finding a procedure or guide for accomplishing an objective or task. In the ViRbot planner subsystem there are two planning layers, the upper is the actions planner, based on a rule base system, and the lower later the movements planner, based on the Dijkstra algorithm. Action planning requires searching in a state-space of configurations to find a set of the operations that will solve the problem.

3.1   Actions Planner

The Robot is able to perform operations like grasping an object, moving itself from on place to another, etc. Then the objective of action planning is to find a sequence of physical operations to achieve the desired goal. These operations can be represented by a state-space graph.

We use the rule base system CLIPS developed by NASA [10], as an inference engine that uses forward state-space search that finds a sequence of steps that leads to an action plan. Actions planning works well when there is a detailed representation of the problem domain. In the ViRbot architecture the actions planning module uses conceptual dependency as the base for representing the problem domain. After a command is spoken, a CD representation of it is generated. The rule base system takes the CD representation, and using the state of the environment will generate other subtasks represented by CDs and micro-instructions to accomplish the command. The micro-instructions are primitive operations acting directly on the environment, such as operations for moving objects.

For example when the user says **"Robot, go to the kitchen"**, the following CD is generated:

(PTRANS (ACTOR Robot) (OBJECT Robot) (FROM Robot's-place) (TO Kitchen))

It is important to notice is that the user could say more words in the sentence, like **"Please Robot, go to the kitchen now, as fast as you can"** and the CD representation would be the same. That is, there is a transformation of several possible sentences to a one representation that is more suitable to be used by an actions planner.

All the information required for the actions planner to perform its operation is contained in the CD. The planner just needs to find the best global path between the Robot's place and the Kitchen, see figure 2, thus the rule base system issue the following command to the movements planner:

(MOVEMENTS-PLANNER get-best global-path Robot's-place to Kitchen)
And the answer of the movements planner is the following:

(best-global-path $Robot's - place\ place_1\ place_2...place_n\ Kitchen$)

Now a new set of PTRANS are generated asking the robot to move to each of the places issued by the planner:

(PTRANS (ACTOR Robot) (OBJECT Robot) (FROM Robot's-place) (TO $place_1$))

(PTRANS (ACTOR Robot) (OBJECT Robot) (FROM $place_1$) (TO $place_2$))

.

.

.

(PTRANS (ACTOR Robot) (OBJECT Robot) (FROM $place_i$) (TO $place_j$))
and finally
(PTRANS (ACTOR Robot) (OBJECT Robot) (FROM $place_n$) (TO Kitchen))

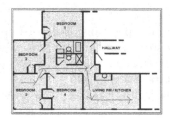

Fig. 2. The movements planner finds a global path from one of the bedrooms to the kitchen

For a more complex example, such as the user saying **"Robot, give the newspaper to the Father."**

First, the following CD representation is generated:

(ATRANS (ACTOR Robot) (OBJECT newspaper) (TO Father) (FROM newspaper's-place))

Then the actions planner finds a sequence of actions to perform the requested order, that is, the Robot needs to go for the object, to pick it up, and to deliver it to the place in which the Father is. These actions are represented by the following CDs:

(PTRANS (ACTOR Robot) (OBJECT Robot) (TO newspaper's-place) (FROM Robot's-place))

(GRASP (ACTOR Robot) (OBJECT newspaper) (TO Robot's-hand) (FROM newspaper's-place))

(PTRANS (ACTOR Robot) (OBJECT Robot) (TO father's-place) (FROM newspaper's-place))

After these CDs are issued some rules of the rule base system are fired, issuing new actions, until the Robot reaches the place where the object is supposed to be. When the Robot finds the newspaper, and is in the same room of the recipient, it will carry this object to the Father. Then the new position of the object and who has it are updated in the rule base system.

4   Context Recognition

In the previous sections it was shown how CDs can be used to recognize multimodal commands and aid the planning necessary to carry the commands out. The final level of the rule base system recognizes sets of events, or context. Context plays an important role for understanding the information provided in a conversation, so it is used to increase the speech recognition accuracy and the robot performance.

The rule base system receives the best M phrases recognized from the speech recognition module. Starting with the first phrase the rule base system will try to interpret it according to the present context.

Occasionally, the speech system gives recognition errors, so if the top speech output does not make any sense given a particular context, then the rule base system will start looking at the second best phrase, and so on until it finds the phrase that make most sense as in the follows example:

if the user says **"Robot, bring the milk"**, and the speech recognition module returned the following best sentences:

1. Robot, bring the mail
2. Robot, bring the milk
3. Robot, bring the ink

The speech recognition module made an incorrect recognition by failing to distinguish milk from mail. The rule base system can correct this by looking at the context surrounding the sentences. If a user was in the kitchen making breakfast, then from this context is more probable that the second sentence is the correct one; that is, the user is asking for the milk and not for the mail. If this process did not work for all M phrases then the rule base system may ask the user to repeat the phrase or to provide more information.

4.1   Context Representation Using Scripts

One way to represent the context is through scripts [11]. The term script was taken from the theater community and it describes all possible sequences of events that an actor or entity may perform under certain conditions, they include the actions place. Using scripts it is easy to make inferences about events for which there are incomplete information or are ambiguous. Scripts serve to encode common sense knowledge. Scripts are also used to fill the gaps between seemingly unrelated events. Every script has different slots associated with it that are filled with actors, objects and directions when it is instantiated.

The scripts in the ViRbot rule base system are written using CD representations. As it was mentioned before, the spoken sentences are represented using CDs, and these CDs correspond with some of the script's CDs.

The first problem that appears is how to choose the appropriate script given the present conditions. The script is activated when the user says a sentence related to it, or when he performs certain actions related to the script. In either case they are CDs that match the CDs of the first steps of the script to be activated.

For example when the user says the sentence **"I will make breakfast"**, or if he is in the kitchen and he starts picking objects related to making food, then the script for making breakfast is triggered.

Here is the making breakfast script, each of the actions may not occur in the established order, and some actions may not occur at all.

* The actor enters the kitchen:
(PTRANS (ACTOR ?actor) (OBJECT ?actor) (FROM ?actor's-place) (TO kitchen))

The notation ? after a name it means a variable name that will used in all the script.

* The actor ask the robot to come to the kitchen:
(SPEAK (ACTOR ?actor) (OBJECT "Robot come to the kitchen") (TO ιυυυι))

* The robot enters the kitchen:
(PTRANS (ACTOR robot) (OBJECT robot) (TO kitchen) (FROM robot's-place))

First the actor needs to obtain all the food and tools elements to prepared the food. He will pick or ask them to the robot.

* The actor goes to the the the cupboard or fridge:
(PTRANS (ACTOR ?actor) (OBJECT ?actor) (TO cupboard) (FROM ? actors-place-kitchen))
or
(PTRANS (ACTOR ?actor) (OBJECT ?actor) (TO fridge) (FROM ?actors-place-kitchen))

* The actor opens the fridge or cupboard. This action is represented by two primitives, first the actor moves its hand to the cupboard's or fridge's handle, then it applies force to it opening the door:
(MOVE (ACTOR ?actor) (OBJECT ?actor-hands) (TO cupboard's-handle))
or
(MOVE (ACTOR ?actor) (OBJECT ?actor-hands) (TO fridge's-handle))
(PROPEL (ACTOR ?actor)(OBJECT ?cupboard's-handle) (FROM ?actor-hands))
or
(PROPEL (ACTOR ?actor) (OBJECT ?fridge's-handle) (FROM ?actor-hands))

* The actor picks some of the food or tools elements:
(PTRANS (ACTOR ?actor) (OBJECT ?object) (TO ?actor's-hands) (FROM ?object-place))

The objects that are more probable to be used are of the food type and kitchen tools: eggs, onions, milk, pans, glasses, cups, knifes, etc. Some of the food needs to be transfered from one container to another, for example the milk can be transfered from the milk bottle to a glass:

(PTRANS (ACTOR ?actor) (OBJECT ?obj) (FROM ?container-obj) (TO glass))

* The actor asks for the food or tool elements to the robot:
(SPEAK (ACTOR ?actor) (OBJECT "Robot bring the ?object") (TO robot))
This instruction is represented as follows:
(ATRANS (ACTOR robot) (OBJECT ?object) (FROM ?robot) (TO ?actor's-hands))

* After the actor has all the elements to prepare food he will start making it. The representation of this actions requires to many primitives. It is assumed that he finish to prepare food when he says "The food is ready":
    (SPEAK (ACTOR ?actor) (OBJECT "The food is ready") (TO robot)),
    or when he starts taking the food to the dining room table:
    (PTRANS (ACTOR ?actor) (OBJECT food) (FROM kitchen) (TO dining-room-table)),
    or when he asks the robot to do that:
    (SPEAK (ACTOR ?actor) (OBJECT "Bring the food to the table") (TO robot))
    (PTRANS (ACTOR robot) (OBJECT food) (FROM kitchen) (TO dining-room-table))

After this point the script for having breakfast is selected. When one variable of the script is found then it is instantiated whenever it appears in them. In the previous example, if the Mother was the actor preparing the food then whenever the slot ?actor appears is substituted by Mother. In this way is possible that some CD's will be complete even if direct actions did not occur or were not observed.

Using scripts helps to answer questions about information that was not observed when the actions happened. For instance, if the user asks for milk to drink it, he will need to put the milk in a glass, and even if this action was not observed the script for making food specify that he did that. Then, it can be asked if the user used a glass for the milk, and the answer should be yes.

Using scripts helps to make the speech recognition more reliable, because it rejects words that does not fit into the currently active script. In the example of the previous section in which the user says **"Robot, bring the milk"**, the speech recognition module found the following sentences:

1. Robot, bring the mail
2. Robot, bring the milk
3. Robot, bring the ink

In the script of making food the second sentence is more probable than the first one, because in this script the objects that are more likely are the ones related to food.

5   Experiments and Results

The Human-Robot interaction subsystem was tested under two situations. In the first situation it was tested how well the robot performed for sentences of the type: "Robot, go with the Father", "Robot, go to the kitchen", "Robot, give the newspaper to the Mother", "Robot, where is the newspaper?", etc. The environment represented a house, set in our laboratory, based on the layout proposed by the Robocup@Home competition[1]. In this environment the robot TPR was used, see figure 3.

Fig. 3. Test Robot TPR8

Fig. 4. In this virtual kitchen, the user's actions can be detected

In the second situation it was used a virtual reality mobile robot simulator that recreated a virtual house with a kitchen included. In this virtual environment, see figure 4, it was tested the script of making breakfast. This simulator was able to give the position and the orientation of the user's head and hand. With this information it could be detected which object the user was pointing at or grabbing and thus the user's actions could be related to the CDs of the making breakfast script.

The speech recognition system alone, under noise conditions, found correctly the following categories, in which the sentence components are classified in the spoken sentences, with the following frequencies: Actors with 94%, Objects with 90%, Recipients 73%, Verbs with 83%, Questions with 86%, and words that were not part of the sentence (insertion words) 70%. The insertion words, words incorrectly added by the speech recognition system, may cause the robot to fail to perform the asked command, which means that the robot could perform correctly, the commands only 30% of the time. Combining all the information provided by the Microsoft Speech Recognizer with Conceptual Dependency techniques, the system was able to overcome this problem and it found a representation that it was executed correctly by the robot 75% of the time.

6   Conclusions and Discussion

Conceptual Dependency representation helped to increase the recognition rate and also helped the actions planner to perform the desired user's command. The use of Conceptual Dependency to represent spoken command given to a robot helped to the actions planner in the ViRbot system to find a set of steps to accomplish the required commands. Conceptual Dependency is useful to represent basic commands and simple dialog with a robot, but is not useful to represent more complex spoken commands. The most significant contribution in this research is the successful combination of a commercially available speech recognition system and AI techniques together to enhance the speech recognition performance. At this time, the system has the following features: speaker independent, medium size vocabulary, loose grammar and context dependent. For future research we will have more scripts of common daily activities. The ViRbot system was tested in the Robocup@Home [1] category in the RoboCup competition at Bremen, Germany in 2006 and in Atlanta in 2007, where our robot TPR8, obtained the third place in this category. The scripts module will be used in the test in which a service robot helps a person to prepare a cooking recipe in the Robocup@Home competition, at the RoboCup 2008 in China.

References

1. Robocup@Home (2006), http://www.ai.rug.nl/robocupathome/
2. Savage, J., Billinhurst, M., Holden, A.: The ViRbot: a virtual reality robot driven with multimodal commands. In: Expert Systems with Applications, vol. 15, pp. 413–419. Pergamon Press, Oxford (1998)
3. Schank, R.C.: Conceptual Information Processing. North-Holland Publishing Company, Amsterdam (1975)
4. Microsoft Speech SDK (2006), http://www.microsoft.com/speech/
5. Rabiner, L., Biing-Hwang: Fundamentals of Speech Recognition. Prentice Hall, Englewood Cliffs (1993)
6. Savage, J.: A Hybrid System with Symbolic AI and Statistical Methods for Speech Recognition. PhD Dissertation University of Washington (August 1995)
7. Dominey, P.F., Weitzenfeld, A.: A Robot Command, Interrogation and Teaching via Social Interaction. In: IEEE-RAS International Conference on Humanoid Robots, December 6-7, Tsukuba, Japan (2005)
8. Lytinen Steven, L.: Conceptual Dependency and Its Descendants. Computer Math. Applic. 23(2-5), 51–73 (1992)
9. Cohen, P.R.: The Role of Natural Language in a Multimodal Interface. In: Proceedings UIST 1994, pp. 143–149. ACM Press, New York (1992)
10. CLIPS Reference Manual Version 6.0. Technical Report Number JSC-25012. Software Technology Branch, Lyndon B. Johnson Space Center, Houston, TX (1994)
11. Schank, R.C., Leake, D.: Computer Understanding and Creativity. In: Kugler, H.-J. (ed.) Information Processing 1986, pp. 335–341. North-Holland, New York (1986)

An Omnidirectional Camera Simulation for the USARSim World

Tijn Schmits and Arnoud Visser

Universiteit van Amsterdam, 1098 SJ Amsterdam, The Netherlands
{tschmits,arnoud}@science.uva.nl

Abstract. Omnidirectional vision is currently an important sensor in robotic research. The catadioptric omnidirectional camera with a hyperbolic convex mirror is a common omnidirectional vision system in the robotics research field as it has many advantages over other vision systems. This paper describes the development and validation of such a system for the RoboCup Rescue League simulator USARSim.

After an introduction of the mathematical properties of a real catadioptric omnidirectional camera we give a general overview of the simulation method. We then compare different 3D mirror meshes with respect to quality and system performance. Simulation data also is compared to real omnidirectional vision data obtained on an 4-Legged League soccer field. Comparison is based on using color histogram landmark detection and robot self-localization based on an Extended Kalman filter.

Keywords: RoboCup, USARSim, Omnidirectional Vision, Simulation, Catadioptric Omnidirectional Camera, Landmark Detection, Kalman Filter.

1 Introduction

Agents operating in a complex physical environment more often than not benefit from visual data obtained from their surroundings. The possibilities for obtaining visual information are numerous as one can vary between imaging devices, lenses and accessories. Omnidirectional vision, providing a 360° view of the sensor's surroundings, is currently popular in the robotics research area and is currently an important sensor in the RoboCup.

Omnidirectional views can be obtained using multiple cameras, a single rotating camera, a fish-eye lens and or a convex mirror. A catadioptric vision system consisting of a conventional camera in front of a convex mirror with the center of the mirror aligned with the optical axis of the camera, is the most generally applied technique for omnidirectional vision. Mirrors which are conic, spherical, parabolic or hyperbolic all are able to provide omnidirectional images [1].

An omnidirectional catadioptric camera has some great advantages over conventional cameras, one of them being the fact that visual landmarks remain in the field of view much longer than with a conventional camera. Also does the imaging geometry have various properties that can be exploited for navigation

L. Iocchi et al. (Eds.): RoboCup 2008, LNAI 5399, pp. 296–307, 2009.

or recognition tasks, improving speed and efficiency of the tasks performed by the robot. The main disadvantage of omnidirectional cameras over conventional ones is the loss of resolution in comparison with standard images [2].

Omnidirectional vision has played a major part in past and present research. At the University of Amsterdam, the Intelligent Autonomous Systems Group uses omnidirectional vision for Trajectory SLAM and for appearance based self-localization using a probabilistic framework [3,4]. Related to the RoboCup, Heinemann et al. use a novel approach to Monte-Carlo localization based on images from an omnidirectional camera [5]. Lima et al. have used a multi-part omnidirectional catadioptric system to develop their own specific mirror shape which they use for soccer robot self-localization [6].

At the RoboCup Rescue Simulation League one of the simulators which is used and developed is USARSim [7], the 3-D simulation environment of Urban Search And Rescue (USAR) robots and environments, built on top of the Unreal Tournament game and intended as a research tool for the study of human-robot interaction and multi-robot coordination [8]. When we started this project, USARSim did not offer a simulation model of an omnidirectional vision sensor. The Virtual Robots Competition, using USARSim as the simulation platform, aims to be the meeting point between researchers involved in the Agents Competition and those active in the RoboCup Rescue League. As the omnidirectional catadioptric camera is an important sensor in the robotics field, we decided to develop this camera for the USARSim environment. In this paper we will describe the elements which simulate a catadioptric omnidirectional camera in USARSim.

2 Method

This section will describe the model of the real omnidirectional vision system on which our simulation model is based and it will explain the rationale behind our simulation method. Figure 1 is an image of the real omnidirectional vision system next to a simulated version of that camera.

Fig. 1. The catadioptric omnidirectional camera, real and simulated

2.1   The Real Camera

The virtual omnidirectional camera model is based on the catadioptric omni-directional vision system shown in Figure 1. It is employed by the Intelligent Systems Lab Amsterdam (ISLA) and consists of a Dragonfly® 2 camera made by Point Grey Research Inc. and a Large Type Panorama Eye® made by Accowle Company, Ltd. The Dragonfly® 2 camera is an OEM style board level camera designed for imaging product development. It offers double the standard frame rate, auto-iris lens control and on-board color processing[1]. It captures omnidirectional image data in the reflection of the Panoramic Eye® hyperbolic convex mirror.

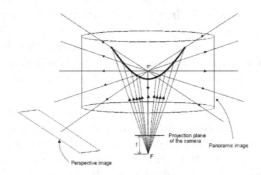

Fig. 2. A schematic of the Single Effective Viewpoint

The hyperbolic surface is designed to satisfy the Single Viewpoint Constraint (SVC). For a catadioptric system to satisfy the SVC, all irradiance measurements must pass through a single point in space, called the effective viewpoint. This is desirable as it allows the construction of geometrically correct perspective images which can be processed by the enormous amount of vision techniques that assume perspective projection: as the geometry of the catadioptric system is fixed, the directions of rays of light measured by the camera are known a priori. This allows reconstruction of planar and cylindrical perspective images, as is depicted in Figure 2 [1,9,10]. Baker and Nayar constructed surface equation which comprises a general solution of the single viewpoint constraint equation for hyperbolic mirrors [1]:

$$\left(z - \frac{c}{2}\right)^2 - r^2 \left(\frac{k}{2} - 1\right) = \frac{c^2}{4}\left(\frac{k-2}{k}\right)(k \geq 2) \qquad (1)$$

with c denoting the distance between the camera pinhole and the effective viewpoint, r is defined by $r = \sqrt{x^2 + y^2}$ and k is a constant.

---

[1] See http://www.ptgrey.com/products/dragonfly2/ Last accessed: December 20th, 2007.

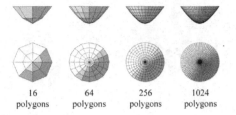

Fig. 3. Mirror Surfaces with increasing polygon counts

2.2   The Virtual Camera Mesh

To recreate this catadioptric camera in simulation first the hyperbolic convex has to build with small polygons as building blocks. The 3D mesh was modeled to scale in Lightwave® and imported into the Unreal Editor™. To investigate the influence of the mirror polygon count, four mirror meshes were created. The mirror surfaces are defined by Equation 1 with $k = 11.546$ and $c = 2.321$, and have 2^4, 2^6, 2^8 and 2^{10} polygons respectively (Figure 3).

2.3   Cube Mapping in Unreal

A mirror should reflect its surroundings. Unfortunately, the Unreal Engine is only capable of rendering a limited number of planar reflective surfaces. Instead of using many reflecting polygons, a limited number of virtual cameras are placed at the effective viewpoint. Multiple virtual cameras are needed to see the surroundings (omnidirectional). This idea is equivalent to the approach of Beck et al., who developed a catadioptric camera meeting the SVC for the Gazebo simulator. They mapped images of the environment to the faces of a cube and then applied this texture to the surface of a three-dimensional object, i.e. a mesh object resembling the surface of a hyperbolic mirror [11]. In the USARSim environment a similar approach is followed based on CameraTextureClients.

Virtual Camera Placement. CameraTextureClients are Unreal Tournament objects which project virtual camera views on textures. Originally designed to create security camera monitor screens in the Unreal Tournament game, they can be used to obtain images of the environment which then can be mapped on the surface of a hyperbolic mirror. If done correctly, the mapping creates a perspective distortion typically for catadioptric cameras. To create the effect of a proper cube mapping, five virtual cameras with a 90° field of view (FOV) need to be placed on the Effective Viewpoint in 90° angles of each other as depicted in Figure 4. A 90° FOV for all the cameras will result in the projection planes of the cameras touching each other at the edges, which means that all viewing angles are covered by exactly one camera. For this particular application five cameras are needed instead of six, as a camera looking at the top side of the cube will register image data which should not be reflected by the virtual mirror. It can therefore be omitted to save computational requirements. Figure 5 shows

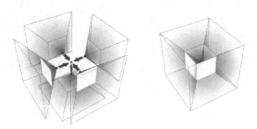

Fig. 4. Placement of the 5 virtual cameras

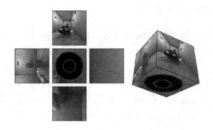

Fig. 5. 5 Virtual Camera views and a cube mapping

five images produced by virtual cameras placed in this position and how they relate to the mapping cube.

Mapping the Camera Textures. For a real catadioptric omnidirectional camera satisfying the SVC using a hyperbolic mirror, the relation between the effective single viewpoint pitch incidence angle θ and the radius of the correlating circle in the omnidirectional image r_i is known to be defined by the following equation

$$r_i = sin(\theta)\frac{h}{(1 + cos(\theta))} \qquad (2)$$

where h is the radius of the 90° circle [12] and, as the shape of the mirror defines the location of the 90° circle, h is directly related to c and k in Equation 1.

The proper UV-mapping of the simulated mirror surface produces omnidirectional image data displaying the same relation between θ and r_i as defined by Equation 2. A point projected UV-mapping of the mapping cube from the Effective Viewpoint to the 3D hyperbolic mirror surface, depicted in Figure 6, has been verified to produce data which concurs with Equation 2. A simulation rendering of the relation between θ and r_i is depicted in Figure 7.

2.4 Simulation Architecture

To set up the simulation of the catadioptric camera, three groups of elements need to be added to a robot configuration in the USARBot.ini file.

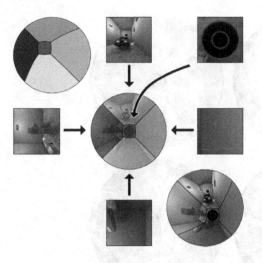

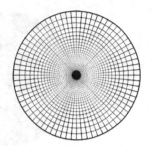

Fig. 7. A simulation mirror rendering of the relation between incidence angles and image pixel locations, correlating to Equation 2. The red line depicts the 90° incidence angle and each line depicts a five degree difference.

Fig. 6. UV Mapping the Virtual Camera Views

First, the static mesh needs to be added. When this is spawned with the robot, it provides a surface on which a mirror reflection is projected. The Static Mesh placement coordinates relate to the SVC center of projection.

Second, the CameraTextureClients and virtual cameras need to be added which create the projections on the dynamic textures, mapped on the mirror surface, effectively simulating the reflection. The virtual cameras need to be placed on the SVC center of projection in a manner depicted in Figure 4.

Finally, a camera needs to be added with a fixed relative position and orientation to capture the reflection and provide actual image data in an Unreal Client.

3 Experimental Results

The performance of the simulation method is demonstrated from two perspectives. First, sensor data quality is compared to the influence of the mirror polygon count on system performance. Second, self localization on a RoboCup 2006 four-legged league soccer field is performed within the simulated environment, using color histogram landmark detection and an extended Kalman filter.

3.1 Mirror Polygon Size Reflection Influence

As the UV-texture mapping method linearly interpolates texture placement on a single polygon, improper distortions occur when mirror polygons are too big. Figure 8 shows omnidirectional images obtained from mirror surfaces described in subsection 2.2. The first two reflection images, produced by the mirrors based on the 16 and 64 polygon surfaces respectively, show heavy distortion of the reflection image due to the linear interpolation. The data provided by these surfaces does not relate to a proper reflection simulation. The 256 polygon surface does provide a proper reflection image, though Figure 9 shows a detail of the two

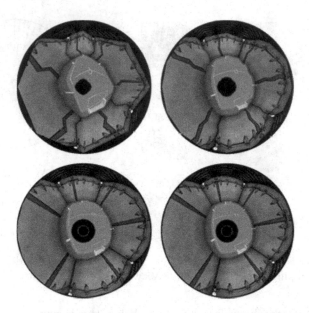

Fig. 8. 360 × 360 Pixel mirror surface view comparison. Top left: 16 polygons. Top right: 64 polygons. Bottom left: 256 polygons. Bottom right: 1024 polygons.

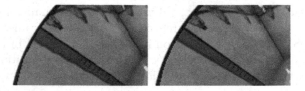

Fig. 9. Mirror Surface Comparison Detail. Left: 256 polygons. Right: 1024 polygons.

surfaces with 256 and 1024 polygons respectively, which depicts a slight quality difference: The 256 polygon mirror still shows minor distortion, jagging of a straight line, which in the 1024 polygon image mirror does not occur. Augmenting the number of polygons to 4096 resulted in images of a quality identical to those produced by the 1024 polygon surface, making the 1024 polygon surface the most proper surface for this simulation method.

Figure 10 shows the influence of mirror surface polygon count on system performance. All four mirror surfaces were mounted on a P2AT robot model and spawned in two RoboCup maps:

- DM-Mapping_250.utx and
- DM-spqrSoccer2006_250.utx.

The test was run on a dual core 1.73 GHz processor 1014 MB RAM system with a 224 MB Intel GMA 950 video processor. It is clear that FPS is influenced by the presence of an omnidirectional camera, though this influence does not

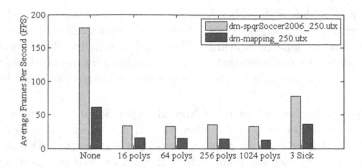

Fig. 10. Mirror surface polygon count related to USARSim performance

depend on the number of polygons. The influence is of the omnidirectional camera is comparable with another often used sensor, the LMS200 laser scanner.

3.2 Real Data Comparison

To compare the simulated omnidirectional camera to the real omnidirectional system on which it was based, we decided to do color histogram landmark detection on cylindrical projections of the omnidirectional camera data and perform self-localization using an Extended Kalman Filter in these two environments:

- on a RoboCup 2006 Four-Legged League soccer field of the Lab of Intelligent Autonomous Systems Group, Faculty of Science, University of Amsterdam, using the omnidirectional camera described in Subsection 2.1;
- in the USARSim DM-spqrSoccer2006_250.utx map [13], using the developed omnidirectional camera simulation.

Four 3-dimensional RGB color histograms were created, one for each landmark color (i.e. *cyan*, *magenta* and *yellow*) and one for all non-landmark environment colors (*neg*). These four histograms H_{cyan}, $H_{magenta}$, H_{yellow}, and H_{neg} were used to define three pixel classifiers based on the standard likelihood ratio approach [14], labeling a particular $rgb \in RGB$ value a specific landmark color if

$$\frac{P(rgb|landmark\ color)}{P(rgb|neg)} \geq \Theta \tag{3}$$

where Θ defines a threshold value which optimizes the balance between the costs of false negatives and false positives.

The Extended Kalman Filter (EKF), a recursive filter which estimates a Markov process based on a Gaussian noisy model, is based on formulas provided in [15]. The robot position μ_t is estimated on the control input u_t and landmark measurement m_t, as defined by Equations 4 to 6 respectively.

$$\mu_t = (x_t, y_t, \phi_t) \tag{4}$$

$$u_t = (v_t, \delta\phi_t) \tag{5}$$

$$m_t = (r_t, \gamma_t) \tag{6}$$

where x_t and y_t define the robot location in world coordinates, ϕ_t defines the robot direction, v_t defines the robot velocity, $\delta\phi_t$ defines robot rotation angle, γ_t defines the landmark perception angle. User input u is defined by noisy odometry sensor data and the measured distance to a landmark, r_t, is calculated based on the inverse of Equation 2, $\theta \leftarrow f(r_i)$, and camera effective viewpoint height.

Real Setting. In the real setting a Nomad Super Scout II, mounted with the omnidirectional vision system described in 2.1, was placed at a corner of the soccer field and it was driven on a clockwise path along the edges of the field until it reached its starting point, where it then crossed the field diagonally. The real omnidirectional camera suffers from many image degradation artifacts (defocus and motion blur) not yet modeled in simulation. The most important factors to reliable detect landmarks based on color histograms appear to be the constant lighting conditions and a clean laboratory. Under realistic circumstances detection algorithm also detects a large amount of false positives. For a fair comparison with the clean and constant simulation world, landmark locations in the recorded omnidirectional images were determined manually. Figure 11, left, shows omnidirectional image data obtained during this run.

Figure 12 and 13 on the left show results obtained in this run. The robot starts in the lower left of Figure 12 and moves up. The EKF estimate is relies on the user input error (odometry) and mostly ignores the observations of the landmark ahead. This is the result of propagation of the high certainty of the initial robot location formulated by an a priori estimate error covariance matrix containing solely zero valued entries combined with a strong deviation of the user input, which leads to sensor readings to be discarded by the validation gate [16]. But as uncertainty about robot location increases, in the 23rd timestep measurements start to pass the validation gate, which leads to adjustments of the gain matrix K. The difference between the predicted and actual measurements of landmarks becomes now important, and the estimation error significantly lowers as can be seen in Figure 13, left. When the robot returns to the lower left corner, estimation error increases again as user input deviation to the left without enough landmark observations to correct this error. The estimation error drops again after timestep 80 as the robot rotates to face the center of the soccer field.

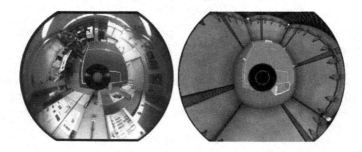

Fig. 11. Omnidirectional Image Data, Real and Simulated

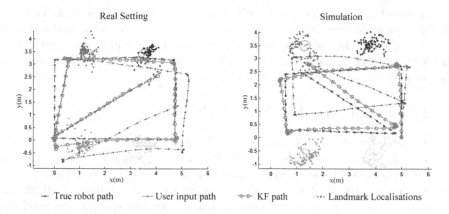

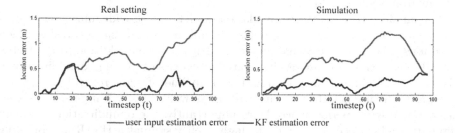

Fig. 12. 2006 Four Legged League Soccer Field Self Localization Results in a real and simulated environment

Fig. 13. Absolute differences between true and predicted robot locations in meters

Simulation. In the simulated environment, a P2AT robot model mounted with the 1024 polygon mirror omnidirectional camera was spawned in an opposite where the real robot started its path and it followed a similar path on the simulated Four Legged League soccer field, though counter-clockwise. Landmark locations were determined using the color histogram classifier, and the EKF algorithm was run using the same parameters as with the real run. Figure 11, right, shows omnidirectional image data obtained during this run.

In Figure 12 on the right, the run in the USARSim environment, several differences can be observed. First of all, landmark measurements are less accurate, due to the fact that these were produced by the automatic color histogram based landmark location method. Secondly, the initial pose estimate is better as landmark measurements pass the validation gate instantly. This can be seen because the KF estimation and the user input differ. In the first five steps KF estimation error is higher than with the user input, though as user input error increases the KF estimation error remains well below 0.5 meters over the course of the whole run. The most striking difference is the diagonal path at timesteps 84 to 100 in which the KF estimate describes a parallel path to the true path with an average location error of only 0.349 meters.

A general similarity between both runs can be observed as well. The EKF location estimate error is accurate within 0.6 meters for both the real and simulation run, while the location error based on user input which rises well above 1 meter in at several timesteps. The simulation results are not perfect, but show the same variations as can be obtained in a real setting.

4   Discussion and Further Work

Based on the theory described in Section 2 and regarding the results in the previous section we can conclude that USARSim can now simulate a Catadioptric Omnidirectional Camera which meets the Single Viewpoint Constraint. Like any other camera in the USARSim environment, it does not simulate all effects that can degrade image quality. The result is that a color histogram based pixel classifier can be used in the simulated environment to accurately locate landmarks where this classifier fails to do so in a real setting. A valid simulation model should implement these degradation models and USARSim as a whole could benefit from such an addition.

The omnidirectional camera simulation model does provide image data which has been verified to concur with realistic omnidirectional image transformation equations. The omnidirectional camera is also a valuable addition to the RoboCup Rescue League as test results show the benefits of utilizing omnidirectional vision. As it is known that landmarks are in line of sight regardless of robot orientation, it implies that victims will be as well. This will improve robot efficiency and team performance. More importantly, the omnidirectional camera simulation adds to the realism of the high fidelity simulator USARSim, bringing the Simulation League one step closer to the Rescue League.

Further work could also be done to create omnidirectional vision systems based on other real-life cameras used in the RoboCup. Not only catadioptric systems with different mirror shapes, but also omnidirectional cameras based on for instance fish-eye lenses, could be explored.

Acknowledgments

The authors would like to thank Bas Terwijn for helping out with providing the real omnidirectional image data at the Intelligent Systems Laboratory Amsterdam. A part of the research reported here is performed in the context of the Interactive Collaborative Information Systems (ICIS) project, supported by the Dutch Ministry of Economic Affairs, grant nr: BSIK03024.

References

1. Baker, S., Nayar, S.K.: A theory of single-viewpoint catadioptric image formation. International Journal of Computer Vision 35, 1–22 (1999)
2. Gaspar, J., Winters, N., Santos-Victor, J.: Vision-based navigation and environmental representations with an omnidirectional camera. IEEE Transactions on Robotics and Automation 16, 890–898 (2000)

3. Bunschoten, R., Kröse, B.: Robust scene reconstruction from an omnidirectional vision system. IEEETransactions on Robotics and Automation 19, 351–357 (2003)
4. Booij, O., Terwijn, B., Zivkovic, Z., Kröse, B.: Navigation using an appearance based topological map. In: Proceedings of the International Conference on Robotics and Automation ICRA 2007, Roma, Italy, pp. 3927–3932 (2007)
5. Heinemann, P., Haass, J., Zell, A.: A combined monte-carlo localization and tracking algorithm for robocup. In: Proceedings of the 2006 IEEE/RSJ International Conference on Intelligent Robots and Systems (IROS 2006), pp. 1535–1540 (2006)
6. Lima, P., Bonarini, A., Machado, C., Marchese, F.M., Marques, C., Ribeiro, F., Sorrenti, D.G.: Omnidirectional catadioptric vision for soccer robots. Journal of Robotics and Autonomous Systems 36, 87–102 (2001)
7. Carpin, S., Lewis, M., Wang, J., Balakirsky, S., Scrapper, C.: Bridging the Gap Between Simulation and Reality in Urban Search and Rescue. In: Lakemeyer, G., Sklar, E., Sorrenti, D.G., Takahashi, T. (eds.) RoboCup 2006: Robot Soccer World Cup X. LNCS, vol. 4434, pp. 1–12. Springer, Heidelberg (2007)
8. Jacoff, A., Messina, E., Evans, J.: A standard test course for urban search and rescue robots. In: Proceedings of the Performance Metrics for Intelligent Systems Workshop, pp. 253–259 (2000)
9. Geyer, C., Daniilidis, K.: Catadioptric projective geometry. International Journal of Computer Vision 45, 223–243 (2001)
10. Grassi Jr., V., Okamoto Jr., J.: Development of an omnidirectional vision system. Journal of the Brazilian Society of Mechanical Sciences and Engineering 28, 58–68 (2006)
11. Beck, D., Ferrein, A., Lakemeyer, G.: A simulation environment for middle-size robots with multi-level abstraction. In: Visser, U., Ribeiro, F., Ohashi, T., Dellaert, F. (eds.) RoboCup 2007: Robot Soccer World Cup XI. LNCS (LNAI), vol. 5001, pp. 136–147. Springer, Heidelberg (2008)
12. Nayar, S.K.: Catadioptric Omnidirectional Camera. In: IEEE Conference on Computer Vision and Pattern Recognition (CVPR), pp. 482–488 (1997)
13. Zaratti, M., Fratarcangeli, M., Iocchi, L.: A 3D Simulator of Multiple Legged Robots Based on USARSim. In: Lakemeyer, G., Sklar, E., Sorrenti, D.G., Takahashi, T. (eds.) RoboCup 2006: Robot Soccer World Cup X. LNCS, vol. 4434, pp. 13–24. Springer, Heidelberg (2007)
14. Fukunaga, K.: Introduction to statistical pattern recognition, 2nd edn. Academic Press Professional, Inc., San Diego (1990)
15. Thrun, S., Burgard, W., Fox, D.: Probabilistic Robotics. MIT Press, Inc., Cambridge (2005)
16. Kristensen, S., Jensfelt, P.: An Experimental Comparison of Localisation Methods, the MHL Sessions. In: Proc. of the IEEE/RSJ International Conference on Intelligent Robots and Systems (IROS 2003), pp. 992–997 (2003)

Introducing Image Processing to RoboCupJunior

PALB VISION – A First Implementation of Live Image Processing in RCJ Soccer

Christoph Siedentop, Max Schwarz, and Sebastian Pfülb

RoboCupJunior Team C-PALB
CJD-Christophorusschule, Königswinter, Germany*
christophsiedentop@gmail.com,
Max@x-quadraht.de,
mail@sebastianpfuelb.de
http://www.c-palb.de

Abstract. For a long time, teams in the RoboCup Junior competition have relied on the same basic sensorical appliances. It is time to evolve. We believe that, the introduction of image processing to RoboCupJunior will take Soccer to a new level of intelligent and less aggressive gameplay. We found it especially challenging to design a system that could not only successfully detect the goal, but also any obstructions, i.e. robots from the opponent's team, in order to score goals more precisely. This work aims to prove that the implementation of image processing does not need to be as much work as one might assume. Our project PALB VISION is designed to serve as an example for other teams in RCJ who plan to use camera-based detection software on their future robots.

We use a CMUcam3 with onboard processing connected to an ATmega 2560. The code for visual detection is written in C and runs directly on the CMUcam. The camera only takes into account the small area between the upper and lower edge of the wall and applies three simple filters for each pixel on a horizontal line.

We have created a quick and reliable vision system which processes ten frames per second and is very resistant to changing illumination. By implementing a special calibration mode, the pre-game setup is reduced to less than one minute. PALB VISION has turned out to be an improvement to the game of RoboCupJunior Soccer and may provide a robust framework for other teams who wish to adopt live image processing into their strategy.

---

* Special thanks go to Dr Winfried Schmitz a great mentor whose continous commitment to all different kinds of Robotics activities are outstanding and unmatched. We are also greatful to our understanding parents for their support throughout the years we spent working on robots. Also we would like to thank those not mentioned but who helped us indirectly and expanded our minds in school and at the various RoboCup competitions.

L. Iocchi et al. (Eds.): RoboCup 2008, LNAI 5399, pp. 308–317, 2009.

1   Challenge

In the RCJ Soccer 2vs2[1] discipline the field spans 122 cm by 183 cm[2] and is surrounded by a wall of 14 cm in height. "The walls are painted matte black."[3] An opening of 45 cm in width and a depth of 7 cm in each of the two shorter sides of the field represent the goals. The entire goal is painted matte grey[4]. The robots are 22 cm in diameter and 22 cm high [5]. The weight limit in 2vs2-Secondary is 2.5 kg[6] and the robot must act completely autonomously.

2   Setup

PALB VISION is based on the CMUCam3[7] developed by Carnegie Mellon University. Before we opted for the CMU Cam, we used a simple USB webcam connected to an ARM 9-based board[8] with software developed in C++. However, this setup was abandoned due to its insufficient processing speed.

The robot itself is powered by an Atmel ATmega2560[9] to which the CMUCam is connected via RS232.

The camera is mounted just above the dribbler at a height of 12 cm from the ground, facing forwards. (See Figure 1)

Fig. 1. Robot opened for camera to be visible

---

[1] Note: Since only our 2vs2 Striker is equipped with a camera all following information applies to 2vs2 soccer.
[2] [Rules, § 1.1.2]
[3] [Rules, § 1.3]
[4] "75% matte white and 25% matte black"[Rules, § 1.4.3]
[5] [Rules, § 2.1]
[6] Secondary refers to games where students are between 15 and 19 years old.
[7] [CMUCam3]
[8] portux920T
[9] [ATmega 2560]

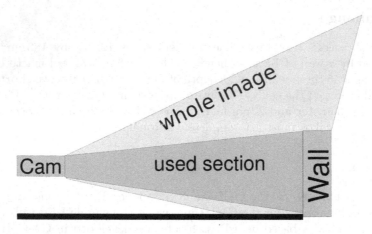

Fig. 2. Vision field

3   Filter

In order to make PALB VISION as simple and error-free as possible and to ensure that the wall is always visible, the CMUcam only processes a cropped version of the original picture (Figure 2).

The image data is loaded as a single line (color) and 14 lines (greyscale). Greyscale values are calculated with the formula of Listing 1.1.

There are five filter passes per frame, which are covered in the following paragraphs.

3.1   Pass 1: Goalie Detection

The first filter pass is the goalkeeper detection which consists of two sub-passes:

1. The filter renders the vertical greyscale variance for each horizontal position. It analyses the seven pixels above and below the middle line and checks whether the values vary. Goalies significantly vary vertically compared to the black wall.
2. The filter renders color information of the picture. Since the wall and the goal have virtually no saturation (grey / black), high saturation areas belong to an enemy goalie.

The results of this filter pass are stored and used by pass 2.

3.2   Pass 2: Division into Sections

For later use we define three types of areas in the image:

0: CLASS_WALL
1: CLASS_GOAL
2: CLASS_GOALIE

Fig. 3. Colorful robot with many edges

In the second pass, the entire horizontal line is divided into multiple sections by searching for edges in the image. The edge detection uses all of the 14 lines of greyscale and looks for significant changes of luminance in at least 10 lines. The previously calculated goalie areas (see Pass 1) are premarked as CLASS_GOALIE and no edge detection is rendered in those areas because goalies typically have many edges (See Figure 3).

3.3   Pass 3: Classification

Pass 3 classifies the remaining section by comparing their brightness level with previously calibrated reference luminances (Listing 1.2).

3.4   Pass 4: Simplification

Small (within a preset range) "gaps" between sections of the same class are erased. Sections which are considered too short, are also erased.

If neighboring sections are of the same class, they are merged. The result is that sequential sections always have different classes.

Listing 1.1. Calculating greyscale values

```
1  uint8_t pv_pixel_bw(uint8_t *pixel) //Calculates ←
       the black/white (bw) pixel for PalbVision
2  {
3  return    ((double)0.299) * ((double)pixel[0])
4      + ((double)0.584) * ((double)pixel[1])
5      + ((double)0.114) * ((double)pixel[2]);
6  }
```

3.5   Pass 5: Result

The fifth and final pass sends the width and position of the longest section which is classifed as CLASS_GOAL to the robot's controller. Additionally the "goalie" section closest to this "goal" section is transmitted. (Fig. 4)

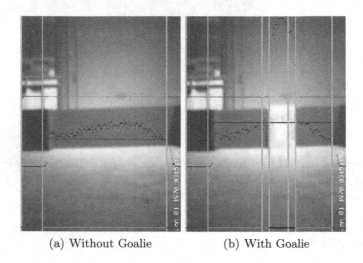

(a) Without Goalie   (b) With Goalie

Fig. 4. The sections are distinguished by vertical lines. The overlayed scatter plot represent greyscale values. The photos are taken from an older version of the system but with the same algorithm. See Figure 6 to 7 for newer images.

4   Application in Strategy

The use of real-time image processing opens up many new ways for a more strategic play in the RCJ. At the moment, our robot acts according to the following principles:

Listing 1.2. Comparing average brightness of a section with goal and wall reference

```
1  if(abs(goal_ref - section->avg) < abs(wall_ref - ↩
       section->avg)) {
2      section->class = CLASS_GOAL;
3  } else {
4      section->class = CLASS_WALL;
5  }
```

Whenever the robot is in possession of the ball, the camera is used to aim towards the goal and, if necessary, evade the goalie. As soon as a sufficiently large section of goal is detected in the center of the visual field, the kicker is triggered.

If the goal is too narrow to aim at, PALB VISION decides whether this is due to a goalie or a mispositioning of the robot. In the first case, the robot moves sideways until the goal is wide enough. In the second case, the robot rotates until the goal is wide enough.

5 Calibration

There are six values which have to be calibrated in order to make PALB VISION operate properly. `goal_ref` and `wall_ref` define reference luminance for goal and wall (See Listing 1.3. Because lighting changes at every table, these two have to be recalibrated before every game. `goalie_sigma_thr` and `goalie_sat_thr` define the thresholds above a pixel is thought of as a goalie based on variance (sigma)[10] and saturation. However, other than the goal and wall references these two settings do not depend on the lighting. Saturation and variance are relatively consistent, even with different lighting conditions. The last values to calibrate are two values that are used in pass 4 (simplification). `goalie_min_size` describes the minium size of a goalie in the picture. `goalie_max_gap` determines the maximal length of a goal section lieing between two goalie sections may be before being merged.

Listing 1.3. Default values for calibration

```
1   // color values 0-255
2   int goal_ref = 50;
3   int wall_ref = 35;
4   int goalie_sigma_thr = 70;
5   int goalie_sat_thr = 70;
6
7   //in pixel, picture width 176
8   int goalie_min_size = 1;
9   int goalie_max_gap = 10;
```

Fig. 5. Easy Calibration with external display

---

[10] We initially worked with the standard deviation but used the variance later because it is simpler to calculate.

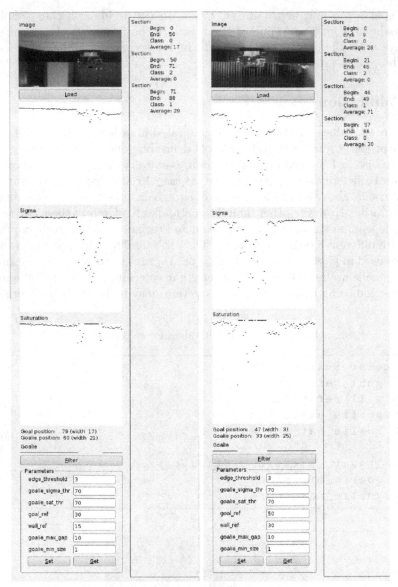

(a) Goal is off-centered   (b) Robot is far away from goal

Fig. 6. Different results of the filter

5.1   Calibration Before Game Play

Since there is usually little time before and in between matches left for calibration, we had to make sure PALB VISION has a quick and easy to use calibration system. It is based on a graphic display together with a numpad connected to the main controller (Figure 5). By aiming the robot towards goal and wall and

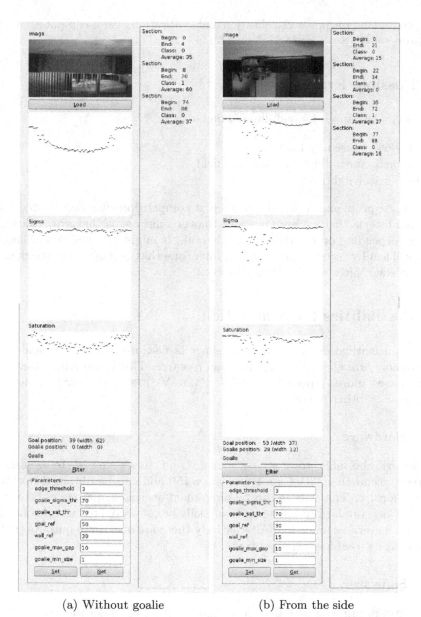

(a) Without goalie   (b) From the side

Fig. 7. Different results of the filter

holding down a specified button, the controller reads several values from the camera, averages them and stores them as the new reference. This procedure takes less than a minute. It is also possible to adjust those values manually, e.g. for tweaking. All values are stored in the EEPROM of the main controller board.

For programing purposes visualisation and calibration over a computer interface is also possible (See Figure 6 - 7 on 314f)

6 Possible Errors

Errors can be caused by bad illumination, e.g.

- Shadows caused by humans
- Inhomogenous illumination of the field
- Insufficient lighting[11]

Since these problems do not occur often at competition sites, we ignored them.

One obstacle that we were not able to overcome yet, is that greyscale values change depending on the distance to the wall. It might be possible to eliminate this problem by using a different goal filter, one that is more adaptive than the current static filter used in PALB VISION.

7 Possibilities for Adaptation

The main motivation we wrote this paper is because RoboCup – and Robo-CupJunior – are about sharing ideas and research. This is something that could be endorsed more, especially in RCJ. PALB VISION is designed to be easily adoptable by other teams.

7.1 Hardware

Teams will be able to use any camera and processing system, although we recommend the CMU Cam 3. It costs € 150 and can be obtained worldwide [CMUCam3][12]. The camera can be programed over a regular serial connection through any standard computer and can easily be connected to most μ-controllers through its serial connection. Alternatively there are also servo outputs (PWM) which can be used for various outputs.

7.2 Software

The software supplied with the CMU Cam is published under the Apache license [CMUCam3, CMUcam.org] and comes with some example programs of computer vision. RoboCupJunior teams will also be able to download the PALB VISION source code from C-PALB's website [www.c-palb.de]. The source code is licensed under GPL license.

---

[11] The system works fine with illumination over 290 lux. Best results are obtained at 580 lux.

[12] For countries outside Europe, North America or Singapore Elektronikladen [Elektronikladen] will send it.

8   Conclusion

In the future, our image processing system could be employed to detect where an opposing robot is moving. This along with features like detecting how far away the ball is might be very useful on the goalie as well.

We think that the entire RoboCupJunior community can profit from using real-time image processing and even though PALB VISION at the moment resembles a rather rudimentary system, the introduction of camera vision to RoboCupJunior in our opinion is a huge step forward.

References

[Rules]   RoboCupJunior Soccer Rules (2007),
                      http://rcj.sci.brooklyn.cuny.edu/rcj2007/
                      soccer-rules-2007.pdf
[CMUCam3]   http://www.cmucam.org
[Robotikhardware]   http://www.robotikhardware.de
[ATmega 2560]   Atmega 2560 from Atmel, http://www.atmel.com/dyn/
                      products/product_card.asp?part_id=3632,
                      http://www.atmel.com/dyn/resources/prod_documents/
                      doc2549.pdf
[www.c-palb.de]   http://www.c-palb.de
[Elektronikladen]   http://elmicro.com/de/cmucam3.html

Multi-robot Range-Only SLAM by Active Sensor Nodes for Urban Search and Rescue

Dali Sun[1], Alexander Kleiner[1], and Thomas M. Wendt[2]

[1] Department of Computer Sciences
University of Freiburg
79110 Freiburg, Germany
{sun,kleiner}@informatik.uni-freiburg.de
[2] Department of Microsystems Engineering
University of Freiburg
79110 Freiburg, Germany
wendt@imtek.uni-freiburg.de

Abstract. To jointly map an unknown environment with a team of autonomous robots is a challenging problem, particularly in large environments, as for example the area of devastation after a disaster. Under such conditions standard methods for Simultaneous Localization And Mapping (SLAM) are difficult to apply due to possible misinterpretations of sensor data, leading to erroneous data association for loop closure. We consider the problem of multi-robot range-only SLAM for robot teams by solving the data association problem with wireless sensor nodes that we designed for this purpose. The memory of these nodes is utilized for the exchange of map data between multiple robots, facilitating loop-closures on jointly generated maps. We introduce RSLAM, which is a variant of *FastSlam*, extended for range-only measurements and the multi-robot case. Maps are generated from robot odometry and range estimates, which are computed from the RSSI (Received Signal Strength Indication). The proposed method has been extensively tested in USARSim, which serves as basis for the *Virtual Robots* competition at RoboCup, and by real-world experiments with a team of mobile robots. The presented results indicates that the approach is capable of building consistent maps in presence of real sensor noise, as well as to improve mapping results of multiple robots by data sharing.

1   Introduction

To jointly map an unknown environment with a team of autonomous robots is a challenging problem, particularly in large areas as for example the area of devastation after a disaster. In USAR (Urban Search and Rescue) the mapping problem is generally harder, due to difficult operating conditions, such as unstructured environments and existing constraints for communication. However,

---

[1] This research was partially supported by DFG as part of the collaborative research center SFB/TR-8 Spatial Cognition R7.

L. Iocchi et al. (Eds.): RoboCup 2008, LNAI 5399, pp. 318–330, 2009.
© Springer-Verlag Berlin Heidelberg 2009

the ability to map environments jointly is an essential requirement for coordinating robots and humans, such as first responders, within this domain. In urban environments GNSS (Global Navigation Satellite System) positioning is affected by the *multipath propagation* [6]. Buildings in the vicinity of the receiver reflect GNSS signals, resulting in secondary path propagations with longer propagation time, causing erroneous position estimates. Furthermore, long-range communication is perturbed by building structures made of reinforced concrete. Conventional methods for Simultaneous Localization And Mapping (SLAM), which require reliable data association for loop closure, are difficult to apply due to possible misinterpretations of sensor data. For example, bad visibility caused by smoke and fire affects vision-based tracking methods, and arbitrarily shaped structures from collapsed buildings require laser range finder-based data association to be fully carried out in 3D.

In this paper, we consider the problem of multi-robot SLAM for large robot teams in USAR. The data association problem is solved by using the unique IDs of deployed sensor nodes as features which are detectable in presence of low visibility via an omni-directional 2.4 GHz antenna. Furthermore, the memory of these nodes is used for indirect communication, e.g. for the exchange of map data between multiple robots. More precise, robots subsequently store their position estimates of known sensor nodes into the memory of nodes within writing range. This information can be utilized by other robots for updating their individual map representations, facilitating loop-closures on jointly generated maps. In contrast to other approaches that require a priori a subset of sensor node locations that have to be surveyed in advance (so called anchor nodes), our method learns this locations stepwise from robots exploring the terrain. Moreover, new sensor nodes can arbitrarily be deployed during robot exploration, e.g. while searching for victims in an unknown environment.

The proposed method, named RSLAM, is a variant of *FastSlam* [16], extended for the multi-robot case and range-only measurements that are derived from the RSSI (Received Signal Strength Indication). We use a voting scheme for determining initial node locations from pairwise intersections of the signal strength measured at different robot locations [17]. The method is computationally efficient, i.e. applicable in real-time, since it applies fast computable updates from robot odometry and rare range observations only. Each update is carried out in $O(nk)$, where n is the number of sensor nodes and k is the number of robot trajectories considered at the same time.

For our experiments we developed sensor nodes [20] meeting the *ZigBee* specification [8]. They are equipped with three sensors, measuring air pressure, temperature, and node orientation. The first two sensors might be utilized for user applications, such as monitoring the temperature and the air pressure during crisis management. Measurements of the node's orientation are useful for detecting the alignment of the antenna, which might be used for improving the RSSI in order to increase the accuracy of distance measurements.

The proposed method has been extensively tested in the USARSim [1] simulation environment which serves as basis for the *Virtual Robots* competition at RoboCup. We modified the simulator in that it provides range readings of virtual sensor nodes with respect to a model for signal path attenuation [18], which has been parameterized according to the developed hardware. The simulation results show that our method is capable to successfully map diverse types of environments by robot teams, also indoors, where signal strength has been heavily perturbed by walls. Finally, we present results from a real-world experiment demonstrating the capability of our system to handle communications of multiple nodes at the same time, as well as to deal with real sensor noise.

The remainder of this paper is structured as follows. In Section 2 related work is discussed, and in Section 3 the real system and the simulation system utilized for experiments are described. In Section 4 the sensor model, and in sections 5 the introduced SLAM method are discussed. Finally, we present experimental results in Section 6 and draw conclusions in Section 7.

2   Related Work

Inspired by the fundamental work of Smith et al. [19], early work on SLAM was mainly based on the Extended Kalman Filter (EKF) [3]. In connection with radio transmitters, the SLAM problem has been addressed as "range-only" SLAM [4,10] since the bearing of the radio signal cannot accurately be determined. RFIDs have been successfully utilized for localizing mobile robots [7] and emergency responders [9,15]. Hähnel et al. [7] successfully utilized Markov localization for localizing a mobile robot in an office environment. Their approach deals with the problem of localization in a map previously learned from laser range data and known RFID positions, whereas the work presented in this paper describes a solution that performs RFID-based localization and mapping simultaneously during exploration. Also sensor networks-based Markov localization for emergency response has been studied [9]. In this work, existing sensor nodes in a building are utilized for both localization and computation of a temperature gradient from local sensor node measurements. Miller and colleagues examined the usability of various RFID systems for the localization of first responders in different building classes [15]. During their experiments, persons were tracked with a Dead Reckoning Module (DRM). In former work we introduced RFID-SLAM, an extension of graph-based SLAM for networks of RFIDs [11]. Furthermore, we demonstrated the efficient deployment of sensor nodes for multi-robot coordination and exploration [21].

3   Test Platform

The test platform utilized for experiments is based on a team of 4WD differentially steered *Zerg* robots, as depicted in Figure 1(a). Each robot is equipped with a *Sick* S30-B Laser Range Finder (LRF), an Inertial Measurement Unit (IMU), over-constrained odometry for wheel-slippage detection, and a mobile sensor node.

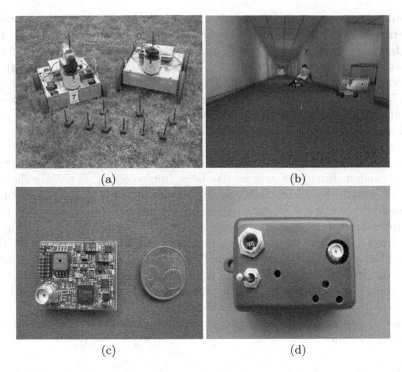

Fig. 1. (a) A team of *Zerg* robots equipped with antennas for sensor node communication, and (b) the corresponding simulation model in USARSim (b). (c-d) The designed sensor node: (a) the PCB, and (b) with housing.

The simulated counter part (see Figure 1(b)) has been designed for the USAR-Sim simulator developed at the University of Pittsburgh [2]. USARSim, which serves as a basis for the RoboCup Rescue virtual competition, allows realistic simulations of raw sensor data and robot actuators, which can directly be accessed via a TCP/IP interface. Sensors, such as odometry and LRF, can be simulated with the same parameters as they are found on real robots. We modified the simulator in that it provides range readings of virtual sensor nodes with respect to a model of signal path attenuation [18] (see Section 4). The model considers obstacles, such as walls and trees, that are located between two transceivers. Obstacles are detected by a ray tracing operation that is applied each time two transceivers are within communication range. Both platforms, the real and the simulated robot, achieved the first place during the RoboCup 2005 *Rescue Autonomy competition*, and the RoboCup 2006 *Virtual Robots competition*, respectively [12,13].

We developed wireless sensor nodes based on the *TI CC2420* transceiver [8] that meets the *ZigBee* specification. The transceiver implements an anti-collision protocol allowing the simultaneous reading of multiple nodes within range. The printed circuit board (PCB) is shown in Figure 1(c), and the sensor node with housing and omni-directional antenna is shown in Figure 1(a,d). An on board eight-bit micro controller [14] monitors the sensors on the PCB and establishes the communication link via the transceiver. The communication protocol is tailored for power minimization and transparent access to the radio link. Furthermore, the sensor node is equipped with three sensors, measuring air pressure, temperature, and antenna orientation, which might be utilized for user applications. Measurements of the antenna orientation, for example, can be considered to improve the reliability of communication. RSSI and Link Quality Indication (LQI) are directly read from the transceiver chip and are directly transmitted to the reader.

In order to enable communication between several sensor nodes at the same time, a random broadcast scheme has been implemented. If the air channel is occupied heavily by radio traffic, nodes are sleeping for a random amount of time before re-sending their data. Alternatively, we experimented with a polling mechanism triggered by a master node that requests data packages within fixed cycles from each node, which led to a high latency time of the system. With the random broadcast scheme, we measured an average data rate of 28.6 Hz for one, 27.5 Hz for four, and 16.5 Hz for eight nodes transmitting at the same time. This data rate is sufficient also for mobile platforms since we assume a node density of maximally 4 nodes being in communication range at any location. Finally, we determined experimentally a maximal range of up to 35 meters for performing reliable communications.

4   Sensor Model

The Transceiver-Receiver (TR) separation, i.e. the distance between an observed sensor node and the detector, can generally be estimated from the power of the signal known as RSSI (Received Signal Strength Indication). The sensor model describes the relation between the RSSI and range estimates denoted by distance r an variance σ_r. Signal propagation, particularly in indoor environments, is perturbed by damping and reflections of the radio waves. Since these perturbations can depend on the layout of the building, the construction material used, and the number and type of nearby objects, modeling the relation between signal path attenuation and TR separation is a challenging problem.

Seidel and Rapport introduced a model for path attenuation prediction that can also be parameterized for different building types and the number of walls between transceiver and receiver [18]. The model relates the signal power P to distance d in the following way:

$$P(d)[dB] = p(d_0)[dB] - 10n \log \frac{d}{d_0} + X_\sigma[dB], \tag{1}$$

where $p(d_0)$ is the signal power at reference distance d_0, n is the mean path loss exponent that depends on the structure of the environment, and X_σ is the standard deviation of the signal.

During practical experiments we noticed that signal strength measurements of the used transceiver contain a high amount of outliers. Therefore, we applied

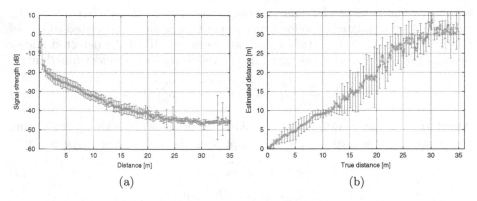

Fig. 2. (a) Measured signal strength at varying distance. (b) Relation between estimated distance and real distance. Each data point has been averaged from ten measurements, green points denote the computed mean and red bars the standard deviation.

RANSAC [5], which is an iterative method to estimate model parameters from a set of observations containing outliers. The parameter we are interested in is the average signal strength given a set of measurements. Therefore, a subset of measurements is randomly sampled, the average computed, and finally all the other measurements are tested against this average. If measurements fit well to the average, they are added to the subset, and rejected otherwise. The procedure is repeated iteratively until the subset contains sufficient data points.

We conducted several measurements for determining the model parameters of the described 2.4 GHz sensor nodes in the testing environment, which are $p(d_0) = -12dB$, and $n = 2$. The standard deviation X_σ is taken from a look-up table that has been determined experimentally, as shown in Figure 2(a). Figure 2(a) depicts the mean and variance of the measured signal strength at different distances averaged over 10 measurements at each location. Figure 2(b) compares the estimate of the parametrized model from Equation 1 with ground truth.

5   RFID SLAM

We utilize *FastSlam* [16] in order to compute simultaneously the locations of the robot and of the sensor nodes. FastSlam estimates simultaneously the robot path $l_1, l_2, ..., l_t$ and beacon locations $b_1, b_2, ..., b_N$, where $l_i = (x_i, y_i, \theta_i)$, and $b_j = (x_j, y_j)$. The path of the robot is estimated by using a particle filter with M particles, whereas each particle possesses its own set of Extended Kalman Filters (EKFs). EKFs are estimating beacon locations independently from each other conditioned on the robot path. Hence, there are $M \times N$ low-dimensional EKFs updated by the system.

We denote each EKF's state vector by $s_i = (b_x, b_y)^T$, and the corresponding 2×2 covariance matrix by Σ_{s_i}, where $(b_x, b_y)^T$ is the location of beacon i. From an observation $z = r$ of a landmark at range r with covariance σ_z (see Section 4), each particle is updated as follows: first, the observation is associated to one of the EKFs stored in the particle based on the unique ID of the detected nodes. Second, the observation is predicted by the following measurement function:

$$h_i(s) = \left(\sqrt{(r_x - b_x)^2 + (r_y - b_y)^2} \right). \tag{2}$$

Third, the associated state vector is updated from the observation by computing the *innovation* v and covariance σ_v by applying the law of error propagation:

$$v_i = z - h_i(s) \tag{3}$$

$$\sigma_{v_i} = \nabla h_s \Sigma_s \nabla h_s^T + \sigma_z, \tag{4}$$

where ∇h_s is the Jacobian matrix:

$$\nabla h_s = \left(-\frac{\Delta x}{d}, -\frac{\Delta y}{d} \right), \tag{5}$$

where $\Delta x = r_x - b_x$, $\Delta y = r_y - b_y$, and $d = \sqrt{\Delta x^2 + \Delta y^2}$. Then, the following Kalman update is applied:

$$K = \Sigma_{s_i} \nabla h_s^T \sigma_{v_i}^{-1} \tag{6}$$

$$s_{i+1} = s_i + K v_i \tag{7}$$

$$\Sigma_{s_{i+1}} = \Sigma_{s_i} - K \sigma_{v_i} K^T. \tag{8}$$

Finally, we update from each observation an importance weight for each particle based on the Mahalanobis distance between the measurement and the prediction:

$$w_{i+1} = w_i - v_i^T \sigma_{v_i}^{-1} v_i. \tag{9}$$

During each cycle, these weights are utilized to stochastically sample robot paths that best explain the measured ranges. According to the particle filtering framework, these particles are furthermore propagated based on the motion model of the robot if steering commands occur [16].

5.1   EKF Initialization

One difficulty in range-only SLAM is to estimate the initial beacon locations. This is particularly important since beacon observations do not contain bearing information as required by SLAM methods in general. We use a voting scheme for determining initial node locations from pairwise intersections of range measurements at different robot locations [17]. This is carried out by maintaining a grid for each beacon, where each cell represents the probability that the beacon is located at the corresponding location. The grid is updated by a likelihood function $f(o_i, z_i)$ that assigns to each cell a probability with respect to the robot's pose estimate from odometry o_i and range observation z_i. Given range measurements, the likelihood function generates circular probability distributions centered at the current pose of the robot, and width according to the confidence of the observation. After integrating several observations, the most likely beacon location can be determined by taking the maximum over all grid cells. The location at this maximum is taken to initialize one EKF for each particle.

5.2 Multi-robot Mapping

The memory of the sensor nodes is used for indirect communication, e.g. for the exchange of map data between multiple robots. If sensor node k is within writing distance of robot R, its current map estimate, consisting of n_R known beacon locations, is stored into the local memory M_k of the node:

$$M_k \leftarrow M_k \cup \langle j, b_j, \Sigma_{b_j}, R \rangle, \quad for\, all\, j \in n_R \tag{10}$$

The stored data is utilized by other robots for updating their individual maps. This is carried out by calculating the local offset between individual robot maps from the relative pose displacement to common sensor nodes. Note that we assume that the IMU angle of each robot is aligned to magnetic north leading to displacements without angular components.

Map data from other robots can significantly accelerate the convergence of the map from a single robot. First, the initialization procedure described in Section 5.1 is not required for nodes that are observed the first time, if their location estimate is already known. Second, known sensor node locations can directly be updated by the EKFs, facilitating loop closure on jointly generated maps. Note that the update procedure automatically degrades from *SLAM* to *localization* if many of the communicated location estimates have tight covariance bounds, i.e. are close to ground truth.

6   Results

RSLAM has been tested in various USARSim environments generated by the National Institute of Standards and Technology (NIST). They provide both indoor and outdoor scenarios of the size bigger than $1000m^2$, reconstructing the situation after a disaster. Furthermore, we conducted real-world experiments showing the capability of the developed sensor nodes to handle communications of multiple clients at the same time, as well as the capability of the proposed approach to deal with real sensor noise.

Figure 3 compares visually the mapping result from wheel-odometry and RSLAM, which has been carried out by aligning measurements from the LRF according to the pose estimates. The map generated from odometry is unusable for navigation, whereas the map generated by the introduced method is close to ground truth. For this experiment the map *RC07-Mapping* has been used, which is also listed in Table 1.

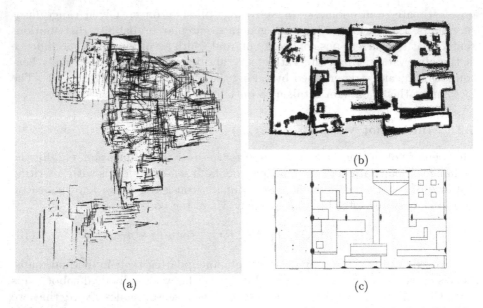

Fig. 3. Indoor map generated from wheel odometry (a), and multi RSLAM (b). The ground truth image has been provided by NIST (c).

Table 1. Avg. Cartesian positioning errors: Comparison between wheel odometry, single-robot RSLAM, and multi-robot RSLAM applied on different USARSim maps. Values are averaged from the total trajectory of each robot. Ground truth has been directly taken from the simulator.

Map name (size in $[m^2]$)	Robot	Odometry [m]	RSLAM [m] (single)	RSLAM [m] (multi)	# RFID
RC07-Plywood (109)	1st	4.7 ± 3.4	0.3 ± 0.9	0.3 ± 0.9	7
	2nd	3.5 ± 2.6	0.8 ± 1.3	0.6 ± 1.1	
RC07-Mapping (250)	1st	7.0 ± 4.9	0.5 ± 1.1	0.5 ± 1.1	8
	2nd	4.1 ± 3.0	0.6 ± 1.4	0.3 ± 0.8	
RC06-Day4a (846)	1st	8.2 ± 3.2	0.4 ± 0.6	0.4 ± 0.6	14
	2nd	13.2 ± 5.8	2.0 ± 1.9	0.8 ± 1.5	
RC07-Factory (1975)	1st	17.4 ± 8.3	2.1 ± 2.8	2.1 ± 2.8	58
	2nd	28.2 ± 9.6	4.6 ± 3.6	1.6 ± 2.5	
RC07-Mobility (3819)	1st	12.9 ± 6.9	1.4 ± 2.7	1.4 ± 2.7	58
	2nd	13.3 ± 6.3	1.7 ± 3.0	1.3 ± 2.3	
RC06-Day4b (4529)	1st	5.0 ± 3.3	0.5 ± 0.9	0.5 ± 0.9	60
	2nd	6.7 ± 4.3	1.2 ± 1.2	0.3 ± 1.0	
	3rd	5.2 ± 2.7	0.4 ± 0.8	0.1 ± 0.4	
	4th	12.7 ± 3.4	0.8 ± 1.0	0.2 ± 0.6	

Table 2. Avg. Cartesian positioning errors: Comparison between wheel odometry and RSLAM applied on real robots. Ground truth has been manually generated by geo-referencing.

	Odo. [m]	RSLAM [m] (single)	RSLAM [m] (multi)	Speed [m/s]	Dist [m]	# RFID
1st robot	8.6 ± 6.0	1.1 ± 0.6	1.1 ± 0.6	1.0	410	
2nd robot	5.0 ± 4.0	2.1 ± 1.9	1.3 ± 0.9	1.2	348	9
3rd robot	7.1 ± 4.5	2.2 ± 1.7	1.3 ± 1.0	1.3	320	
4th robot	5.3 ± 3.2	1.9 ± 0.9	1.5 ± 0.6	1.5	282	

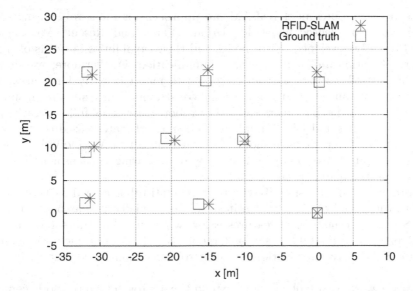

Fig. 4. Real-robot experiment: Comparison of the locations of the deployed real sensor nodes with manually measured ground truth

Table 1 summarizes the results from applying the method on simulated indoor and outdoor maps. Maps starting with *RC06*, and *RC07* were introduced by NIST for RoboCup 2006 and 2007, respectively. Ground truth, i.e. the true trajectories of the robots, is directly provided by the simulator. The results show that single RSLAM successfully improves maps generated from odometry. Moreover, multi RSLAM, i.e. taking advantage of map data stored by other robots into node memories, yields even better results than single RSLAM.

Table 2 summarizes the results from real-robot experiments. During this experiment real sensor nodes have been distributed at fixed height and with fixed orientation in an outdoor environment. Four robots were manually steered [1] through the environment while computing map estimates, i.e. the robot path and sensor node location, online. This information was used by subsequent robots

---

[1] Note that two robots have been steered two times sequentially.

for initializing their particle filters. Also during this experiment, maps generated from wheel odometry have been significantly improved by applying RSLAM. As can be seen by the third row, map data provided by other robots helps to increase this improvement. Due to limited accuracy of range readings computed from signal strength, a minimal possible positioning error around one meter has been reached. Figure 4 depicts the true positions and the positions finally estimated by the robots. During all experiments, an average computation time of 20 ms on a *Pentium4 1 GHz* has been measured.

7   Conclusions and Future Works

In this paper we presented RSLAM, an approach for multi-robot range-only SLAM based on sensor nodes for Urban Search and Rescue. We demonstrated recently developed sensor nodes and their capabilities in terms of signal strength detection and multi-sensor communication. Furthermore, we showed that RSLAM allows it to a team of robots to map efficiently large areas under severe communication and operational constraints. Our results from simulation and real-word experiments show that robots can successfully correct noisy odometry readings and jointly improve their map estimates based on the wireless nodes. Moreover, we have shown that with multi RSLAM individual robots consistently gain better mapping results by data sharing via sensor nodes than carrying out the mapping task on their own.

Sensor nodes greatly simplify the task of multi-robot SLAM in three ways: First, features can be uniquely identified, solving trivially data association problems. Second, the number of features is low w.r.t. visual features and thus the SLAM problem is tractable even for large areas. Third, node memories can be used for indirect communication allowing robots to jointly correct their local maps.

In future work we are planning to extend the approach by inter-node communication allowing nodes to improve their estimates in a decentralized manner, and to synchronize map data with other nodes that are in direct communication range. During our experiments we observed high variations of distance measurements when orientation and height of sensor nodes have been modified. Our future goal is to address these problems by integrating sensor measurements from the nodes, such as orientation and height. This will allow to generate specific sensor models for each node leading to more accurate distance estimations.

References

1. Balakirsky, S., Scrapper, C., Carpin, S., Lewis, M.: USARSim: providing a framework for multi-robot performance evaluation. In: Proceedings of PerMIS 2006 (2006)
2. Carpin, S., Lewis, M., Wang, J., Balakirsky, S., Scrapper, C.: Bridging the gap between simulation and reality in urban search and rescue. In: Lakemeyer, G., Sklar, E., Sorrenti, D.G., Takahashi, T. (eds.) RoboCup 2006: Robot Soccer World Cup X. LNCS (LNAI), vol. 4434, pp. 1–12. Springer, Heidelberg (2007)

3. Dissanayake, M.W.M.G., Newman, P., Clark, S., Durrant-Whyte, H.F., Csorba, M.: A solution to the simultaneous localization and map building (slam) problem. IEEE Transactions on Robotics and Automation 17(3), 229–241 (2001)
4. Djugash, J., Singh, S., Corke, P.: Further results with localization and mapping using range from radio. In: Proc. of the Fifth Int. Conf. on Field and Service Robotics, Pt. Douglas, Australia (2005)
5. Fischler, M.A., Bolles, R.C.: Random sample consensus: A paradigm for model fitting with applications to image analysis and automated cartography. Comm. of the ACM 24, 381–395 (1981)
6. Grewal, M.S., Weill, L.R., Andrews, A.P.: Global Positioning Systems, Inertial Navigation, and Integration. John Wiley & Sons, Chichester (2001)
7. Hähnel, D., Burgard, W., Fox, D., Fishkin, K., Philipose, M.: Mapping and localization with rfid technology. In: Proc. of the IEEE International Conference on Robotics and Automation (ICRA) (2004)
8. Texas Instruments. Datasheet, cc2420 2.4 ghz ieee 802.15.4 / zigbee-ready rf transceiver, rev swrs041 (2006), http://www.chipcon.com
9. Kantor, G., Singh, S., Peterson, R., Rus, D., Das, A., Kumar, V., Pereira, G., Spletzer, J.: Distributed search and rescue with robot and sensor team. In: Proc. of the Int. Conf. on Field and Service Robotics (FSR), pp. 327–332. Sage Publications, Thousand Oaks (2003)
10. Kehagias, A., Djugash, J., Singh, S.: Range-only slam with interpolated range data. Technical Report CMU-RI-TR-06-26, Robotics Institute. Carnegie Mellon University (2006)
11. Kleiner, A., Dornhege, C.: Real-time localization and elevation mapping within urban search and rescue scenarios. Journal of Field Robotics 24(8–9), 723–745 (2007)
12. Kleiner, A., Steder, B., Dornhege, C., Hoefler, D., Meyer-Delius, D., Prediger, J., Stueckler, J., Glogowski, K., Thurner, M., Luber, M., Schnell, M., Kuemmerle, R., Burk, T., Braeuer, T., Nebel, B.: Robocuprescue - robot league team rescuerobots freiburg (germany). In: RoboCup 2005 (CDROM Proceedings), Team Description Paper, Rescue Robot League, Osaka, Japan (2005)
13. Kleiner, A., Ziparo, V.A.: Robocuprescue - simulation league team rescuerobots freiburg (germany). In: RoboCup 2006 (CDROM Proceedings), Team Description Paper, Rescue Simulation League, Bremen, Germany (2006)
14. Silicon Laboratories. Datasheet, c8051f310/1/2/3/4/5/6/7, rev 1.7 08/06 (2006), http://www.silabs.com
15. Miller, L.E., Wilson, P.F., Bryner, N.P., Francis, Guerrieri, J.R., Stroup, D.W., Klein-Berndt, L.: Rfid-assisted indoor localization and communication for first responders. In: Proc. of the Int. Symposium on Advanced Radio Technologies (2006)
16. Montemerlo, M.: FastSLAM: A Factored Solution to the Simultaneous Localization and Mapping Problem with Unknown Data Association. PhD thesis, Robotics Institute. Carnegie Mellon University, Pittsburgh, PA (July 2003)
17. Olson, E., Leonard, J., Teller, S.: Robust range-only beacon localization. Autonomous Underwater Vehicles, 2004 IEEE/OES, pp. 66–75 (2004)
18. Seidel, S.Y., Rapport, T.S.: 914 mhz path loss prediction model for indoor wireless communications in multi-floored buildings. IEEE Trans. on Antennas and Propagation (1992)

19. Smith, R., Self, M., Cheeseman, P.: Estimating uncertain spatial relationships in robotics. Autonomous Robot Vehicles 1, 167–193 (1988)
20. Wendt, T.M., Reindl, L.M.: Reduction of power consumption in wireless sensor networks through utilization of wake up strategies. In: 11th WSEAS international Conference on Systems, Crete Island, Greece (2007)
21. Ziparo, V.A., Kleiner, A., Nebel, B., Nardi, D.: Rfid-based exploration for large robot teams. In: Proc. of the IEEE Int. Conf. on Robotics and Automation (ICRA), Rome, Italy (to appear, 2007)

Analysis Methods of Agent Behavior and Its Interpretation in a Case of Rescue Simulations

Tomoichi Takahashi

Meijo University, Tenpaku, Nagoya, 468-8501, Japan
ttaka@ccmfs.meijo-u.ac.jp

Abstract. The agent-based approach has been proved to be useful for the modeling phenomena of traditional social fields. In order to apply agent-based simulation results to practical usages, it is necessary to show potential users the validity of the simulation outputs that arise out of the agents' behaviors. In cases that involve human actions, it is difficult to obtain sufficient amounts of data on real cases or experimental data to validate the simulation results.

In this paper, we review the metrics that have been used to evaluate rescue agents in the Rescue Simulation Agent competition and propose a method to analyze the simulation output by presenting the agent behavior with a probability model. We present that the analysis results of the method are comparable to a human-readable interpretation and with task-dependent knowledge and discuss its applicability to real-world cases.

1 Introduction

Society consists of multi-heterogeneous entities. The simulation of various phenomena in the society involves human behaviors and social structures. By presenting human behaviors as agents, the agent-based approach has enhanced the potential usage of computer simulations as a tool for modeling and analyzing human behaviors from social scientific viewpoints [3]. When disasters threaten human life, we want to use agent-based social simulation (ABSS) to predict the evacuation behavior of person and to estimate damage of disasters. Like a triage system that patients will be sorted according to need when medical conditions are insufficient for all patients to be treated, the emergency centers of local governments will prepare their prevention plans against disasters, so that the damage from the disaster will become the least with their limited rescue resources.

When the ABSS is applied to practical usages, it is necessary to demonstrate the validity of the ABSS's outputs to the potential users. In scientific and engineering fields, the following principle has been repeatedly used to increase the fidelity of simulations: *Guess → Compute consequence → Compare experiment results with simulation results* [2]. Social phenomena are in contrast to physical phenomena that can be explained using the laws of natural science. The social phenomena are not so much objectively measured but subjectively interpreted

L. Iocchi et al. (Eds.): RoboCup 2008, LNAI 5399, pp. 331–342, 2009.

by humans [6]. It makes difficult to systematically analyze the social phenomena and to evaluate the performances of agents. In most cases, it is difficult to obtain data on real cases or conduct experiments to verify the results of ABSS and the analysis results.

Earthquake disasters and rescue simulations are composed of disaster simulators and human-behavior simulations. While the components of disaster simulators, such as fire and building collapse, have been programmed on the basis of models developed in civil engineering fields, agent technology that is used to simulate human behaviors with these disaster simulations does not have such theoretical background. This makes it difficult to compare the results with the real data and follow the principles, because we cannot conduct physical experiments on disasters on a real scale, involve humans in the experiments, and the results of agent-based approach simulation have emergent properties.

In this paper, considering the disaster and rescue simulation system as an example of ABSS, we propose a method to analyze the simulation outputs without any metrics related with the application domain and show the interpretation of the analysis results. In section 2, the background of this study and related studies are described. Section 3 discusses the performance metric that has been used in RoboCup Rescue Agent Competition. Section 4 shows the agent behavior presentation based on a probability model and experiment results. The discussion on future research topics and a summary are presented in section 5.

2   Background and Related Studies

There are several human behaviors that lead to life-threatening situations. One example is the crowd behavior induced by panic. Studies have been conducted for simulating models of collective behaviors of pedestrians. For, example, D. Helbing et al. demonstrated an optimal strategy for escaping from a smoke-filled room by modeling pedestrians as Newton particles and the interaction among them as a generalized force model [1]. The methods are macro level simulations and the wills of individuals, for example some people stop not to push others in crowded places, are not presented with the force model. It is a future issue to implement the wills of agents and to observe how human behavior will change when announcing proper indications to the crowds are given in the panic.

The other example is rescue and evacuation movements during natural disasters. When disasters take place, the collapse of buildings will hurt civilians and block roads with debris. Fires burn houses, and smoke from these burning houses impedes the activities of firefighters. Kitano et al. proposed the RoboCup Rescue simulation (RCRS) that integrates various disaster simulation results and agent actions [5]. The wills of agent are implemented to represent the behaviors of civilians, firefighters, the human rescuers etc. Their microscopic behaviors are summed to the output of RCRS such as the number of living civilians, the area of houses that are not burnt, the recovery rate of blocked road and so on. They are metrics used in disaster related documents.

With these metrics, some attempts have been done to apply RCRS to practical applications. Schurr et al. applied the RCRS to train human rescue officers [8]. In their system, the human officers command the rescue agents in the RCRS as they would do to real rescuers. How many damages are decreased at the end of simulations by their commands is used to estimate how well the officers direct them at the disaster situations. We showed that simulation results of the RCRS have a good correlation with data in the fire department report and also reported comments from local governments when they were questioned about the possibility of using the RCRS as their tools [9]. The comments are as follows:

- There are no precedents that ABSS is used to design the prevetion plans and no theoretical backgrounds for the models.
- The simulation size is quite different from a real one; for example, the number of agents is smaller than that in a real world situation.

These comments indicate that the validity of the ABSS is required to persuade users who have the will to use the ABSS as their tools. The requirement is not limited to only the RCRS but is also to the cases when the ABSS is applied to fields such as

- real-life data which could validate the results of simulations are often nonexistent or difficult to obtain,
- domains are too dynamic or complex to make models.

3   Studies on Evaluation of Agents' Performance from the Past Competitions

Given initial disaster conditions, RCRS simulates disaster and rescue situations. The rescue agents save people or extinguish fires, and their behaviors change the disaster situations. RCRS outputs the number of casualties, the amount of damage from fires, and other metrics. The metrics are intricately linked with each other, and sometimes conflict each other. For example, the decision of rescue headquarters to save victims may increase the damage from fires because they have limited rescue resources. It is difficult to set a multi-attribute metric that measures the behaviors of rescue agent properly.

Ranking index: The RCRS has been used the following formula to rank the performance of teams since 2001 competition.

$$V = (P + \frac{H}{Hint}) \times \sqrt{\frac{B}{Bmax}}$$

where P is the number of living civilian agents, H denotes the health point values (i.e., how much stamina the agents have after rescue operations) of all agents, and B is the area of houses that are not burnt. Hint and Bmax are values at start; the score decreases as the disasters spread. Teams with higher scores have been evaluated to perform better rescue performance.

Table 1. Changes in metrics for two sensing conditions

sensing C.	metrics	initial V.	Team X	Team Y	Team Z
Base	P	120	79	98	90
	H	1200	574	606	582
	B	14270	14213	14176	13707
	V	12100	78.9	97.7	88.2
Half-vision	P	120	79	64	65
	H	1200	565	464	600
	B	14270	14213	4434	13681
	V	12100	78.9	35.7	83.3

Robustness to changes: Robustness is another property of the performance of agents. Robust agents will perform little more than before when the conditions to the agents become worse. In RoboCup 2004, they had a challenge session to test the robustness of agents [7]. A team that was consisted of 13 fire brigades, 6 ambulances and 12 police agents performed rescue operations under varying sensing conditions in the same situations of disaster. The relative variations of metrics, such as $|V_{Base} - V_{Half-vision}|/V_{Base}$, are used to estimate the robustness of agents' performances. Table 1 shows the results of three teams for two sensing conditions, Base and Half-vision. Agents could see objects that are within 100m from them in Base sensing condition and the radius of visibility is set to 50m in Half-vision case. From Table 1,

- Team X is the most robust one. Although the performance on H decreases a little at Half-vision case, the scores of V are practically equal.
- In the Base case, teams Y and Z saved more lives than team X, while more houses were burnt down.
- The decreases in P and B of team Y were significant compared to the others. Especially, the decrease of B in Half Vision case leads that team Y was not robust.

Time change of rescue: The factor of time change is important, when rescue rates are taken into consideration from the viewpoint that the initial actions are important for rescue. Figure 1 shows Team Y's time sequence changes of the metric for two simulations. The vertical axis is the relative values to the initial and the horizontal axis shows time steps. The left figure shows the rate of P is constant in the Base condition, while it decreases as time proceeds in the Half-vision condition (right figure).

Attributes such as P, H and B are typical metrics that are present in disaster reports. These evaluation using P, H, B and V are helpful to estimate the overall results of the simulations, however the macro-level properties are complex collective outputs that arise out of agent behaviors and their interactions. They do not have the resolution to analyze what behaviors of agents cause the differences of these metrics and do not give a clue to explain why the performance of one agent team is better than the others.

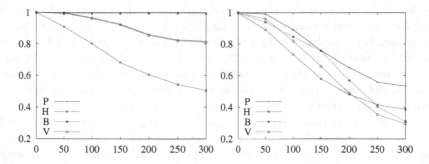

Fig. 1. Changes of Team Y's metrics during simulation steps (Left:Base condition, Right:Half vision condition)

4   Analysis of Agents Behaviors Based on Probability Model

It is necessary to interpret objectively the behaviors of the specific agents from the simulation results. We propose a method to analyze the agents' behaviors.

4.1   Problem Formulation and Behavior Presentation

Multi agent system (MAS) is simply formalized as follows [10].

$$action : S^* \to A, \quad env : S \times A \to \mathcal{P}(S),$$

where $A = \{a_1, a_2, \ldots, a_l\}$ is a set of agent actions, $S = \{s_1, s_2, \ldots, s_m\}$ is a set of environment states and $\mathcal{P}(S)$ is the power set of S. An agent at s_i plans possible actions and executes an action, a_i, according to their knowledge and data that they have gained. The action changes the situations of the environment. The interaction between an agent i and the environment leads to the next state. It is represented as a history h_i:

$$h_i : s_0 \xrightarrow{a_0} s_1 \xrightarrow{a_1} s_2 \xrightarrow{a_2} s_3 \cdots .$$

The set of all agents' histories, $\mathcal{H} = \{h_1, \ldots, h_n\}$ where n is the number of agents, is the information displayed on a screen.

The users of MAS [1] are not necessarily aware of the agents' states, s_i, and their will that causes actions, a_i. The users set the parameters of simulations and check the visual changes in $\mathcal{H}$ that are assumed to display the changes in the states of agents, s_i^o. The observable changes are not equal to the changes of agents' states, s_i, that are not displayed on the screen. The users observe how

---

[1] In this section, the "user" of MAS are referred to people at rescue centers of local governments. They are not programmers, so they are not familiar with the programs and only observe the output of ABSS.

the situations change on monitors, create gradually their own models of agent behaviors, and understand how the ABSS mirrors the social behaviors.

Our idea is to implement the process of the users by extracting the features of agent behavior from $\mathcal{H}$ based on the probability models. Behaviors of agents with the same aim will select similar situations to achieve their goals. However, agents do not necessarily act in the same way because their internal states, the history of their actions, and prior information are different. The behaviors of agents at state i are presented as a set $\{p_{ij}\}$ where p_{ij} is the probability that the agents at state i will take an action that causes it will be at state j at the next time. The probability describes rescue agent behavior by assuming the following:

Assumption 1. Agents select their actions to attain their goal efficiently and promptly.

Assumption 2. States that agents visit more often are more important for the agents than other states.

Assumption 3. Differences between actions with good and bad performances are represented as differences in the history of actions.

A stochastic matrix $\mathbb{P} = \{p_{ij}\}$, where p_{ij} is the probability that the agents at state i will be at state j after one time step, has the following conditions $p_{ij} \geq 0$ and $\sum_i p_{ij} = 1$ for all j. The transition probability can be approximated by taking the ensemble average of how many times agents change states from i to j from the ABSS output. When applying a stochastic matrix directly to interpret agent behaviors, there are the following problems:

- Let m be the size of states, the size of $\mathbb{P}$ is $m \times m$. Knowing m itself indicates that the states of the agents are known before analyzing their behaviors.
- $\sum_i p_{ij} = 1$ for all j implies that transitions to any state can occur. In simulations, there may be some states that do not appear. In such states, $\sum_i p_{ij}$ is zero.

We use a frequency matrix, $\mathbb{F} = \{f_{ij}\}$ instead of $\mathbb{P}$. $\mathbb{F}$ is a variant of $\mathbb{P}$ where p_{ij} are normalized as $\sum_{ij} f_{ij} = 1$. The following properties are indicated by considering $\mathbb{F}$ as a directed graph with f_{ij} s associated with the correspondent edges.

Property 1. $\mathbb{F}$ indicates the behavior of corresponding agents. The rank of $\mathbb{F}$ is proportional to the range of agent motions.

Property 2. Large elements of dominant eigenvectors correspond to states that are important in social simulations.

Property 3. When agents behave two patterns, the eigenvectors corresponding to the patters are independent each other.

4.2   Exemplifications of Matrix Representation

Two illustrative cases are shown to explain the matrix presentation of agents' behavior.

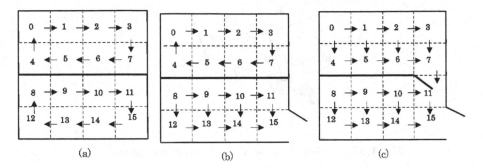

Fig. 2. Three cases of human movements; the room is divided into 4×4 cells and 16 cells correspond to the visual states

Crowds evacuation behavior: Figure 2 shows human movements when they escape from a smoke-filled room. The room is divided into two parts and the figures correspond to the following cases.

(a) Unfortunately, people are unfamiliar with the room and move around in circles to find an exit.
(b) People in the lower part know there is an exit. When persons come at cell 15, they notice the exit and exit.
(c) People in the upper room know that there is a door between two parts. When people come at cell 7, they notice there is the door and move to the lower part through it.

Their behaviors are observed as the time sequence of their locations. The latter part of a 16×16 matrix for case (b) is shown below. It show the elements of a sub matrix from 8th to 11th row and 8th to 15th column when the agent at cell 8 moves to whether cell 9 or cell 12 with probability 0.5 for both.

$$
\begin{array}{c}
\quad\quad 8 \quad\ 9 \quad\ 10 \quad 11 \quad 12 \quad 13 \quad 14 \quad 15 \\
\begin{array}{c} 8 \\ 9 \\ 10 \\ 11 \end{array}
\left(
\begin{array}{cccccccc}
0.5 & & & & 0.5 & & & \\
 & 0.5 & & & & 0.5 & & \\
 & & 0.5 & & & & 0.5 & \\
 & & & & & & & 1
\end{array}
\right)
\end{array}
$$

The rank of the matrix corresponding to case (a), (b) and (c) are 16, 14, and 12, respectively, and the components of eigenvectors are separated according to the patterns of the movements.

Rescue agents' behavior: Figure 3 illustrates a situation where a rescue agent comes from n_0 and extinguishes the fires. The agent knows that there are two fires near n_2 and n_3. At crossing n_1, the agent determines which fire to extinguish and selects whether it will go forward, turn right or perform some other actions. The rescue agent behavior at n_1 is presented as $P_1 = \{p_{10}, p_{11}, p_{12}, p_{13}, p_{14}, p_{15}\}$. p_{1j} is the probability that the agents will be at n_j at the next simulation step.

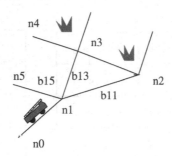

Fig. 3. Illustration of rescue agent behavior model at n_1

Table 2. Metrics of behavior of fire brigade agents in stochastic matrix presentation

Sensing Condition	Team X			Team Y			Team Z		
	Size/Rank	ratio of E.V.s 2/1	3/1	Size/Rank	ratio of E.V.s 2/1	3/1	Size/Rank	ratio of E.V.s 2/1	3/1
Base	770/632	0.87	0.67	583/520	0.85	0.39	434/390	0.61	0.22
Half vision	840/767	0.37	0.25	338/250	0.30	0.30	404/364	0.46	0.16

2/1 (3/1): the ratio of 2nd (3rd) eigen value to the dominant one.

n_j covers all nodes in the map, because the agents can pass one or more nodes in one simulation step.

4.3　Analysis of Rescue Simulation Results and Interpretation

The observed states, $\{s^o\}$, are properties that are recognized visually. Assuming the positions of agents represent their states, p_{ij}s are calculated from how frequent the agents visit to the places.

Analysis from static properties of the matrix: Table 1 at Section 3 shows the results of 2004 Challenge sessions that were performed on a virtual city map (Figure 4). The map has 1065 roads, 1015 nodes, and 953 buildings. They correspond to the states of environments. Table 2 shows the size, rank, and eigenvalues of $\mathbb{F}$ for the behaviors of fire brigades.

Table 2 shows interesting features that cannot be obtained from Table 1. The size of $\mathbb{F}$ is the number of points traversed by the fire brigades. Under the Base sensing condition, the fire brigades of teams X and Y visited 770 and 583 points, respectively. The difference shows that the fire brigades of two teams move differently. At the Half-vision sensing condition, the agents of team X move a larger area than the Base condition, although the metrics of team X in Table 1 are similar values, 78.9, for two different sensing condition cases. The agents of team Y and Z move across smaller areas than the Base condition. This shows team X's behavior is different from the others.

The sizes of the eigenvalues correspond to the importance of the places that the agents visited. When there are two major places, the values of the corresponding

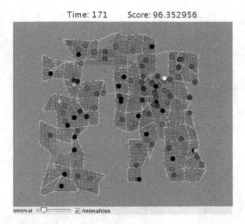

Fig. 4. Map used in 2004 Challenge Game

eigenvalues are of the same order and the ratio of the 1st and 2nd eigenvalues is close to 1.0. In case the agents visit one place, the 1st eigenvalue becomes dominant, the eigenvalues diminish in size, and the ratios decrease to 0.0. Columns 2/1 and 3/1 in Table 2 show the ratios of the dominant eigenvalue to the 2nd and the 3rd values, respectively. It is interesting that the ratios become less at the Half-vision condition than at the Base condition. It is inferred that sensing ability restricts the agents to search and rescue at new places.

Analysis from time sequence changes: Figure 5 shows the snapshots of simulations on the Kobe, Japan. The map consists of 820 roads, 766 nodes and 754 buildings. Thirteen firefighter agents, 7 ambulances, 11 police persons, and 85 civilians are involved in this simulation. The followings can be interpreted from the snapshots.

(a) Initially, three fires break out at B1, B2 and B3.
(b) At 50 steps, fires at B1 and B2 are extinguished, while the fire at B3 continues to spreads.
(c) Agents gather to extinguish the fire at B3.
(d, e) After 150 steps, the spread of the fire is prevented by fire fighting actions.

Table 3 shows the time sequence changes of the burning rate $(1 - B/Bmax)$ and the properties of $\mathbb{F}$ for the fire brigade agents. The properties are the ones in Table 2 and the positions of the biggest components of the dominant, the second and the third eigenvectors.

- The rank of the matrix increases with the simulation steps. This indicates that the range of the fire brigade agents becomes wider with time.
- The first eigenvalue becomes dominant over time. It indicates that, initially, the agents move separately and later they behave in a similar manner as the simulation proceeds.

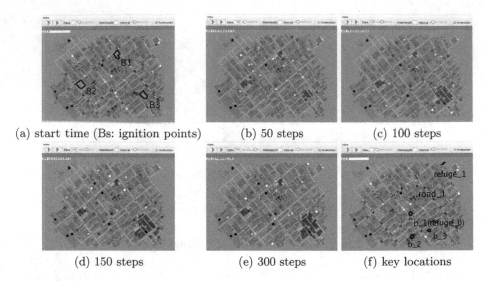

(a) start time (Bs: ignition points)   (b) 50 steps   (c) 100 steps

(d) 150 steps   (e) 300 steps   (f) key locations

Fig. 5. Time sequences of disaster simulations (Black and blue squares indicate the houses that are burnt down and the houses where the fire is extinguished, respectively. The other squares indicate the burning houses. Red, yellow, white, and black circles indicate fire brigades, civilians, ambulances and dead agents, respectively.)

Table 3. Time sequence of burning rate, size of matrix and eigenvectors

step	burning rate	matrix			key locations corresponds to component of dominant e.vectors		
		size/rank	ratio of E.V.s		1st	2nd	3rd
			2/1	3/1			
50	2.6%	155/135	0.38	0.38	b_1	b_2	road_1
100	4.0%	246/217	0.11	0.10	b_1	road_1	b_2
150	4.9%	300/271	0.06	0.06	b_1	road_1	b_2
200	5.0%	355/325	0.05	0.04	b_1	road_1	b_2
250	5.0%	422/388	0.04	0.04	b_1	road_1	b_3
300	5.1%	484/442	0.03	0.03	b_1	b_3	road_1

(f) The figure shows the locations that correspond to the key components of dominant vectors in Table 3.

- Refuge 0 (b_1) is the key building in all the steps,
- from 100 and 250 step, a place (road 1) near the fire (B1) is an important place that corresponds to the second dominant eigen vector,
- after from 250 step, a place($b\,3$) near the other fire (B3) appears to be the key place.

The interpretations correspond well to the interpretations from the snapshots.

5   Discussion and Summary

MAS provides tools to simulate and analyze complex social phenomena. To put ABSS into practical use, the following features are required.

- human behaviors are involved.
- the simulation results are validated.

The validations of ABSS have been done by comparing the simulation data and real-world data. The real-world data of disasters are obtained from the reports that are published from governments or insurance companies. The real-world data are macro level ones and it is difficult to take account of microscopic agent behaviors.

In this paper, we review how the rescue agents have been evaluated in the RCRS agent competitions by the macro-level properties, and propose a method of analyzing agents' behavior based on probability model. The method presents the agent behaviors with a stochastic matrix and analyzes the behaviors by the matrix's properties such as size, rank, eigenvalues and eigenvectors. The method is applied to disaster and rescue simulations, and the experimental results show interpretations that are acceptable from the rescue task knowledge.

The probability model has been widely used to model the variations in the parameters of simulation and agent characters. A large number of simulations with changing them statistically are used to verify simulation results. For example, Johnson et al. have simulated the evacuation of public buildings and proposed risk assessment techniques [4]. Our approach uses a stochastic approach to represent the states of the agents and analyze their behaviors. It shows the range of agents' activities and key locations of their rescue actions and gives clues to interpret the results of ABSS.

The analysis method is based on mathematical quantity such as rank, eigenvalue, etc. They are independent of domain specific evaluations and suggest the potential of the method that will be applied to other fields. For example, when the motions of individual in a crowd will be video tracked, the method can assess the validity of ABSS by comparing its outputs and the behaviors in the real-world data. And the ABSS approach will be used to support decision-making in the future, the proposed method will provide a function to explain the results.

References

1. Farkas, I., Helbing, D., Vicsek, T.: Simulating dynamical features of escape panic. Nature 407, 487–490 (2000)
2. Feynman, R.P.: The Character of Physical Law. MIT Press, Cambridge (1967)
3. Jennings, N.R.: Agent-based computing: Promise and perils. In: Proc. IJCAI 1999, pp. 1429–1436 (1999)
4. Johnson, C.W.: The application of computational models for the simulation of large scale evacuations following infrastructure failures and terrorist incidents. In: Proc. NATO Research Workshop on Computational Models of Risk to Infrastructure, pp. 9–13 (May 2006)

5. Kitano, H., Tadokoro, S., Noda, I., Matsubara, H., Takahashi, T., Shinjou, A., Shimada, S.: Robocup rescue: Search and rescue in large-scale disasters as a domain for autonomous agents research. In: IEEE International Conference on System, Man, and Cybernetics (1999)
6. Moss, S., Edmonds, B.: Towards good social science. Journal of Artificial Societies and Social Simulation 8(4) (2005)
7. RoboCup2004 (2004), http://robot.cmpe.boun.edu.tr/rescue2004/
8. Schurr, N., Marecki, J., Kasinadhuni, N., Tambe, M., Lewis, J.P., Scerri, P.: The defacto system for human omnipresence to coordinate agent teams: The future of disaster response. In: AAMAS 2005, pp. 1229–1230 (2005)
9. Takahashi, T., Ito, N.: Preliminary study to use rescue simulation as check soft of urban's disasters. In: Workshop: Safety and Security in MAS (SASEMAS) at AAMAS 2005, pp. 102–106 (2005)
10. Weiss, G.: Multiagent Systems. MIT Press, Cambridge (2000)

Spiral Development of Behavior Acquisition and Recognition Based on State Value

Yasutake Takahashi[1], Yoshihiro Tamura[1], and Minoru Asada[1,2]

[1] Dept. of Adaptive Machine Systems, Graduate School of Engineering
Osaka University
[2] JST ERATO Asada Synergistic Intelligence Project
Yamadaoka 2-1, Suita, Osaka, 565-0871, Japan
{yasutake,yoshihiro.tamura,asada}@ams.eng.osaka-u.ac.jp
http://www.er.ams.eng.osaka-u.ac.jp

Abstract. Both self-learning architecture (embedded structure) and explicit/implicit teaching from other agents (environmental design issue) are necessary not only for one behavior learning but more seriously for life-time behavior learning. This paper presents a method for a robot to understand unfamiliar behavior shown by surrounding players through the collaboration between behavior acquisition and recognition of observed behavior, where the state value has an important role not simply for behavior acquisition (reinforcement learning) but also for behavior recognition (observation). That is, the state value updates can be accelerated by observation without real trials and errors while the learned values enrich the recognition system since it is based on estimation of the state value of the observed behavior. The validity of the proposed method is shown by applying it to a soccer robot domain.

1 Introduction

Reinforcement learning has been studied well for motor skill learning and robot behavior acquisition in both single and multi-agent environments. Especially, in the multi-agent environment, observation of surrounding players make the behavior learning rapid and therefore much more efficient [1,3,7]. Actually, it is desirable to acquire various unfamiliar behavior with some instructions from others in real environment because of huge exploration space and enormous learning time to learn. Therefore, behavior learning through observation has been more important. Understanding observed behavior does not mean simply following the trajectory of an end-effector or joints of demonstrator. It means reading his/her intention, that is, the objective of the observed behavior and finding a way how to achieve the goal by oneself regardless of the difference of the trajectory. From a viewpoint of the reinforcement learning framework, this means reading rewards of the observed behavior and estimating sequence of the value through the observation.

Takahashi et al.[6] proposed a method of not only to learn and execute a variety of behavior but also to recognize observed behavior executed by surrounding

L. Iocchi et al. (Eds.): RoboCup 2008, LNAI 5399, pp. 343–354, 2009.

players supposing that the observer has already acquired the values of all kinds of behavior the observed agent can do. The recognition means, in this paper, that the robot categorizes the observed behavior to a set of its own behavior acquired beforehand. The method seamlessly combines behavior acquisition and recognition based on "state value" in reinforcement learning scheme. Reinforcement learning generates not only an appropriate behavior (a map from states to actions) to accomplish a given task but also a utility of the behavior, an estimated sum of discounted rewards that will be received in future while the robot is taking an appropriate policy. This estimated discounted sum of reward is called "state value." This value roughly indicates closeness to the goal state of the given task if the robot receives a positive reward when it reaches the goal and zero else, that is, if the agent is getting closer to the goal, the value becomes higher. This suggests that the observer may recognize which goal the observed agent likes to achieve if the value of the corresponding task is going higher.

This paper proposes a novel method that enhances behavior acquisition and recognition based on interaction between learning and observation of behavior. Main issues to be attacked here is

- categorization of unfamiliar behavior based on reading rewards, and
- enhancement of the behavior recognition and acquisition.

A learning/recognizing robot assumes that all robots and even the human player share reward models of the behavior. For example, all robots and the human player receive a positive reward when the ball is kicked into the goal. This assumption is very natural as we assume that we share "value" with colleagues, friends, or our family in our daily life. The reading rewards during the observation of unfamiliar behavior gives a hint to categorize the observed behavior into the one that is meaningful to the observer. For the second issue, a robot learns its behavior through not only trials and errors but also reading rewards of the observed behavior of others (including robots and humans). Fig. 1 shows a rough idea of our proposed method. $V(s)$ and $\hat{V}(s)$ are the state value updated by oneself and the state value estimated through observation, respectively. Takahashi et al.[6] showed the capability of the proposed method mainly in case that the observer has already acquired a number of behavior to be recognized beforehand. Their case study showed how this system recognizes observed behavior based on the

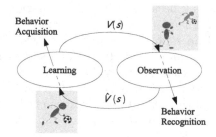

Fig. 1. Interaction between Learning and Observation of Behavior

state value functions of self-behavior. This paper shows how the estimated state value of observed behavior, $\hat{V}(s)$, gives feedback to learning and understanding unfamiliar observed behavior and this feedback loop enhances the performance of observed behavior recognition. The validity of the proposed method is shown by applying it to a soccer robot domain including a human player.

2   Outline of the Mechanisms

The reinforcement learning scheme, the state/action value function, recognition of observed behavior based on state value function, and the modular learning system for various behavior acquisition/recognition are explained, here.

2.1   Behavior Learning Based on Reinforcement Learning

An agent can discriminate a set S of distinct world states. The world is modeled as a Markov process, making stochastic transitions based on its current state and the action taken by the agent based on a policy π. The agent receives reward r_t at each step t. State value V^π, the sum of the discounted rewards received over time under execution of a policy π, will be calculated as follows:

$$V^\pi = \sum_{t=0}^{\infty} \gamma^t r_t \ . \tag{1}$$

For simplicity, it is assumed here that the agent receives a positive reward if it reaches a specified goal and zero else. Then, the state value increases if the agent comes close to the goal by following a good policy π. The agent updates its policy through trials and errors in order to receive higher positive rewards in future. Analogously, as animals get closer to former action sequences that led to goals, they are more likely to retry it. For further details, please refer to the textbook of Sutton and Barto [4] or a survey of robot learning [2].

Here we introduce a model-based reinforcement learning method. A learning module has a forward model which represents the state transition model and a behavior learner which estimates the state-action value function based on the forward model in a reinforcement learning manner.

Each learning module has its own state transition model. This model estimates the state transition probability $\hat{\mathcal{P}}_{ss'}^a$ for the triplet of state s, action a, and next state s':

$$\hat{\mathcal{P}}_{ss'}^a = Pr\{s_{t+1} = s' | s_t = s, a_t = a\} \tag{2}$$

Each module has a reward model $\hat{\mathcal{R}}_s$, too:

$$\hat{\mathcal{R}}(s) = E\{r_t | s_t = s\} \tag{3}$$

All experiences (sequences of state-action-next state and reward) are simply stored to estimate these models. Now we have the estimated state transition

Fig. 2. Robots with a human player in a Soccer Field

probability $\hat{\mathcal{P}}_{ss'}^a$ and the expected reward $\hat{\mathcal{R}}_s$, then, an approximated state-action value function $Q(s, a)$ for a state action pair s and a is given by

$$Q(s, a) = \sum_{s'} \hat{\mathcal{P}}_{ss'}^a \left[\hat{\mathcal{R}}(s') + \gamma V(s') \right] \qquad (4)$$

$$V(s) = \max_a Q(s, a), \qquad (5)$$

where $0 < \gamma < 1$ is a discount factor[1].

2.2   Behavior Recognition Based on Estimated Values

An observer watches a demonstrator's behavior and maps the sensory information from an observer viewpoint to a demonstrator's one with a simple mapping of state variables[2]. Fig. 2 shows two robots, a human player and color-coded objects, e.g., an orange ball, and a goal. The robot has an omni-directional camera on top. A simple color image processing is applied in order to detect the color-coded objects and players in real-time. The mobile platform is based on an omni-directional vehicle. These two robots and the human play soccer such as dribbling a ball, kicking it to a goal, passing a ball to the other, and so on. While playing with objects, they watch each other, try to understand observed behavior of the other, and emulate them. Fig. 3 shows a simple example of this transformation. It detects color-coded objects on the omni-directional image, calculates distances and directions of the objects in the world coordinate of the observer, and shifts the axes so that the position of the demonstrator comes to center of the demonstrator's coordinate. Then it roughly estimates the state information of the demonstrator. A sequence of its state value from the estimated state of the observed demonstrator is estimated. If the state value goes up during an observed behavior, it means that the behavior derived by the state value system is valid for explaining the observed behavior. Fig. 4 shows an example task of navigation in a grid world and a map of the state value of the task. There is a goal state at the top center of the world. An agent can move one of the neighboring square in the grids every step. It receives a positive reward only when

---

[1] The discount factor represents that distant rewards are less important.

[2] For a reason of consistency, the term "demonstrator" is used to describe any agent from which an observer can learn, even if the demonstrator does not have an intention to show its behavior to the observer in this paper.

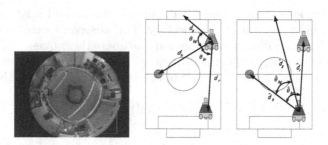

Fig. 3. Estimation of view of the demonstrator. Left : a captured image the of observer, Center : object detection and state variables for self, Right : estimation of view of the demonstrator.

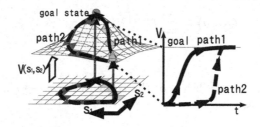

Fig. 4. Behavior recognition based on the change of state value

it stays at the goal state while zero else. There are various optimal/suboptimal policies for this task as shown in the figure. If the agent follows an appropriate policy, the value is going up even if it is not exactly the optimal one.

Here we define recognition reliability g that indicates how much the observed behavior would be reasonable to be recognized as a behavior

$$g = \begin{cases} g + \beta & \text{if } V_t - V_{t-1} > 0 \text{ and } g < 1 \\ g & \text{if } V_t - V_{t-1} = 0 \\ g - \beta & \text{if } V_t - V_{t-1} < 0 \text{ and } g > 0 \ , \end{cases}$$

where β is an update parameter, and 0.1 in this paper. This equation indicates that the recognition reliability g will become large if the estimated state value rises up through time and it will become low when the estimated state value goes down. Another condition is to keep g value in the range from 0 to 1.

2.3   Learning by Observation

In the previous section, behavior recognition system based on state value of its own behavior is described. This system shows robust recognition of observed behavior [5] only when the behavior to be recognized has been well-learned beforehand. If the behavior is under learning, then, the recognition system is not able to show good recognition performance at beginning. The trajectory of the observed behavior can be a bias for learning behavior and might enhance the behavior learning based on the trajectory. The observer cannot watch actions of observed behavior directly and can only estimate the sequence of the state of the observed robot. Let s_t^o be the estimated state of the observed robot at time t. Then, the estimated state value $\hat{V}^o$ of the observed behavior can be calculated as below:

$$\hat{V}^o(s) = \sum_{s'} \hat{P}_{ss'}^o \left[\hat{R}(s') + \gamma V^o(s') \right] \tag{6}$$

where $\hat{P}_{ss'}^o$ is state transition probability estimated from the behavior observation. This state value function $\hat{V}^o$ can be used as a bias of the state value function of the learner V. The learner updates its state-action value function $Q(s, a)$ during trials and errors based on the estimated state value of observed behavior $\hat{V}^o$ as below:

$$Q(s, a) = \sum_{s'} \hat{P}_{ss'}^a \left[\hat{R}(s') + \gamma V'(s') \right] \tag{7}$$

while

$$V'(s) = \begin{cases} V(s) & \text{if } V(s) > \hat{V}^o(s) \text{ or the transition}(s, a) \to s' \text{is well experienced} \\ \hat{V}^o(s) & \text{else} \end{cases}$$

This is a normal update equation as shown in (4) except using $V'(s)$. The update system switches the state value of the next state s' between the state value of own learning behavior $V(s')$ and the one of the observed behavior $\hat{V}^o(s')$. It takes $V(s')$ if the state value of own learning behavior $V(s')$ is bigger than the one of the observed behavior $\hat{V}^o(s')$ or the state transition $(s, a) \to s'$ is well experienced by the learner, $\hat{V}^o(s')$ else. This means the state value update system takes $\hat{V}^o(s')$ if the learner does not estimate the state value $V(s')$ because of lack of experience at the state s' from which it reaches to the goal of the behavior. $\hat{V}^o(s')$ becomes a bias for reinforcing the action a from the state s even though the state value of its own behavior $V(s')$ is small so that it leads the learner to explore the space near to the goal state of the behavior effectively.

2.4   Modular Learning System

In order to observe/learn/execute a number of behavior in parallel, we prepare a number of behavior modules each of which adopts the behavior learning and behavior recognition method as shown in Fig. 5. The learner tries to acquire a number of behavior shown in Table 1. The table also describes necessary state variables for each behavior. A range of each state variable is divided into 11 in order to construct a quantized state space. 6 actions are prepared to be selected by the learning module for each behavior: Approaching the goal, approaching the teammate, approaching the ball, leaving from the ball, turn around the ball clockwise and counterclockwise.

A demonstrator is supposed to show a number of behavior which are not informed directly to the observer. In order to update the estimate values of the

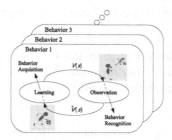

Fig. 5. Modular Learning System for Behavior Acquisition and Recognition

Table 1. List of behavior learned by self and state variables for each behavior

Behavior	State variables
Approaching a ball	d_b
Approaching a goal	d_g
Approaching the teammate	d_r
Shooting a ball	d_b, d_g, θ_{bg}
Passing a ball	d_b, d_r, θ_{br}

behavior the demonstrator is taking, the observer has to estimate which behavior the demonstrator is taking correctly. If the observer waits to learn some specific behavior by observation until it becomes able to recognize the observed behavior well, bootstrap of leaning unfamiliar behavior by observation cannot be expected. Therefore, the observer(learner) maintains a history of the observed trajectories and updates value functions of the observed behavior with

- high recognition reliability or
- high received reward.

The observer estimates the state of the demonstrator every step and the reward received by the demonstrator is estimated as well. If it is estimated that the demonstrator receives a positive reward by reaching to the goal state of the behavior, then, the observer updates the state value of the corresponding behavior even if it has low recognition reliability for the observed behavior. The update strategy enhances to estimate appropriate values of the observed behavior.

3   Behavior Learning by Observation

3.1   Experimental Setup

In order to validate the effect of interaction between acquisition and recognition of behavior through observation, two experiments are set up. One is that the learner does not observe the behavior of other but tries to acquire shooting/passing behavior by itself. The other is that the learner observes the behavior of other and enhances the learning of the behavior based on the estimated state value of the observed behavior. In the former experiment, the learner follows the learning procedure:

1. 15 episodes for behavior learning by itself
2. evaluation of self-behavior performance
3. evaluation of behavior recognition performance
4. goto 1.

On the other hand, in the later experiment, it follows :

1. 5 episodes for observation of the behavior of the other
2. 10 episodes for behavior learning by self-trials with observed experience
3. evaluation of self-behavior performance
4. evaluation of behavior recognition performance
5. goto 1.

Both learners attempt to acquire behavior listed in Table 1. The demonstrator shows one of the behavior one by one but the observer does not know which behavior the demonstrator is taking. In both experiments, the learner follows ϵ-greedy method; it follows the greedy policy with 80% probability and takes a random action else. Performance of the behavior execution and recognition of observed behavior during the learning time is evaluated every 15 learning episodes. The performance of the behavior execution is the success rate of the behavior while the learner, the ball, and the teammate are placed at a set of pre-defined positions. The one of the behavior recognition is the average length of period in which the recognition reliability of the right behavior is larger than 70% during the observation. The soccer field area is divided 3 by 3 and the center of the each area is a candidate of the initial position of the ball, the learner, or the teammate. The performances are evaluated in all possible combinations of the positions.

3.2   Recognition of Observed Behavior

Before evaluating the performance of the behavior execution and behavior recognition of other during learning the behavior, we briefly review how this system estimates the values of behavior and recognizes the observed behavior after the observer has learned behavior. When the observer watches a behavior of the other, it recognizes the observed behavior based on repertoire of its own behavior. Figs. 6 (a) and (b) show sequences of estimated values and reliabilities of the behavior, respectively. The line that indicates the passing behavior keeps tendency of increasing value during the behavior in this figures. This behavior is composed of behavior of approaching a ball and approaching the teammate again, then, the line of approaching a ball goes up at the earlier stage and the line of approaching the teammate goes up at the later stage in Fig. 6(a). All reliabilities start from 0.5 and increase if the value goes up and decrease else. Even when the value stays low, if it is increasing with small value, the recognition reliability of the behavior increases rapidly. The recognition reliability of the

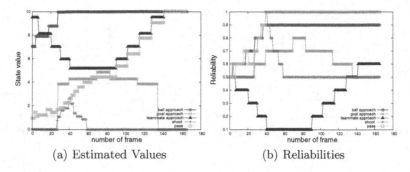

(a) Estimated Values   (b) Reliabilities

Fig. 6. Sequence of estimated values and reliabilities during a behavior of pushing a ball to the magenta player, red line : approaching a ball, green line : approaching the goal, light blue line : passing, blue line : approaching the other, magenta line : shooting

behavior of pushing a ball into the teammate reaches 1.0 at middle stage of the observed behavior. "Recognition period rate" of observed behavior is introduced here to evaluate how long the observer can recognize the observed behavior as a correct one. The recognition period rate is 85% here ,that means, the period in which the recognition reliability of the passing behavior is over 70% is 85% during the observation.

3.3 Performance of Behavior Learning and Recognition

In this section, performances of the behavior execution and behavior recognition during learning the behavior are shown. Fig. 7 shows success rates of the behavior and their variances during learning in cases of learning with/without value update through observation. The success rates with value update of all kinds of behavior grows more rapidly than the one without observation feedback. Rapid learning is one of the most important aspect for a real robot application. The success rate without observation sometimes could not reach the goal of the behavior at the beginning of the learning because there is no bias to lead the robot to learn appropriate actions. This is the reason why the variances of the rate is big. On the other hand, the system with value update through observation utilizes the observation to bootstrap the learning even though it cannot read exact actions of observed behavior.

Recognition performance and recognition period rate of observed behavior and their variances are shown in Figs. 8 and 9, respectively. They indicate a similar aspect with the success rates of the acquired behavior. The performance of the behavior recognition depends on the learning performance. If the learning system has not experienced enough to estimate state value of the observed behavior, it cannot perform well. The learning system with value update with observed behavior rapidly enables to recognize the behavior while the system without value update based on the observation has to wait to realize a good recognition performance until it estimates good state value of the behavior by its own trials and errors.

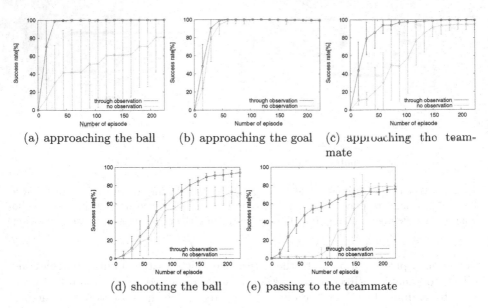

(a) approaching the ball   (b) approaching the goal   (c) approaching the team-
                                                                mate

(d) shooting the ball   (e) passing to the teammate

Fig. 7. Success rate of the behavior during learning with/without observation of demonstrator's behavior

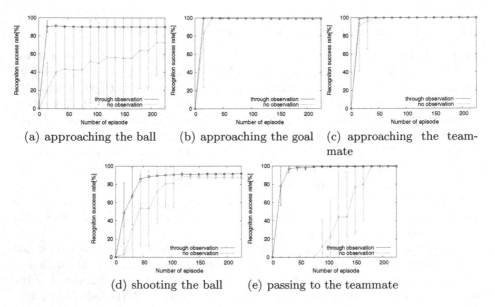

(a) approaching the ball   (b) approaching the goal   (c) approaching the team-
                                                                mate

(d) shooting the ball   (e) passing to the teammate

Fig. 8. Recognition performance of the behavior during learning with/without observation of demonstrator's behavior

Those figures show the importance of learning through interaction between behavior acquisition and recognition of observed behavior.

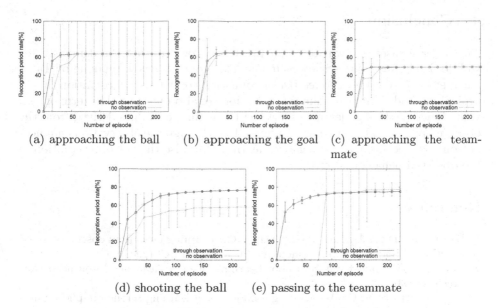

(a) approaching the ball   (b) approaching the goal   (c) approaching the teammate

(d) shooting the ball   (e) passing to the teammate

Fig. 9. Recognition period rate of the behavior during learning with/without observation of demonstrator's behavior

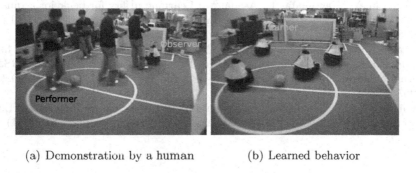

(a) Demonstration by a human   (b) Learned behavior

Fig. 10. Demonstration of a passing behavior by a human and acquired passing behavior after the observation and learning

3.4   Experiments with a Real Robot and a Human Demonstrator

The proposed architecture was applied to the real robot system. The robot observed demonstrations of shooting and passing behavior played by a human player first, then, the robot learned the behavior by itself. The human player showed shooting and passing behavior 5 times for each. The real learning time was half an hour for each behavior. Figs. 10 (a) and (b) show one scene of human demonstration of the passing behavior and acquired passing behavior, respectively. The robot acquired the observed behavior within an hour.

4　Conclusion

The observer uses its own value functions to recognize what the demonstrator will do. Preliminary investigations in a similar context have been done by Takahashi et al. [5] and they showed better robustness of behavior recognition than a typical method. In paper, unknown behavior are also understood in term of own value function through learning based on the estimated values derived from the observed behavior. Furthermore, value update through the observation enhances not only the performance of behavior learning but also the one of recognition of the observed behavior effectively.

References

1. Bentivegna, D.C., Atkeson, C.G., Chenga, G.: Learning tasks from observation and practice. Robotics and Autonomous Systems 47, 163–169 (2004)
2. Connell, J.H., Mahadevan, S.: Robot Learning. Kluwer Academic Publishers, Dordrecht (1993)
3. Price, B., Boutilier, C.: Accelerating reinforcement learning through implicit imitatione. Journal of Articial Intelligence Research (2003)
4. Sutton, R.S., Barto, A.G.: Reinforcement Learning: An Introduction. MIT Press, Cambridge (1998)
5. Takahashi, Y., Kawamata, T., Asada, M.: Learning utility for behavior acquisition and intention inference of other agent. In: Proceedings of the 2006 IEEE/RSJ IROS 2006 Workshop on Multi-objective Robotics, pp. 25–31 (October 2006)
6. Takahashi, Y., Kawamata, T., Asada, M., Negrello, M.: Emulation and behavior understanding through shared values. In: Proceedings of the 2007 IEEE/RSJ International Conference on Intelligent Robots and Systems, pp. 3950–3955 (October 2007)
7. Whitehead, S.D.: Complexity and cooperation in q-learning. In: Proceedings Eighth International Workshop on Machine Learning (ML 1991), pp. 363–367 (1991)

Determining Map Quality through an Image Similarity Metric

Ioana Varsadan, Andreas Birk, and Max Pfingsthorn

Jacobs University Bremen*
Campus Ring 1
28759 Bremen, Germany

Abstract. A quantitative assessment of the quality of a robot generated map is of high interest for many reasons. First of all, it allows individual researchers to quantify the quality of their mapping approach and to study the effects of system specific choices like different parameter values in an objective way. Second, it allows peer groups to rank the quality of their different approaches to determine scientific progress; similarly, it allows rankings within competition environments like RoboCup. A quantitative assessment of map quality based on an image similarity metric Ψ is introduced here. It is shown through synthetic as well as through real world data that the metric captures intuitive notions of map quality. Furthermore, the metric is compared to a seemingly more straightforward metric based on Least Mean Squared Euclidean distances (LMS-ED) between map points and ground truth. It is shown that both capture intuitive notions of map quality in a similar way, but that Ψ can be computed much more efficiently than the LMS-ED.

1 Introduction

One option to evaluate robotic systems is to use competitions such as RoboCup [1], the European Land Robot Trials (ELROB) [2], the AAAI robot competition [3], or the Grand Challenge [4]. In this kind of competitions, the level of system integration and engineering skills of a particular team are evaluated by generating a ranking. According competitions hence have their limits in measuring the performance of a robotic subsystem or a particular algorithm as they concentrate on mission based evaluation of complete systems.

A more fine grain evaluation approach is task-based performance testing [5], similar to observing a mouse in maze. The Intelligent Systems Division (ISD) of the US National Institute of Standards and Technology (NIST) has for example developed a collection of test elements to assess the performance of response robots. The test elements are arranged in so-called arenas [5][6][7][8]. This more fine grain evaluation can of course also be used in competitions; a ranking can simply be produced by summing the results over the different tests up, respectively by letting all participants compete in the same collection of test elements.

---

* Formerly International University Bremen.

L. Iocchi et al. (Eds.): RoboCup 2008, LNAI 5399, pp. 355–365, 2009.

An example for this is the usage of NIST test arenas in the RoboCup Rescue League.

Task based performance is excellent when there are simple ways to determine the quality with which a task was solved. For example, a robot can or can not negotiate over a particular type of obstacle, it can detect a victims in a given time frame under given constraints or not, or it can manipulate an object or not. Through repeated trials and variations of the challenges, it is even possible to statistically measure in which degree the robot is capable to solve the task. But it gets significantly more difficult if solving the task is related to the generation of complex output. Then, the decision of whether and to which degree the robot has solved its task is much more difficult. One perfect example for this problem is the task to generate a map.

Mapping in general is a core problem for mobile robots [9]. It is hence surprising, that the vast amount of publications in this field does not provide any numerical comparisons of map quality. The benefits of new approaches are usually presented through theoretical advantages, run times, and a qualitative assessment of map quality, e.g., by displaying the generated maps and the results of other approaches or a ground truth plan for comparison. Surprisingly often, even the qualitative comparison basis in form of a ground truth plan or maps generated by other approaches is omitted.

Here, a new metric for assessing the quality of the predominantly used form of maps, namely occupancy grids [10][11], is proposed. The metric is very fast to compute and it gives meaningful numerical feedback about the distortion of a map due to common error sources, namely salt and pepper noise as well as translation, rotation and scale error. The metric is based on an image similarity function Ψ [12]. The computation of the metric is embedded in a tool for assessing map quality, the Jacobs Map Analysis Tool. The Jacobs Map Analysis Tool is open source and freely available as download from the Jacobs University Robotics Group.

The map quality metric based on Ψ has some advantages over alternative metrics, which have been proposed before for assessing map quality. The discussion of ψ and the comparison to alternative metrics is done in the following section 2.

2   Metrics for Map Quality

2.1   The Image Similarity Ψ

The method proposed here for evaluating map quality is based on an image similarity metric. The similarity function Ψ [12] is defined as a sum over all colors of the average Manhattan-distance of the pixels with color c in picture a to the nearest pixel with color c in picture a' ($\psi(a, a', c)$) and the average distance vice versa ($\psi(a', a, c)$):

$$\Psi(a, a') = \sum_{c \in C} (\psi(a, a', c) + \psi(a', a, c)) \tag{1}$$

where

$$\psi(a, a', c) = \frac{\sum_{a[p_1]=c} min(md(p_1, p_2)|a'(p_2) = c)}{\#_c(a)} \qquad (2)$$

where

- C denotes the set of colors of constant size, $a[p]$ denotes the
- color c of pixel array a at position $p = (x, y)$,
- $md(p_1, p_2) = |x_1| + |x_2| + |y_1| + |y_2|$ is the Manhattan distance between pixels p_1 and p_2
- and $\#_c(a)$ is the number of pixels of color c in a.

In the context of occupancy grid maps, the color c simply corresponds to occupancy information. This means for each occupied cell in map m, the Manhattan distance to the nearest occupied cell in ground truth is computed and vice versa.

The main advantage of Ψ is that it can be very efficiently computed, namely in linear time. Figure 1 shows the algorithm. It is based on two relaxation steps as illustrated in figure 2.

2.2   Alternative Metrics

An alternative method of assessing map quality for grid maps is the cross entropy [13]. In information theory, the Shannon entropy or information entropy is a measure of the uncertainty associated with a random variable. Cross entropy - also known as Kullback-Leibler divergence - is used to measures the difference

```
for y = 0 to n-1 {
        for x = 0 to n-1 {
                if a'((x,y)) = c
                        d-map_c[x][y] = 0
                else
                        d-map_c[x][y] = ∞
        }
}
for y = 0 to n-1 {
        for x = 0 to n-1 {
                h = Min{d-map_c[x-1][y] + 1, d-map_c[x][y-1] + 1}
                d-map_c[x][y] = Min{d-map_c[x][y], h}
        }
}
for y = n-1 to 0 step -1 {
        for x = n-1 to 0 step -1 {
                h = Min{d-map_c[x+1][y] + 1, d-map_c[x][y+1]+1}
                d-map_c[x][y] = Min{d-map_c[x][y], h}
        }
}
```

Fig. 1. The Algorithm for Computing Ψ

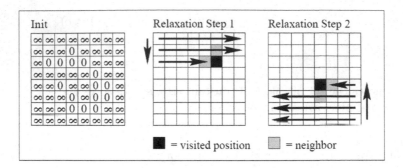

Fig. 2. The working principle of computing ψ. It is a simple relaxation algorithm that just takes one pass for initialization and two passes over the map for processing.

between two probability distributions [14]. This can be employed to measure the correlation between two occupancy grids, especially a map as estimated grid and ground truth as reference. The cross entropy is defined as:

$$CrossEntropy = \sum_{x,y}^{N} (1 + A(x,y)log2(B_(x,y)) + (1 - A_(x,y))log2(1 - B_(x,y)))$$

(3)

where A is the reference grid and B is the estimated grid.

The major disadvantage of cross entropy is that it only takes information from co-located cells in the map and in ground truth into account. The map hence has to be pretty well aligned with the ground truth to give any meaningful result. In the case of any larger disturbances, cross entropy fails to provide information about the map quality. The obvious way to solve this significant disadvantage is to take a measure of distance between similar points in the reference grid and the estimated one. A straightforward choice are Least Mean Squares of Euclidean Distances (LMS-ED) [15]. As we will also see in the experiment section 4, LMS-ED as well as any similar Euclidean neighborhood based metric is expensive to compute. It is hence usually not applied to all cells of the whole occupancy grids, but only to a very limited sets of landmarks [16][17]. This gives the additional burden of having to use some feature extraction technique to find suited landmarks. Furthermore, the feature extraction is usually based on advanced techniques like SIFT [18], which actually compensate some of the error, which is supposed to be measured by the map quality metric.

3   Ground Truth Data Generation

A reason for the scarce research in map evaluation by comparison to a reference map might be the unavailability of ground truth data. But then there is still the option to at least compare the result from on technique to the result from an other. For example, [19] compares positions acquired by the robot while

Fig. 3. An example of a rescue arena. It consists of standardized test elements, which facilitates the generation of a ground truth floor plan.

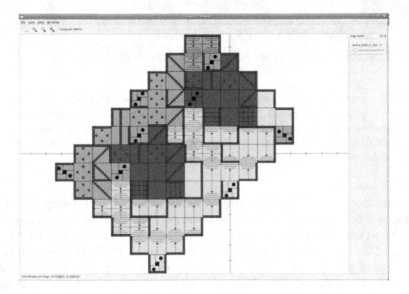

Fig. 4. The generation of a ground truth map with the Rescue Arena Designer as part of the Jacobs Map Analysis Tool. This module of the Jacobs Map Analysis Tool allows to easily generate ground truth data by combining standardized components by drag and drop.

doing a particular SLAM algorithm [20] with positions acquired by Monte Carlo Localization (MCL) [21]. As the MCL is more accurate than the SLAM algorithm and as it is in addition corrected by human supervision, the data from the MCL is considered to be a kind of ground truth.

Here, ground truth data is generated by special feature of the Jacobs Map Analysis Tool, namely the ground truth designer module. As already mentioned

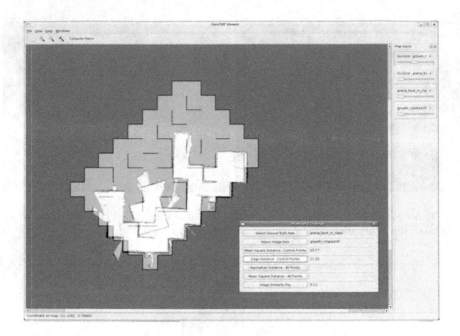

Fig. 5. The overlay module of the Jacobs Map Analysis Tool. It allows to display a map - or even several ones in parallel - over ground truth data. A slider allows to change the transparency of each layer, i.e., map. This makes it very easy for human users to get a qualitative assessment of the quality of a single map, respectively to compare the quality of different maps against each other.

in the introduction, National Institute of Standards and Technology (NIST) has developed a collection of test elements arranged in so-called arenas [5][6][7][8]. The ground truth designer exploits the fact that the test elements have a fixed footprint. The floor plan of a robot test arena is hence determined by the arrangement of the test elements (figure 3). The ground truth designer features symbols for the different test elements, which can be easily placed by drag and drop to generate a ground truth floor plan (figure 4).

The use of the ground truth designer is of course not mandatory for map comparison. It is also possible to import ground truth data into the Jacobs Map Analysis Tool in form of raster data for which the popular GeoTiff standard is supported.

The Jacobs Map Analysis Tool allows features an overlay modus where maps and ground truth can be displayed in parallel over each other (figure 5). A transparency slider is associated with each layer, i.e., map. By adding some transparency to a map lying over ground truth, the true floor can shine through the map. This gives hence some easy possibility for human users to have a qualitative assessment of the quality of the map. Of course, the same mechanisms can be used to compare two different maps to each other.

4   Experimental Results

In the following experiments, the performance of Ψ a map quality metric is compared with the Least Mean Squared Euclidean Distance (LMS-ED). For this purpose, maps were created with a range of different degrees of error for several error types. Concretely, different levels of salt and pepper noise as well as rotation, translation and scale errors are tested. Examples of the different effects of the errors to a map are shown in figure 6.

In the figures 7 to 10, the effects of the different error sources to the different map quality metrics are shown. Please note that for a perfect map, i.e., for a map that is identical to ground truth, both metrics always return a value of Zero. For increasing error levels, both metrics return increasing values. Note

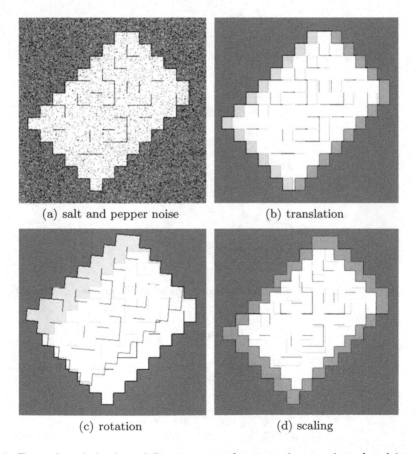

(a) salt and pepper noise   (b) translation

(c) rotation   (d) scaling

Fig. 6. Examples of the four different types of systematic error introduced into the maps. The disturbed maps are displayed over the ground truth; each map is shown in bright black and white over the ground truth data, which shines through as gray scale values.

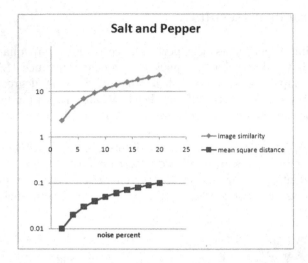

Fig. 7. The influence of different levels of salt and pepper noise to the map quality metrics

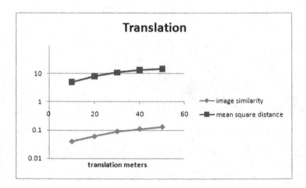

Fig. 8. The influence of different levels of translation to the map quality metrics

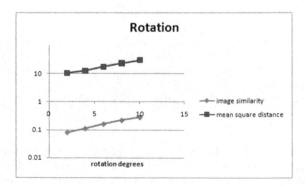

Fig. 9. The influence of different levels of rotation to the map quality metrics

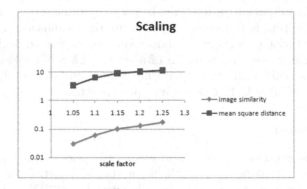

Fig. 10. The influence of different levels of scaling to the map quality metrics

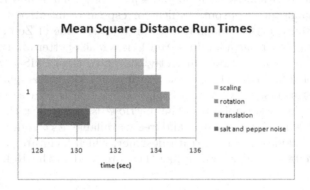

Fig. 11. Average run times for computing Least Mean Squared Euclidean Distances (LMS-ED)

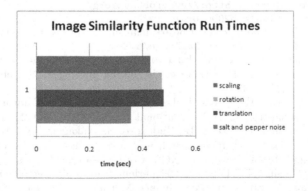

Fig. 12. Average run times for computing the image similarity Ψ

that the shapes of both metrics are very similar for all error types, i.e., there is no observable difference in using either Ψ or LMS-ED for assessing map quality in a numerical way.

But while the outputs of both metrics behave very similar, there is a significant difference in the run times to compute both metrics as shown in figures 11 and 12. The computation of Ψ is always almost four orders of magnitude faster than LMS-ED. While ψ takes a few tenths of a second, LMS-ED takes in the order of two minutes. This is due to the fact that the computation of Ψ can be done in linear time, whereas LMS-ED requires quadratic time.

5   Conclusion

A metric for assessing map quality is proposed. It is based on the image similarity Ψ. It is shown that it behaves much like a standard metric for map quality, namely Least Mean Squared Euclidean Distances (LMS-ED). Concretely, it is shown through different levels of common error sources, namely salt and pepper noise as well as rotation, translation and scale errors, that Ψ and LMS-ED both give similar quantitative outputs, which correspond to intuitive notions of map quality. Especially, perfect maps lead to an output value of Zero; higher levels of error lead to higher output values. But Ψ is a much better alternative as map metric as it can be much more efficiently computed than LMS-ED. Concretely, the runtime of Ψ is linear whereas the runtime of LMS-ED is quadratic. In the concrete experiments, ψ accordingly takes a few tenths of a second, LMS-ED takes in the order of two minutes. The metric Ψ is included in the Jacobs Map Analysis Tool, which is open source and freely available. As an additional feature, the tool also includes a ground truth designer, which facilitates the generation of reference data for NIST arenas like they are used in the RoboCup Rescue League.

References

1. The robocup federation, http://www.robocup.org/
2. Fgan, http://www.elrob2006.org/
3. Association for the advancement of artificial intelligence, www.aaai.org
4. Darpa, www.darpa.mil/grandchallenge/
5. Jacoff, A., Weiss, B., Messina, E.: Evolution of a performance metric for urban search and rescue. In: Performance Metrics for Intelligent Systems (PERMIS), Gaithersburg, MD (2003)
6. Jacoff, A., Messina, E., Evans, J.: Performance evaluation of autonomous mobile robots. Industrial Robot: An International Journal 29(3), 259–267 (2002)
7. Jacoff, A., Messina, E., Weiss, B., Tadokoro, S., Nakagawa, Y.: Test arenas and performance metrics for urban search and rescue robots. In: Proceedings of the Intelligent and Robotic Systems (IROS) Conference (2003)
8. Jacoff, A., Messina, E., Evans, J.: Experiences in deploying test arenas for autonomous mobile robots. In: Performance Metrics for Intelligent Systems (PERMIS), Mexico City (2001)
9. Thrun, S.: Robotic mapping: A survey. In: Lakemeyer, G., Nebel, B. (eds.) Exploring Artificial Intelligence in the New Millenium. Morgan Kaufmann, San Francisco (2002)
10. Elfes, A.: Using occupancy grids for mobile robot perception and navigation. Computer 22(6), 46–57 (1989)

11. Moravec, H., Elfes, A.: High resolution maps from wide angle sonar. In: Proceedings of the IEEE International Conference on Robotics and Automation, pp. 116–121 (1985)
12. Birk, A.: Learning geometric concepts with an evolutionary algorithm. In: Proc. of The Fifth Annual Conference on Evolutionary Programming. MIT Press, Cambridge (1996)
13. Moravec, M.B.H.: Learning sensor models for evidence grids. In: CMU Robotics Institute 1991 Annual Research Review, pp. 8–15 (1993)
14. Kullback, R.A.L.S.: On information and sufficiency. Ann. Math. Stat., 79–86 (1951)
15. Yairi, T.: Covisibility based map learning method for mobile robots. In: Zhang, C., Guesgen, H.W., Yeap, W.-K. (eds.) PRICAI 2004. LNCS, vol. 3157, pp. 703–712. Springer, Heidelberg (2004)
16. Wnuk, K.: Dense 3d mapping with monocular vision (March 2005)
17. Blanco, J.-L., González, J., Fernandez-Madrigal, J.-A.: A new method for robust and efficient occupancy grid-map matching. In: Martí, J., Benedí, J.M., Mendonça, A.M., Serrat, J. (eds.) IbPRIA 2007. LNCS, vol. 4478, pp. 194–201. Springer, Heidelberg (2007)
18. Lowe, D.G.: Distinctive image features from scale-invariant keypoints. International Journal of Computer Vision 60(2), 91 (2004)
19. Nuchter, A., Wulf, O., Hertzberg, J., Wagner, B.: Benchmarking urban 6d slam. In: IEEE/RSJ International Conference on Intelligent Robots and Systems (2007)
20. Nuechter, A., Lingemann, K., Hertzberg, J., Surmann, H.: 6d slam – mapping outdoor environments. In: IEEE International Workshop on Safety, Security, and Rescue Robotics (SSRR). IEEE Press, Los Alamitos (2006)
21. Dellaert, F., Fox, D., Burgard, W., Thrun, S.: Monte carlo localization for mobile robots. In: Proceedings of the IEEE International Conference on Robotics and Automation, vol. 2, pp. 1322–1328 (1999)

Real-Time Spatio-Temporal Analysis of Dynamic Scenes in 3D Soccer Simulation

Tobias Warden[1], Andreas D. Lattner[2], and Ubbo Visser[3]

[1] Center for Computing Technologies – TZI,
Universität Bremen, Germany
warden@tzi.de
[2] Institute of Computer Science, Information Systems and Simulation
Goethe-Universität Frankfurt am Main, Germany
lattner@cs.uni-frankfurt.de
[3] Department of Computer Science, University of Miami, USA
visser@cs.miami.edu

Abstract. We propose a framework for spatio-temporal real-time analysis of dynamic scenes. It is designed to improve the grounding situation of autonomous agents in (simulated) physical domains. We introduce a knowledge processing pipeline ranging from relevance-driven compilation of a qualitative scene description to a knowledge-based detection of complex event and action sequences, conceived as a spatio-temporal pattern matching problem. A methodology for the formalization of motion patterns and their inner composition is introduced and applied to capture human expertise about domain-specific motion situations. We present extensive experimental results from the 3D soccer simulation that substantiate the online applicability of our approach under tournament conditions, based on 5 Hz a) precise and b) noisy/incomplete perception.

1 Introduction

Autonomous agents that reside in and interact with a dynamic, physical real-time environment require a conceptualization thereof suitable with regard to their respective tasks. For individual tasks such as path planning or object tracking it is often considered effectual to rely on a sophisticated quantitative scene representation based primarily on geometrical location of mission-relevant scene residents in a certain frame of reference as well as spatial relations among them. For agents with retrospection- and optionally some foresight capabilities, a "film roll" of temporally ordered, static snapshots of the ambient scene gathered at successive, discrete points in time seems to constitute an adequate metaphor for their knowledge representation. In the RoboCup domain, agents are not exclusively concerned with individual tasks. On the field, they also seek to participate successfully in simulated soccer matches. These matches involve technically challenging group activities both cooperative and adversarial in nature. Due to its idiosyncratic game characteristics and fixed laws of the game, soccer presets a distinct contextual setting for the decisive interpretation of game-relevant motion situations in the ambient dynamic scene.

L. Iocchi et al. (Eds.): RoboCup 2008, LNAI 5399, pp. 366–378, 2009.

Our basic assumption is that this interpretation is feasible based on the data already inaccessibly encoded in the "film roll", employing the means of spatio-temporal analysis of dynamic scenes. Agents which are rendered "conscious" of their contextual setting and equipped with domain expertise via explicit formalizations of domain-specific motion patterns can arrive at a broader grounding situation as they are to some extent aware of being engaged in a soccer match and grasp temporally extended dynamic elements of the game.

We discuss three main challenges: First, the existing qualitative observation must be transferred into a compact qualitative scene description which generalizes from particular situation idiosyncrasies. Attention focusing caters for the selection of the game-relevant observation sub-sampling for further processing. Second, a logical representation for motion situations that accommodates human conceptualization with spatial and temporal dimension structured in an ontology with taxonomic and partonomic aspects must be specified. Third, a knowledge-based detection procedure is required whose design criteria comprise the adaption of existing proved inference techniques, an exploitation of the motion ontology for computational efficiency and a hierarchical bottom-up detection which immediatly leverages preceding intra-cyclic detection yield.

2   Related Work

Miene et al. proposed a comprehensive method for incremental assembly of quantitative time series data via monotonicity and threshold-based classification into spatio-temporal ground facts and a basic, domain-agnostic, hierarchically organized logical formalization for motion situations (situations, events and (multi-agent) action). A prototypical a-posteriori detection of soccer-related motion incidences was implemented and evaluated extensively with focus on detection quality and robustness in the context of the 2D soccer simulation [1,2].

Wendler et al. proposed an approach for action detection geared towards an efficient OO implementation in the context of a comprehensive behavior recognition/prediction framework [3]. The detection is incremental as incidence hypotheses are proposed at an early stage of action progression and henceforth tracked, being eventually verified or discarded. A limiting factor for scalability is the fact that any motion pattern, regardless of its complexity, must be built from a fixed set of ground predicates. Also, the real-time aptitude of the approach is in parts due to constraints in the formalization of the temporal configuration of motion pattern constituents. In [4], Herzog et al. opposed declarative (design-time) and procedural (run-time) motion pattern specifications and found that generally speaking the former can be mapped to a set of procedural alternatives.

Further relevant work in the area of game analysis aims at support for coaching activities such as the offline automated team analyst ISAAC by Raines et al. [5] which is based on offline learning, data mining and statistical techniques. Intille and Bobick present a probabilistic framework for the classification of single, highly structured offensive plays in American football [6]. More recently, Beetz et al. introduced the Football Interaction and Process Model (FIPM) [7] that classifies motion situations into four crisp top-level actions and can estimate potential failures of actions based on a decision tree trained with a database of previous games and a domain-specific feature set.

None of the presented approaches handles our complete set of requirements for scalability, real-time aptitude and robustness against low-quality input.

3   Compilation of a Qualitative Scene Description

The first step in our analysis workflow comprises qualitative abstraction from an existing quantitative scene representation [1,8,9]. We consider a subset of a complete scene representation, namely *location* and *translational velocity* of scene actors conceived as multivariate time series. Raw Cartesian coordinates are converted to polar form. Based here upon time series for *planar location, elevation, speed, acceleration, motion direction and rise/fall* are compiled. We also consider pairs of objects by forming time series for *planar distance* and *spatial relation* among actors. Each time series is then associated with a qualitative predicate P_i with a finite set of symbolic states $\mathcal{SYM}_i$ [10].

We implement categorical classifiers with flexible bounds based work by Steinbauer et al. [11]. In order to deal with perception noise during classification, Steinbauer proposed the concept of predicate hysteresis for binary classifiers as a means to mitigate noise-induced, rapid interval oscillation in the vicinity of class boundaries. We generalize their concept for classifiers with arbitrary numbers of target-classes, operating over unbounded or cyclic value co-domains. We also use a two-tier classification for multivariate time series. We initially employ univariate classifiers separately along each input dimension obtaining an intermediate symbol vector which is consequently mapped to a single target class. Using this concept it is possible to construct a bivariate region classifier that handles composite regions that form homogeneous local neighborhoods [10].

While the classification yields the value of the considered predicate instances for successive mapping cycles (e.g. $velocity(\mathsf{pl}_1, \mathsf{slow})$ instance-of P_{vel}) the mandatory next step is the association of qualitative predicates with a temporal validity interval to obtain fluent facts $\mathrm{FACT}(p, i) \in \mathbb{F}_{\mathrm{atom}} : p$ instance-of P_j as basic substrate for pattern detection. In [2], Miene proposed a fact assembly scheme based on explicit validity extension in successive cycles. Yet, to put less strain on updating the KB we use an alternative scheme where new facts are initially asserted with a right-open interval $i = \langle s, \infty \rangle \in \mathcal{I}_{\vdash}$ and later revised with a closed interval $j = \langle s, e \rangle \in \mathcal{I}_{\parallel}$ when the state of the predicate instance changes thus that the validity duration is now known. Once the ground facts have been updated we compile derived facts $\mathbb{F}_{\mathrm{infer}}$ which are based on tracking sets of ground facts (e.g. the ball-player distances for ball ownership).

For an efficient qualitative mapping we implemented a domain-specific focus heuristic based on suggestions by Retz-Schmidt et al. [12] that filters the available quantitative input in a preprocessing step such that only relevant scene aspects are considered. Although the focus is in principle suited for multiple key objects, we were concerned only with the detection of ball-centered motion situations.

Thus, we found it sufficient to treat the ball as single focus point. It defines the attention focus by a fixed radius of 13 m such that a) all unary relations of scene actors within the radius and b) binary relations from the ball to the other focused actors are considered.

4   Formalization of Motion Situations

Besides a suitable pool of ground fact instances $\dot{\mathbb{F}}_{\mathrm{infer}} \cup \dot{\mathbb{F}}_{\mathrm{atom}}$, our knowledge-based pattern detection requires background knowledge of the targeted motion situations. We distinguish two principal categories of extensive motion situations, namely events and actions/action sequences. Beginning with events, we formally specify event/action classes and their internal composition.

Definition 1 (Event Class). Let id denote a unique identifier, $\mathcal{ACT}$ the set of actors in a dynamic scene. Let further $f_a : \mathbb{N}_n \to \mathcal{A} \equiv (a_1, a_2, \ldots, a_n)$ denote an ordered set of elements a_i over a co-domain $\mathcal{A}$ with arity $n \in \mathbb{N}$. An equivalence class of events, $ev \in \mathbb{E}$ is then defined as first-oder logic term as follows:

$$id(args) . args := ref \circ div, \quad ref : \mathbb{N}_{|n} \to \mathcal{ACT}, \quad div : \mathbb{N}_{|n} \to \mathcal{SYM}. \qquad \square$$

The ordered set $ref \subseteq args$ (reference) specifies the scene actors involved in an event incidence, e.g. as a kick initiator, while $div \subseteq args$ (diversification) comprises event-specific parameters, e.g. the kick height, which define particular traits of certain event incidences. We adopted the inclusion of the diversification following suggestions by Wendler [13], yet we forwent the use of parameters with sub-symbolic co-domain in order to retain a homogeneous qualitative description throughout. An event equivalence class constitutes a succinct class signature such as $kick(player, dir, height, type) . ref = (player), div = (dir, height, type)$ for the kick event.

An event incidence $\dot{ev} \in \dot{\mathbb{E}}$ is expressed in terms of its class signature with a distinct assignment of its arguments and is associated with a closed validity interval $i \equiv \langle s, e \rangle \in \mathcal{I}_{\|}.s, e \in \mathbb{N}^{+0}$. Following notational suggestions by Allen, a particular kick is noted as $\mathrm{OCCUR}(kick(\mathsf{pl}_1, \mathsf{north}, \mathsf{high}, \mathsf{volley}), i) . i \in \mathcal{I}_{\|}$.

Besides via its signature, each equivalence class of events is described by the necessary and sufficient conditions for an actual event occurrence, that is its event pattern (cf. Fig. 1), as follows:

Definition 2 (Event Pattern). Let $ev \in \mathbb{E}$ denote an equivalence class of events and let $i, j \in \mathcal{I}_{\|}$. The event pattern associated with ev is defined as:

$$\mathrm{OCCUR}(ev, i) \Leftrightarrow Pat(i, ev) \wedge \forall j . In(i, j) : \neg Pat(j, ev)$$

$Pat(i, ev)$ is defined as a formula in first-order logic with the codomain:

$$dom(Pat(i, ev)) = \mathbb{F}_{\mathrm{assert}} \cup \mathbb{F}_{\mathrm{infer}} \cup \mathbb{E} \cup \mathcal{IR} \cup \mathcal{SR} \cup \mathcal{M}. \qquad \square$$

Event patterns comprise as valid constituents asserted ground facts $\mathbb{F}_{\mathrm{assert}}$, inferred facts $\mathbb{F}_{\mathrm{infer}}$ and subordinate events $\mathbb{E}$. Temporal (semi-)interval relations $\mathcal{IR}$ as proposed by Allen and Freksa specify their temporal configuration. Additional

$\textsc{Occur}(kick(pl, dir^{g:8}, height, \mathsf{volley}), \langle s, e \rangle) \Leftarrow$

$\quad \textsc{Fact}(acceleration(\mathsf{ball}, \mathsf{increasing}), \langle s, \mathsf{inf} \rangle)$

$\quad \wedge \textsc{Fact}(freeBall(), \langle s_1, \mathsf{inf} \rangle)$

$\quad \wedge Older(\langle s_1, \mathsf{inf} \rangle, \langle s, \mathsf{inf} \rangle)$

$\quad \wedge \textsc{Fact}(motionDir(\mathsf{ball}, dir^{g:8}), \langle s, \mathsf{inf} \rangle)$

$\quad \wedge \textsc{Fact}(motionDir(\mathsf{ball}, dir_2^{g:8}), \langle s_2, e_2 \rangle)$

$\quad \wedge Meets(\langle s_2, e_2 \rangle, \langle s, \mathsf{inf} \rangle)$

$\quad \wedge (dir^{g:8} \neq dir_2^{g:8})$

$\quad \wedge \textsc{Fact}(distance(\mathsf{ball}, pl, \mathsf{veryClose}), \langle s_3, e_3 \rangle)$

$\quad \wedge TailToTailWith(\langle s_2, e_2 \rangle, \langle s_3, e_3 \rangle)$

$\quad \wedge \textsc{Holds}(zpositionTrend(\mathsf{ball}, trend), \langle s, e \rangle)$

$\quad \wedge Translate\uparrow_{ztrend}^{height}(trend, height)$

$\textsc{Occurring}(ballTaming(player), \langle s, e \rangle)) \Leftarrow$

$\quad \textsc{Occur}(receive(player), \langle s, e_1 \rangle)$

$\quad \wedge \textsc{Fact}(xBallControl(player), \langle s, e_2 \rangle)$

$\quad \wedge \textsc{Occur}(retreat(player, \mathsf{ball}), \langle s_3, e_3 \rangle)$

$\quad \wedge Meets(\langle s, e_2 \rangle, \langle s_3, e_3 \rangle)$

$\quad \wedge \textsc{Fact}(freeBall(), \langle s_3, e_4 \rangle)$

$\quad \wedge \textsc{Occur}(receive(player), \langle s_5, e \rangle))$

$\quad \wedge Meets(\langle s_3, e_4 \rangle, \langle s_5, e \rangle)$

Fig. 1. Examples for specified motion patterns. *left:* the volley kick event pattern, *right:* the ball taming event pattern.

relations ($\mathcal{SR}$) allow for an enforcement of spatial constraints. For convenience and expressiveness reasons, we also adopted two meta-logical functions from the Prolog domain in the $\mathcal{M}$ set, namely $setOf()$ and negation as failure.

The terms in $dom(Pat(i, ev))$ are assembled to a valid formula of first-order logic using the logical connectives $\wedge$, $\vee$. The composition of $Pat(i, ev)$ is flexible as it allows to use subordinate event classes in the specification of superordinate event patterns. We use a simplified notation for event patterns which conforms to $\textsc{Occur}(ev, i) \Leftarrow Pat(i, ev)$. The notation of the event pattern thus contributes to a clarification of the pattern usage.

We define action classes $ac \in \mathbb{A}$ as terms in first-order logic with a composition analogous to events. Let $pass(source, target, dir, height, force, succ)$ where $ref = (source, target)$, $div = (dir, height, force, succ)$ denote the equivalence class for passes. Based upon this concrete action incidence expressed as

$$\textsc{Occurring}(pass(\mathsf{pl}_1, \mathsf{pl}_2, \mathsf{forward}, \mathsf{low}, \mathsf{weak}, \mathsf{succ}), i) \in \dot{\mathbb{A}} . i \in \mathcal{I}_{\|}$$

where $\dot{\mathbb{A}}$ is the set of concrete action instances. Action classes are described in terms of their respective necessary and sufficient incidence conditions $Pat(i, ac)$: $i \in \mathcal{I}_{\|}$, $ac \in \mathbb{A}$.

Definition 3 (Action Pattern for ac $\in \mathbb{A}_{atom} \subset \mathbb{A}$). Let $a \in \mathbb{A}_{atom}$ denote an equivalence class of actions and let $i, j \in \mathcal{I}_{\|}$. Then, the action pattern is defined as:

$$\textsc{Occurring}(ac, i) \Leftrightarrow Pat(i, ac) \wedge \forall j . In(i, j) : \neg Pat(j, ac)$$

where $dom(Pat(i, ac)) = dom(Pat(i, ev)) \cup \mathbb{A}$. $\qquad \square$

The fact that $Pat(i, ac)$ may also contain subordinate action classes allows for the following two scenarios: *specialization* of basic actions (ball transfer $\mapsto$ pass) and specification of *action sequences* with a non self-similar progression (one-two pass). $dom(Pat(i, ac))$ thus enables scalable concise specification of action with growing complexity. Following Miene et al. [2] we exploit the natural hierarchy of the considered actions. The more complex the action patterns, the more complex constituents may contribute to $Pat(i, ac)$. Yet, at the same time, the incorporation of less complex constituents down to simple facts remains a valid option with regard to precise detail specification (cf. Fig. 1).

So far, only action sequences with a self-similar character ($\mathbb{A}_{atom} \subset \mathbb{A}$) have been treated. An examination of the space of conceivable action sequences found in dynamic scenes unveils an additional class of action sequences. These are comprised of a homogeneous, finite, yet in its length a priori indeterminate concatenation of a particular basic action. The dribbling is a paradigmatic example for this class of action sequences, denoted as $\mathbb{A}_{loop}$, since dribble sequences are composed of simple atomic dribblings where an agent forwards the ball once.

Let $\text{OCCURRING}(\dot{a}c, i) : \dot{a}c$ instance-of $ac \in \mathbb{A}_{loop}$, $i \in \mathcal{I}_{\parallel}$ denote a self-similar action sequence, then it is possible that $\exists j : In(i, j) \land \text{OCCURRING}(\dot{a}c, j)$.

Definition 4 (Action Pattern ac $\in \mathbb{A}_{loop} \subset \mathbb{A}$). Let $ac_{sup} \in \mathbb{A}_{atom} \cup \mathbb{A}_{loop}$. Let further $i, j, k \in \mathcal{I}_{\parallel}$. The action pattern is then defined as:

$$\text{OCCURRING}(ac, i) \Leftrightarrow \text{OCCURRING}(ac_{sup}, i)$$
$$\lor (\text{OCCURRING}(ac, j) \land \text{OCCURRING}(ac_{sup}, k)$$
$$\land Meets(j, k) \land Equals(i, j + k) \land Cont(ac, ac_{sup})) \qquad \square$$

In the latter case, $Cont(ac, ac_{sup}) : dom(Cont(ac, ac_{sup})) = \mathcal{SR}$ enforces spatial continuity constraints to be met by both constituents that are combined into the longer action sequence. To conclude, relying using Def. 3 and Def. 4, it is now possible to specify simple actions as well as both types of actions sequences.

5   Knowledge-Based Pattern Detection

The formalization for motion patterns corresponds to first-order definite clauses where the premise $Pat(i, mo) : mo \in \mathbb{E} \cup \mathbb{A}$ specifies the inner composition of the respective pattern while the consequence specifies the event/action class (cf. Fig. 1). Therefore, a transfer to an operational logic programming language (Prolog) is possible in a natural way. Once the semantics of the spatial and temporal relations $\mathcal{SR} \cup \mathcal{IR} \subset dom(Pat(i, mo))$ are realized in the target language, spatio-temporal pattern matching is enabled that can recurse to standard logical inference techniques, i.e. sophisticated backward chaining. We describe an efficient deployment strategy of logical inference for pattern matching.

A first step in an efficient incremental detection strategy is to determine the detection order elements in a pool of motion patterns, to be applied in each detection pass. We capitalize on the ontological relations among the desired motion classes captured in their respective $Pat(i, mo)$. We use a detection hierarchy based on the maximum level of complexity of the substantial constituent terms $\mathcal{T}_{mo}$ in $Pat(i, mo)$ (i.e. $\mathcal{T}_{mo} \subset \mathbb{F} \cup \mathbb{E} \cup \mathbb{A}$) as follows: First, (inferred) atomic facts ($fact \in \mathbb{F}_{infer} \cup \mathbb{F}_{atom}$) constitute a fundamental level of complexity $l(fact) = 0$. The hierarchy level of particular extensive motion patterns then evaluates to $l(mo) = 1 + \text{argmax}(l(t_i))$ where $t_i \in \mathcal{T}_{mo}$.

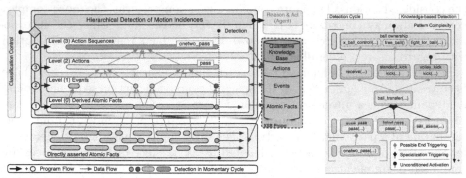

(a) Schematic overview of the bottom-up character of the detection

(b) Trigger relations among motion classes

Fig. 2. Details of the knowledge-based detection of motion incidences

Each consecutive detection pass d_i can then be implemented as a bottom-up search for new motion incidences with respect to this pattern hierarchy. This search starts with patterns of $l(mo) = 1$ that can be matched against the momentary pool of atomic facts $\dot{\mathbb{F}}_i$ alone. When new motion incidences are detected, they are fed immediatly into the qualitative KB ($\dot{\mathbb{F}}_i \cup \dot{\mathbb{E}}_i \cup \dot{\mathbb{A}}_i$) which is crucial as it enables immediate intra-cycle reuse of the partial detection yield achieved so far. Motion classes whose pattern comprise subordinate events/actions can be matched subsequently against the updated KB contents (cf. Fig. 2(a)).

Regarding real-time efficiency, the goal of a knowledge-based detection strategy is to try and match for a certain motion pattern iff the momentary detection cycle has already yielded necessary and sufficient for either guaranteed or quite probable detection success. We propose to derive appropriate decision heuristics, *specialization* (SP) and *possible end triggering* (PE), based on the observation that the motion incidences are in general related via certain relationships to subordinate incidences such that the detection of the latter during a detection cycle triggers the detection of superordinate motion incidences in the first place.

First, motion patterns can be modeled as direct specializations of subordinate patterns. Since specialized patterns, being located higher in the pattern hierarchy, are detected subsequent to their generalization, it is evident that a detection attempt is only promising once the detection of the generalized pattern has already yielded a concrete result. Otherwise, it is sound to skip the detection attempt for specialized concepts as they are bound to fail.

A second relationship is derived from the internal composition of $Pat(i, mo)$ which is applicable for a greater subset of patterns. For a particular pattern, one or more closing events/actions may exist, denoted here as atomic *PE*-triggers. Within a pass, the detection of a pattern associated with a single *PE*-trigger is attempted once a trigger incidence has already been added to the momentary, partial detection yield. For the detection of a pattern associated with multiple triggers, the fulfillment policy determines that a detection is attempted either if instances for at least a single (existential policy) or all specified triggers (universal policy) have been added to the detection yield. We found that those types can be employed for a considerable subset of all engineered motion patterns (cf. 2(b)), thus leading to a considerably increased ratio of successful detection attempts.

6 Experiments and Results

To date we have prototypically implemented the complete scene analysis workflow [10]. Our qualitative scene description is based on a minimal set of nine ground predicate types directly asserted incrementally by our qualitative abstraction module and three additional derived ball ownership predicates. We have formally specified ten event, eight action, and two sequence patterns and transferred them to Prolog rules as foundation for our knowledge-based detection. Necessary spatial and temporal relations have been implemented as rules such that we can use Prolog as backend for spatio-temporal pattern matching. We simulated 15 regular 3D Soccer Simulation League (SSL) matches with Virtual Werder playing 3 times respectively against 5 top-rated sphere simulation teams, namely *Fantasia*, *SEU*, *WrightEagle*, *FC Portugal* and *Aeolus*. An extended version of the *3D Soccer Server 0.5.3* developed by the *UTUtd 3D* team [14] which features support for an online coach was used for the experiments. All games were conducted with the server configuration acknowledged as standard for official RoboCup competitions and simulated in a non-distributed setting on a single host machine (cf. [10]). Both, the Virtual Werder players and our online coach compiled log files during the simulation runs encoding their respective perceptions as S-expressions. Moreover, our coach performed online analyses for all conducted test games.

To evaluate general feasibility and robustness of our analysis module, we conducted a two-tiered experiment using quantitative scene perceptions of varying quality as input data. The basic test was based on the complete, precise, allocentric perception of our coach agent (update freq.: 5 Hz). We compiled a sound ground truth as baseline for quality comparison by hand-tagging a subset of six pass-oriented motion situations covered by our prototype. The constraint in the coverage of considered motion situations is due to the difficulty to manually estimate the temporal extension of ball-ownership-related facts and to discern ball deflections from attempted receptions in the soccer monitor. Other patterns such as mutual kicks or one-two passes occurred too infrequently to be statistically significant in both hand-tagged half times against *FC Portugal* and *SEU*.

We performed the quality analysis offline, seizing perception logs compiled by the coach agent in the batch simulation of the test games. Table 1(a) outlines statistics of our offline analysis based on coach perception logs. Precision values beyond 90% accordant to low false-positive classifications indicate a sound manual engineering of the considered motion patterns. The recall beyond 85% for event and 70% for action incidences suggests that our motion patterns were not yet expressive enough to cover all conceivable incidence traits. Thereby, the lower recall for actions is due to the hierarchical bottom-up detection scheme which propagates recall failures for base concepts to superordinate levels. The

Table 1. Basic quality assessment of the proposed analysis approach

(a) Analysis quality under full perception

Motion Situation		P_{Sys}	P_{Tru}	$\cap$	m_{prec}	m_{rec}
VW3D vs. FC Portugal: 1500 analysis cycles						
kick	ev	62	67	62	100.0	92.5
receive	ev	78	76	76	97.4	100.0
ball transfer	ac	59	65	57	96.6	87.7
pass	ac	17	19	16	94.1	84.2
pass (fail)	ac	23	28	23	100.0	82.1
self assist	ac	18	18	18	100.0	100.0
VW3D vs. SEU: 1500 analysis cycles						
kick	ev	63	71	63	100.0	88.7
receive	ev	83	83	79	95.1	95.1
ball transfer	ac	60	69	57	95.0	82.6
pass	ac	36	41	36	100.0	87.8
pass (fail)	ac	16	20	14	87.5	70.0
self assist	ac	8	8	7	87.5	87.5
Accumulated Results $\equiv$ complete game						
kick	ev	125	138	125	100.0	90.5
receive	ev	161	159	155	96.3	97.4
ball transfer	ac	119	134	114	95.7	85.0
pass	ac	53	60	52	98.1	86.6
pass (fail)	ac	39	48	37	94.8	77.1
self assist	ac	26	26	25	96.1	96.1

(b) Comparing analysis systems

Motion Situation	m_{prec}	m_{rec}
Analysis by Miene et al. [2] $\rightarrow$ 2D Soccer Simulation		
pass	95.5	93.3
pass (fail)	89.8	92.8
self assist	100.0	95.5
$avg_{\text{Mien.}}$	93.8	93.2
Analysis by the authors $\rightarrow$ 3D Soccer Simulation		
pass	98.1	86.6
pass (fail)	94.8	77.1
self assist	96.1	96.1
$avg_{\text{Ward.}}$	96.3	86.6
Δ_{avg}	2.5	-6.6
avg_{i}: concept average		

Symbol Description	
m_{prec}, m_{rec}	precision, recall
P_{Sys}	system detections
P_{Tru}	ground truth
$\cap$	mutual detections

data in Table 1(a) also suggests correlations between analysis quality and the idiosyncratic style of play of the considered opponents. For instance the *SEU* team consequently plays a rapid kick'n'rush soccer with minimal ball control times while *FC Portugal* features a rather slow game development. As we did not anticipate volley play effectively enough, results are suboptimal in this case. We consciously chose a set of motion situations for our evaluation with a direct equivalent in the evaluation conducted by Miene in [2]. Thus, it was possible to compare results obtained in the 2D and 3D soccer simulation leagues. The data in Table 1(b) shows that although the 3D SSL demands for more sophisticated motion patterns due to added realism, simulation dynamics and the treatment of motion traits, the general analysis quality is competitive.

Having substantiated basic feasibility of our analysis, we proceeded with a second quality assessment in order to identify the analysis tolerance when using agent perception logs as input. Agents in the sphere-based 3D SSL need to cope with noisy, egocentric perception with a 180° field-of-view (update freq. 5 Hz). The Virtual Werder agents employ a ball-oriented poke-around strategy, integrate successive percepts and use a particle filter to deduct noise [15]. In our offline test, agent perception refers to the best estimate of the real state of affairs by the agent, rather than raw perception. Tab. 2 presents the results of our comparative assessment where our analysis was employed with perception logs

Table 2. Influence of incomplete, noisy perception on analysis quality

(a) Quality with Coach Perception

Motion Sit.		P_{Sys}	P_{Tru}	$\cap$	m_{prec}	m_{rec}
kick	ev	50	53	47	94.0	88.7
receive	ev	64	64	60	93.8	93.8
ball transf.	ac	44	49	39	88.6	79.6
pass	ac	20	21	19	95.0	90.5
pass (fail)	ac	20	23	17	85.0	73.9
self assist	ac	5	5	3	60.0	60.0

(b) Quality with Agent Perception

Motion Sit.		P_{Sys}	P_{Tru}	$\cap$	m_{prec}	m_{rec}
kick	ev	43	53	43	100.0	81.1
receive	ev	62	64	60	96.7	93.8
ball transf.	ac	38	49	35	92.1	71.4
pass	ac	20	21	17	85.0	81.0
pass (fail)	ac	13	23	13	100.0	56.5
self assist	ac	6	5	5	83.3	100.0

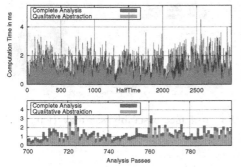
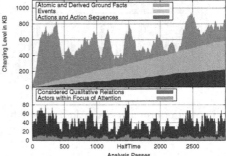

Fig. 3. Basic Evaluation for a complete simulation match of our analysis prototype, showing computation time (left) and knowledge base charging (right)

from our coach agent and a player acting as a midfielder for a single half time of a match against the *Aeolus 3D* team. It shows that regardless of degrading input the precision of our analysis remains high while the recall declines mildly about 10% due to a growing number of non-detections, especially of kicks. The latter seems to be in part due to both our choice of granularity in the qualitative abstraction, present neglect of the location of the observer agent relative to the relevant scene part and the low frequency of perception updates.

The real-time aptitude of our approach was investigated in the next step where we evaluated runtime key measures compiled by our coach agent in our simulation test series. The average measured computation time for a complete analysis pass varies team-dependent between an encouraging 2.04 and 2.63 ms. While the maxima for the respective teams are notably higher ($\langle 7.47, 16.26 \rangle$ ms), the associated Q_{95} quantiles and the location of μ_t halfway between Q_5 and Q_{95} suggest that isolated outliers are responsible for these results. This hypothesis is substantiated by Fig. 3 (left) which shows a paradigmatic graph of the required computation time for our analysis plotted for a complete match. Besides real computation time, we also considered the simulation-specific distribution of

Table 3. Evaluation of time consumption for computation of a single analysis cycle by our coach agent in ms of real time. 3-game average results.

VW3D vs.	Aeolus	Fantasia	Wright Eagle	SEU	FC Portugal
Median	2.211	2.035	2.491	2.626	2.264
$\langle Q_5, Q_{95} \rangle$	$\langle 1.02, 4.07 \rangle$	$\langle 0.99, 3.61 \rangle$	$\langle 1.20, 4.42 \rangle$	$\langle 1.28, 4.70 \rangle$	$\langle 0.75, 4.27 \rangle$
$\langle \min, \max \rangle$	$\langle 0.41, 8.71 \rangle$	$\langle 0.30, 7.47 \rangle$	$\langle 0.35, 16.26 \rangle$	$\langle 0.38, 9.28 \rangle$	$\langle 0.30, 8.86 \rangle$
Mean $\mu_t \pm \sigma_t$	2.324 ± 0.957	2.152 ± 0.812	2.012 ± 1.056	$2.754 + 1.113$	2.379 ± 1.083
Acts./cycle	8.147 ± 2.732	6.871 ± 2.52	8.564 ± 2.62	7.77 ± 2.11	8.339 ± 2.81
Rels./cycle	49.88 ± 16.39	42.2 ± 15.1	52.38 ± 15.73	47.61 ± 12.65	51.03 ± 16.83

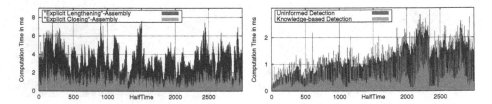

Fig. 4. Performance variation caused by choice of fact assembly strategy (left) and effect of knowledge-based detection strategy

consumed sim steps per analysis cycle. Given the fact that a full *percept-reason-act* cycle for agents, both the coach and players, within the 3D SSL spans 20 sim steps, the consumption of no more than a single complete sim step is definite proof for real-time aptitude with regard to deployment in soccer simulation.

The comparatively high σ_t for the complete analysis as well as mapping/detection fractions ($\sim \frac{1}{3}\mu_t$ to $\frac{1}{2}\mu_t$) which seems conspicuous at first in Tab. 3, is in fact a distinctive trait of the implemented analysis. Regarding qualitative abstraction, cycles with higher computational load, e.g. when the ball is moving swift such that spatial binary relations change frequently alternate with cycles of relative calmness where spatial relations remain largely unaltered. Furthermore, due to the guided recognition approach, the detection fraction of the analysis is comparatively expensive in passes where sequences of extensive motion incidences are detected in succession, bottom-up, building upon each other, e.g. when a one-two pass is completed. In other passes, no motion incidences are detected at all e.g. when the ball is flying or rolling freely and only a small subset of pattern matchings are attempted (cf. Fig. 3).

The time consumption ratio between scene interpretation and qualitative abstraction is nearly balanced with light prevalence to the latter. While both aspects of our analysis approach benefited significantly from real-time optimizations (cf. Fig. 4), larger gains were rendered possible by the proposed fact assembly strategy based on open validity intervals.

7   Conclusion

Spatio-temporal analysis of dynamic scenes in real-time environments is challenging by nature. We presented an approach for relevance-based temporal tracking of single object's motion traces and spatial relations among them. We extended prior formalizations for motion situations to cover self-similar action sequences and concept traits and proposed a knowledge-based detection strategy. Applying our idea to soccer simulation, we showed that our approach benefits an online coach as well as regular field players as solid foundation for opponent modelling or plan recognition. We have performed extensive tests which approve analysis quality with regard to precision and recall both using precise, complete input data and, as a primer, noisy input from competition agents. Also, unconstrained real-time aptitude of our approach has been substantiated. Looking forward, we plan to perform a careful review of our qualitative scene representation to examine its cognitive plausibility and task adequacy. To improve the coverage of possible motion situations with the set of modeled motion classes we plan to examine to what extent manual knowledge engineering can be automated with pattern mining approaches [16]. Due to its domain-agnosticism we also pursue a transfer of our analysis to new domains such as autonomous logistics.

Acknowledgements. The presented research was partially funded by the *Senator für Bildung und Wissenschaft, Freie Hansestadt Bremen* ("FIP RoboCup", Forschungsinfrastrukturplan) and the CRC 637 "Autonomous Logistic Processes".

References

1. Miene, A., Visser, U., Herzog, O.: Recognition and Prediction of Motion Situations Based on a Qualitative Motion Description. In: Polani, D., Browning, B., Bonarini, A., Yoshida, K. (eds.) RoboCup 2003. LNCS (LNAI), vol. 3020, pp. 77–88. Springer, Heidelberg (2004)
2. Miene, A.: Räumlich-zeitliche Analyse von dynamischen Szenen. Dissertation, Universität Bremen. Number 279 in DISKI. Akademische Verlagsgesellschaft Aka GmbH, Berlin, Germany (2004)
3. Wendler, J., Bach, J.: Recognizing and Predicting Agent Behavior with Case Based Reasoning. In: Polani, D., Browning, B., Bonarini, A., Yoshida, K. (eds.) RoboCup 2003. LNCS (LNAI), vol. 3020, pp. 729–738. Springer, Heidelberg (2004)
4. Herzog, G.: Utilizing Interval-Based Event Representations for Incremental High Level Scene Analysis. Technical Report 91, Universität des Saarlandes (1995)
5. Nair, R., Milind, T., et al.: Automated Assistants for Analyzing Team Behaviors. Journal of Autonomous Agents and Multi-Agent Systems 8(1), 69–111 (2004)
6. Intille, S., Bobick, A.: Recognizing Planned, Multiperson Action. Computer Vision and Image Understanding: CVIU 81(3), 414–445 (2001)
7. Beetz, M., Kirchlechner, B., Lames, M.: Computerized Real-Time Analysis of Football Games. IEEE Pervasive Computing 4(3), 33–39 (2005)
8. Sprado, J., Gottfried, B.: What Motion Patterns Tell Us about Soccer Teams. In: Iocchi, L., Matsubara, H., Weitzenfeld, A., Zhou, C. (eds.) RoboCup 2008: Robot Soccer World Cup XII. LNCS (LNAI), vol. 5399, pp. 614–625. Springer, Heidelberg (2009)

9. Gottfried, B.: Representing Short-Term Observations of Moving Objects by a Simple Visual Language. Journal of Visual Languages and Computing 19(3), 321–342 (2007)
10. Warden, T.: Spatio-Temporal Real-Time Analysis of Dynamic Scenes in the RoboCup 3D Soccer Simulation League. Master's thesis, Universität Bremen (2007)
11. Fraser, G., Steinbauer, G., et al.: Application of Qualitative Reasoning to Robotic Soccer. In: 18th International Workshop on Qualitative Reasoning, USA (2004)
12. Retz-Schmidt, G.: Recognizing Intentions, Interactions, and Causes of Plan Failures. In: User Modeling and User-Adapted Interaction, vol. 1, pp. 173–202. Kluwer Academic Publishers, The Netherlands (1991)
13. Wendler, J.: Automatisches Modellieren von Agenten-Verhalten: Erkennen, Verstehen und Vorhersagen von Verhalten in komplexen Multi-Agenten-Systemen. PhD thesis, Humboldt-Universität zu Berlin (2003)
14. Aghaeepour, N., Disfani, F.M., et al.: Ututd2006-3D Team Description Paper for 3D Development Competition. Technical report, University of Teheran (2006)
15. Lattner, A.D., Rachuy, C., Stahlbock, A., Visser, U., Warden, T.: Virtual Werder 3D Team Documentation. Technical Report 36, TZI - Center for Computing Technologies, Universität Bremen (2006)
16. Lattner, A.D.: Temporal Pattern Mining in Dynamic Environments. Dissertation, Universität Bremen. Number 309 in DISKI. Akademische Verlagsgesellschaft Aka GmbH, Berlin, Germany (2007)

Coaching Robots to Play Soccer via Spoken-Language

Alfredo Weitzenfeld[1], Carlos Ramos[1], and Peter Ford Dominey[2]

[1] ITAM, San Angel Tizapán, México DF, CP 0100
alfredo@itam.mx
http://www.cannes.itam.mx/Alfredo/English/Alfredo.htm
[2] Institut des Sciences Cognitives, CNRS, 67 Blvd. Pinel, 69675 Bron Cedex, France
dominey@isc.cnrs.fr
http://www.isc.cnrs.fr/dom/dommenu-en.htm

Abstract. The objective of this paper and our current research is to develop a human-robot interaction architecture that will let human coaches train robots to play soccer via spoken language. This work exploits recent developments in cognitive science, particularly notions of grammatical constructions as form-meaning mappings in language, and notions of shared intentions as distributed plans for interaction and collaboration between humans and robots linking perceptions to action responses. We define two sets of voice-driven commands for human-robot interaction. The first set involves action commands requiring robots to perform certain behaviors, while the second set involves interrogation commands requiring a response from the robot. We then define two training levels to teach robots new forms of soccer-related behaviors. The first level involves teaching new *basic* behaviors based on action and interrogation commands. The second level involves training new *complex* behaviors based on previously learnt behaviors. We explore the two coaching approaches using Sony AIBO robots in the context of RoboCup soccer standard platform league previously known as the four-legged league. We describe the coaching process, experiments, and results. We also discuss the state of advancement of this work.

1 Introduction

We expect interaction between humans and robots to be as natural as interaction among humans. As part of this work we are developing a domain independent language processing system that can be applied to arbitrary domains while having psychological validity based on knowledge from social cognitive science. In particular our architecture exploits models on: (i) language and meaning correspondence relevant to both neurological and behavioral aspects of human language developed by Dominey et al. [1], and (ii) perception and behavior correspondence based on the notion of shared intentions or plans developed by Tomasello et al. [2, 3]. In Weitzenfeld and Dominey [4] we describe preliminary results using two prior robotic platforms, the Event Perceiver System and AIBO robots under simpler tasks. The current paper reports more advanced yet preliminary results in coaching robots to play soccer in the context of RoboCup [5] standard platform league where ITAM's Eagle Knights team regularly competes at the top level [6, 7]. In the standard platform league two teams of four fully autonomous robots in red or blue uniform play soccer on a 6m by 4m carpeted soccer field using Sony's AIBO robots (up until 2008).

L. Iocchi et al. (Eds.): RoboCup 2008, LNAI 5399, pp. 379–390, 2009.

Robots have as main sensor a color-based camera with main actuators the four robot legs. The AIBO also includes wireless communication capabilities to interact with the game controller and other robots in the field. The field includes two colored goals and two colored cylinders used for localization. In order to win robots need to score as many goals as possible by kicking the orange ball into the opposite team's goal. As in human soccer, teams need to outperform the opponents by moving faster, processing images more efficiently, localizing and kicking the ball more precisely, and having more advanced individual and team behaviors. In general, approaches to robot programming vary from direct programming to learning approaches that are usually intended to optimize individual robot tasks. Our human-robot interaction approach is intended to coach robots to play soccer initially as individuals and then as a team as real human coaches currently do. The following sections describe the human-robot interaction architecture, the coaching approach, current experimental results, and conclusions.

2 Human-Robot Interaction Architecture

In this section we describe the human-robot interaction spoken-language architecture we have developed using the CSLU Speech Tools Rapid Application Development (RAD) system [8] for: (a) scene processing for event recognition, (b) sentence generation from scene description and response to questions, (c) speech recognition for posing questions, (d) speech synthesis for responding, and (e) sending and receiving textual communications with the robot.

2.1 Learning <Sentence, Meaning> Bindings

Dominey and Boucher [9, 10] describe a system that can adaptively acquire a limited grammar by training with human narrated video events. An image processing algorithm extracts the meaning of the narrated events translating them into action descriptors, detecting physical contacts between objects, and then using the temporal profile of contact sequences in order to categorize the events (see [11]). The visual scene processing system is similar to related event extraction systems that rely on the characterization of complex physical events (e.g. give, take, stack) in terms of composition of physical primitives such as contact (e.g. [12, 13]). Each narrated event generates a well formed *<sentence, meaning>* pair that is used as input to a model that learns the sentence-to-meaning mappings as a form of template where nouns and verbs can be replaced by new arguments in order to generate the corresponding new meanings. Each grammatical construction corresponds to a mapping from sentence to meaning. This information is also used to perform the inverse transformation from meaning to sentence. These templates or grammatical constructions (see [14]) are identified by the configuration of grammatical markers or function words within the sentences [15]. The construction set provides sufficient linguistic flexibility, for example, when a human commands a robot to push a ball towards the goal *push(robot, ball)*.

2.2 Learning <Percept, Response> Bindings

Ad-hoc analysis of interaction among humans during teaching-learning reveals the existence of a general intentional plan that is shared between teachers and learners. In a generalized independent learning platform is described where new *<percept, response>* constructions can be acquired by binding together perceptual and behavioral capabilities. Three components are involved in these *<percept, response>* constructions: (i) the percept, either a verbal command or an internal system state, originating from vision or another sensor; (ii) the response to this percept, either a verbal response or a motor response from the existing behavioral repertoire; and (iii) the binding together of the *<percept, response>* construction and its subsequent validation that it was correctly learned. The system then links and saves the *<percept, response>* pair so that it can be used in the future.

Having human users control and interrogate robots through spoken language results in the ability to naturally teach robots individual action sequences conditional on perceptual values or even more sophisticated shared intention tasks involving multiple robots such as passing the ball between robots when one of them is blocked or far away from the goal. In Dominey and Weitzenfeld [15] we describe a general approach for action, interrogation and teaching robots simple tasks.

3 Coaching Robots to Play Soccer

In order to demonstrate the generalization of the human-robot interaction architecture we have begun a series of experiments in the domain of RoboCup Soccer a well documented and standardized robot environment that provides a quantitative domain for evaluation of success. While no human intervention is allowed during a game, in the future humans could play a decisive role analogous to real soccer coaches adjusting in real-time their team playing characteristics according to the state of the game, individual or group performance, or the playing style of the opponent. Furthermore, a software-based coach may become incorporated into the robot program analogous to the RoboCup simulated coaching league where coaching agents can learn during a game and then advice virtual soccer agents how to optimize their behavior accordingly (see [16, 17]).

3.1 Coaching System

The coaching system developed by our group is illustrated in Figure 1. The spoken language interface is provided by the CSLU-RAD while communication to the Sony AIBO robots is done in a wireless fashion via UDP messages. Messages communicate the EK2007 system with MsgToAIBO module in the remote computer. The UDP messages contain a data structure that can be used to send textual commands to the robot or interrogate the robot about internal states. The MsgToAIBO module broadcasts the UDP messages containing information about commands and interrogations.

The human coach uses voice commands to control the behavior of the AIBO. These voice commands are translated into regular text commands by the CSLU-RAD system and then sent transmitted to the EK2007 system controlling the behavior of the robot.

The EK2007 system is shown in more detail in Figure 2. It contains the following modules: *EKMain* controlling robot behavior; *EKMotion* controlling robot movement; *EKVision* processing camera images to identify objects of interest; *EKLoc* computing

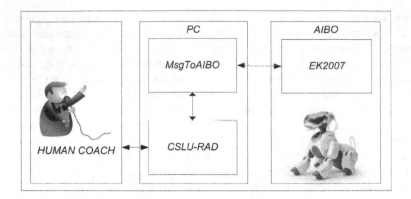

Fig. 1. Coaching system. A human coach interacts with the CSLU-RAD system to command or interrogate the AIBO robot. Textual messages are sent between the EK2007 system controlling the AIBO and the MsgToAIBO module in the remote computer.

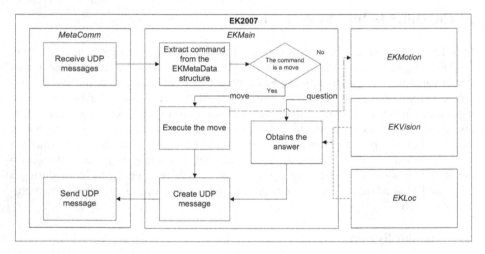

Fig. 2. EK2007 system. Commands are sent and receive via MetaComm. EKMain processes all commands and questions interacting with EKMotion, EKVision and EKLoc accordingly.

robot position in the field; and *MetaComm* sending and receiving UDP messages between robots and computer. The *MsgToAIBO* module takes as parameter a *Meta-Data* containing information that includes an action or interrogation command and a destination robot identifier. When a UDP messages is received by the robot, the *MetaComm* module processes the *MetaData* verifying that the message is intended for the current robot, in which case it is delivered to the *EKMain* module. If the command refers to an action command, an action control is sent to *EKMotion*; otherwise if it refers to a question, information is requested from either *EKVision* or *EKLoc*. EK-Main responds with a UDP message sent back to the computer via the *MetaComm* module. Action commands finish immediately while interrogation commands wait for an answer.

3.2 Action and Interrogation Commands

In order to demonstrate the human-robot coaching architecture we have defined and developed a number of basic action and interrogation commands in the context of soccer playing robots. These actions and interrogations are executed as voice commands using the CSLU-RAD software.

Table 1 describes action commands showing corresponding *voice command, CSLU node* (see Figure 3), *command id*, and *behavior*. Action commands include *Stop, Walk, Turn Right, Turn Left, Kick Ball, Hold Ball, Turn Right with Ball, Turn Left with Ball, Go to Ball*, and *Block Ball*. Note that certain actions such as *Go to Ball* depend on perceptions, in this case seeing the ball.

Table 1. Action commands showing voice command, corresponding CSLU node, command id and behavior

Voice Command	CSLU node	Command Id	Behavior
Stop	Stop	0	Stop
Walk	Walk	1	Walk
Right	TurnR	9	Turn Right
Left	TurnL	10	Turn Left
Kick	Kick	11	Kick Ball
Hold Ball	Hold	12	Hold Ball
Right with Ball	TRH	13	Turn Right with Ball
Left with Ball	TLH	14	Turn Left with Ball
Go to Ball	Go -> Ball	20	Go to Ball and Stop in Front of Ball
Block	Block	22	Block Ball

Table 2. Interrogation commands showing voice command, corresponding CSLU node, command id and response

Voice Command	CSLU node	Command Id	Response - Id
Do you see the ball?	Ball?	50	sees the ball – 1, otherwise - 0
Do you see the yellow goal?	YGoal?	51	sees the yellow goal – 1, otherwise - 0
Do you see the cyan goal?	CGoal?	52	sees the cyan goal – 1, otherwise - 0
Is the ball near?	BallN?	59	sees the ball near – 1, otherwise - 0

Table 2 describes the interrogation commands showing corresponding *voice command, CSLU node* (see Figure 3), *command id*, and *response*. Interrogation commands include *Do you see the ball?, Do you see the yellow goal?, Do you see the cyan goal?*, and *Is the ball near?*

CSLU-RAD defines a directed graph where each node in the graph links voice commands to specific behaviors and questions sent to the robot as shown in Figure 3. The *select* node separates action and interrogation commands. Action commands are represented by the *behaviors* node ('behavior' voice command) while interrogation

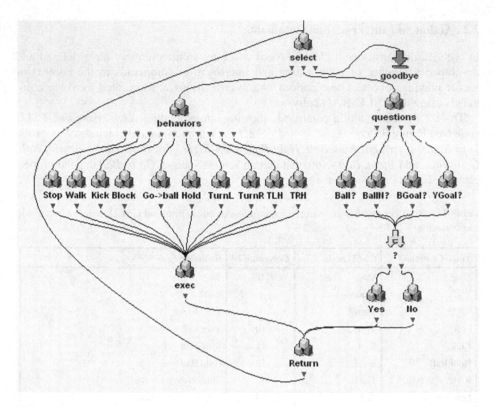

Fig. 3. The CLSU-RAD diagram describes the basic set of behaviors and questions that can be sent as voice commands to the robot

commands are represented by the *questions* node ('question' voice command). Behavior nodes include 'Stop', `Walk', 'Kick', 'Go->ball', 'Hold, 'TurnL', 'TurnR', 'TLH', and TRH; while question nodes are 'Ball?' ('Do you see the ball?'), 'BallN? ('Is the ball near?'), 'CGoal' ('Do you see the cyan goal?') and 'YGoal´ ('Do you see the yellow goal?'). Behavior commands are processed by the *exec* node while questions are processed by the *?* node that waits for a 'Yes' or 'No' reply from the robot. Finally, the *Return* node goes back to the top of the hierarchy corresponding to the *select* node. The *goodbye* node exits the system.

3.3 Teaching New Behaviors

Action and interrogation commands form the basis for teaching new behaviors in the system. In particular, we are interested in teaching soccer-related tasks at two levels: (i) basic behaviors linking interrogations to actions such as *if you see the ball* then *go to the ball,* or *if you are near the ball* then *shoot the ball;* and (ii) complex behaviors composed of previously learnt behaviors such as *"Go to the ball and shoot the ball".*

The teaching process is based on individual action and interrogation commands. To achieve teaching, we have extended the CSLU-RAD model shown in Figure 3 to

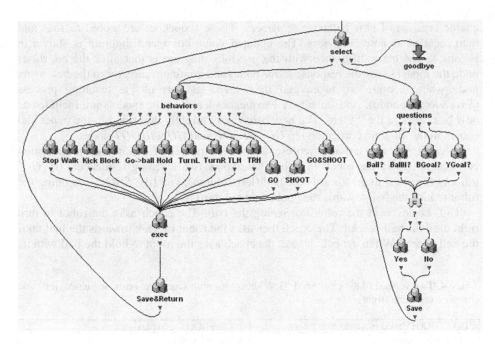

Fig. 4. The CLSU-RAD diagram describes the extended model for training the robot to Go and Shoot the ball

Table 3. Individual 'Go' (left column) and 'Shoot' (right column) training sequences

'GO' training sequence	'SHOOT' training sequence
RAD: Select option	RAD: Select option
User: Question	User: Question
User: Do you see the ball?	User: Do you see cyan goal?
RAD: No	RAD: No
User: Turn right	User: Turn left
RAD: Select option	RAD: Select option
User: Question	User: Question
User: Do you see the ball?	User: Do you see cyan goal?
RAD: Yes	RAD: Yes
User: Go to the ball	User: Kick
RAD: Select option	RAD: Select option
User: Question	User: Goodbye
User: Is the ball near?	
RAD: No	
User: Go to the ball	
RAD: Select option	
User: Question	
User: Is the ball near?	
RAD: Yes	
User: Hold the ball	
RAD: Select option	
User: Goodbye	

enable creation of new behavior sequences. These sequences are stored as files and then recalled as new behaviors. The updated voice command diagram is shown in Figure 4. The main difference with the previous diagram is that after the *questions* node the model saves the response (*Save* node) and continues directly to the *behaviors* node where actions are taken and then saved as part of the teaching process (*Save&Return* node). Additionally, all sequences learnt in the models are included as new behaviors in the system. The new behaviors currently described in the paper and shown in the diagram correspond to *GO*, *SHOOT*, and *GO&SHOOT* nodes.

A teaching conversation is represented by a sequence of action and interrogation commands sent to the robot. In Table 3 we show two teaching sequences: (i) '*GO*' training the robot to go towards the ball (left column); and (ii) '*SHOOT*' training the robot to kick the ball towards the goal (right column).

'GO' begins with the robot not seeing the ball. The coach asks the robot to turn right until the ball is seen. The coach then asks the robot to walk towards the ball until the ball is near. When the ball is near, the coach asks the robot to hold the ball with its

Table 4. Two version of the combined 'Go&Shoot' training sequence: basic sequence (left) and complex sequence (right)

'GO&SHOOT' basic sequence	'GO&SHOOT' complex sequence
RAD: Select option	RAD: Select option
User: Question	User: Behavior
User: Do you see the ball?	User: Go
RAD: No	RAD: Select option
User: Turn right	User: Behavior
RAD: Select option	User: Shoot
User: Question	RAD: Select option
User: Do you see the ball?	User: Goodbye
RAD: Yes	
User: Go to the ball	
RAD: Select option	
User: Question	
User: Is the ball near?	
RAD: No	
User: Go to the ball	
RAD: Select option	
User: Question	
User: Is the ball near?	
RAD: Yes	
User: Hold the ball	
RAD: Select option	
User: Question	
User: Do you see cyan goal?	
RAD: No	
User: Turn left	
RAD: Select option	
User: Question	
User: Do you see cyan goal?	
RAD: Yes	
User: Kick	
RAD: Select option	
User: Goodbye	

front legs and head. 'SHOOT' begins with the robot holding the ball. The coach asks the robot to keep turning right until it sees the cyan goal. The coach then asks the robot to kick the ball towards the goal. Note how we group interactions together into *interrogation-response-action* subsequences, e.g. (1) "User: Do you see the ball?", (2) "RAD: No", and (3) "User: Turn right". These groups of three commands may be repeated several times as part of a single teaching conversation as in 'GO' and 'SHOOT' cases. New behaviors are then recalled as "User: Go" and "User: Shoot".

There are two ways to build 'GO&SHOOT" training sequences: (i) use the *basic* sequence shown in Table 4 (left column) consisting of all the training commands used for 'GO' and 'SHOOT' added together; or (ii) use the *complex* sequence combining the learnt 'GO' and 'SHOOT' sequences as shown in Table 4 (right column). In the basic case, training proceeds as before by combining questions, responses and action commands. In the complex case, training involves simply recalling the already taught sequences i.e. 'GO' sequence followed by 'SHOOT'.

4 Coaching Results

We tested the coaching architecture on the two basic behaviors described in the previous section: 'GO' and 'SHOOT', and then combined them into 'GO&SHOOT' in its basic and complex form. Table 5 shows the average user training time in seconds for

Table 5. Average training time for 'GO','SHOOT', and 'GO&SHOOT' basic and complex sequences

Average Training Time (sec)			
GO	SHOOT	GO&SHOOT - basic	GO&SHOOT – complex
63	29	93	40

Table 6. Test time for 'GO&SHOOT' behaviors with the robot and ball located at different initial positions. Both basic and complex 'GO&SHOOT' training sequences performed similarly.

Test	Robot Initial field position	Ball field position	Time (sec)				Total	Kick Result
			GO		SHOOT			
			looking for the ball	going to the ball	looking for the goal	kicking		
1	Left	Left	0	8	2	4	14	Goal
2	Center	Left	0	10	2	4	16	Goal
3	Right	Left	4	14	2	3	23	Fail
4	Left	Center	2	10	1	4	17	Fail
5	Center	Center	0	10	0	4	14	Goal
6	Right	Center	7	14	0	3	24	Goal
7	Left	Right	2	15	4	3	24	Goal
8	Center	Right	6	11	5	4	26	Goal
9	Right	Right	0	10	5	3	18	Goal

the four sequences. Training time includes all voice interactions between human and robot as well as the time it takes the robot to perform the actual task. Note that 'GO&SHOOT' in its basic form is a direct sum of the individual 'GO' and 'SHOOT' training sequences. On the other hand, the average training time for 'GO&SHOOT' in its complex form is less than half since the task is taught through a more compact set of human-robot interactions based on the already learnt 'GO' and 'SHOOT' sequences. In other words, basic sequences take much longer to train than complex sequences since many more human-robot interactions are required to achieve the intended behavior, i.e. questions, response and action commands. The numbers shown in Table 5 are based on two to four trainings for basic behaviors and single trainings for complex behaviors.

Table 6 shows the result from testing the 'GO&SHOOT' task in the standard platform soccer field under real lighting conditions. Both basic and complex forms performed comparably. The test involves going for the ball and shooting it into the goal with different initial positions for both ball and robot. We test on half of the field. The part of the task that takes the longest is going to the ball. Note how time varies depending on the robot and ball relative initial position, i.e. takes more time to complete the task if robot and ball are on different sides of the goal. Additionally, tests take longer if the robot does not see the ball immediately. We also include in the table the kick results, i.e. goal or fail. Note that scoring the goal depends on the robot correct orientation and kicking.

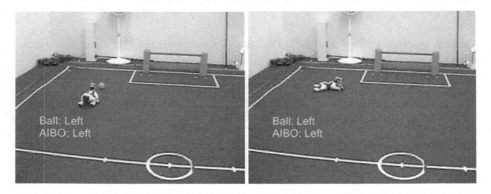

Fig. 5. The images show the behavior of the AIBO robot during 'GO&SHOOT' testing task. The left image shows part of the 'GO' task while the right image shows part of the 'SHOOT' task. The task was tested with ball and robot at different initial positions.

Figure 5 shows two images taken from 'GO&SHOOT' testing tasks. The figure in the left shows the robot going to the ball, part of 'GO' task, while the figure in the right shows the robot shooting the ball, part of the 'SHOOT' task. The full video for training and testing sets can be found in [18].

5 Conclusions and Discussion

We have described in this paper results from our current research in the development of a generalized approach to human-machine interaction via spoken language in the context of robot soccer. The coaching architecture previously described exploits recent developments in cognitive science - particularly notions of grammatical constructions as form-meaning mappings in language, and notions of shared intentions as distributed plans for interaction and collaboration binding perceptions to actions. With respect to social cognition, shared intentions represent distributed plans in which two or more collaborators have a common representation of an action plan in which each plays specific roles with specific responsibilities with the aim of achieving some common goal. In the current study, the common goals were well defined in advance (e.g. teaching the robots new relations or new behaviors), and so the shared intentions could be built into the dialog management system.

To demonstrate the interaction model we described in the current work how to coach a robot to play soccer by teaching new behaviors at two levels: (i) basic behaviors trained from a sequence of existing actions and interrogations, and (ii) complex behaviors trained from newly trained sequences. We tested the architecture on 'GO' and 'SHOOT' tasks using Sony AIBOs in the context of RoboCup soccer standard platform league where our ITAM Eagle Knights team consistently competes at the top level. This work has demonstrated our technical ability to teach robots new behaviors, yet we need to provide with a more natural interaction between humans and robots. In particular the dialog pathways are somewhat constrained, with several levels of hierarchical structure in which the user has to navigate the control structure with several single word commands in order to teach the robot a new relation, and then to demonstrate the knowledge, rather than being able to do these operations in more natural single sentences. In order to address this issue, we are reorganizing the dialog management where context changes are made in a single step. Also, in order to focus the interactions, we are working around scenarios in which the human can interact with several robots at once collaborating around a shared goal such as passing the ball between themselves as part of more sophisticated game playing. In such scenario the human coach will be eventually be able to transmit knowledge in the form *"if blocked pass the ball to player behind"*. Such a command will modify an internal robot database with *"if possess(ball) and goal(blocked) then pass(ball)"*.

Our long term goal in human-robot coaching is to be able to positively affect team performance during a real game. Eventually, these coaching capabilities will be done by an agent coaches inside the robot.

Acknowledgements

Supported by French-Mexican LAFMI, ACI TTT Projects in France, UC-MEXUS CONACYT, CONACYT #42440, and "Asociación Mexicana de Cultura" in Mexico.

References

1. Dominey, P.F., Hoen, M., Lelekov, T., Blanc, J.M.: Neurological basis of language in sequential cognition: Evidence from simulation, aphasia and ERP studies. Brain and Language 86(2), 207–225 (2003)
2. Tomasello, M.: Constructing a language: A usage-based theory of language acquisition. Harvard University Press, Cambridge (2003)

3. Tomasello, M., Carpenter, M., Call, J., Behne, T., Moll, H.: Understanding and sharing intentions: The origins of cultural cognition. Behavioral and Brain Sciences (2006)
4. Weitzenfeld, A., Dominey, P.: Cognitive Robotics: Command, Interrogation and Teaching in Robot Coaching. In: Lakemeyer, G., Sklar, E., Sorrenti, D.G., Takahashi, T. (eds.) RoboCup 2006: Robot Soccer World Cup X. LNCS, vol. 4434, pp. 379–386. Springer, Heidelberg (2007)
5. Kitano, H., Asada, M., Kuniyoshi, Y., Noda, I., Osawa, E.: Robocup: The robot world cup initiative. In. Proceedings of IJCAI 1995 Workshop on Entertainment and AI/ALife (1995)
6. Weitzenfeld, A., Martínez, A., Muciño, B., Serrano, G., Ramos, C., Rivera, C.: EagleKnights 2007: Four-Legged League, Team Description Paper, ITAM, Mexico (2007)
7. Martínez-Gómez, J.A., Weitzenfeld, A.: Real Time Localization in Four Legged RoboCup Soccer. In: Proc. 2nd IEEE-RAS Latin American Robotics Symposium, Sao Luis Maranhao Brasil, September 24-25 (2005)
8. CSLU Speech Tools Rapid application Development (RAD),
   http://cslu.cse.ogi.edu/toolkit/index.html
9. Dominey, P.F., Boucher, J.D.: Developmental stages of perception and language acquisition in a perceptually grounded robot. Cognitive Systems Research 6(3), 243–259 (2005)
10. Dominey, P.F., Boucher, J.D.: Learning to talk about events from narrated video in a construction grammar framework. AI 167(1-2), 31–61 (2005)
11. Kotovsky, L., Baillargeon, R.: The development of calibration-based reasoning about collision events in young infants. Cognition 67, 311–351 (1998)
12. Siskind, J.M.: Grounding the lexical semantics of verbs in visual perception using force dynamics and event logic. Journal of AI Research (15), 31–90 (2001)
13. Steels, L., Baillie, J.C.: Shared Grounding of Event Descriptions by Autonomous Robots. Robotics and Autonomous Systems 43(2-3), 163–173 (2002)
14. Goldberg, A.: Constructions. U Chicago Press, Chicago (1995)
15. Bates, E., McNew, S., MacWhinney, B., Devescovi, A., Smith, S.: Functional constraints on sentence processing: A cross linguistic study. Cognition (11), 245–299 (1982)
16. Riley, P., Veloso, M., Kaminka, G.: An empirical study of coaching. In: Distributed Autonomous Robotic Systems, vol. 6. Springer, Heidelberg (2002)
17. Kaminka, G., Fidanboylu, M., Veloso, M.: Learning the Sequential Coordinated Behavior of Teams from Observations. In: Kaminka, G.A., Lima, P.U., Rojas, R. (eds.) RoboCup 2002. LNCS, vol. 2752, pp. 111–125. Springer, Heidelberg (2003)
18. Videos coaching AIBO robots to play soccer,
    http://robotica.itam.mx/EKCoaching/

Player Positioning in the Four-Legged League

Henry Work, Eric Chown, Tucker Hermans,
Jesse Butterfield, and Mark McGranaghan

Department of Computer Science
Bowdoin College
8650 College Station
Brunswick, ME, 04011, USA
{hwork,echown,thermans,mmcgrana}@bowdoin.edu
jbutterf@cs.brown.edu
http://robocup.bowdoin.edu

Abstract. As RoboCup continues its march towards the day when robots play soccer against people, the focus of researchers' efforts is slowly shifting from low-level systems (vision, motion, self-localization, etc) to high-level systems such as strategy and cooperation. In the Four-Legged League (recently renamed to the Standard Platform League), teams are still struggling with this transition. While the level of play has consistently risen each year, teams continue to remain focused on low-level tasks. Surprisingly few of the 24 Four-Legged teams that competed at RoboCup 2007 were able to self-position at the beginning of the game, despite penalties incurred for not doing so. Things considered to be standard in 'real' soccer – positioning, passing, overall strategies – are still, after 10 years of research, far from a given within the league, and are arguably in their infancy compared to other RoboCup leagues (Small-Sized, Mid-Sized, Simulation). Conversely, for the top teams, many of these low-level systems have been pushed far enough that there is little to be gained in soccer performance from further low-level system work. In this paper we present a robust and successful player positioning system for the Four-Legged League.

1   Introduction

In soccer, possession of the ball is paramount to success. The team controlling the ball has a higher likelihood of scoring goals, and keeping the other team from doing so. Following other teams [2], we firmly believe that to maximize possession and thus win games our team must be the fastest to the ball. For 2007, our positioning system was designed with the sole focus of getting to the ball quickly and insuring possession.

We begin this paper with some background and then describe our positioning system, including its behavior tree, how our players position around the ball, how we play with inactive robots, how out-of-bounds balls are handled, and our potential fields system. We gauge our results with possession statistics from RoboCup 2007. Finally, we conclude with an outlook on future high-level positioning and strategy within the league.

L. Iocchi et al. (Eds.): RoboCup 2008, LNAI 5399, pp. 391–402, 2009.
© Springer-Verlag Berlin Heidelberg 2009

2   Background

The 2007 Four-Legged League [1] robot is the Sony Aibo ERS-7, which all teams must use without any type of hardware modification. Although the Aibo is a fairly cheap and robust robot, its sensors are limited and present numerous challenges [6].

As for higher-level systems, few teams at the 2007 competitions were able to accurately position their robots on the field. While considerable research has been conducted on opponent recognition [3] [4] and localization [5], no team uses such technology in games, to our knowledge. Passing challenges in 2006 and 2007 encouraged teams to invest more research into passing the ball, however, few deliberate passes were completed during games at RoboCup 2007. These limitations are more a factor of the Aibo's sensor limitations than anything else.

Overall, while the quality of play has consistently risen each year, the league is just beginning to move into high-level strategy and positioning.

3   Positioning

The intention of our behavioral system is to facilitate a coordinated system that allows for high-level team play. The framework we devised is a behavior tree that is strongly influenced by traditional soccer: we have strategies, formations, roles, and sub-roles. It is similar to, but was developed independently from other research in the Four-Legged League [10] [2].

First, the system defines overall team characteristics; a team may be more offensive than defensive. Second, we define formations dependent on the position of the ball on the field. For example, when the ball is near the opponent's goal, our robots are positioned to a) score, b) get second chances on missed shots, and c) have a defender bracing for a counter-attack. Third, the system assigns each player robot roles(e.g. defender or chaser) which defines its behavior and position, typically dependent on ball location. Finally, we define sub-roles for each role; if one player is a defender, there are different positioning rules depending on whether the ball is on the offensive or defensive side of the field.

3.1   Strategies

A strategy specifies a general manner of play; a team should be able to play more defensively sometimes, more offensively at others. While some teams have experimented with strategies [5] [11] [2], a more effective approach is to change strategies on the fly, triggered by events during the game. For example, if our team is down a goal with 2:00 minutes left in the game, our team should scramble after the ball in a last ditch effort to score. Further, our team should gather statistics as the game progresses and try to become more or less offensive in order to match the strengths and weaknesses of the opposing team.

We define a strategy as a set of formations. For example, we could build a defensive strategy which would require its component formations to include two

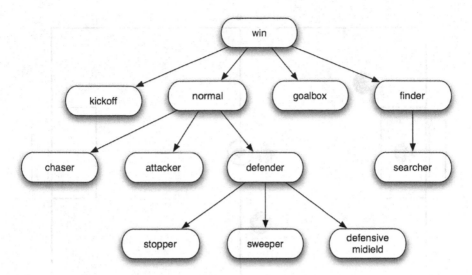

Fig. 1. An overview of our behavior tree. The top level is our single strategy for 2007. The second level are formations, third are roles. The final level is sub-roles, and this chart only shows a small selection of the Defender Role.

defenders. The decision-making process for determining which strategies should be used is a complex problem and is a target of current research within our team. While the strategy framework was in place for the 2007 competitions, we were not able to implement any strategies described above. By competition, our team had only one strategy: *win*. We consider dynamic strategies to be one of our biggest research opportunities in 2008.

3.2 Formations

Unlike human soccer, each player in the Four-Legged League has the exact same skill level and thus is able to switch roles whenever a situation calls for it. A robot may be the primary attacker for half the game, and then a defender for the other. The details of our role switching system are described more thoroughly in [9].

Formations act as a layer above the role switching system to allow for different roles to be specified for selection by the agents as well as sometimes dictating specific roles when well known situations arise. An example of one of our formations is *Kickoff*.

The kickoff is the one situation in a soccer game which is relatively constant; the ball is in the center of the field, the players are not moving, etc. Thus we can ensure that robot A will always be set in position to start playing defense, robot B will be in the center ready to become the chaser, and robot C will be on the wing ready to become the attacker.

This formation lasts for a specified time period or until something unexpected occurs, such as the ball goes out of bounds or a player is penalized. The style

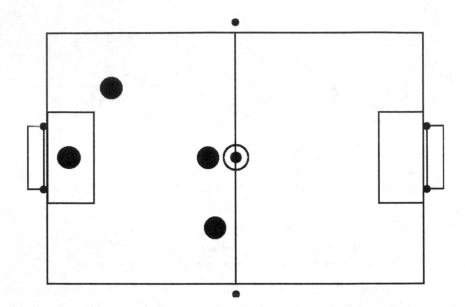

Fig. 2. Offensive Kickoff Position. Immediately after kickoff, the robot closest to the ball (the smaller black filled circle in the center of the field) will become *Chaser*, the one on top will migrate into the center of the field to become a *Defender*, and the one on the bottom will assume the *Attacker* role.

of play in the *Kickoff* formation does not look any different than our typical style of play; the formation simply reinforces the method of soccer we wish to be played.

3.3   Roles

Roles define the fundamental activity a robot should be performing from frame to frame and directly influence which behavioral states in the player's FSM are to be used. The five roles used at RoboCup 2007 were *Chaser*, *Attacker*, *Defender*, *Searcher*, and *Goalie*. These roles are fairly broad in scope and thus we define a number of sub-roles to further control the team dynamic.

3.4   Sub-roles

Sub-roles are divisions of *Roles* which are dynamically assigned by the position of the ball on the field. Each sub-role carries an assignment of a specific point or line on the field for positioning.

Smooth transitions between sub-roles is very important. If the ball were to land on the midfield line, we may have two sub-roles for a *Defender*: one for when the ball is on the offensive side of the field and another for when it is on the defensive side of the field. Because of the particularly noisy ball estimates in our league, there may be oscillation between these two sub-roles. Even slight hesitations in decision-making may give opponents an advantage, so reducing their frequency is important.

Our major strategy for coping with the uncertainty in role switching involves buffering the decision-making process. A robot must be 'sure' of a sub-role decision for about a third of a second before it decides to switch to another sub-role. This means that the robot must have estimates that place the ball's location inside a different sub-role's zone for a constant period of time before it decides to switch. Further, we overlap the sub-role ball zones so that once a ball is in one zone it must leave it convincingly before being considered outside the zone. These features significantly reduce hesitation, and thus improve our odds of maintaining possession of the ball.

3.5   An Example: The Defender

In soccer, the defender's job is prevent the other team from scoring by stopping the ball when on the defensive end of the field. Because we have only one strategy, *win*, the defender's high-level behavior does not change as the game matures (it remains equally defensive during the match). Most of our formations, *normal*, *kickoff*, *goalbox* require a defensive presence and so the *Defender* role is used heavily in games. The *Defender* role consists of three sub-roles: *Stopper*, *Sweeper*, and *Defensive Midfield*. The basic rules we use to define our defender are: a) it should never cross half-field nor enter its own goalbox, and b) it should always position itself between the ball and its own goal to prevent shots.

The main *Defender* sub-role is *Stopper*, which attempts to stay between the ball and the goal. We activate this sub-role when the ball is in the middle of the field. This position relies on ball localization estimates; it is calculated 100 centimeters from the ball on a line between the ball and the back of our own goal. The *Stopper* sub-role was integral to our success at competition.

When the ball is at the opponent's end of the field, the *Defender* switches to a *Defensive Midfield* sub-role, which clips its position at half-field. The robot still positions itself between the ball and its own goal, however it will not chase the ball into opponent territory.

Conversely, the *Sweeper* sub-role is active when the ball is very deep within its own territory. The *Sweeper* position is a static (x,y) coordinate just above its goalbox which makes sure the defender does not get in the way of a teammate chasing the ball. If the ball rolled into the goalbox, the *Goalbox* formation would take over and this sub-role would cease.

4   Positioning on the Ball

To go along with our 'possession is king' approach, one important feature of positioning is that all robots need to be facing the ball at all times. While always having visual contact with the ball would be ideal, this is not possible. First, the frequency of occlusion and obstruction of the ball is quite high: other robots get in the way; referees step on the field to remove penalized robots; and the ball occasionally 'teleports' when it gets placed back on the field after going out-of-bounds. Further, to stay properly localized, positioning robots must continually scan for landmarks.

Our positioning robots split their time between tracking the ball and self-localizing. If they detect any significant velocity of the ball, the robots track the ball until it stops moving. Because we share ball information throughout our team, having positioning robots stay well localized improves ball localization for other robots who may have an obstructed view (particularly the goalie).

When they are positioning, robots also try to keep the center of their bodies facing the ball, ready to become the *Chaser*. Because local data is more trustworthy, when a robot sees the ball, it uses relative estimates to align itself; when it does not see the ball, it uses its global estimate of the ball's position.

One important note is that the *Chaser* robot always keeps its eye on the ball. Unfortunately, this negatively affects the *Chaser*'s self-localization (it must rely purely on odometry). But keeping track of the ball is paramount to possession; teams that choose to do quick head scans to re-localize when chasing the ball increase their chances of losing the ball. We have seen many instances of a robot looking away, only to lose sight of the ball when it looks back.

5   Inactive Teammates

It is crucial for a team to adjust its strategies according to how many robots are actually on the field. In the Four-Legged League there is no stoppage of play; teams are forced to play shorthanded whenever a penalty is called or a robot shuts off. The detection of the operability of teammates has lots of useful extensions for general multi-agent systems research. For our purposes, if our team does not recognize that we have inactive defender, we may give up a goal.

The news of a 30-second penalty must propagate across the entire team as quickly as possible. In our implementation, whenever a robot is penalized, the offending robot will immediately broadcast a packet to its teammates. A penalized robot continues to broadcast its status until the penalty is over. Similarly, when a robot literally shuts off, the 'dead' robot broadcasts an emergency packet to its teammates on shutdown. When the robot is turned on and starts sending packets again, its teammates know that it is alive and well.

We design our formations and roles to consider all cases of teammate inactivity. When we have one robot inactive, we have no attacker. If two robots are inactive we don't have a chaser that goes beyond half-field. There were many times during competition where this work saved us from getting scored upon (especially when our defender received a penalty).

6   Ball Out-of-Bounds

A crucial part of our positioning system is taking into consideration the ball going out-of-bounds as per the 2007 Four-Legged League rules [8]. There is no pausing of the game when the ball leaves the field; referees place the ball back onto the field according to rules designed to penalize the team which knocked the ball out. We don't believe in trying to insure that the ball stays inbounds; we assume that the ball will be bumped out-of-bounds constantly. In our three matches during the round of eight, the ball went out of bounds on average more than 42 times per match, which translates to an average of more than twice per minute of play.

When the ball is closer to any edge of the field, our positioning anticipates the ball leaving the field. One teammate covers the ball's placement if our team kicks it out, another is positioned if the other team does. Our third robot, the *Chaser*, goes for the ball. If we have one inactive robot, we always cover the more defensive out-of-bounds placement (with two, we just have a *Chaser*).

Where out-of-bounds strategies are particularly useful is with goal-line situations. The worst out-of-bounds penalty in the game comes when a robot shoots or accidentally kicks the ball out-of-bounds along the opposing team's goalline. This penalty results in the ball being moved half the field length and occurred quite often in our games (an average of 13 times per game during our three round of eight games in 2007). Our central strategy is to have our *Defender*, who never crosses midfield, always 'anticipate' the side of the field on which the ball is more likely go out. The *Defender* keeps track of side-to-side position of the ball down field and tries to stay parallel with its motion. As the ball moves from side-to-side on the field, so does the *Defender*. Ideally, if the ball goes out-of-bounds, it amounts to a de facto pass to our defender. Top teams employ similar strategies, but ours were particularly effective: of the 39 times it occurred in our final three games of RoboCup 2007, we retrieved the ball 32 times. Picking up the ball in this situation is critical to keeping the ball in the opponent's territory.

This half-field recovery situation also presents an interesting problem due to the large displacement involved: ball capture. From the robot's perspective, this situation seems very odd. At one moment, the ball is near the opponent's goalline and has some velocity. Then suddenly, the ball disappears (as it is picked up by a referee), and reappears in a very different place with zero velocity. This ball capture problem is particularly acute with our localization system, an Extended Kalman Filter [12]. To the filter, this situation seems a lot like noise at first-steady readings that radically oscillate. And so our defenders often appeared sluggish in their responses to the ball (despite their good positioning). Instead of capitalizing on their positioning and scooping up the ball and sending it back into enemy territory, they behaved tentatively. The solution to this problem was to conditionally change the parameters of the filter. When a robot is in a defensive role and is near midfield, its ball model is run with different parameters such that it will respond to the ball far more quickly. This change, implemented right before our final game, brought an immediate boost in performance.

7   Potential Fields

To get our robots to specific, strategic positions, we need each one to avoid running into other teammates, running off the field, running into their own goalbox, or blocking teammates' shots on goal. The system we implemented for these purposes is flexible and avoids unmanageable increases in the number of cases as our decision-making becomes more sophisticated. Inspired by other work in the league [7] [3], we use a potential field positioning system.

7.1   Charges Overview

At the core of the potential field system are 'charges'. Charges either attract robots to or repel robots from certain points or line segments in the playing field. By placing several of these charges around the playing field, we can easily influence the movement and positioning of the robots.

A charge influences a robot through the charge's force function, which we can also think of as the heights of the potential field the charge creates. For example, we say that a repulsive point charge generates a 'potential hill' with a peak at its point location and height that decays with increasing distance from the point. A robot using potential fields to navigate in the presence of such a charge will tend to move away from the charge point towards locations with lower potential charges, thus keeping the robot away from the point as desired.

To model situations in which the robot must consider many such objects when deciding how to move, we use multiple paramaterized charges. For example, we might place repulsive line charges around the sidelines, a repulsive charge at the location of each of the other robots on the field, and an attractive charge at the point to which we would like the robot to move. With such a potential field set up, we could aggregate the positioning influences of all of the charges to determine how to move.

7.2   Charges Implementation

Following [7], we use exponential functions to model the height of potential fields and the partial derivative of these height functions with respect to the x and y location of the robot to determine how it should move.

7.3   Equilibrium Detection

In the course of normal play, robots will often reach points of equilibrium in the potential field, for example in a potential 'trough' or 'cup'. It is important that as the robots approach and move past such positions that they recognize them as such and stop moving instead of oscillating repeatedly across the equilibrium. To implement equilibrium detection we find the value of the height function at some small distance away from the robot in each of the 4 cardinal directions as well as at its current location. The heights obtained can then be used to determine if the robot is at an effective equilibrium point. For example, if the height to the left and the height to the right are both greater than the current height, we say that the robot is at equilibrium with respect to its movement in x direction.

Fig. 3. An example of the output of the potential fields visualizer. Lighter green indicate regions of higher potential, red lines indicate expected paths of the robots.

7.4 Implementation

We start with a basic set of charges for the field: charges for the sidelines, and particularly for our own goalbox (see above). Second, every teammate has a charge in the system, and every packet received from a teammate updates the location of their charge to their self-localization estimates. When one player has possession of the ball, we expand its charge to a wider area so that all other teammates do not disrupt that player's advancement of the ball. Further, we create a line-charge between that player's (x, y) estimate and the back of the opponent's goal. This keeps robots from getting in the way of their teammates' shooting chances.

The potential fields system sits at the bottommost rung of the positioning hierarchy. When the positioning system has chosen a formation, role, sub-role, and produces an (x, y) coordinate on the field to move to, the potential fields system takes this desired position, considers the other charges in the system, and then suggests the best direction of movement. Potential fields were an integral part of our overall positioning efforts.

8 Results

RoboCup provides obvious metrics to measure the total quality of the team: wins, losses, goals scored, and goals allowed. However, to effectively measure

Table 1. Time in minutes in which the ball was in each of the two halves of the field. The first column refers to the opponent's defensive side, the second column to our defensive side.

Opponent	Opponent's Side	Our Side
Team A	11:37	8:23
Team B	14:35	5:25
Team C	12:13	7:47
Totals	38:25	21:35

Table 2. Analysis of how often our team got to the ball first on different out-of-bounds situations. The first number in each column is how many times our team got to the ball first, the second number is the total number of times the ball went out of bounds.

Opponent	Sideline	Endline	Total
Team A	18 / 30	11 / 15	29 / 45
Team B	17 / 27	7 / 8	24 / 35
Team C	18 / 31	14 / 16	32 / 47
Totals	53 / 88	32 / 39	85 / 127

Table 3. Number of traps attempted, and successes, by our team and our opponents by game. The first two columns represent the figures for our team, the next two columns for our opponents.

Opponent	Attempts	Success	Attempts	Success
Team A	93	67	49	32
Team B	123	71	39	19
Team C	100	67	57	45
Total	316	205	145	96

possession, we have chosen three statistics: the time the ball spent in either half of the field, out-of-bounds situation handling, and number of attempted grabs. In order to avoid skewing our results we have limited our analysis to games in the final three rounds of the tournament where only the top competitors remained. The results presented here were gleaned by human analysis of video taken during the matches. As a side note, we feel it would be beneficial for the RoboCup community to begin to develop performance metrics other than goals scored.

How much time the ball spent on either half of the field is a good indicator of general possession. Each match is exactly twenty minutes. The results are summarized in Table 1. As you can see, the ball spent more time on the opponent's side of the field in each of our games, particularly against Team B (when our win margin, not shown, was the highest).

Another important metric is our ability to regain possession after a ball gets kicked out-of-bounds. The results are summarized in Table 2. Overall we got to the ball first more than twice as often as our opponents when the ball was replaced after going out-of-bounds. When the out-of-bounds occurred over an endline this shot up to a ratio better than four to one.

In our league robots generally attempt to trap the ball against their chests before getting ready to kick. It stands to reason that if our goal is to get to the ball more quickly then a good measure of success is the number of times such 'traps' are attempted. We have analyzed how often each team attempted to trap (and how often the traps were successful). Since robot speeds are relatively equal, this is another case where positioning should be the deciding factor. Once again we will limit our results to matches played from the quarterfinals on. The results are summarized in Table 3. The results show that our team attempted traps more than twice as often as our opponents. We also trapped successfully at a similar ratio to our opponents. The high trap ratio reflects the fact that many of the basic soccer skills in RoboCup are close to being optimized by the top teams.

9 Conclusion

While hardly the most sophisticated system on paper, our positioning system has been proven to work under the adverse conditions found in competition. This reflects one of our primary research goals - to build systems that work as well in the real world as they do in the lab.

Over the next few years, we firmly believe that improvements to the overall quality of play will come from smarter and more situationally-aware positioning, faster information propagation, and from integrated positioning systems such as potential fields. We hope that our research encourages teams to continue to focus on high-level behaviors and coordination even as the league moves to a new platform.

References

1. The RoboCup Federation (2005), http://www.robocup.org
2. Quinlan, M., Henderson, N., Nicklin, S., Fisher, R., Knorn, F., Chalup, S., King, R.: The 2006 NUbots Team Report. Technical Report, University of Newcastle (2006)
3. Quinlan, M., Nicklin, S., Hong, K., Henderson, N., Young, S., Moor, T., Fisher, R., Douangboupha, P., Chalup, S., Middleton, R., King, R.: The 2005 NUbots Team Report. Technical Report, University of Newcastle (2005)
4. Röfer, T., Brose, J., Carls, E., Carstens, J., Gohring, D., Juengel, M., Laue, T., Oberlies, T., Oesau, S., Risler, M., Spranger, M., Werner, C., Zimmer, J.: The German National RoboCup Team. Technical Report, Darmstadt University, Bremen University, Humboldt University, Berlin (2006)
5. Hebbel, M., Nistico, W., Schwiegelshohn, U.: Microsoft Hellhounds Team Report 2006. Technical Report, Dortmund University (2006)
6. Nistico, W., Rofer, T.: Improving Percept Reliability in the Sony Four-Legged Robot League. In: Lakemeyer, G., Sklar, E., Sorrenti, D.G., Takahashi, T. (eds.) RoboCup 2006: Robot Soccer World Cup X. LNCS, vol. 4434. Springer, Heidelberg (2007)

7. Knorn, F.: Algorithm Development and Testing for Four Legged League Robot Soccer Passing. Dissertation (2006)
8. The 2007 RoboCup Four-Legged League Rules (2007),
   http://www.tzi.de/4legged/pub/Website/Downloads/Rules2007.pdf
9. Work, H., Chown, E., Hermans, T., Butterfield, J.: Robust Team-Play in Highly Uncertain Environments. In: Proceedings of the Seventh International Conference on Autonomous Agents and Multiagent Systems (2008)
10. Morioka, N.: Road To RoboCup 2005: Behaviour Module Design and Implementation System Integration. Technical Report, The University of New South Wales (2005)
11. Röfer, T., Burkhard, H., von Stryk, O., Schwiegelshohn, U., Laue, T., Weber, M., Juengel, M., Gohring, D., Hoffmann, J., Altmeyer, B., Krause, T., Spranger, M., Brunn, M., Dassler, M., Kunz, M., Oberlies, T., Risler, M., Hebbel, M., Nistico, W., Czarnetzki, S., Kerkhof, T., Meyer, M., Rohde, C., Schmitz, B., Wachter, M., Wegner, T., Zarges, C.: German Team: RoboCup 2005. Technical Report, Technical Report, Dortmund University, Darmstadt University, Bremen University, Humboldt University, Berlin (2005)
12. Kalman, R.E.: A new approach to linear filtering and prediction problems. Transactions of the ASME - Journal of Basic Engineering 82, 35–45 (1960)

Humanoid Robot Gait Generation
Based on Limit Cycle Stability

Mingguo Zhao, Ji Zhang, Hao Dong,
Yu Liu, Liguo Li, and Xuemin Su

Department of Automation
Tsinghua University, Beijing, P.R. China
mgzhao@mail.tsinghua.edu.cn,
{ji-zhang03,donghao00,liuyu,
lilg06,suxm06}@mails.thu.edu.cn
www.au.tsinghua.edu.cn/robotlab/rwg

Abstract. This paper presents the gait generation and mechanical design of a humanoid robot based on a limit cycle walking method- *Virtual Slope Control*. This method is inspired by Passive Dynamic Walking. By shortening the swing leg, the robot walking on level ground can be considered as on a virtual slope. Parallel double crank mechanisms and elastic feet are introduced to the 5 DoF robot leg, to make the heelstrike of the swing leg equivalent to the point-foot collision used in *Virtual Slope Control*. In practical walking, the gait is generated by connecting the two key frames in the sagittal and lateral plane with sinusoids. With the addition of leg rotational movement, the robot achieves a fast forward walking of 2.0leg/s and accomplishes omnidirectional walking favorably.

1 Introduction

Zero-Moment Point is a well known method wildly used on humanoid robots [1,2,3,4] including Asimo [5], HRP [6], and Qrio [7]. In the RoboCup Humanoid League, there is the leading team Darmstadt Dribblers [8,9] etc. In ZMP walking, stability is ensured by constraining the stance foot to remain in flat contact with the ground at all times [10]. Such an artificial constraint is somewhat too restrictive in that it inherently limits the performance of the gait [11]. The robots are under-achieving in terms of speed, efficiency, disturbance handling, and natural appearance compared to human walking [12].

Limit Cycle Walking is a new conception for biped walking, where the periodic sequence of the steps is stable as a whole but not locally stable at every instant in time [13]. Compared with ZMP, Limit Cycle Walking has fewer artificial constraints and more freedom for finding efficient, fast and robust walking motions [11]. One typical paradigm of Limit Cycle Walking is Passive Dynamic Walking, which presents human-like natural motions when the robot walks on a shallow slope without any actuation, powered only by gravity [14,15,16,17]. Among the existing 2D and 3D robots with active joints, Limit Cycle Walking and Passive

L. Iocchi et al. (Eds.): RoboCup 2008, LNAI 5399, pp. 403–413, 2009.
© Springer-Verlag Berlin Heidelberg 2009

Dynamic based robots have more advantages in terms of energy efficiency and disturbance rejection ability [12,18,19,20,21].

To realize Passive Dynamic based powered walking on level ground, our idea is to create a virtual slope for the robot by bending the knee of the swing leg and extending the knee of the stance leg. For a fixed gait, only three conditions are needed to keep the walking stable, all of which are easily satisfied. Therefore such a method could become predominant for real-time applications such as the RoboCup humanoid competition. We name this powered walking method *Virtual Slope Control*.

According to the principle of *Virtual Slope Control*, a humanoid robot Stepper_3D was designed using parallel double crank mechanisms and elastic feet. For practical walking, we designed a simple gait, which has only two key frames in the sagittal plane. The gait has eight parameters, all of which have strongly intuitive meanings that make parameters tuning effortless. With only a few hours of hand tuning, the robot presented good walking results.

The remainder of this paper is organized as follows. In Section 2, the *Virtual Slope Control* method is introduced. In Section 3, the mechanical design of our humanoid robot Stepper_3D is presented. In Section 4, we illustrate the gait for practical walking. Section 5 presents the experiment results and Section 6 the conclusion and future work.

2   Virtual Slope Control Method

2.1   Principium of Virtual Slope Control

In Passive Dynamic Walking, a robot walks along a down hill slope without any actuation, as shown in Fig. 1, the gravity potential energy provided by the slope turns into walking kinetic energy and gets lost at heelstrike [14]. If the slope inclination angle is appropriate, the complementary gravity potential energy E_s equals the heelstrike releasing energy E_r and a stable gait can be synthesized [16].

In level walking, our idea is to make the robot walk as on a virtual slope. As shown in Fig. 2, we suppose that the robot leg length can be infinitely shortened. During each walking step, the swing leg is shortened by a fixed ratio. In this way, the body of the robot experiences a virtual slope. As in Passive Dynamic Walking, if the inclination angle of the virtual slope is appropriate, a stable gait could be synthesized.

In practical walking, the leg length of the robot can not be infinitely shortened, so in the swing phase, we actively extend the stance leg that has been shortened in the last walking step (Fig. 3). The leg shortening and extending is actualized by bending and unbending the knee joint. When extending the stance leg, an amount of energy E_c is added into the walking system. If this energy equals the gravity potential energy provided by the virtual slope E_s and the heelstrike releasing energy E_r (Fig. 4), a stable gait can be synthesized on level ground.

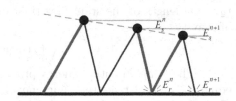

Fig. 1. Passive Dynamic Walking

Fig. 2. Virtual Slope Walking by Shortening Legs

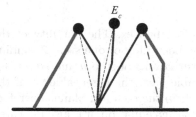

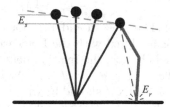

Fig. 3. Actively Extending the Stance Leg and Complementary Energy E_c

Fig. 4. Gravity Potential Energy Provided by Virtual Slope E_s and Heelstrike Releasing Energy E_r

2.2 Basic Conditions for Powered Stable Walking with Virtual Slope Control

The theoretical analysis of the *Virtual Slope Control* method is based on the model of two massless legs and a point mass trunk. For a fixed robot gait, whose joint angular velocities at heelstrike are zero, only the three following conditions have to be satisfied for stable walking.

(1) The angular velocity of the stance leg should be a small positive value, otherwise the stance leg would leave the ground by the pulling of centrifugal force. In the following inequation, θ/T is the mean of the angular velocity of the stance leg in one waking step; α_0 is the initial angle of the stance leg (Fig. 5); θ is the angle between the two legs at heelstrike; T is the walking period; r is the

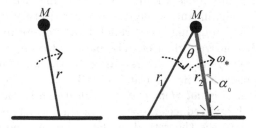

Fig. 5. Variables used in Basic Walking Conditions, Left: Swing; Right: Heelstrike

length of the stance leg; r_1 is the length of the swing leg right after heelstrike; r_2 is the length of the stance leg right after heelstrike; And g is the acceleration of gravity.

$$0 < \frac{\theta}{T} \leq \sqrt{(g\cos(\theta - \alpha_0) + \min r'')/r_1} \tag{1}$$

(2) The complementary energy provided by extending the stance leg should be equal to the heelstrike releasing energy, while E_c is the complementary energy; E_r is the heelstrike releasing energy; M is the mass of the trunk (Fig. 5); ω_* is the angular velocity of the stance leg right after heelstrike; And t is the time variable.

$$\begin{aligned} E_c &= \int_0^T Mg\cos(\alpha_0 - \frac{\theta}{T}t)r'\mathrm{d}t - \frac{1}{2}M\frac{\theta}{T^2}^2(r_1^2 - r_2^2) \\ &= E_r = \frac{1}{2}Mr_2^2\omega_*^2\tan^2\theta \end{aligned} \tag{2}$$

(3) The gait should be robust to small disturbances. The stability of the gait is analyzed with the Poincaré map at the beginning of one step. Assuming linear behavior, the relation between the perturbations of step n and step $n+1$ is $[\Delta\omega_{n+1}\Delta t_{n+1}]^T = J[\Delta\omega_n \Delta t_n]^T$. For a stable gait, the eigenvalues λ of the Jacobi matrix J should be within the unit circle in the complex plane. Therefore the following equation should be satisfied. From such equation, it is also indicated that extending the stance leg late could enhance walking stability.

$$\int_0^T \sin(\alpha_0 - \frac{\theta}{T}t)r'\mathrm{d}t < 0 \quad r' \geq 0 \tag{3}$$

3   Mechanism of Stepper_3D

3.1   Overall Introduction

Stepper_3D (Fig. 6) is the lower body of the humanoid robot which will be used in RoboCup 2008 by the Tsinghua Hephaestus team. Its structure is designed under the competition rules of the RoboCup Humanoid League [22]. The robot is 0.44m in height and 2.51kg in weight. Dimensional parameters are shown in Fig. 7. The robot has 11 DoF: 1 yaw DoF on top of the trunk, 3 orthogonal DoF at each hip, 1 pitch DoF at each knee, and 1 roll DoF at each ankle. All the DoF are actuated by Robotis digital servo motors.

The uniqueness of Stepper_3D comes from adopting the parallel double crank leg mechanism shown in Fig. 6. This design arises from the principle of *Virtual Slope Control*, and its foundation-Limit Cycle Walking, where the state of the robot at heelstrike is significant, while the exact trajectory in the swing phase is less relevant [11]. By adopting the parallel double crank mechanisms, we are able to mechanically constrain the feet to be parallel to the ground at heelstrike, and avoid the angular cumulative errors caused by the joint motors in normally designed legs. Therefore, the flat foot at heelstrike can be approximated by the point-foot model as in *Virtual Slope Control*. Additionally, this design saves the two ankles' pitch DoF and reduces the weight of the legs. The mass distribution of the robot is closer to the theoretical model of two massless legs and a mass point trunk (Fig. 5).

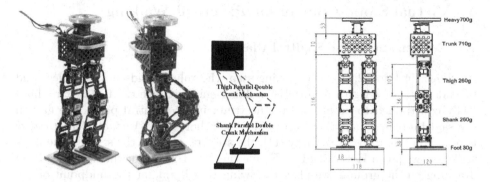

Fig. 6. Stepper_3D with the parallel double crank leg mechanism

Fig. 7. Stepper_3D Structure and Body Parameters

3.2 Details of Mechanism Design

The leg of Stepper_3D is composed of two sets of parallel double crank mechanisms (Fig. 6), forming the thigh and shank respectively. Since this mechanism has only one DoF, any of the four vertexes can be used as the active rotating axis, while the other three rotate passively. On Stepper_3D, the active rotating axes are placed at the front vertexes of the knee joint (Fig. 8), while the hip and ankle vertexes act as passive ones (Figs. 9, 10).

During the swing phase, there might be a slight angle between the sole of the stance leg and the ground. To reduce its impact on the walking stability, we used flexible materials on the bottom of the feet. As shown in Fig. 10, the soft sponge cushion is added under the elastic aluminum sheet which is placed under the rigid baseboard.

The batteries and processing boards are all located inside the trunk box (Fig. 11) rather than on the legs or feet, decreasing the mass of the legs to better approach the theoretical model. An iron weight approximating the upper body's mass and inertia is placed above the trunk (Fig. 11) to simulate the completed robot walking conditions.

Fig. 8. Knee   **Fig. 9.** Hip   **Fig. 10.** Ankle   **Fig. 11.** Trunk

4   Virtual Slope Control for Practical Walking

4.1   Movement in the Sagittal Plane

As shown in Fig. 12, in each walking step, the robot bends its swing leg, and extends its stance leg to restore the walking energy. Since the feet are mechanically constrained to be parallel to the trunk in the sagittal plane, the gait in the sagittal plan equals that of a point-foot robot. Since *Virtual Slope Control* is based on Limit Cycle Stability [11], the sticking point is the state of the heelstrike, so we put a Heelstrike Key Frame therein. To prevent the swing leg from dragging on the ground, we placed a Swing Key Frame at the midpoint of the swing phase. The two key frames are described by four parameters as in Fig. 12. Since the exact trajectories of the joint angles are of little concern in limit cycle walking [11], we simply use smooth sinusoids to connect the key frames as described by Eq. 4 and Eq. 5. The connected trajectories are shown in Fig. 14 and the definitions of the joint angles are shown in Fig. 13.

In the four gait parameters shown in Fig. 12, θ is related to the step length. In practical walking, it is equally divided between the two hip joints. α is related to the complementary energy, the larger α is, the more walking energy is complemented. β is related to the lifting height of the swing leg, and T is the step period. It is intuitive that a walk with a higher step frequency (small T) or

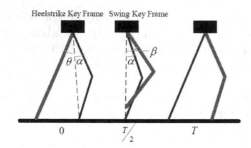

Fig. 12. Sagittal Movement

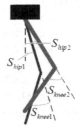

Fig. 13. Definitions of Joint Angles in Sagittal Plane

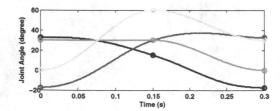

Fig. 14. Trajectories of Sagittal Joint Angles in One Walking Step. $T = 0.3s, \theta = 35°, \alpha = 15°, \beta = 30°$, The cycles denote the key frames. Blue: Hip1; Red: Hip2; Green: Knee1; Yellow: Knee2.

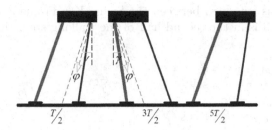

Fig. 15. Lateral Swing Movement

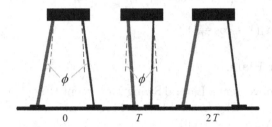

Fig. 16. Lateral Step Movement

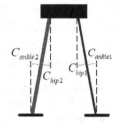

Fig. 17. Definitions of Joint Angles in Lateral Plane

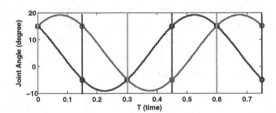

Fig. 18. Trajectories of Lateral Joint Angles in 2.5 Walking Steps. $T = 0.3s, \varphi = 5°, \gamma = 10°, \phi = 10°$; Blue: Hip1, Ankle1; Red: Hip2, Ankle2;Brown: Lateral Swing Key Frames; Orange: Lateral Step Key Frames.

Fig. 19. Rotational Movement

a larger step length (large θ) costs more energy. So the faster the walking is, the larger α should be. Let $\theta = 0, \alpha = 0$, the robot march in place.

This gait can easily satisfy the three basic walking conditions. First by giving a bigger T and a smaller θ, condition (1) can be satisfied. Second, by changing α, the complementary energy can be adjusted to satisfy condition (2). Finally, the robot stance leg is extended between the Swing Key Frame and the Heelstrike Key Frame, which is in the second half of the walking step, thus condition (3) can be satisfied.

$$
\begin{cases}
S_{hip1} = \frac{\theta}{2}\cos\frac{\pi t}{T} + \alpha \\
S_{hip2} = -\frac{\theta}{2}\cos\frac{\pi t}{T} + \frac{\beta}{2}(1 - \cos\frac{2\pi t}{T}) \\
S_{knee1} = 2\alpha \\
S_{knee2} = \beta(1 - \cos\frac{2\pi t}{T})
\end{cases}
\quad 0 < t \le T/2
\tag{4}
$$

$$
\begin{cases}
S_{hip1} = \frac{\theta}{2}\cos\frac{\pi t}{T} + \frac{\alpha}{2}(1 - \cos\frac{2\pi t}{T}) \\
S_{hip2} = -\frac{\theta}{2}\cos\frac{\pi t}{T} + \alpha + \frac{\beta-\alpha}{2}(1 - \cos\frac{2\pi t}{T}) \\
S_{knee1} = \alpha(1 - \cos\frac{2\pi t}{T}) \\
S_{knee2} = 2\alpha + (\beta - \alpha)(1 - \cos\frac{2\pi t}{T})
\end{cases}
\quad T/2 < t \le T
\tag{5}
$$

4.2   Movement in the Lateral Plane

In the lateral plane, we define two movements: Lateral Swing Movement (Fig. 15) and Lateral Step Movement (Fig. 16). The Lateral Swing Movement serves in every gait while the Lateral Step Movement contributes to lateral walking and omnidirectional walking. Each movement includes two key frames. In the key frames shown in Fig. 15, γ indicates the lateral swing amplitude. φ indicates the angle between the two legs and in part determines the lateral swing period [15]. So adjusting φ could synchronize the lateral movement and sagittal movement. In the key frames shown in Fig. 16, ϕ indicates the lateral step length. The feet are controlled to be parallel to the trunk. The key frames are connected by sinusoids as Eq 6. The connected trajectories are shown in Fig. 18 and the definitions of the joint angles are shown in Fig. 17.

$$
\begin{cases}
C_{hip1} = C_{ankle1} = \varphi - \frac{\gamma}{2}\sin\frac{\pi t}{T} + \phi\cos\frac{\pi t}{T} \\
C_{hip2} = C_{ankle2} = \varphi + \frac{\gamma}{2}\sin\frac{\pi t}{T} + \phi\cos\frac{\pi t}{T}
\end{cases}
\tag{6}
$$

4.3   Rotational Movement

To the rotational movement, we put two key frames as in Fig. 19 and use sinusoids for the connection, while λ indicates the rotation amplitude.

5   Experiment Results

5.1   Forward Walking

By tuning the four parameters of the sagittal movement θ, α, β, T (Fig. 12) and the two parameters of the lateral swing movement φ, γ (Fig. 15), Stepper_3D presented a fast and stable walking and reached the speed of 0.5m/s and the relative speed of 2.0leg/s. By giving different θ and α, the robot could change its speed continuously from 0m/s to 50cm/s. The video frames of forward walking are shown in Fig. 20. The comparison of our result to those of the other robots in RoboCup 2008 is shown in Fig. 23.

Fig. 20. Forward Walking Video Frames

Fig. 21. Lateral Walking Video Frames

Fig. 22. Rotation Video Frames

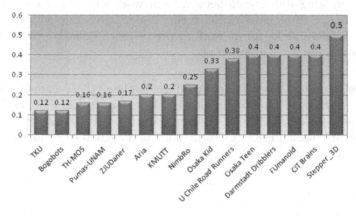

Fig. 23. Comparison of Walking Speed (m/s). All the robots on RoboCup 2008 whose speed exceed 0.1m/s are presented in the figure. Stepper_3D reached the speed of 0.5m/s.

5.2 Lateral Walking and Rotation

By setting $\theta = 0, \alpha = 0$, tuning β, T (Fig. 12) and φ, γ (Fig. 15), the robot could march in place. Then by giving a slight value to the lateral step parameter ϕ (Fig. 16) or rotation parameter λ (Fig. 19), it could walk sideward or rotate as shown in Figs. 21, 22.

5.3   Omnidirectional Walking

While lateral walking and rotation could be realized by adding the lateral step movement and rotation movement to the gait of marching in place respectively, omnidirectional walking can be realized by adding the two movements to forward walking. In this way, Stepper_3D accomplished walking and rotating simultaneously. All the videos about the walking experiments could be found on our website http://www.au.tsinghua.edu.cn/robotlab/rwg/Robots.htm.

6   Conclusion and Future Work

In this paper, we present the mechanical design and gait generation of our humanoid robot Stepper_3D based on a limit cycle walking method-*Virtual Slope Control*. The parallel double crank mechanism is adopted and the elastic materials are used on the bottom of the feet. The gait is extraordinarily simple with strongly intuitive parameters, which results in a very high walking speed and also accomplishes omnidirectional walking favorably. The experiment results indicate that the mechanical design and the gait of *Virtual Slope Control* are suitable for RoboCup Competitions.

Currently, we are working on the theoretical analysis of energy supplementation and walking stability. Hopefully, such work will start to appear in our papers in two or three months. Since this paper only deals with a fixed gait, other possible avenues for future work involve sensor feedback introduction, gait adjustment, and active gait switching.

Acknowledgments. The authors would like to thank the Tsinghua Hephaestus team. This work was supported in part by the Open Project Foundation of National Industrial Control Technology Key Lab (No. 0708003) and Open Project Foundation of National Robotics Technology and System Key Lab of China(No. SKLRS200718).

References

1. Vukobratovic, M., Juricic, D.: Contribution to the synthesis of biped gait. In: Proc. IFAC Symp. Technical and Biological Problem on Control, Erevan, USSR (1968)
2. Juricic, D., Vukobratovic, M.: Mathematical Modeling of Biped Walking Systems (ASME Publ., 1972) 72-WA/BHF-13 (1972)
3. Vukobratovic, M., Stepanenko, Y.: Mathematical models of general anthropomorphic systems. Mathematical Biosciences 17, 191–242 (1973)
4. Vukobratovic, M.: How to control the artificial anthropomorphic systems. IEEE Trans. System, Man and Cybernetics SMC-3, 497–507 (1973)
5. Hirose, M., Haikawa, Y., Takenaka, T., et al.: Development of humanoid robot ASIMO. In: Proc. IEEE/RSJ International Conference on Intelligent Robots and Systems, Workshop 2, Maui, HI, USA, October 29 (2001)
6. Kaneko, K., Kanehiro, F., Kajita, S., et al.: Humanoid robot HRP-2. In: Proc. IEEE International Conference on Robotics and Automation (ICRA), New Orleans, LA, USA, April 26 - May 1, pp. 1083–1090 (2004)

7. Nagasaka, K., Kuroki, Y., Suzuki, S., et al.: Integrated motion control for walking, jumping and running on a small bipedal entertainment robot. In: Proc. IEEE International Conference on Robotics and Automation (ICRA), New Orleans, LA, USA, April 26 - May 1, vol. 4, pp. 3189–3194 (2004)
8. Friedmann, M., Kiener, J., Petters, S., et al.: Versatile, high-quality motions and behavior control of humanoid soccer robots. In: Proc. 2006 IEEE-RAS International Conference on Humanoid Robots, Genoa, Italy, December 4, pp. 9–16 (2006)
9. Hemker, T., Sakamoto, H., Stelzer, M., et al.: Hardware-in-the-Loop Optimization of the Walking Speed of a Humanoid Robot. In: Proc. CLAWAR 2006, Brussels, Belgium, September 12-14 (2006)
10. Vukobratovic, M., Frank, A., Juricic, D.: On the Stability of Biped Locomotion. IEEE Transactions on Biomedical Engineering 17(1) (1970)
11. Hobbelen, D., Wisse, M.: Limit Cycle Walking. In: Humanoid Robots: Human-like Machines, ch. 14, p. 642. I-Tech Education and Publishing, Vienna (2007)
12. Collins, S., Ruina, A., Tedrake, R., et al.: Efficient Bipedal Passive-Dynamic Walkers. Science 307(5712), 1082–1085 (2005)
13. Hurmuzlu, Y., Moskowitz, G.: Role of Impact in the Stability of Bipedal Locomotion. International Journal of Dynamics and Stability of Systems 1(3), 217–234 (1986)
14. McGeer, T.: Passive Dynamic Walking. International Journal of Robotics Research 9, 62–82 (1990)
15. Garcia, M.: Stability, Scaling, and Chaos in Passive Dynamic Gait Models. PhD Thesis, Cornell University, Ithaca, NY (1999)
16. Wisse, M.: Essentials of Dynamic Walking: Analysis and design of two-legged robots. PhD Thesis, Delft University of Technology, Netherlands (2004)
17. Tedrake, R.: Applied Optimal Control for Dynamically Stable Legged Locomotion. PhD Thesis, MIT, MA (2004)
18. Pratt, J., Chew, C.-M., Torres, A., et al.: Virtual Model Control: An Intuitive Approach for Bipedal Locomotion. The International Journal of Robotics Research 20(2), 129–143 (2001)
19. Chevallereau, C., Abba, G., Aoustin, Y., et al.: RABBIT: a testbed for advanced control theory. IEEE Control Systems Magazine 23(5), 57–79 (2003)
20. Geng, T., Porr, B., Wörgötter, F.: Fast Biped Walking with a Sensor-driven Neuronal Controller and Real-Time Online Learning. The International Journal of Robotics Research 25(3), 243–259 (2006)
21. Morimoto, J., Cheng, G., Atkeson, C.G., et al.: A simple reinforcement learning algorithm for biped walking. In: Proceedings of IEEE International Conference on Robotics and Automation, vol. 3 (2004)
22. Kulvanit, P., Stryk, O.: RoboCup Soccer Humanoid League Rules and Setup for the, competition in Suzhou, China (2008), https://lists.cc.gatech.edu/mailman/listinfo/robocup-humanoid

Playing Creative Soccer: Randomized Behavioral Kinodynamic Planning of Robot Tactics

Stefan Zickler and Manuela Veloso

Computer Science Department
Carnegie Mcllon University
Pittsburgh, Pennsylvania, USA
{szickler,veloso}@cs.cmu.edu

Abstract. Modern robot soccer control architectures tend to separate higher level tactics and lower level navigation control. This can lead to tactics which do not fully utilize the robot's dynamic actuation abilities. It can furthermore create the problem of the navigational code breaking the constraints of the higher level tactical goals when avoiding obstacles. We aim to improve such control architectures by modeling tactics as sampling-based behaviors which exist inside of a probabilistic kinodynamic planner, thus treating tactics and navigation as a unified dynamics problem. We present a behavioral version of Kinodynamic Rapidly-Exploring Random Trees and show that this algorithm can be used to automatically improvise new ball-manipulation strategies in a simulated robot soccer domain. We furthermore show how opponent-models can be seamlessly integrated into the planner, thus allowing the robot to anticipate and outperform the opponent's motions in physics-space.

1 Introduction

Multi-agent control architectures as encountered in robot soccer tend to follow a common paradigm: controls are layered in a hierarchical structure with higher level decision-making controls located at the top, flowing to more concrete, lower level controls as we move towards the bottom. One particularly successful approach is the "Skills, Tactics, and Plays" (STP) model [1]. Here, a multi-agent *playbook* performs role-assignment at the top of the hierarchy. These roles are then executed by the individual agents and modeled in the form of *tactics* which again consist of a state machine that can invoke even lower level *skills*. Navigational controls are modeled on the skill level and can range from very simplistic reactive controls to more elaborate motion planning techniques.

While this STP approach has proved very successfully in several real-world robot domains, there exists at least one remaining limitation. The lower level motion planning code is generally unaware of the higher level strategic goals of the game. Similarly, the higher level strategy is mostly unaware of the inherent dynamics of the physical world that the robot is acting within. While the navigational control might for example perform obstacle-avoidance to prevent collisions, it might at the same time also violate the assumptions of the higher

L. Iocchi et al. (Eds.): RoboCup 2008, LNAI 5399, pp. 414–425, 2009.

level tactic, such as being able to dribble the ball successfully towards the goal. A related issue is the problem of predictability. There might exist several tactics that the agent can chose from, but each of them will typically perform its task deterministically, given the current state of the world. There is little to none creative variation in the motions that the robot executes, which can ultimately not only lead to missed opportunities, but also to the opponent team quickly adapting to one's strategy.

Our work aims to solve these problems by embedding the skills and tactics components of STP into a kinodynamic planning framework. By treating soccer as a physics-based planning problem, we are able to automatically generate control sequences which achieve higher level tactical goals while adhering to any desired kinodynamic navigation constraints. To generate such control sequences, we introduce Behavioral Kinodynamic Rapidly-Exploring Random Trees: a randomized planning algorithm that allows us to define tactics as nondeterministic state machines of sampling-based skills. Finally, we show that our approach is able to creatively improvise new ball-manipulation strategies in a simulated soccer domain and compare its performance with traditional static tactic approaches.

1.1   Related Work

Several multi-agent control architectures exist that are applicable to robot soccer [1,2,3,4,5]. Our presented approach aims to improve the single-robot control components of such layered approaches and is able to work underneath any higher level multi-agent role assignment methodology, such as playbooks [1], state machines, or even market-based methods [6]. In terms of single agent control, our work adopts the concept of tactics as a state machine of skills as presented in [1]. However, in our case, both the tactics and skills are modeled inherently probabilistically and act as search axioms inside of a kinodynamic planner. We are currently unaware of any other approach that has unified planning, strategy, and lower-level dynamics in such a way.

Our approach builds heavily upon kinodynamic randomized planning. Introduced by LaValle [7], Rapidly-Exploring Random Trees (RRT) are a planning technique which allows rapid growth of a search tree through a continuous space. The algorithm has been refined by Kuffner and LaValle [8] and has been used for collision free motion planning in many robotics applications, such as in the RoboCup Small Size team by Bruce and Veloso [9,10]. LaValle and Kuffner [11] successfully showed that RRT can be adapted to work under Kinematic and Dynamic planning constraints.

Internally, our planning algorithm uses a rigid-body simulator to perform its state transitions. The task of robustly and accurately simulating rigid body dynamics is well-understood [12] and there exist several free and commercial simulators, such as the Open Dynamics Engine (ODE), Newton Dynamics, and Ageia PhysX.

2   Modeling Soccer as a Kinodynamic Planning Problem

Instead of artificially separating our control model into higher level tactics and lower level navigation, we treat soccer as a general control planning problem. Given an initial state x_{init} in the state-space $\mathbf{X}$, our algorithm is to find an acceptable sequence of control parameters which lead us to a final state x_{final} located within a predefined goal region $X_{goal} \subseteq \mathbf{X}$. Additional constraints might exist on the intermediate states of our action sequence, such as collision avoidance or velocity limits. A very common tool to solve such problems is randomized planning.

Randomized planning techniques are search algorithms which apply probabilistic decision making. One advantage of randomized approaches is that we are able to shape their behavior by modifying the underlying sampling distributions. More importantly, randomized techniques tend to work well in continuous environments as they are less likely to "get stuck" due to some predefined discretization of the set of possible actions. Rapidly-Exploring Random Trees (RRT) is a popular example of modern randomized planning techniques. Standard RRT as introduced by LaValle [7] is outlined in algorithm 1. We start with a tree T containing a single node x_{init} representing our initial state. We then continuously sample random points x_{sample} located anywhere within our domain. We locate the node x_{source} in T that is closest to x_{sample} using a simple nearest neighbor lookup. We extend node x_{source} towards x_{sample}, using a predefined constant distance, thus giving us a new node, x_{new}. Finally, we add x_{new} to T with x_{source} being its parent. This process is repeated until we either reach the goal or give up. A visual example of a typical RRT search is given in figure 1.

Since robot soccer is a problem which can contain kinematic and dynamic constraints, we want our planner to operate in second order time-space. In the planning literature this is commonly referred to as kinodynamic planning [13]. RRT can be adapted to work in a kinodynamic environment as shown by LaValle and Kuffner [11].

```
T.AddVertex(x_init);
for k ← 1 to K do
    x_sample ← SampleRndState();
    x_source ← NNeighbor(T,x_sample);
    x_new ← Extend(x_source,x_sample);
    if IsValidExtension(x_new) then
        T.AddVertex(x_new);
        T.AddEdge(x_source,x_new);
        if x_new ∈ X_goal then
        |   return x_new ;
        end
    end
end
return Failed;
```

Algorithm 1. Standard RRT

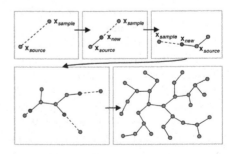

Fig. 1. A visual demonstration of the standard RRT algorithm

Finally, it is important to point out that planning robot soccer tactics is significantly more challenging than a pure path planning problem. The crucial difference is that in robot soccer, our agent's motions and the higher level planning goal are actually allowed to be disjoint subsets of our state space. For example, our planning goal is typically defined in terms of the ball being delivered to a particular target, whereas our actions can only be applied to the robot, but not the ball itself. Thus, the only way to manipulate the ball, is through the interactions defined by our physics simulator, such as rigid body dynamics. This prevents us from using a simple goal-heuristic such as modifying RRT's sampling distribution. Without the bias of a useful goal heuristic, our search would become infeasible.

2.1   Behavioral Kinodynamic Planning for Robot Tactics

To guide our search through dynamics space, we embrace and extend the concept of tactics and skills as presented by Browning and colleagues in [1]. For our work, tactics and skills still follow their traditional role to act as informed controllers of our robot by processing the input of a particular world state and providing an action for our robot to execute. In other words, they act as an implementation of a robot's behavior (such as "Attacker" or "Goalie") and contain some kind of heuristic intelligence that will attempt to reach a particular goal state.

However, the significant difference is that in our case, both the tactic and its skills reside within a kinodynamic planner. That is, instead of greedily executing a single tactic online, we use planning to simulate the outcome of many different variations of a tactic in kinodynamic space. Our agent then selects one particularly good solution for execution. To achieve this variation in tactic execution, we modify the traditional definition of a tactic to be modeled probabilistically as

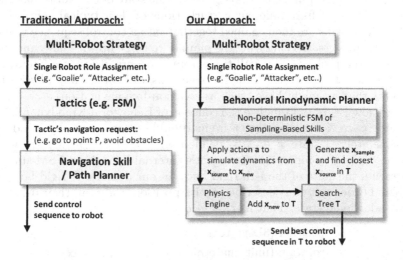

Fig. 2. A diagram of a traditional STP-style approach (left) in comparison to our behavioral kinodynamic planning approach (right)

a Nondeterministic Finite State Machine. Similarly, skills are sampling-based, and thus deliver nondeterministic actions. A diagram showing this structural difference is shown in figure 2.

In the nondetermininistic FSM, transitions are modeled probabilistically. The probabilities are defined by the programmer and act as a behavioral guideline for how likely the tactic is to chose a particular sequence of state-transitions. By being modeled in this nondeterministic fashion, each execution of the tactic can result in a different sequence of state transitions, thus reducing predictability of our robot's actions. However, we are in full control over how much "creative freedom" we want to provide to the tactic by adjusting the transition probabilities.

To further increase the freedom of choice in our agent's behavior, we also modify our Skills to act non-deterministically. In traditional STP, a skill would normally perform an analytic reactive control given the current state of the world. For example, a traditional "shoot-on-goal" skill would aim for the center of the largest opening in the goal, and take a shot. This not only leads to very high predictability, but it also ignores the fact that there might exist other shots that could be more likely to succeed, given a particular dynamics state of the world. In our new non-deterministic version of a "shoot-on-goal" skill, we instead use sampling to select randomly from a set of target points throughout the goal area. We then rely on the fact that our planner will execute this skill many times, always selecting a different sample point and simulating the skill's outcome in kinodynamic space.

Given this non-deterministic model of tactics and skills, we can now use search to cover many different simulated executions of the tactic in our kinodynamic search space. The goal of this methodology is to be able to search over a variety of executions and find one particular instance that leads us successfully to the goal state. The inherent creativity of this approach arises from the fact that skills no longer execute deterministic motions based on the state of the world, but instead can chose randomly from different perturbations of their typical behavior. The way that these skills are then chained together into a control sequence is again a probabilistic process, thus allowing the planner to come up with new, and unique solutions to the problem. Since all the planning occurs inside of a kinodynamic framework, we can guarantee that the solutions will satisfy any physics-based constraints that we might have (such as collision-free navigation).

We are now ready to formalize the concept of Behavioral Kinodynamic RRT. The State Space $\mathbf{X}$ describes the entire modifiable space of our physical domain. More concretely, a state $x \in \mathbf{X}$ is defined as $x = [t, fsm, r_0, \ldots, r_n]$ where t represents time, fsm represents the agent's internal behavioral FSM-state, and r_i represents the state of the i-th rigid body in our domain. A rigid body state in a second order system is described by its position, rotation, their derivatives, and any additional state. That is, $r_i = [p, q, v, \omega, o]^T$ where

p : position (3D-vector)

q : rotation (unit quaternion or rotation matrix)

v : linear velocity (3D-vector)

ω : angular velocity (3D-vector).

o : optional additional state-variables (e.g. robot's actuators).

It is important to point out that r may include both active and passive rigid bodies. Active rigid bodies are the ones which we can directly apply forces to as an execution of our planning solution, such as robots. All other bodies which can solely be manipulated through rigid body dynamics (e.g. the ball) are considered passive. For simplicity, we will assume that the total number of rigid bodies in our domain stays constant.

The Action Space $\mathbf{A}$ (also known as control space) is defined by the set of possible actions of our agent. An action $a \in \mathbf{A}$ of an agent consists of force, torque, and possibly actuation commands, such as kick or dribble. Robot-dependent kinematic constraints are modeled by the possible torques and forces that can be applied to the robot.

The Sampling Space $\mathbf{S}$ is what we draw our random samplings from. It is a subset of the state-space, i.e. $\mathbf{S} \subseteq \mathbf{X}$.

```
T.AddVertex(x_init);
for k ← 1 to K do
    s_random ← SampleRandomState();
    x_source ← KinematicNearestNeighbor(T,s_random);
    fsm_state ← FsmTransition(x_source.fsm);
    if IsValidFSMstate(fsm_state) then
        a ← ApplyFsmBehavior(fsm_state,x_source,s_random);
        x_new ← Simulate(x_source,a,Δt);
        x_new.t ← x_source.t + Δt;
        x_new.fsm ← fsm_state ;
        if IsValidPhysicsState(x_new) then
            T.AddVertex(x_new);
            T.AddEdge(x_source,x_new,a);
            if x_new ∈ X_goal then
                | return x_new ;
            end
        end
    end
end
return Failed;
```

Algorithm 2. Behavioral Kinodynamic RRT

Using this terminology, we can now define Behavioral Kinodynamic RRT as shown in algorithm 2. We start with a tree T containing an initial state $x_{init} \in \mathbf{X}$. We then enter the main RRT loop. The function `SampleRandomState` uses an internal probability distribution to provide us with a sample s_{random} taken from the sampling space $\mathbf{S}$. It is important that the sampling space $\mathbf{S}$ and the probability distribution are carefully chosen to match the particular robot domain.

For a typical omni-directional robot movement, the sampling space **S** might consist of three dimensions, representing a point and orientation in space. Similarly, the sampling distribution might be spread uniformly throughout the confines of the domain. The `KinematicNearestNeighbor` function defines a distance metric between a sample s_{random} and an existing state x from our tree T.

The definition of a consistent distance metric is crucial for a correct operation of the algorithm. Since we are planning in second order timespace, using an Euclidean nearest neighbor distance approach will fail. This is because the Euclidean distance between two nodes' positions ignores the fact that the nodes have velocities attached to them. Thus, distance is not a good indicator of how much time it will take to reach a particular node. A much more reliable approach is to define a metric based on estimated time to get from $x_i \in T$ to s_{random}. This can be achieved by using a simple acceleration-based motion model to compute the minimal estimated time for the robot to reach its target position and orientation. While this certainly will not always constitute an accurate prediction of the time traveled (in particular because we are not taking potential obstacles into account), it is still a good heuristic for the nearest neighbor lookup.

Once we have located the nearest neighbor x_{source} towards s_{random}, we now call `FsmTransition` to perform a nondeterministic FSM state transition giving us the new internal FSM state fsm_state. We then make sure that this state is valid by ensuring that the FSM has not reached any end-condition. Finally, we apply the behavior associated with fsm_state by calling `ApplyFsmBehavior`. The actual FSM behavior can be any skill, ranging from a simple "full stop" to much more elaborate controls. It is important to note, that each FSM behavior is able to make full use of the sample s_{random}. For example, we might imagine a skill called "move towards point" which will accelerate our robot towards the sample s_{random}. This in fact implies, that the traditional kinodynamic RRT algorithm is actually a subset of the introduced Behavioral Kinodynamic RRT. If we imagine a behavioral FSM with a single behavior that implements the typical RRT extend-operator then this behavioral RRT would behave algorithmically identical to standard RRT.

The result of `ApplyFsmBehavior` will be an action a describing a vector in the action-space **A**. We can now access our physics-engine, load the initial state x_{source}, apply the forces and torques defined by a, and simulate Δt forward in time which will provide us with a new state x_{new}. We then query the physics engine to ensure that the simulation from x_{source} to x_{new} did not violate any of our dynamics constraints, such as collisions. If accepted, x_{new} is added to T. This entire process is repeated until we either reach the goal, or give up. Once the goal has been reached, we simply backtrack until we reach the root of T and reverse that particular action sequence.

The presented pseudocode will stop as soon as a solution has been found. Because we are using a randomized planning scheme, there are no immediate guarantees about the optimality of the solution. However, one could easily enhance the algorithm to keep planning until a solution of a desired quality has been found. Various metrics, such as plan length or curvature could be used to compare multiple solutions.

2.2 Domain and Robot Modeling

To plan in dynamics space, an accurate physics model of the domain is required. Modern physics simulators allow the construction of any thinkable rigid-body shapes and are sufficiently reliable in simulating motions and collisions based on friction and restitution models. In our approach, this domain description is treated as a module and can be interchanged independently of the planner, and its tactics and skills. Skills are similarly designed with platform independence in mind, allowing an abstracted description of motion, sensing, and actuation which will then be translated into physics-forces defined by a modular platform-dependent robot-model.

2.3 Opponent Modeling

Since we are dealing with a robot soccer domain, we need to assume that we are planning against an adversary. Note, that our planner stores the entire physics state x for any node when performing its search through our dynamics time-space. This conveniently allows us to integrate an opponent model into the search-phase. Similar to our own behavior, such opponent model can be reactive to the given state of the world as defined in a particular x_{source}. For example, we can model an opponent which will attempt to steal the ball from us, by defining a simple deterministic skill. Using this model, our planner will anticipate and integrate the opponent's motions during the planning phase and only deliver solutions which will out-maneuver the predicted opponent's motions. In algorithm 2, this opponent model would apply its actions for any given plan-state transition, similar to `ApplyFsmBehavior`. The `Simulate` function will then not only simulate the actions of our own agent, but also the ones of the opponent. Obtaining an accurate opponent model is certainly a completely different challenge that we will not cover in this paper. Nevertheless, even simple assumptions, such as extrapolating the opponent's linear motion, or assuming a simple "drive to ball" strategy, should be significantly preferable over the assumption of a static opponent.

3   Experimental Results

We present results in a simulated environment that resembles the RoboCup Small-Size League. The simulated robot models represent our actual Small-Size RoboCup robots. Our simulated model has been proved successfully for developing RoboCup code and models the motion, sensing, and ball-manipulation capabilities of our robots with good accuracy. The planning framework was implemented in C++. Ageia PhysX was chosen as the underlying physics engine.

Fig. 3. A behavioral nondeterministic finite state machine used for generating dribbling sequences

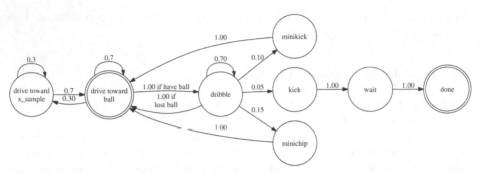

Fig. 4. An example of planning an attacker-tactic with chip and flat kicks. The left two images show the growth of the Behavioral Kinodynamic RRT through the domain (robots and ball are not displayed). Yellow nodes represent the position of the ball. Green nodes represent our agent, any other nodes represent the three defenders respectively. The right three images highlight the selected solution and show the attacker executing it.

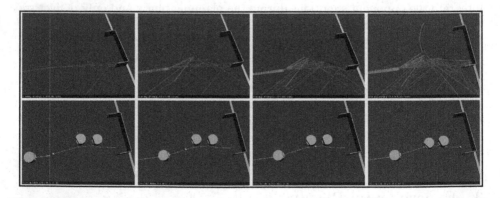

Fig. 5. An example of planning an attacker-tactic which has been limited to flat kicks only. The top row shows the growth of the Behavioral Kinodynamic RRT through the domain (robots and ball are not displayed). Blue nodes represent the position of the ball. Green, purple, and red represent positions of attacker, defender, and goalie respectively. The bottom row highlights the selected solution and shows the attacker executing it.

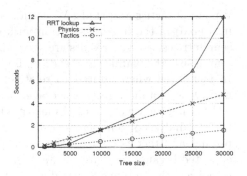

Method	Tree Size	% Success
Linear Tactic	n/a	30%
BK-RRT	1000	25%
BK-RRT	2500	40%
BK-RRT	5000	50%
BK-RRT	10000	65%

Fig. 6. Average success rate of a simulated "attacker vs. two defenders" scenario. Each row was computed using 20 randomly initialized trials.

Fig. 7. A performance analysis of the different planning components throughout growing tree sizes

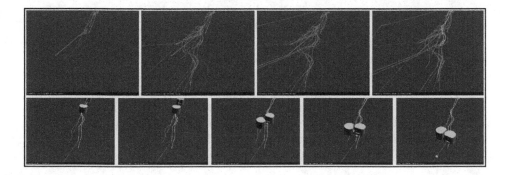

Fig. 8. An example of planning a tactic to out-dribble an opponent and score a goal. The top row shows the growth of the Behavioral Kinodynamic RRT through the domain (robots and ball are not displayed). Blue nodes represent the position of the ball. Green and purple nodes represent the positions of the attacker and opponent respectively. The bottom row highlights the selected solution and shows the attacker executing it.

The results were computed on a Pentium 4 processor running Linux. The action timestep Δt used in all tests was 1/60th of a second.

The first experiment used a probabilistic "Attacker" tactic to shoot a goal against an opponent defense. The skills within this tactic were a sampling-based flat kick and a sampling-based chip-kick. Kick-strength and kick-aiming for both of those skills were modeled as free planning variables using uniform sampling distributions. The decision between the two skills was also a free planning variable because it was modeled using the nondeterministic state machine. The opponent defense was modeled using a simple RoboCup defender code, which would try to block the ball from going into the goal. Visual results of planning an attacker strategy can be seen in figures 4 and 5.

To evaluate the qualitative performance of our planning approach to traditional control methods, we ran samples of a linear execution of a deterministic version of tactics and skills. We then compared the success rate (trials resulting in a goal) of the linear execution with our approach using different maximum search tree sizes. The results are shown in figure 6. We can see that with growing search size, the planner also ends up finding successful solutions more frequently. In fact, even when limited to relatively small search tree sizes (such as 2500 nodes), we obtain a greater average success rate than linear executions.

Planning however, does come with the trade-off of computational time. A timing analysis of our planner can be seen in figure 7. The total time spent on physics and tactics computations scales linearly with tree size, which is to be expected because they are constant time operations applied per node. For larger tree sizes, the significant slowdown is the RRT nearest neighbor lookup which scales exponentially as the tree size increases. However, for smaller trees, such as 5000 nodes, the physics engine is still the major bottleneck. This is good news in a way, as processing power will undoubtedly increase and dedicated physics acceleration hardware is becoming a commodity.

Our second experiment was designed to emphasize the creative power of our approach. We created an opponent model which attempts to steal the ball from our robot, whereas our robot attempts to out-dribble the opponent and score a goal. The tactic used a state machine similar to the one outlined in figure 3. Note, that all of the ball-manipulation skills within this state machine contain free sampling-based planning variables to increase the variety of created solutions. It is also interesting to note that some of the FSM transition probabilities actually depend on the current state x_{source}. Some visual results from this experiment can be seen in figure 8. Note that additional videos are available at http://www.cs.cmu.edu/~szickler/papers/robocup2008/.

4 Conclusion and Future Work

We presented a randomized behavioral kinodynamic planning framework for robot control generation. In particular, we introduced the Behavioral Kinodynamic RRT algorithm and demonstrated how it can be used to effectively search the dynamics space of a robot-soccer domain. We demonstrated that our planning-based approach is able to execute attacker-tactics with a greater success rate than linear approaches, due to its ability to sample from different possible control-sequences.

One obvious goal of future work will be to apply this approach to real robotic hardware, bringing along the challenges of model accuracy and computational performance. Another aspect that should be addressed in the future is the ability of re-planning. As the environment tends to change very quickly in a domain such as the RoboCup Small Size league, it might be a better strategy to generate shorter, but more densely sampled plans, instead of predicting the environment too far into the future. Furthermore, the question how to re-use the previously computed result during re-planning needs to be answered. Finally, it would also be interesting to see how to integrate multi-agent control processes into the planning framework. Currently, the presented framework computes the control strategy for each robot independently, possibly using serial prioritized planning, which can be globally non-optimal. Allowing the planner to devise a joint control policy for sets of robots will be an interesting research problem, as it increases the dimensionality of the planning problem significantly.

References

1. Browning, B., Bruce, J., Bowling, M., Veloso, M.: Stp: Skills, tactics and plays for multi-robot control in adversarial environments. IEEE Journal of Control and Systems Engineering 219, 33–52 (2005)
2. Parker, L.: ALLIANCE: an architecture for fault tolerant multirobot cooperation. IEEE Transactions on Robotics and Automation 14(2), 220–240 (1998)
3. Behnke, S., Rojas, R.: A Hierarchy of Reactive Behaviors Handles Complexity. In: Balancing Reactivity and Social Deliberation in Multi-Agent Systems: From Robocup to Real-World Applications (2001)
4. D'Andrea, R.: The Cornell RoboCup Robot Soccer Team: 1999-2003, pp. 793–804. Birkhauser Boston, Inc., New York (2005)
5. Laue, T., Rofer, T.: A Behavior Architecture for Autonomous Mobile Robots Based on Potential Fields. In: Nardi, D., Riedmiller, M., Sammut, C., Santos-Victor, J. (eds.) RoboCup 2004. LNCS, vol. 3276, pp. 122–133. Springer, Heidelberg (2005)
6. Bernadine Dias, M., Zlot, R., Kalra, N., Stentz, A.: Market-based multirobot coordination: A survey and analysis. Proceedings of the IEEE 94(7), 1257–1270 (2006)
7. LaValle, S.: Rapidly-exploring random trees: A new tool for path planning. Computer Science Dept, Iowa State University, Tech. Rep. TR, 98–11 (1998)
8. Kuffner Jr, J., LaValle, S.: RRT-connect: An efficient approach to single-query path planning. In: Proceedings of IEEE International Conference on Robotics and Automation 2000. ICRA 2000, vol. 2 (2000)
9. Bruce, J., Veloso, M.: Real-time randomized path planning for robot navigation. In: Proceedings of IROS 2002, Switzerland (October 2002)
10. James Bruce and Manuela Veloso: Safe Multi-Robot Navigation within Dynamics Constraints. Proceedings of the IEEE, Special Issue on Multi-Robot Systems (2006)
11. LaValle, S.M., Kuffner Jr., J.: Randomized Kinodynamic Planning. The International Journal of Robotics Research 20(5), 378 (2001)
12. Baraff, D.: Physically Based Modeling: Rigid Body Simulation. In: SIGGRAPH Course Notes, ACM SIGGRAPH (2001)
13. Donald, B., Xavier, P., Canny, J., Reif, J.: Kinodynamic motion planning. Journal of the ACM (JACM) 40(5), 1048–1066 (1993)

A Robot Referee for Robot Soccer[*]

Matías Arenas, Javier Ruiz-del-Solar, Simón Norambuena, and Sebastián Cubillos

Department of Electrical Engineering, Universidad de Chile
{marenas,jruizd}@ing.uchile.cl

Abstract. The aim of this paper is to propose a robot referee for robot soccer. This idea is implemented using a service robot that moves along one of the field sides, uses its own cameras to analyze the game, and communicates its decisions to the human spectators using speech, and to the robot players using wireless communication. The robot uses a video-based game analysis toolbox that is able to analyze the actions at up to 20fps. This toolbox includes robots, ball, landmarks, and lines detection and tracking, as well as refereeing decision-making. This robot system is validated and characterized in real game situations with humanoid robot players.

1 Introduction

One of the RoboCup main goals is allowing robots to play soccer as humans do. A natural extension of this idea is having robots that can referee soccer games. Refereeing task are very similar to playing task, but differentiate in the fact that a referee has to correctly interpret every situation, a single wrong interpretation can have a large effect in the game result. The main duty of a robot referee should be the analysis of the game, and the real-time refereeing decision making (referee decisions can not be delayed). A robot referee should be able to *follow the game*, i.e. to be near the most important game actions, as human referees do. In addition it should be able to communicate its decisions to the human or robot players, assistant referees, and spectators. This communication can be achieved using speech, gestures, data networks or visual displays, depending on the distance of the message's receptor, and the available communications mechanisms. The robot referee should primarily use its own visual sensors to analyze the game. In large fields or in games where the ball moves very fast or travels long distances, the robot referee could use external cameras, in addition to assistant referees. Thus, a robot referee should have 3 main subsystems: (i) video-based game analysis, (ii) self-positioning and motion control, and (iii) interfaces to communicate decisions. Interestingly, the video-based game analysis subsystem, in addition to be used for refereeing decision-making, can be used to obtain game statistics (% of the time that the ball is in each half of the field, % of ball possession of each team, number of goals of each team player, etc.), as well as for video annotation and indexing, which could be later used to retrieve an automated summary or a semantic description of the game. Besides, the robot referee could be used as commentator of robot soccer games.

---

[*] This research was partially supported by FONDECYT (Chile) under Project Number 1061158.

L. Iocchi et al. (Eds.): RoboCup 2008, LNAI 5399, pp. 426–438, 2009.

In this context, the aim of this paper is to propose a robot referee for robot soccer. This robot referee is specially intended to be used in the RoboCup SPL 2-legged league, and in the RoboCup humanoid league. The referee is a service robot that moves alongside one of the field lines (see figure 1 and 2), uses its own cameras to analyze the game, and communicates its decisions to the human spectators using speech and to the robot players using wireless communication. The robot video-based game analysis subsystem is based on the one proposed in [27], but the robot detection module has been largely improved by the use of statistical classifiers. One interesting feature of the robot referee is its ability to express facial gestures while refereeing, like being angry when a foul is committed or happy when a goal is scored, which makes him very attractive to human spectators. To the best of our knowledge a similar system has not been proposed in the literature. This paper is organized as follows. In section 2 some related work is presented. The here-proposed robot referee is presented in section 3. In section 4 some experimental results of the application of this system are presented. Finally, in section 5 conclusions of this work are given.

Fig. 1. The robot referee in a typical game situation

2 Related Work

Computer vision based analysis of sport videos has been addressed by many authors (e.g. [1]-[21]), and nowadays is a hot topic within the multimedia video analysis community. There are also successful video analysis programs that are being used by TV sport channels (e.g. Hawkeye [22] and Questec [23]). Applications have been developed in almost all massive sports such as tennis ([11][12][14][18]), soccer ([2][4][5][15][16][20][21]), baseball ([8][13]), basketball ([10]), and American football ([9]). However, to the best of our knowledge the automatic analysis of robot sports has not been addressed in the literature, except for [27].

The major research issues for the automatic analysis of human sports include (see a survey in [18]): ball and players tracking, landmarks detection (goal mouth, oval, side lines, corners, etc.), tactic analysis to provide training assistance, highlight extraction (ace events in tennis, score events in soccer and basketball, etc.), video summarization (automatic generation of game summaries), content insertion (e.g. commercial banner replacement in the field), and computer-assisted refereeing (e.g. offside detection). Some of these research issues are still open as for example fully autonomous tactic analysis, soccer offside detection considering player's intention, or video summarization; current solutions are semi-automatic and provide assistance to human referees or video operators. We believe that many of the accumulated knowledge in the automatic analysis of human sports can be employed for the automatic analysis of robot sports. Probably the main obvious difference being that in the robot sport case, robot identification is a hard problem, because usually all players look the same, because they all correspond to the same robot model (e.g. RoboCup SPL league). However, the robots can be individualized using the team uniform color [24] and the player's number [26].

The here-proposed robot referee makes use of the accumulated knowledge in automatic analysis of human sports, and in the RoboCup soccer leagues. The robot video-based game analysis subsystem is based on [27], however, the robot detection (one of the core referee functionality) is improved, instead of using SIFT descriptors, a cascade of boosted classifiers allows the robust detection of robots.

3 Proposed Robot Referee

3.1 Hardware

As robot referee we use our service robot *Bender*. This personal/social robot was originally designed to be used for the RoboCup @Home league. The main idea behind its design was to have an open and flexible testing platform. The robot has shown to be adequate for the @Home league (it won the *RoboCup @Home 2007 Innovation Award*), and has been used to provide multimedial and ubiquitous Web interfaces [29], and as speaker (lecturer) in activities with children (see pictures in [34]). Bender's main hardware components are:

- A robot's chest that incorporates a tablet PC as the main processing platform of the robot. We use a HP TC4200, powered with a 1.86 GHz Intel Centrino, with 1 GB DDR II 400 MHz, running Windows XP Tablet PC edition. The tablet includes 802.11bg connectivity. The screen of the tablet PC allows: (i) the visualization of relevant information for the user (a web browser, images, videos, etc.), and (ii) entering data thanks to the touch-screen capability.
- A robot's head with pan and tilt movements, two CCD cameras, one microphone and two loudspeakers. The head incorporates a face with the capability of expressing emotions. The head is managed by a dedicated hardware, which is controlled from the tablet PC via USB.
- A robot's arm with 3 degrees of freedom (DOF), two in the shoulder and one in the elbow. The arm is powered with a 3-finger hand. Each finger has 2 DOF. The arm is managed by a dedicated hardware, which is controlled from the tablet PC via USB.
- A mobile platform where all described structures are mounted. The platform provides mobility (differential drive), and sensing skills (16 infrared, 16 ultrasound, and 16 bumpers). The whole platform is managed by a dedicated hardware, which is controlled from the tablet PC via USB.

One interesting feature of Bender is that the relative angle between the mobile platform and the robot body can be manually adjusted. In the robot standard configuration

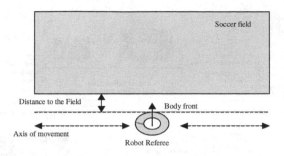

Fig. 2. Robot referee positioning

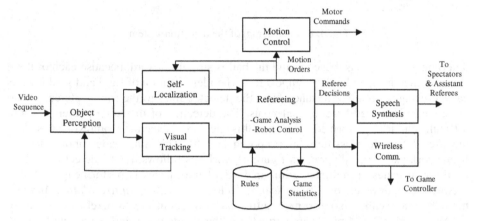

Fig. 3. Block diagram of the robot referee controller

this angle is set to 0 degrees, which allows the normal robot movement. For the task of refereeing, the angle is set to 90 degrees. This allows the robot to have a frontal view of the field while moving along the line, even though it has a differential drive configuration (see figure 2).

3.2 Robot Controller

The block diagram of the proposed robot referee controller is shown in figure 3. The system is composed by seven main modules *Object Perception*, *Visual Tracking*, *Self-localization*, *Refereeing*, *Motion Control*, *Speech Synthesis*, and *Wireless Communications*, and makes use of two databases: *Rules* (input) and *Game Statistics* (output).

The *Object Perception* module has two main functions: object detection and object identification. First, all the objects of interest for the soccer game (field carpet, field and goal lines, goals, beacons, robot players, ball) are detected using color segmentation and some simple rules, similar to the ones employed in any RoboCup soccer robot controller. No external objects, as for example, spectators or legs of assistant referees or team members are detected (in some leagues assistant referees and team members can manipulate the robots during a game). The identification (identity

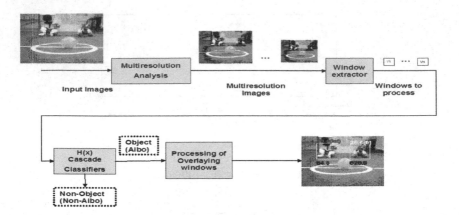

Fig. 4. Block diagram of the detection system

determination) of goals, beacons and the ball is straightforward, because each of them has a defined form and color composition. The identification of field and goal lines is carried using the relative distance from the detected lines to the robot referee, and to the already identified beacons and goals. The detection of the robot players is more difficult, and it is performed using a multiscale robot detection framework based on the use of boosted classifiers, as proposed in [28]. This multiscale robot detection framework (see block diagram in figure 4) works as follows: (i) To detect the robots at different scales, a multiresolution analysis of the images is performed, by down-scaling the input image by a fixed scaling factor --e.g. 1.2-- (*Multiresolution Analysis* module). This scaling is performed until images of about 24x24 pixels are obtained. (ii) Windows of 24x24 pixels are extracted in the *Window Extraction* module for each of the scaled versions of the input image. (iii) The windows are analyzed by a nested cascade of boosted classifier (*Cascade Classification Module*). (iv) In the *Overlapping Detection Processing* module, the windows classified as positive (they contain a robot) are fused (normally a robot will be detected at different scales and positions) to obtain the final size and position of the detections. Real-time robot processing is achieved thanks to the use of cascades of classifiers, and because this analysis is carried out only at the beginning of each robot tracking sequence (see the feedback connection from *Visual Tracking* to *Object Perception* in figure 3).

The *Visual Tracking* module is in charge of tracking the moving objects, i.e. the ball and the robot players. The implemented tracking system is built using the mean shift algorithm [30], applied over the original image (not the segmented one). The seeds of the tracking process are the detected ball and robot players. As in [32], a Kalman Filter is employed to maintain an actualized feature model for mean shift. In addition, a fast and robust line's tracking system was implemented (see description in [27]). Using this system, it is not necessary to detect the lines in each frame. Using the described perception and tracking processes, the system is able to track in near real time (up to 20 fps) all game moving objects and the lines.

The *Self-localization* module is in charge of localizing the robot referee. As in the case of the robot players, this functionality is achieved using the pose of the landmarks (goals and beacons) and the lines, and odometric information. The only difference being that in the case of the robot referee, the movements are not executed inside the field, but outside, along one of the field sides (see figure 2).

The *Refereeing* module is in charge of analyzing the game dynamics and the actions performed by the players (e.g. kicking or passing), and detecting game relevant events (goal, ball out of the field, illegal defender, etc.). This analysis is carried out using information about static and moving detected objects, and the game rules, which are retrieved from the *Rules* database. In addition, this module is in charge of the referee positioning. The module should keep the referee outside of the field, but at a constant distance of the field side (the referee should move along one of the field sides), it should control de robot's head and body movement to allow the robot to correctly *follow the game*, by avoiding obstacles and without leaving the field area (in case that the ball or a player leave the field.). It is important that the referee always perceives and follows the main elements of game-play. In the present this is done by following the ball, and estimating the position of the players in the field. In future implementations we plan to use several cameras to have more information of the activities in the field. The outputs of this module are refereeing decisions (e.g. goal was scored by team A) that are sent to the *Speech Synthesis* and *Wireless Communication* modules, motion orders that are sent to the *Motion Control* module, and game statistics (e.g. player 2 from team A score a goal) that are stored in the corresponding database.

The *Motion Control* module is in charge of translating motion orders into commands for the robot motors. These commands allow the control of the robot pose, the robot head pose, the robot facial expressions, and the robot arm.

Finally, the *Speech Synthesis* and *Wireless Communication* modules communicate the referee decisions to robot players and human assistant referees and spectators. Wireless communication is straightforward, while speech synthesis is achieved using the CSLU toolkit [35].

3.3 Refereeing

This module is in charge of analyzing the game dynamics, determining the actions performed by the players, and the game relevant events. This analysis is carried out using the information of the static and moving objects detected, and the game rules, which depend on the specific RoboCup soccer leagues.

Most of the game situations can be analyzed using information on the position of the ball and the robot players in the field, the time during which the ball and the robot players stay in a given field area, the localization of the field lines and goal lines, and for some specific cases, the identity of the last robot that touches the ball. For instance, to detect a goal event, a ball crossing one of the two goal lines should be detected. The correct positioning of a given robot depends on its own position and in some cases the position of the other players and the ball. The identity of the scoring robot is the identity of the last robot that touches the ball. Thus, using simple rules many situations such as "goal detection", "ball leaving the field", "game stuck", "robot kickoff positioning", "robot leaving the field" or "robot falling" can be detected. There are complex situations that depend on the exact relative position between two players or between a player and the ball that cannot be robustly detected. Some exemplar situations are "ball holding", "goalie/player pushing" and "robot obstruction".

For example, in the case of the RoboCup humanoid league, from which rules will be used to run our experiments, the following situations (definitions taken from [31]) can be analyzed:

- *Goal*: "A goal is scored when the whole ball passes over the goal line, between the goal posts and under the crossbar, provided that no infringement of the rules has been committed previously by the team scoring the goal".
- *Robots kickoff positioning*: "All players are in their own half of the field. The opponents of the team taking the kick-off are outside the center circle until the ball is in play. The ball is stationary on the center mark. The referee gives a signal. The ball is in play when it is touched or 10 seconds elapsed after the signal.
- *Ball In and Out Play*: "The ball is out of play when it has wholly crossed the goal line or touch line whether on the ground or in the air or when play has been stopped by the referee. The ball is in play at all other times, including when it rebounds from a goalpost, crossbar, corner pole, or human and remains in the field of play".
- *Global Game Stuck*: "the referee may call a game-stuck situation if there is no progress of the game for 60s".
- *Illegal defense and attack*: "Not more than one robot of each team is allowed to be inside the goal or the goal area at any time. If more than one robot of the defending team is inside its goal or goal area for more than 10s, this will be considered illegal defense. If more than one robot of the attacking team is inside the opponent's goal or goal area for more than 10s, this will be considered illegal attack".

However, there are some other situations that are much harder to analyze, because it is required either to judge the intention of the players (e.g. robot pushing) or to solve visual occlusions that difficult the determination of the relative position of the robot legs or the ball (e.g. ball holding). For the moment we have not implemented the detection of those situations.

The game statistics that can be computed in our current implementation are: % of the time that the ball is in each half of the field, number of goals of each team and team player, number of direct kicks to the goal by each team and each team player, number of times that each team and each team player sent the ball out of the field, number of illegal defense or attack events of each team, number of times that each team player leaves the field, number of times that each player fall down, number of global game stuck events, and time required for automatic kickoff positioning by each team.

3.4 Robot Perceptor

To detect the playing robots a cascade of boosted classifier was used (based on Adaboost). This detector requires a diversified database of images with the robots that are going to be detected, as well as a number of images not containing any robots. The referee here implemented was mainly focused on the RoboCup Humanoid league, so the training database was made from videos obtained in the league website (videos submitted by the league's teams) and using our humanoid robot (Hajime H18) in our laboratory.

During the training of the cascades, validation and training sets are used. The procedure to obtain both sets is analogous, so only the training dataset is explained. To obtain the training set used at each layer of the cascade classifier, two types of databases are needed: one of cropped windows of positive examples (Humanoid) and one of images not containing the object to be detected. The second type of database is used during the bootstrap procedure to obtain the negative examples (this comes from our implementation of the Adaboost algorithm). The training dataset is used to train the weak classifiers, and the validation database is used to decide when to stop the training of a given layer and to select the bias values of the layer (see details in [28]). To obtain positive examples (cropped windows) a rectangle bounding the robot was annotated and a square of size equal to the largest size of the rectangle was cropped and downscaled to 24x24 pixels. The "positive" database contains several thousand 24x24 images containing Humanoids in different poses, environments, illuminations, etc. The training process was repeated several times in order to obtain a good detection rate with a low quantity of false positives. Each time the process was repeated more images were added to the database to increase the variety of the images included in it (so the classifier becomes more general). The number of images used in the final version of the Robot detector is shown in Table 1.

4 Experimental Results

To prove the usefulness of the proposed system, some experiments were carried out. The results of these tests are shown in the following sections. The quantitative results were obtained from a series of video sequences (with different configurations) taken in our laboratory. In total 5,293 frames were taken for a preliminary analysis of our system. These videos are from short play sequences in the field, some examples can be seen in figures 5, 6 and 7. The idea was to capture different playing sequences, but due to the huge amount of possible situations during a game, only a few of them were captured.

4.1 Object Detection and Tracking

The first evaluation was the Robot Detector module. To do this, the system was programmed to detect robots every 5 to 10 frames, and to track the detections in every frame. Then we counted all the robots that the detector correctly found versus the number of robots that appear on all the frames. We also counted the false detections that appeared in some frames. The tracking systems are much faster than the detection systems, especially the mean shift module used to track the detected robots. This allows the system to work close to real time and the possibility to run faster or slower depending on the possible applications and desired results. Table 2 shows the robot detection results. These results show that the robots were correctly detected in almost all frames, with a false detection every 15 or so frames although, usually, false detections appeared in consecutive frames. These results are quite good and show that the detector is working as intended. Once the robot is correctly detected the tracking system works remarkably well (the mean shift system has no problem tracking object in these environments).

Table 1. Summary of the databases used for training

Class	# Positive examples		# Negative images	
	(Training)	(Validation)	(Training)	(Validation)
Humanoids	3,693	3,693	6,807	3,410

Table 2. Summary of the results for the Robot Detector module

Number of Frames	Number of Robots	Robot Detection Rate	Number of false positive windows
5,293	3,405	98.7%	334

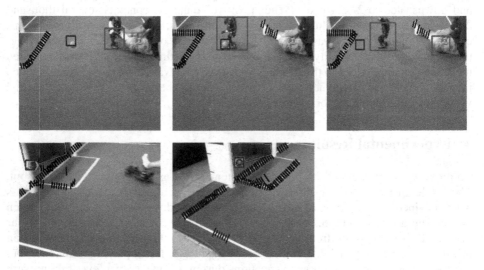

Fig. 5. Selected frames from a robot scoring sequence. Robot/ball tracking window in red/blue. The goal detection event is shown by a red window out of the ball blue window.

Line detection was also evaluated, but only partially. Many lines of many sizes appear in all images and they can appear entirely, partially or barely. The "correct" detection of all lines is difficult to put in numbers because of this, and the many criteria that can exist (a "correctly detected" line can mean that a part of the line was detected or that some parts of the line were detected or that all the line was detected, etc.). Our line detector worked fine on side lines that appeared entirely (or almost), having problems only with those that were farther away. It also detected some of the lines inside the field but had problems with smaller lines (the ones that limit the space for the goalie). Some examples can be seen in figures 5 and 6. The back lines were detected in almost all cases, while the goal area of the lines was a bit more difficult to tag correctly. Our system does a good job at tagging a zone in front of the goal but can sometimes over- or under-estimate the portion of the line that represents the goal line. Beacons, goals and ball detection could be evaluated, but these systems have already been tested before (they are very similar to the ones currently used by the

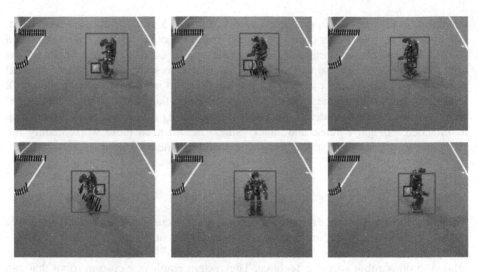

Fig. 6. Selected frames from ball tracking and robot detection

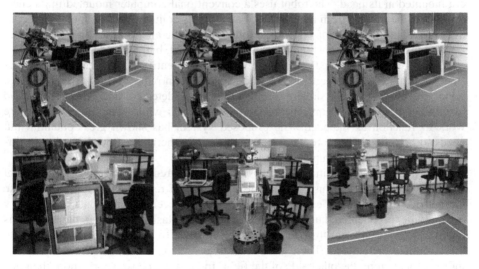

Fig. 7. Selected images of a Refereeing situation

robots in RoboCup league), and give very good results. An example of this is figure 6 which shows the ball detection system and how it works even with total occlusion of the ball (thanks to a Kalman filter).

4.2 Event Detection

Same video sequences as in 4.1 were employed. As these events are less frequent than object apparitions, a statistical evaluation is more complicated. Furthermore these events are harder to detect because a correct detection of all the elements present in the event must be present. For example for goal detection the system needs a good detection on the lines, goal, goal line and ball. Figures 5 and 7 show goal detections, one directly from the camera used by the robot, and the other from an external camera filming the setup of the refereeing robot. Our system correctly detects most of the goal situations, although some false positives appear in some cases (where our goal line is over-estimated and the ball was leaving the field instead of entering the goal). Other situations like "game stuck" or "player leaving the field" where correctly detected but they have to undergo more real games testing to see if it is really working as intended. "Illegal defense and attack" situation still need more work, mainly because of the difficulties of setting the goalie area with line detections.

5 Conclusions

This paper proposes a robot referee for robot soccer. A custom robot is used alongside the soccer field to referee, analyze and possibly comment the current game. It is currently implemented for the RoboCup Humanoid and SPL leagues, but it can be extended to other robot soccer categories. The system employs a custom robot that can move along the soccer field and follow the action by watching the game with a camera mounted in its head. The robot uses a conventional computer mounted in its chest to process the images from the game and analyze its content. The computer also controls the robot moving parts in order to follow the game and to notify the surrounding persons the important events ongoing in the game (goal, ball leaving the field, etc).

Currently, the developed system can detect and identify all humanoid-league defined field objects, and perform the tracking of the moving objects in real-time. A good part of the defined situations can be correctly detected, and other show good results but need more testing in diverse environments while some other need more work to function as intended. The proposed system has shown great potential but needs to be refined for more complex situations. Furthermore it needs testing in different environments to completely prove its usefulness.

The system could be criticized because is it not assured that it always take the right decision (e.g. due to occlusion problems), but human referees also not always take correct decisions. This is especially true in robot soccer environments, where very often non-experimented humans assume refereeing tasks. Future implementations plan to use more cameras to cover the whole field, and to be able to correctly evaluate more complex situations. These cameras can be fixed over the field or in another moving robot along the other side of the field. In most collective sports more than one referee is used to make sure all rules are obeyed, so it is logical to include more sources of information in this case too.

References

1. Assfalg, J., Bertini, M., Colombo, C., Del Bimbo, A.: Semantic Annotation of Sports Videos. IEEE MultiMedia 9(2), 52–60 (2002)
2. Assfalg, J., Bertini, M., Colombo, C., Del Bimbo, A., Nunziati, W.: Semantic Annotation of soccer videos: automatic highlights identification. Computer Vision and Image Understanding 92(2-3), 285–305 (2003)

3. Babaguchi, N., Kawai, Y., Kitahashi, T.: Event Based Video Indexing by Intermodal Collaboration. IEEE Transactions on Multimedia 4(1), 68–75 (2002)
4. Bertini, M., Del Bimbo, A., Nunziati, W.: Soccer Video Highlight Prediction and Annotation in Real Time. In: Roli, F., Vitulano, S. (eds.) ICIAP 2005. LNCS, vol. 3617, pp. 637–644. Springer, Heidelberg (2005)
5. D'Orazio, T., Ancona, N., Cicirelli, G., Nitti, M.: A ball detection algorithm for real soccer image sequences. In: Proc. Int. Conf. Patt. Recog. – ICPR 2002, Canada, August 11-15 (2002)
6. Duan, L.Y., Xu, M., Chua, T.S., Tian, Q., Xu, C.S.: A mid-level representation framework for semantic sports video analysis. In: Proc. of ACM MM 2003, Berkeley, USA, November 2-8, pp. 33–44 (2003)
7. Ekin, A., Tekalp, A.M.: Automatic soccer video analysis and summarization. IEEE Trans. on Image Processing 12(7), 796–807 (2003)
8. Han, M., Hua, W., Xu, W., Gong, Y.H.: An integrated baseball digest system using maximum entropy method. In: Proc. of ACM MM 2002, pp. 347–350 (2002)
9. Li, B., Sezan, M.I.: Event detection and summarization in American football broadcast video. In: Proc. SPIE Conf. on Storage and Retrieval for Media Databases, vol. 4676, pp. 202–213 (2002)
10. Nepal, S., Srinivasan, U., Reynolds, G.: Automatic detection of 'Goal' segments in basketball videos. In: Proc. ACM MM 2001, Ottawa, Canada, pp. 261–269 (2001)
11. Pingali, G., Jean, Y., Carlbom, I.: Real time tracking for enhanced tennis broadcasts. In: Proc. IEEE Conf. Comp. Vision and Patt. Rec. – CVPR 1998, pp. 260–265 (1998)
12. Pingali, G., Opalach, A., Jean, Y.: Ball tracking and virtual replays for innovative tennis broadcasts. In: Proc. Int. Conf. Patt. Recog. – ICPR 2000, Barcelona, Spain, pp. 4146–4152 (2000)
13. Rui, Y., Gupta, A., Acero, A.: Automatically extracting highlights for TV Baseball programs. In: Proc. of ACM MM 2000, pp. 105–115 (2000)
14. Sudhir, G., Lee, J., Jain, A.K.: Automatic classification of tennis video for high-level contentbased retrieval. In: Proc. of IEEE Int. Workshop on Content-based Access of Image and Video Database, pp. 81–90 (1998)
15. Tovinkere, V., Qian, R.J.: Detecting semantic events in soccer games: Towards a complete solution. In: Proc. ICME 2001, pp. 1040–1043 (2001)
16. Wan, K., Yan, X., Yu, X., Xu, C.S.: Real-time goalmouth detection in MPEG soccer video. In: Proc. of ACM MM 2003, Berkeley, USA, pp. 311–314 (2003)
17. Wan, K., Yan, X., Yu, X., Xu, C.S.: Robust goalmouth detection for virtual content insertion. In: Proc. of ACM MM 2003, Berkeley, USA, pp. 468–469 (2003)
18. Wang, J.R., Paramesh, N.: A scheme for archiving and browsing tennis video clips. In: Proc. of IEEE Pacific-Rim Conf. on Multimedia - PCM 2003, Singapore (2003)
19. Wang, J.R., Parameswaran, N.: Survey of Sports Video Analysis: Research Issues and Applications. In: Conferences in Research and Practice in Information Technology (VIP 2003), Sidney, vol. 36, pp. 87–90 (2004)
20. Yow, D., Yeo, B.L., Yeung, M., Liu, B.: Analysis and presentation of soccer highlights from digital video. In: Li, S., Teoh, E.-K., Mital, D., Wang, H. (eds.) ACCV 1995. LNCS, vol. 1035. Springer, Heidelberg (1996)
21. Yu, X., Xu, C.S., Leong, H.W., Tian, Q., Tang, Q., Wan, K.W.: Trajectory-based ball detection and tracking with applications to semantic analysis of broadcast soccer video. In: Proc. of ACM MM 2003, Berkeley, USA, pp. 11–20 (2003)
22. http://www.hawkeyeinnovations.co.uk/
23. http://www.questec.com/

24. Röfer, T., et al.: German Team 2005 Technical Report, RoboCup 2005, Four-legged league (February 2006), http://www.germanteam.org/GT2005.pdf
25. Quinlan, M.J., et al.: The 2005 NUbots Team Report, RoboCup 2005, Four-legged league (February 2006), http://www.robots.newcastle.edu.au/publications/NUbotFinalReport2005.pdf
26. Loncomilla, P., Ruiz-del-Solar, J.: Gaze Direction Determination of Opponents and Teammates in Robot Soccer. In: Bredenfeld, A., Jacoff, A., Noda, I., Takahashi, Y. (eds.) RoboCup 2005. LNCS, vol. 4020, pp. 230–242. Springer, Heidelberg (2006)
27. Ruiz-del-Solar, J., Loncomilla, P., Vallejos, P.: An automated refereeing and analysis tool for the Four-Legged League. In: Lakemeyer, G., Sklar, E., Sorrenti, D.G., Takahashi, T. (eds.) RoboCup 2006: Robot Soccer World Cup X. LNCS, vol. 4434, pp. 206–218. Springer, Heidelberg (2007)
28. Arenas, M., Ruiz-del-Solar, J., Verschae, R.: Detection of Aibo and Humanoid Robots using Cascades of Boosted Classifiers. In: Visser, U., Ribeiro, F., Ohashi, T., Dellaert, F. (eds.) RoboCup 2007: Robot Soccer World Cup XI. LNCS, vol. 5001, pp. 449–456. Springer, Heidelberg (2008)
29. Ruiz-del-Solar, J.: Personal Robots as Ubiquitous-Multimedial-Mobile Web Interfaces. In: 5th Latin American Web Congress LA-WEB 2007, Santiago, Chile, October 31 – November 2, pp. 120–127 (2007)
30. Comaniciu, D., Ramesh, V., Meer, P.: Kernel-Based Object Tracking. IEEE Trans. on Pattern Anal. Machine Intell. 25(5), 564–575 (2003)
31. RoboCup 2008 Humanoid official rules (January 2008), http://mail.fibo.kmutt.ac.th/robocup2008/HumanoidLeagueRules2008draft-2008-01-20.pdf
32. Peng, N.S., Yang, J., Liu, Z.: Mean shift blob tracking with kernel histogram filtering and hypothesis testing. Pattern Recognition Letters 26, 605–614 (2005)
33. Lowe, D.G.: Distinctive Image Features from Scale-Invariant Keypoints. Int. Journal of Computer Vision 60(2), 91–110 (2004)
34. Bender robot official website, http://bender.li2.uchile.cl/
35. CSLU toolkit Official Website, http://cslu.cse.ogi.edu/toolkit/

Detection of Basic Behaviors in Logged Data in RoboCup Small Size League

Koshi Asano, Kazuhito Murakami, and Tadashi Naruse

Faculty of Information Science and Technology
Aichi Prefectural University,
Nagakute-cho, Aichi, 480-1198 Japan

Abstract. This paper describes a method that extracts the basic behaviors of robots such as kicking and passing from the history data of the positions and velocities of the robots and the ball in RoboCup Small Size League (SSL). In this paper, as a first step, we propose an offline method that extracts the basic behaviors of robots from the logged data which is a record of the positions and velocities of the robots and the ball as the time series data. First, paying attention to the ball movement, we extract the line segments in the ball trajectory which satisfy our proposed conditions. These segments arise from the kicking actions. Then we classify the extracted line segments into the detailed kicking actions by analysing the intention of the kicked robot. We also propose algorithms that detect and classify the covering actions. Experimental results show that 98% of the kicking actions are correctly detected and more than 80% of the detected kicking actions are correctly classified, and that 90% of the covering actions are also correctly classified.

1 Introduction

Cooperative plays and higher strategies are studied actively in the RoboCup Small Size League (SSL), since the global cameras can achieve the reliable image processing in comparison to the image processing using the local cameras, which is developed mainly in the humanoid, middle-size and 4-legged leagues. They are also studied actively in the simulation league[1,2]. However, it seems difficult to transfer the technology that was developed in the simulation league into the SSL directly, since the real robot has the uncertainty of the motion.

It is expected, in the highly skilled robotic soccer, that the best strategy should be adaptively chosen by learning the behaviors of the opponent robots. To do so in the SSL, it is necessary to detect the opponent behaviors in the history data of the positions and velocities of the robots and the ball, which are obtained by processing images captured by the over-field cameras every 1/30 or 1/60 seconds. Many teams in the SSL use only the history data of the robots and the ball for computing the positions that the robots should go next. On the other hand, in human soccer, each player recognizes the other players' actions, predicts their next actions and decides his action. To realize human-like robotic

L. Iocchi et al. (Eds.): RoboCup 2008, LNAI 5399, pp. 439–450, 2009.

soccer in the SSL, it is necessary to recognize the opponent behaviors from the history data.

It has been studied to analyze, search and edit the human soccer videos automatically [3,4]. the main purpose of these studies was, however, the recognition of a player, the tracking of the ball and the detection of events based on cinematic features. In contrast, the recognition of behaviors is still remained as a future work. In the RoboCup simulation League, it has been studied to learn when and in which situation the specific events such as the passing and shooting often happen [5]. In the simulation league, since the players and the ball move according to the given physical formula, it is rather easy to detect the behaviors of each player. On the other hand, in the SSL, positions of the robots and the ball in the history data are not always exact, since they are computed by the image processing. To analyse the behavior of players, we have to detect and classify actions from the history data. In this paper, we propose an algorithm that detects kickings and covering actions from the history data and classifies them into a passing, a shooting, a clearing, an interception and so forth. Moreover, we show an experimental result of the detection and classification. It shows that the proposed algorithm is available for our purpose.

2   Classification of Robot Actions

In this section, we classify the actions of robot in robotic soccer based on the logged data which is a history data of the robots and the ball. The basic actions of robot in the SSL are "kicking" and "covering". We classify them into the following detailed actions;

- "Kicking" actions:
  - "Shot",
  - "Pass",
  - "Clear".
- "Covering" actions:
  - "covering for preventing passing",
  - "covering for preventing shooting",
  - "covering for intercepting ball".

Besides the basic actions classified above, there are other actions such as "chip kick" and "dribbling", however, these actions are not considered here[1].

We detect each action classified above in the logged data by using our proposed algorithm described in the next section and after. The logged data is a record of the SSL's competition including following data as the history data;

- Time stamp,
- Referee signal,
- Position, direction, velocity and angle velocity of each robot,
- Position and velocity of ball,

---

[1] These actions are the challenges for the next step.

- Reliability of each recognized object by image processing,
- Camera number that each object is captured.

These data are recorded every $1/60$ seconds.

Hereafter, we use the term "object" meaning a ball, a teammate robot, or an opponent robot.

3   Detection and Classification of Kicking Actions

The detection of the kicking actions discussed here means to extract the segment that the ball starts linear motion and ends it. There is no need that the ball must stand still before the ball starts linear motion. First, we extract the linear motion segment from the trajectory of the ball, then check the segment whether it is caused by the kicking of a robot or not. Second, we classify the action of the detected segment into detailed kicking action by an intention analysis. Third, we classify the effect of the kicking action.

3.1   Algorithm for Detecting Kicking Actions

The detection algorithm of the kicking segment consists of two algorithms, i.e. to extract the maximal linear segment in the ball trajectory and to classify the extracted segment. The algorithms are shown below.

In the algorithms, we use the following notations.
(X_i, Y_i) is the position of the ball at time i,
$\vec{a_i}$ is a vector from (X_i, Y_i) to (X_{i-1}, Y_{i-1}),
$\vec{b_i}$ is a vector from (X_i, Y_i) to (X_{i+1}, Y_{i+1}),
θ_i is an angle which is given by the following equation,

$$\theta_i = \left| 180 - \frac{180}{\pi} \cdot \cos^{-1}\left(\frac{\vec{a_i} \cdot \vec{b_i}}{|\vec{a_i}||\vec{b_i}|} \right) \right| \tag{1}$$

$\overline{\theta_i}$ is an average of θ_i between time i and $i + n - 1$, which is given by,

$$\overline{\theta_i} = \frac{1}{n} \sum_{m=i}^{i+n-1} \theta_m \tag{2}$$

V_i is an average velocity of the ball between time i and $i+n-1$, which is given by,

$$V_i = \frac{1}{n} \sum_{m=i}^{i+n-1} \left| \vec{b_m} \right| \tag{3}$$

f is a function that returns a value in proportion to the inverse of input value, i.e. if $B = f(A)$ then $B \propto A^{-1}$, T_α, T_β and T_ω are threshold values, which are given beforehand, In the above, n is a given number for smoothing noise.

Using above notations, first, we describe an algorithm to detect the linear motion segment.

Algorithm 1. detection of linear motion segment

Extract a linear motion segment in the trajectory of the ball,

Step 1. Set $i \leftarrow 1$

Step 2. Compute

$$\{\overline{\theta_i} < f(V_i)\} \wedge \{V_i > T_\alpha\}. \tag{4}$$

If Eq. (4) holds, then set time i as s the starting time of the linear motion segment, set $j \leftarrow i + 1$ and go to Step 3. If not, set $i \leftarrow i + 1$ and repeat Step 2.

Step 3. Compute $\overline{\theta_j}$ and V_j, and compute

$$\{\overline{\theta_j} < f(V_j)\} \wedge \{\theta_{j+n} < T_\beta\}. \tag{5}$$

If Eq. (5) holds, then set $j \leftarrow j + 1$ and repeat Step 3. Otherwise, set time $j + n$ as e the ending time of the linear motion. If the end of trajectory is reached, then exit, otherwise, set $i \leftarrow j + n + 1$ and go to Step 2.

When the segment is detected, a classification of the kicking action is done by the following "algorithm 2".

Algorithm 2. Classification of kicking action

For the detected segment, do the following.

Step 1. Let s and R_s be the starting time of the linear motion segment and the robot which is closest to the ball at time s, respectively. Let D_k be the distance between the robot R_s and the ball at time k, where $s \leq k \leq s + n$. Compute the followings;

$$\{D_s < T_\gamma\} \wedge \{(D_s < D_{s+1}) \wedge \cdots \wedge (D_{s+n-1} < D_{s+n})\} \tag{6}$$

$$\max_{0 \leq k \leq n-2} \{(D_{s+k+2} - D_{s+k+1}) - (D_{s+k+1} - D_{s+k})\} > T_\omega \tag{7}$$

Equation (6) shows that the ball goes away from the kicked robot R_s and equation (7) shows that the maximal acceleration in the segment should be greater than the threshold T_ω if the ball is kicked by the robot R_s.

If Eq. (6) is true and Eq. (7) holds, the linear motion segment is caused by the kicking action. Then, search the past time. If the previous linear motion segment ends at some time l ($s - m \leq l \leq s$), the kicking is a **direct play**[2]. This leads that the value of m is about 5.

Step 2. If Eq. (6) is true but Eq. (7) does not hold, the linear motion segment is caused by the ball bouncing off the robot. If Eq. (6) is false, it is caused by the unknown reason or action.

---

[2] The direct play is an action that kicks the ball immediately after receiving it[6].

3.2   Intention of Kicking Actions

It is important for the behavior recognition to clarify the intention of the kicking action of the robot. For the kicking actions satisfying both Eqs. (6) and (7), the intention of the kicking actions is judged by the following algorithm.

Algorithm 3. Intention of kicking action
Step 1. Find the team (teammate/opponent) to which the robot R_s belongs. It is easily found in the logged data.
Step 2. Calculate the half line L that begins at the starting point (X_s, Y_s) of the linear motion and goes through the ending point (X_e, Y_e).
Step 3. If the half line goes across one of the objects, where objects are end-line (EL), goal-line (GL), side-line (SL) and teammate robots of R_s at time s, then find the closest object to the robot R_s and its intersection c.
Step 4. From the intersection c, the intention of the kicking action is classified as shown in Table 1.

Table 1. Classification of kicking action

c / R_s	TR's EL	OR's EL	TR's GL	OR's GL	SL	TR	OR
TR	clear	shot	unknown	shot	clear	pass	
OR	shot	clear	shot	unknown	clear		pass

TR means teammate robot.
OR means opponent robot.

3.3   Effect of Kicking Actions

The effect of the kicking action is often different from its intention. Therefore, we realized an algorithm to classify the effect of the kicking action.

Algorithm 4. Effect of kicking action
Step 1. Find the team (teammate/opponent) to which the robot R_s belongs. It is easily found in the logged data.
Step 2. Calculate the line segment L that begins at the starting point (X_s, Y_s) of the linear motion and ends in the ending point (X_e, Y_e). Extend the segment for the length of D_L over the end point.
Step 3. If the extended line segment goes across one of the objects, where objects are end-line (EL), goal-line (GL), side-line (SL) and teammate and opponent robots of R_s at time e, then find the closest object to the robot R_s and its intersection c.
Step 4. From the intersection c, the effect of the kicking action is classified as shown in Table 2.

Table 2. Effect of kicking actions

R_s \ c	TR'sEL	OR'sEL	TR'sGL	OR'sGL	SL	TR	OR
TR	clear	shot miss	own goal	goal	clear	pass	interception
OR	shot miss	clear	goal	own goal	clear	interception	pass

TR means teammate robot.
OR means opponent robot.

4 Detection and Classification of Covering Actions

4.1 Definition of Covering Action

Typical covering actions in the SSL can be classified into the following three actions.

"Covering for preventing passing" action. "Covering for preventing passing" action is an action at free kick or throw-in that covers an opponent robot who will receive the ball and prevent a passing between opponent robots.

"Covering for preventing shooting" action. "covering for preventing shooting" action is an action at free kick or throw-in that covers an opponent robot who will shoot the ball just after receiving it from an opponent's teammate and prevent a shooting. In this case, the covering robot stands on the shooting line. Typical example appears in the case that a robot does the direct play.

"Covering for getting ball" action. "covering for getting ball" action is an action in in-play or at free kick or throw-in that covers an opponent robot holding the ball, and tries to get the ball or prevents a passing or a shooting.

These three covering actions might be done by two or more robots.

By contrast, the action of the robot that defends the goal with keeping a constant distance from the goal is not regarded as the covering action.

4.2 Algorithm for Detecting Covering Action

The purpose of the following algorithms is to detect an opponent robot which covers a teammate robot.

Intuitive features such as the distance between two robots or the angle between the moving direction vectors of them do not work well for the detection of the covering action. We utilize the new feature; minimal of the distances between one robot and each line connecting the ball and the other robots.

"Covering for Preventing Passing" Action. The opponent robot which acts as preventing passing always moves to the position that it can take the ball away. It stands near the receiving robot and tends to move on the line connecting the ball and the receiving robot. Considering this observation, we get the following algorithm that detects the "covering for preventing passing" action.

First, we define the notation used in the algorithm,

R is the robot which receives the passed ball.
(x_i, y_i) is the position of the robot R at time i.
$R_j : j = \{0, 1, \cdots, 4\}$ are the opponent robots.
(x_{ij}, y_{ij}) is the position of the robot R_j at time i.
$L : ax + by + c = 0$ is the line connecting the ball and the robot R.
T is a given threshold.

In the above, n is a given number in the range of 5 to 10 used for smoothing noise.

Algorithm 5. Detection of the "covering for preventing passing" action
Step 1. Set $i \leftarrow 1$.
Step 2. For each j, calculate

$$E1_{ij} = \sqrt{(x_i - x_{ij})^2 + (y_i - y_{ij})^2}, \tag{8}$$

$$E2_{ij} = \frac{|ax_{ij} + by_{ij} + c|}{\sqrt{a^2 + b^2}}, \tag{9}$$

$$E_{ij} = \alpha \cdot E1_{ij} + \beta \cdot E2_{ij}. \tag{10}$$

$E1_{ij}$ shows a distance between the robots R and R_j. $E2_{ij}$ shows a length of the perpendicular from the robot R_j to the line L.

If the foot of the perpendicular on the line L is outside of the line segment beginning from the ball and ending in the robot R, following equation,

$$E2_{ij} = \frac{|ax_{ij} + by_{ij} + c|}{\sqrt{a^2 + b^2}} \cdot \gamma, \tag{11}$$

is used instead of Eq. (9), where γ is a penalty.
    Averaging Eq. (10) in the time interval $(i, i + n - 1)$, we get

$$\overline{E_j} = \frac{1}{n} \sum_{k=i}^{i+n-1} E_{kj}. \tag{12}$$

Finally, we define $CoveringByRobot_j$ by,

$$CoveringByRobot_j = \begin{cases} 0 & \left(\overline{E_j} > T\right) \\ 1 & \left(\overline{E_j} \leq T\right) \end{cases} \tag{13}$$

If $CoveringByRobot_j = 1$, we decide that the robot R_j covers R for preventing passing from time i to $i + n - 1$.
    Set $i \leftarrow i + n$ and repeat step 2 until the end of the logged data.

Detection of "Covering for Preventing Shooting" Action. While the detection of the "covering for preventing passing" action is calculated based on the distance between the passing line and the covering robot, the detection of "covering for preventing shooting" action is calculated based on the distance between the shooting line and the covering robot.

Therefore, we can use the algorithm 5 only replacing the passing line into the shooting line. In our definition, the shooting line is a line connecting the robot R and the center of the goal mouth.

Detection of "Covering for Intercepting Ball". "Covering for intercepting ball" is an action to cover the robot which holds the ball. Therefore, the covering robot will keep its position on the line L' that connects the ball and the robot which holds the ball. We can use the algorithm 5 as well, however, we should replace the R into R_{ball} which is a ball holding robot, and also replace the line L into the line L' in the algorithm 5.

5   Experimental Study

In this section, we show an experimental result of the action detection given above. In the experiment, we applied our algorithm to the logged data that were recorded in the RoboCup 2007 competition.

5.1   Experimental Examples

We show some examples that are the result of the detection and the classification of kicking action.

Detection of Kicking Action. Figure 1 shows an example of the trajectory of the ball. It started from the throw-in on the side line by one robot, then the other robot received the ball and kicked it immediately toward the goal, however the opponent goalkeeper blocked and the ball went out the field and bounced off the outside wall.

Figure 2 shows the velocity (saw-teeth-like line) and moving direction ($\overline{\theta_i}$, dotted line) of the ball calculated from the part shown in Fig. 1 of the logged data. D_k is shown as well. From the figure, it is shown that the three kicking actions by two teammate robots and one goalkeeper are detected correctly using the algorithm 1. Table 3 shows the result of the classification of the intention and the effect of the kicking actions by the algorithms 2 and 3. It is clear from the table that three kickings are classified into correct actions.

Detection of Covering Action. Figure 3 shows the trajectories of five opponent robots and one teammate robot that will receive the ball in some time interval. An arrow beside each trajectory shows the moving direction of corresponding robot.

Figure 4 shows the value of $\overline{E_j}$, Eq. (12) in algorithm 5, between the teammate robot R and the opponent robot R_j for the time interval of trajectories in Fig. 3.

From Fig. 4, we can get $\overline{E_2}$ which is always below threshold T and find it a "covering for preventing passing" action in this time interval.

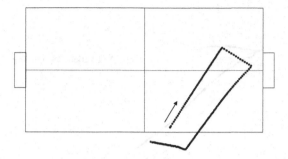

Fig. 1. Ball trajectory

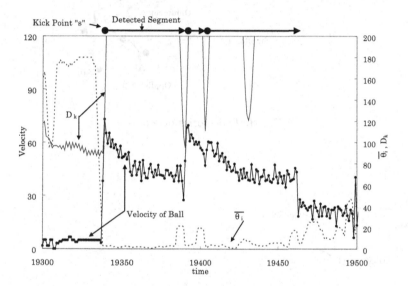

Fig. 2. Kicking action detection

Table 3. Classification of intention of kicking actions

s(time)	e(time)	R_s	Type of kick	Intention	Effect
19338	19390	Yellow-2	Kick	Pass	Pass
19391	19403	Yellow-1	Direct play	Shot	Interception
19404	19463	Blue-4	Kick	Clear	Clear

Yellow : teammate robots
Blue : opponent robots

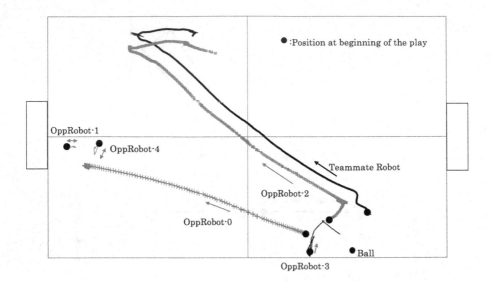

Fig. 3. Trajectories of robots

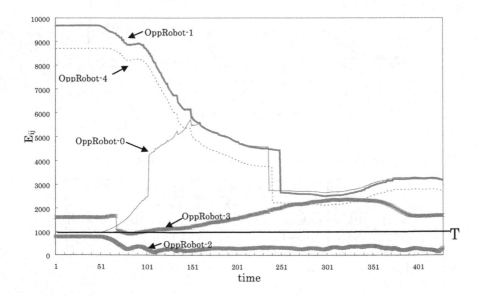

Fig. 4. E_{ij} calculation of Figure 3

5.2   Experimental Result

We got the 11 minutes logged data in RoboCup 2007 competitions. From them, one of the authors got extracted the kick actions and the "covering for preventing passing" actions. We use them as the supervisory data. We applied the algorithms described in the previous sections to the 11 minutes logged data and extracted the actions. Then, we compared the supervisory data and the computer extracted data.

Table 4 and Table 5 show the results.

Table 4. Kick actions

detected kick actions by supervisor	48
totally detected kick actions by computer	54
correctly detected kick actions by computer	47 (97.9%)
undetected kick actions by computer	1
misdetected kick actions by computer	7
correctly detected kicked robot in 47 kicks	46 (97.8%)
correctly classified kicking intention in 47 kicks	38 (80.8%)
correctly classified kicking effect in 47 kicks	39 (82.9%)

Table 5. "covering for preventing passing" actions

detected actions by supervisor	20
detected actions by computer	18
undetected actions by computer	2
correctly detected actions	18
misdetected actions by computer	0

5.3   Discussion

From Table 4, we can get correct classification in over 80% of the 47 correct detected kicks. Failed examples of intention classification are mainly due to misclassification of the shooting action into the passing action. These happen when the other teammate robot stands near the shooting line when robot shoots. Failed examples of the kicking effect classification are mainly due to misclassification of the passing action into the interception action. These happen when many opponent robots are around the ball[3] and one of them is detected as a ball getter while a true getter is a teammate robot.

From Table 5, 90% of the covering action for passing are correctly detected. Undetected covering action happens when a robot runs faster than a covering

---

[3] This is often the case in the SSL.

robot due mainly to the performance difference between competing teams[4], since, in the case, the covering robot does not satisfy Eq. (13).

6   Concluding Remarks

In this paper, we proposed new algorithms that detect and classify the robot actions in the logged data in the SSL, and showed that the algorithms well-detect and well-classify the robot actions. Since this is a sort of time series analysis, programs can run in real time. We developed an off-line analysis, however it can be easily extended to an on-line analysis, since the delay element of the algorithms is only an "n", a smoothing parameter, and its value is small.

As the further study, there remains,

- to confirm the effectiveness of the detection and classification algorithms in the real game,
- to design the system that learns and forecasts the actions of the opponent behaviors by using the detected actions.

References

1. Khojasteh, M.R., Meybodi, M.R.: Evaluating Learning Automata as a Model for Cooperation in Complex Multi-Agent Domains. In: Lakemeyer, G., Sklar, E., Sorrenti, D.G., Takahashi, T. (eds.) RoboCup 2006: Robot Soccer World Cup X. LNCS, vol. 4434, pp. 410–417. Springer, Heidelberg (2007)
2. Lattner, A., Miene, A., Visser, U., Herzog, O.: Sequential Pattern Mining for Situation and Behavior Prediction in Simulated Robotic Soccer. In: Bredenfeld, A., Jacoff, A., Noda, I., Takahashi, Y. (eds.) RoboCup 2005. LNCS, vol. 4020, pp. 118–129. Springer, Heidelberg (2006)
3. Ekin, A., Tekalp, A.M., Mehrotra, R.: Automatic soccer video analysis and summarization. IEEE Transactions on Image Processing 12(7), 796–807 (2003)
4. Nakagawa, Y., Hada, H., Imai, M., Sunahara, H.C.: Automation of the Soccer Game Analysis. IPSJ SIG 2002-CSEC-16, pp. 193–198 (in Japanese) (2002)
5. Kaminka, G., Fidanboylu, M., Chang, A., Veloso, M.: Learning the sequential coordinated behavior of teams from observation. In: Kaminka, G.A., Lima, P.U., Rojas, R. (eds.) RoboCup 2002. LNCS (LNAI), vol. 2752, pp. 111–125. Springer, Heidelberg (2003)
6. Nakanishi, R., Bruce, J., Murakami, K., Naruse, T., Veloso, M.: Cooperative 3-robot passing and shooting in the RoboCup Small Size League. In: Lakemeyer, G., Sklar, E., Sorrenti, D.G., Takahashi, T. (eds.) RoboCup 2006: Robot Soccer World Cup X. LNCS, vol. 4434, pp. 418–425. Springer, Heidelberg (2007)

---

[4] Authors think that this is a would-be covering action and should classify as a covering action.

Using Different Humanoid Robots for Science Edutainment of Secondary School Pupils

Andreas Birk, Jann Poppinga, and Max Pfingsthorn

Jacobs University Bremen*
Campus Ring 1, 28759 Bremen, Germany
a.birk@jacobs-university.de
http://robotics.jacobs-university.de

Abstract. Robotics camps that involve design, construction and programming tasks are a popular part of various educational activities. This
paper presents the results of a survey that accompanied the *Innovationscamp*, a one week intensive workshop for promoting science and
engineering among secondary school pupils through humanoid robots.
Two very different types of platforms were used in this workshop: LEGO
mindstorms, which are widely used for educational activities, and Bioloid
humanoids, which are more commonly used for professional research.
Though the workshop participants were robotics novices, the survey indicates through several statistically significant results that the Bioloid
robots are preferred by the pupils over the LEGO robots as educational
tools.

1 Introduction

In the last decade, robots have become increasingly popular as an integral part of
a wide range of educational activities. There is accordingly a significant amount
of literature on the field of robotics in education, ranging from books, e.g., [1],
over special issues of journals, e.g., [2,3], to a vast amount of general journal
and conference contributions. The list of ongoing projects has a wide span in
several dimensions. It includes for example various age groups, from kids in
their first school years over K-12 to graduate university students [4][5][6][7][8]. It
also includes various teaching goals, ranging from highly specialized topics like
control theory or software engineering to the aim of waking a general interest in
science and technology [9][10][11][12][13][14][15][16][17]. Last but not least, there
is a wide range of organizational forms ranging from short sneak workshops on
a single afternoon to intensive long term engagements in events like RoboCup
Jr.[8][18][19].

The *Innovationscamp* project [20] presented here targets at waking, respectively raising a general interest in science and technology among pupils through
hands on workshops. The data presented here was collected at a Humanoid
Workshop as part of the first *Innovationscamp* camp in October 2007 in Bremen, Germany. The camp lasted one week. It was conducted in a Youth Hostel,

---

* Formerly International University Bremen.

L. Iocchi et al. (Eds.): RoboCup 2008, LNAI 5399, pp. 451–462, 2009.
© Springer-Verlag Berlin Heidelberg 2009

i.e., participation included accommodation. The Humanoid Workshop addressed secondary school pupils. It was accompanied by a study based on questionnaires; one set distributed before the camp and one set distributed right after. The surveys were anonymous and voluntarily. The study indicates through a statistically significant increase in several indicators in the self assessment of the pupils that the main goal of the camp, namely increasing interest in science and technology, is indeed achieved. This is a result, which has been reported for many similar activities before. Here, we focus on the aspect that humanoids were used in the Workshop. Concretely, two very different types of humanoids - LEGO Mindstorms and Bioloid - were used. The participants roughly spend the same amount of time with both types of robots. The participants had before the workshop no or at most very few contact with robots or robotics kits, i.e., they were robotics novices. A surprising result of the survey conducted after the camp is that there is statistically significant evidence that the participants see the Bioloid robot as a better educational tool than the LEGO mindstorms humanoid. Concretely, this includes the assessment of its suitability to get an introduction to humanoids, to learn about mechanics, electronics, respectively programming, to offer many options for own experiments, and to be a robot the participant would like to also continuously work with at home.

The rest of the paper is structured as follows. Section 2 provides background information about the humanoid workshop at the *Innovationscamp*. In section 3, the statements from the survey questionnaire, which deal with the humanoids are presented. Results including a statistical analysis are given in section 4. Section 5 concludes the paper.

2   The Humanoid Workshop at the Innovationscamp

The *Innovationscamp* in October 2007 featured two different workshops: one addressing pupils in the age group 9 to 13 years and dealing with interactive devices, and one addressing pupils in the age group 14 to 17 years and dealing with humanoids. Both workshops were held in parallel. The humanoid workshop, which is the topic of interest of this paper, was attended by 15 participants. Figure 1 shows the distribution of the sex, the age, and the school year of the participants. The 15 participants were divided in 3 groups of 5. Each group was supported by a tutor in form of a senior university student. The assignment to the groups was self-organized by the participants in an acquaintance session at the beginning of the workshop. The groups were working in a single room and they could interact with each other. Staying in a group over the whole week was not enforced, in contrary, exchange of ideas and also of group members was encouraged. One interesting anecdotic observation is that the female participants preferred to form a group of their own.

Two very different types of humanoid robots were used in the workshop: LEGO Mindstorms Humanoid [21] and Bioloid Humanoid [22]. Their most apparent difference is the complexity of their joints. The LEGO Mindstorms humanoid has 3 DOF, one in each leg and one to turn its head. The Bioloid

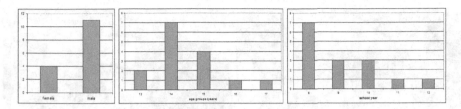

Fig. 1. The distribution of sex, age and the school year of the participants of the humanoid workshop

humanoid in contrast has 18 DOF, 5 per leg, 3 per arm, 1 for the head, and one in the lower torso. The LEGO Mindstorms kits in general and the LEGO humanoid platform in particular are promoted as educational tools for a wide age range, whereas the Bioloid platform is usually considered as a more "professional" platform, which is for example widely used by research institutions in the RoboCup Humanoid League. The main motivation for using the different robot types was to start with a well established, supposedly simply to use platform in form of the LEGO Mindstorm humanoid and then to use the Bioloid for an outlook of what is possible with humanoid robots. Two LEGO Mindstorms Humanoid kits and one Bioloid humanoid were available for the participants. The groups regularly changed between using the different robots; each participant spend roughly the same amount of time with each of the robots.

The LEGO humanoid robots were provided as regular kits to the participants (figure 2). Work with the LEGO robots hence included the mechanical assembly of the robots. This was seen as a feature. It was expected that the mechanical assembly was entertaining and at the same time supporting some of the educational goals. Also, youngsters can be expected to be familiar with LEGO building blocks for mechanical construction. The assembly of a Bioloid robot is in contrast quite complex and very unfamiliar. It requires the fixing of dozens of miniature screws, a very tedious and time consuming task. The Bioloid robot (figure 3) was hence pre-assembled by the organizers before the workshop.

Fig. 2. The LEGO humanoid was provided as basic kit at the camp. Work with this robot hence included assembly as well as programming.

Fig. 3. The Bioloid humanoid was pre-assembled before the camp. The participants could hence concentrate on programming the robot.

Fig. 4. The dance show at the end of the camp (left) where the results of the activities were presented in a playful manner to the parents and friends of the participants (right)

The workshop mainly consisted of hands on sessions but it also included some short lecture style teaching sessions. The classical teaching material based on slides and multi media material covered the following topics in a very high level, overview manner:

- robotics and AI with a particular focus on humanoids
- DC-motors and gears
- feedback loops, control, and servos
- kinematics
- walking machines
- introduction to the LEGO and the Bioloid humanoid

This classical lecture part was mainly done on the first day of the camp. The two following days were fully devoted to hands on sessions where the participants assembled LEGO robots, and programmed the LEGO, respectively Bioloid robots. In doing so, the tutors encouraged experiments and provided background information to link the work to the previously presented teaching material. The fourth day was mainly devoted to developing a concept for a show at the final day of the camp. No constraints were given to the participants; they were asked to freely come up with ideas of how to entertain their parents and friends with a performance of their robots. The participants decided to develop a dance performance. A significant part of the fourth day and the fifth day were devoted to the implementation of the show, which took place on the late afternoon of the fifth day (figure 4). The evening program during the week included watching robot movies as well as playing games and the option for sports activities.

3   The Humanoid Questionnaire

As mentioned before in the introduction, the *Innovationscamp* was accompanied by a survey. The survey was voluntarily and anonymous. All participants of the camp also took part in the survey. One part of the related questionnaires deals with the particular aspects of the humanoids used. Concretely, following questions are asked (please note that the statements are translated):

1. The LEGO humanoid
   (a) is suited to get a first impression of humanoid robots
   (b) is suited to learn something about mechanics
   (c) is suited to learn something about electronics
   (d) is suited to learn something about programming
   (e) offers many options for own experiments
   (f) is a robot I would like to also continuously work with at home
2. The Bioloid humanoid
   (a) is suited to get a first impression of humanoid robots
   (b) is suited to learn something about mechanics
   (c) is suited to learn something about electronics
   (d) is suited to learn something about programming
   (e) offers many options for own experiments
   (f) is a robot I would like to also continuously work with at home
3. Please assess the following statements
   (a) I was aware of LEGO Mindstorm as general robot construction kits already before this camp.
   (b) I also already knew about the LEGO humanoids before this camp.
   (c) I worked already with LEGO Mindstorm before this camp.
   (d) I was aware of the Bioloid humanoid before this camp.
   (e) I knew about humanoids similar to the Bioloid before this camp.
   (f) I worked already with Bioloid or similar humanoids before this camp.

For each statement, the participants had the option to mark four tickboxes numbered from 1 to 4. The numbers correspond to the degree of agreement with each statement, ranging from 1 as the strongest agreement to 4 as the strongest disagreement. To ease this assessment, following additional guidelines were given to the participants on the questionnaire. Please note that also here translations are given:

1. yes, this statement is perfectly correct in my case
2. this statement somewhat applies to me
3. this statement tends not to apply to me
4. no, this statement is not at all correct in my case

4   Results

Figures 5 and 7 show the distributions of the raw data, once for the statements related to the LEGO humanoid (Q1.x), and once for the same statements but related to the Bioloid humanoid (Q2.x). Please note that the *distributions* of answers to each question are shown in figure 5, respectively figure 7, i.e., there is no correlation between the number on the x-axis and a particular student. Figures 5 and 7 show the means and the standard deviations of the data. As can be seen, for each statement under Q1.x and Q2.x, there is a stronger level

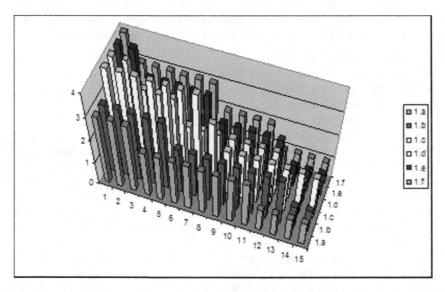

Fig. 5. The distribution of the raw data for the statements under Q1.x dealing with the LEGO Humanoid

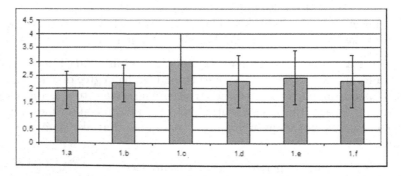

Fig. 6. The means and standard deviations for the statements under Q1.x dealing with the LEGO Humanoid

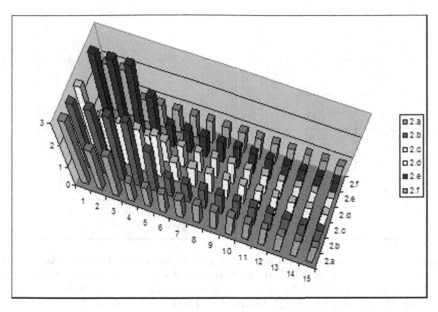

Fig. 7. The distribution of the raw data for the statements under Q2.x dealing with the Bioloid Humanoid

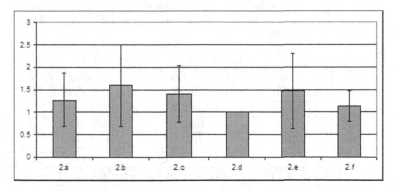

Fig. 8. The means and standard deviations for the statements under Q2.x dealing with the Bioloid Humanoid

of agreement with the statement when it is applied to a Bioloid humanoid than when it is applied to a LEGO humanoid.

This observation is also illustrated in figure 9, which shows the means and the standard deviations of the numerical differences in the agreements with each statement. It is a surprising result that the Bioloid robot is consistently considered to be more suited to get a first impression, to learn something about mechanics, electronics as well as programming, to offer many options for own experiments, and to be a robot the participant would like to also continuously

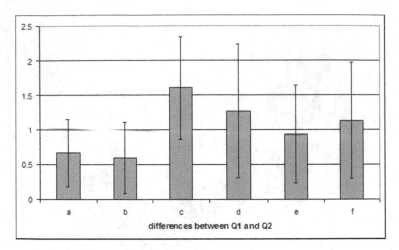

Fig. 9. The mean and standard deviations for the numerical differences between the statements under Q1.x and Q2.x. It can be seen that for each statement under category Q2 dealing with the Bioloid the agreement is much stronger than with the corresponding statement in the category Q2 dealing with the LEGO Humanoid.

Table 1. A t-test analysis of the differences between the answers to statements Q1.x and Q2.x. The analysis shows that the higher levels of agreements with the statements related to the Bioloid are statistically significant. The significance level is set to the conventional value of $\alpha = 0.05$.

	Q1.a Q2.a	Q1.b Q2.b	Q1.c Q2.c
Mean	1.93333333 1.2666667	2.200000 1.600000	3.0000000 1.40000
Variance	0.49523810 0.3523810	0.457143 0.828571	1.0000000 0.40000
t Stat	5.29150262	4.582576	8.4105197
P(T≤t)	0.00005696	0.000213	0.0000004
t Critical	1.76131012	1.761310	1.7613101

	Q1.d Q2.d	Q1.e Q2.e	Q1.f Q2.f
Mean	2.2666667 1.00000	2.400000 1.466666667	2.2666667 1.1333333
Variance	0.9238095 0.00000	0.971429 0.695238095	0.9238095 0.1238095
t Stat	5.1040716	5.136596	5.2642501
P(T≤t)	0.0000802	0.000076	0.0000599
t Critical	1.7613101	1.761310	1.7613101

work with at home. A t-test analysis of the data (table 1) shows that all these results are statistically significant.

Please note that in the teaching material, none of the two robots was given any preference. All fundamental concepts like motors or kinematics were introduced independent of any robot platform. Also in the hands on sessions, the participants simply worked with the robots in a round robin fashion. The tutors explained and motivated all concepts in a similar way by using the different platforms in a fairly equal amount. Whenever for example a question occurred, the tutor used whatever robot was just at hand to answer it. If there was a bias at all, then one may argue that it is in favor of the LEGO robots as two of them were available to illustrate things in contrast to a single Bioloid robot.

One of the most striking surprises is that the Bioloid robot is considered to be more suited to learn something about mechanics though all mechanical assembly was done using LEGO. It seems that the experiments of the participants with balancing the Bioloid robot, with working with its many DOF, and with developing complex motions for it, are considered to be more instructive than the assembly of LEGO parts. Another very surprising result is the fact that the Bioloid robot is considered to be more suited as an device for an introduction to humanoid robots. It was expected that the more widely used LEGO robots could better serve this part.

The numeric analysis of the survey data is also supported by anecdotic evidence. Though the workshop only lasted one week, the participants voiced already well before the end the feeling that they had exhausted the possibilities of the LEGO Mindstorms kit, whereas they kept on discovering new possibilities, especially motion patterns, of the Bioloid humanoid. This is also reflected in the show that was performed at the end of the workshop (figure 4). The workshop participants developed more than a dozen new complex motion patterns for the

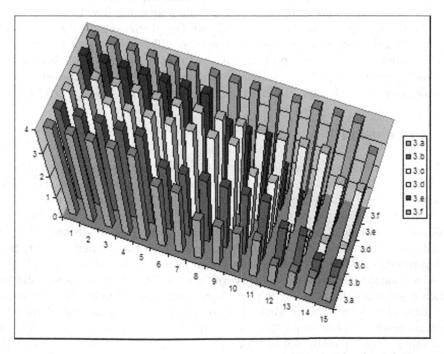

Fig. 10. The distribution of the raw data for the statements under Q3.x dealing with previous exposure to different types of humanoids

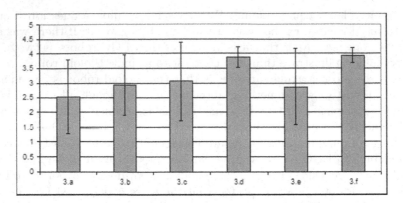

Fig. 11. The mean and standard deviations for the statements under Q3.x dealing with previous exposure to different types of robotics kits and humanoids

Bioloid robot, which were turned into an impressive dance performance. For the LEGO Mindstorms humanoids, they preferred to mainly work on a decoration of the robots than trying to program anything beyond the standard behaviors or to modify the mechanics.

Please note that the participants of the workshop had no or at most very few previous experience with robots or robotics kits. Figures 10 and 11 show the raw data, respectively the means and standard deviations for the statements under Q3.x dealing with previous exposure to different types of robotics kits and humanoids. As can be expected, LEGO Mindstorms were better known than Bioloid robots to the participants before the workshop. But real experiences were not significantly higher for Mindstorms, but quite low for both types of robots.

5   Conclusion

A humanoid workshop for secondary school pupils was presented. The workshop took place in October 2007 as part of the *Innovationscamp* in Bremen, which lasted one week. The workshop was attended by 15 participants who had no or at least very few previous experiences with robots or robot kits. Two very different type of robots were used: LEGO Mindstorms humanoids and a Bioloid humanoid. A survey shows the surprising result that the participants - though being robotics novices - prefer the supposedly more "professional" Bioloid robot as educational tool. For all seven surveyed statements, there is statistically significant evidence that the participants agree stronger with them in relation to the Bioloid. Concretely, the Bioloid is considered to be more suited to get a first impression, to learn something about mechanics, electronics as well as programming, to offer many options for own experiments, and to be a robot the participant would like to also continuously work with at home.

Acknowledgments

Please note the name-change of our institution. The Swiss Jacobs Foundation invests 200 Million Euro in **International University Bremen (IUB)** over a five-year period that started from 2007. To date this is the largest donation ever given in Europe by a private foundation to a science institution. In appreciation of the benefactors and to further promote the university's unique profile in higher education and research, the boards of IUB have decided to change the university's name to **Jacobs University Bremen**. Hence the two different names and abbreviations for the same institution may be found in this paper, especially in the references to previously published material.

References

1. Druin, A., Hendler, J.: Robots For Kids. Morgan Kauffmann, San Francisco (2000)
2. Weinberg, J.B., Yu, X. (eds.): Special Issue on Robotics in Education. Robotics and Automation Magazine, vol. 9, part 1. IEEE, Los Alamitos (2003)
3. Weinberg, J.B., Yu, X. (eds.): Special Issue on Robotics in Education. Robotics and Automation Magazine, vol. 10, part 2. IEEE, Los Alamitos (2003)
4. Buiu, C.: Hybrid educational strategy for a laboratory course on cognitive robotics. IEEE Transactions on Education 51(1), 100–107 (2008)
5. D'Andrea, R.: Robot soccer: A platform for systems engineering. In: American Society for Systems Education, Annual Conference, vol. 2220 (1999)
6. Ahlgren, D.J.: Meeting educational objectives and outcomes through robotics education. In: World Automation Congress, 2002. Proceedings of the 5th Biannual World Automation Congress, 2002. Proceedings of the 5th Biannual, vol. 14, pp. 395–404 (2002)
7. Kolberg, E., Orlev, N.: Robotics learning as a tool for integrating science technology curriculum in k-12 schools. In: Orlev, N. (ed.) Frontiers in Education Conference, 2001. 31st Annual, vol. 1, pp. T2E–12–13 (2001)
8. Lund, H., Pagliarini, L.: Robocup jr. with lego mindstorms. In: Proceedings of the International Conference on Robotics and Automation, ICRA 2000 (2000)
9. Mirats Tur, J., Pfeiffer, C.: Mobile robot design in education. Robotics and Automation Magazine, IEEE 13(1), 69–75 (2006)
10. Howell, A., Way, E., Mcgrann, R., Woods, R.: Autonomous robots as a generic teaching tool. In: Way, E. (ed.) Frontiers in Education Conference, 36th Annual, pp. 17–21 (2006)
11. Williams, A.B.: The qualitative impact of using lego mindstorms robots to teach computer engineering. IEEE Transactions on Education 46(1), 206 (2003)
12. Asada, M., D'Andrea, R., Birk, A., Kitano, H., Veloso, M.: Robotics in edutainment. In: Proceedings of the International Conference on Robotics and Automation (ICRA). IEEE Presse, Los Alamitos (2000)
13. Beer, R.D., Chiel, H.J., Drushel, R.F.: Using autonomous robotics to teach science and engineering. In: Communications of the ACM (1999)
14. Lund, H.: Robot soccer in education. Advanced Robotics Journal 13, 737–752 (1999)
15. Lund, H., Arendt, J., Fredslund, J., Pagliarini, L.: What goes up, must fall down. Journal of Artificial Life and Robotics 4 (1999)

16. Gustafson, D.A.: Using robotics to teach software engineering. In: Frontiers in Education Conference, 1998. FIE 1998. 28th Annual., vol. 2, pp. 551–553 (1998)
17. Mehrl, D., Parten, M., Vines, D.: Robots enhance engineering education. In: Parten, M.E. (ed.) Proceedings Frontiers in Education Conference, 1997. 27th Annual Conference. 'Teaching and Learning in an Era of Change, vol. 2, pp. 613–618 (1997)
18. Sklar, E., Eguchi, A., Johnson, J.: Robocupjunior: learning with educational robotics. In: Kaminka, G.A., Lima, P.U., Rojas, R. (eds.) RoboCup 2002. LNCS (LNAI), vol. 2752, pp. 238–253. Springer, Heidelberg (2003)
19. Kitano, H., Suzuki, S., Akita, J.: Robocup jr.: Robocup for edutainment. In: Proceedings of the International Conference on Robotics and Automation, ICRA 2000 (2000)
20. Innovationscamp: website (2007), http://www.innovationscamp.de
21. LEGO: Mindstorms humanoid (2007),
    http://mindstorms.lego.com/Overview/MTR_AlphaRex.aspx
22. Robotis: Bioloid humanoid (2008),
    http://www.robotis.com/html/sub.php?sub=2&menu=1

Planetary Exploration in USARsim: A Case Study Including Real World Data from Mars*

Andreas Birk, Jann Poppinga, Todor Stoyanov, and Yashodhan Nevatia

Jacobs University Bremen**
Campus Ring 1
28759 Bremen, Germany
<initial>.<lastname>@jacobs-university.de
http://robotics.jacobs-university.de/

Abstract. Intelligent Mobile Robots are increasingly used in unstructured domains; one particularly challenging example for this is planetary exploration. The preparation of according missions is highly non-trivial, especially as it is difficult to carry out realistic experiments without very sophisticated infrastructures. In this paper, we argue that the Unified System for Automation and Robot Simulation (USARSim) offers interesting opportunities for research on planetary exploration by mobile robots. With the example of work on terrain classification, it is shown how synthetic as well as real world data from Mars can be used to test an algorithm's performance in USARSim. Concretely, experiments with an algorithm for the detection of negotiable ground on a planetary surface are presented. It is shown that the approach performs fast and robust on planetary surfaces.

1   Introduction

Planetary exploration is a task where intelligent mobile robots can be valuable tools as impressively demonstrated by the Mars Exploration Rover (MER) mission [1][2][3][4]. Also, the control of the systems still involves a major amount of human supervision [5], i.e., there is still significant need for research to increase the robots' intelligence and autonomy. Furthermore, the preparation of according missions is highly non-trivial. It requires a significant amount of preparation and testing. Here, the use of the Unified System for Automation and Robot Simulation (USARSim) for the purpose of research, testing and planning of planetary exploration missions is evaluated. Concretely, a case study is made were USARSim is used for an approach to terrain classification in the context of planetary exploration.

The Unified System for Automation and Robot Simulation (USARSim) [6] is a high fidelity robot simulator built on top of the Unreal Tournament[7] game engine. Its feature include a commercial physics engine (Karma [8]) and a real-time,

---

* This work was supported by the German Research Foundation (DFG).
** Formerly International University Bremen.

L. Iocchi et al. (Eds.): RoboCup 2008, LNAI 5399, pp. 463–472, 2009.

Fig. 1. The autonomous version of a *Rugbot* with some important on-board sensors pointed out. The SwissRanger SR-3000 and the stereo camera deliver the 3D data for the terrain classification.

Fig. 2. Two Rugbots at the Space Demo at RoboCup 2007 in Atlanta

three-dimensional visualization engine. It is important that these components have been tested for their physical fidelity [9,10,11,12]. The robot model used for the case study in this paper is the Rugbot - from rugged robot - (figure 1), which was first developed for work on Safety, Security, and Rescue Robotics (SSRR). But due to its capabilities to negotiate rough terrain [13][14], it is also an interesting platform for research on planetary exploration (figure 2). The software architecture on the Rugbots is designed to support intelligent functions up to full autonomy [15][16][17].

The case study conducted here deals with terrain classification, especially the detection of drivable ground. This is a very important topic in the space robotics community [18,19,20,21,22] as - despite a human in the loop component - the robots have to move some distances autonomously on their own; the long delay in radio communication simply prohibits pure tele-operation. Here we present an extension of work described in detail in [23], which deals with a very fast but nevertheless quite robust detection of drivable ground. The approach is based on range data from a 3D sensor like a time-of-flight camera like a SwissRanger, respectively a stereo camera. The main idea is to process the range data by a Hough transform with a three dimensional parameter space for representing planes. The discretized parameter space is chosen such that its bins correspond to planes that can be negotiated by the robot. A clear maximum in parameter space hence indicates safe driving. Data points that are spread in parameter space correspond to non-drivable ground. In addition to this basic distinction, a more fine grain classification of terrain types is in principle possible with the approach. An autonomous robot can use this information for example to annotate its map with way points or to compute a risk assessment of a possible path.

The approach has already proven to be useful in in- and outdoor environments in the context of SSRR. The results presented in [23] are based on experiments with datasets with about 6,800 snapshots of range data. Drivability is robustly detected with success rates ranging between 83% and 100% for the SwissRanger and between 98% and 100% for the stereo camera. The complete processing time for classifying one range snapshot is in the order of 5 to 50 msec. The detection of safe ground can hence be done in real-time on the moving robot, which allows using the approach for reactive motion control as well as mapping in unstructured environments. Here, the question of interest is whether the approach is also suited for planetary surfaces and how USARSim can be used to answer this question.

2   Detection of Negotiable Terrain

The terrain classification is based on the following idea. Range images, e.g. from simple 3D sensors in the form of an optical time-of-flight camera and a stereo camera, are processed with a Hough transform. Concretely, a discretized parameter space for planes is used. The parameter space is designed such that each drivable surface leads to a single maximum, whereas non-drivable terrain leads to data-points spread over the space. The actual classification is done by three simple criteria on the binned data arranged in a decision tree like manner (see algorithm 1). In addition to binary distinctions with respect to drivability, more fine grain classifications of the distributions are possible allowing to recognize different categories like plane floor, ramp, rubble, obstacle, and so on in SSRR domains, respectively flat ground, hills, rocks, and so on in planetary exploration scenarios. This transform can be computed very efficiently and allows a robust classification in real-time.

Classical obstacle and free space detection for mobile robots is based on two-dimensional range sensors like laser scanners. This is feasible as long as the robot operates in simple environments mainly consisting of flat floors and plain walls. The generation of complete 3D environment models is the other extreme, which requires significant processing power as well as high quality sensors. Furthermore, 3D mapping is still in its infancy and it is non-trivial to use the data for path planning. The approach presented here lies in the middle of the two extremes.

Algorithm 1. The classification algorithm: First, it checks the bin corresponding to the floor. If it has enough hits, the result "floor" is returned. Otherwise it
uses two simple criteria to verify the usability of the bin with most hits $\text{bin}^{\max}$.
In this case, the class is assigned based on the parameters of $\text{bin}^{\max}$ (line 1).
Otherwise, no plane dominates the Hough space, so an obstacle is reported. $\#S$
is the cardinality of S, PC is the used point cloud. Constants were $t_m = 0.667$,
$t_p = 0.125$, $t_n = 6$, $t_h = 0.15$

1. if $\#\text{bin}_{\text{floor}} > t_h \cdot \#PC$ then
2.   class $\leftarrow$ floor
3. else
4.   if $(\#\{\text{bin} \mid \#\text{bin} > t_m \cdot \#\text{bin}^{\max}\} < t_n)$ and $(\#\text{bin}^{\max} > t_p \cdot \#PC)$ then
5.   class $\leftarrow$ type$(\text{bin}_{\max}) \in \{\text{floor}, \text{plateau}, \text{canyon}, \text{ramp}\}$
6.   else
7.   class $\leftarrow$ obstacle
8.   end if
9. end if

A single 3D range snapshot is processed to classify the terrain, especially with
respect to drivability. This information can be used in various standard ways
like reactive obstacle avoidance as well as 2D map building. The approach is
very fast and it is an excellent candidate for replacing standard 2D approaches
to sensor processing for obstacle avoidance and occupancy grid mapping in non-
trivial environments. More details about the implementation of the approach in
general can be found in [23].

3   Experiments and Results

The terrain classification algorithm is now tested with synthetic and real world
data from Mars in USARSim. The real world data covers the Eagle crater on
Mars, which is modeled in USARSim based on ground truth data from the Mars
Exploration Rover (MER) mission data archives [24] (see also figures 3 and 4).

Three different areas are used, each with 12 samples. Example images for
these terrains can be seen in figure 5. The ground truth is based on visual
assessment by two experienced USARSim users. The results of the classification
are in table 1. It turns out that the algorithm is nearly as successful as in the
original scenario, but significantly faster. This is due to the comparably small
number of points, which nevertheless hardly hinder the success.

In figure 6, you can see exemplary Hough spaces for the four terrains corresponding to the images in figure 5. For the passable Terrain A, the Hough
space is relatively empty except for one maximum. In the histogram in the right
column, it can be seen how one bin with many hits stands out from the others.
In Terrains B and C, a number of planes receive many hits, so an obstacle is
reported. Also note that the algorithm was invariant to the considerably varying
magnitudes in the bins: the maxima were 571 and 527 for A and C, but 1477
for B.

Fig. 3. A Jacobs Rugbot in the RoboCup Virtual Simulator (left), exploring its environment on different planetary surface types (center and right)

Fig. 4. A Rugbot on Mars in the vicinity of the Endurance crater; the environment is modeled based on original data from the opportunity mission

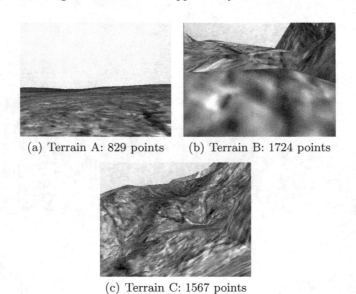

(a) Terrain A: 829 points   (b) Terrain B: 1724 points

(c) Terrain C: 1567 points

Fig. 5. Example CGI for the terrains used in the classification experiments with given number of points in the corresponding point cloud

Table 1. Results of the classification experiments

Terrain	Correctness [%]	median time [msec]	median #points
Terrain A	100	5.124	814
Terrain B	83	8.048	1724
Terrain C	83	8.424	1567

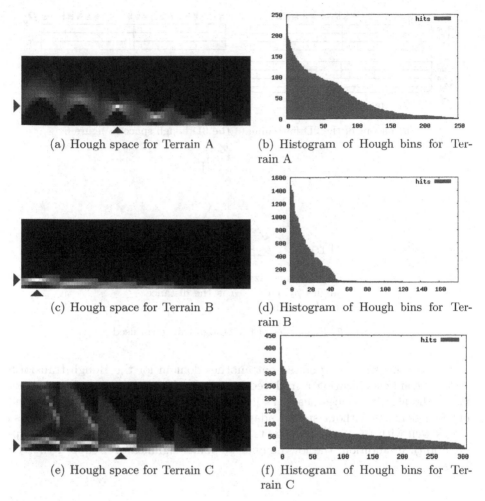

(a) Hough space for Terrain A

(b) Histogram of Hough bins for Terrain A

(c) Hough space for Terrain B

(d) Histogram of Hough bins for Terrain B

(e) Hough space for Terrain C

(f) Histogram of Hough bins for Terrain C

Fig. 6. Results for the scenes in figure 5. In the left column there is a 2D flattening of the 3D hough space (legend in figure 7). The two arrows point at the bin with the maximum number of hits bin^{max}. In the right column there are the bins of the hough space re-ordered by magnitude.

4   Conclusion

We demonstrated the validity of USARSim as a tool for simulation by successfully applying an algorithm that has been shown to work in the real world. This underlines USARSim's usability in the preparation of planetary exploration. A lot of emphasis is put on this phase since many resources are at stake in the actual mission. A low cost software framework like USARSim allows a wider range of companies and research groups to take part in the space effort as it reduces the need for expensive testing environments.

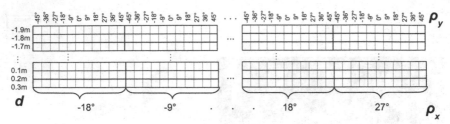

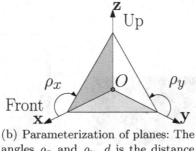

(a) Layout of the 2D flattening of the 3D hough space in figure 6

(b) Parameterization of planes: The angles ρ_x and ρ_y, d is the distance to the origin

Fig. 7. Properties of the Hough transform used

At the same time, we pointed out another domain for the Hough transform based terrain classification introduced in [23]. In the planetary exploration domain, the algorithm does nearly as good as it does in the original indoor and outdoor domains without special adaptations. It was also observed that it also works well with relatively few points (circa 5% of the 25K in the original application). In addition, the low number of points significantly reduced the run time.

Acknowledgments

Please note the name-change of our institution. The Swiss Jacobs Foundation invests 200 Million Euro in **International University Bremen (IUB)** over a five-year period starting from 2007. To date this is the largest donation ever given in Europe by a private foundation to a science institution. In appreciation of the benefactors and to further promote the university's unique profile in higher education and research, the boards of IUB have decided to change the university's name to **Jacobs University Bremen (Jacobs)**. Hence the two different names and abbreviations for the same institution may be found in this paper, especially in the references to previously published material.

Furthermore, the authors gratefully acknowledge the financial support of the *Deutsche Forschungsgemeinschaft* (DFG) for their research.

References

1. Erickson, J.: Living the dream - an overview of the mars exploration project. IEEE Robotics and Automation Magazine 13(2), 12–18 (2006)
2. Biesiadecki, J., Baumgartner, E., Bonitz, R., Cooper, B., Hartman, F.R., Leger, P.C., Maimone, M.W., Maxwell, S.A., Trebi-Ollennu, A., Tunstel, E.W., Wright, J.R.: Mars exploration rover surface operations: driving opportunity at meridiani planum. IEEE Robotics and Automation Magazine 13(2), 63–71 (2006)
3. Lindemann, R., Bickler, D., Harrington, B., Ortiz, G.M., Voothees, C.J.: Mars exploration rover mobility development. IEEE Robotics and Automation Magazine 13(2), 19–26 (2006)
4. Ai-Chang, M., Bresina, J., Charest, L., Chase, A., Hsu, J.C.-J., Jonsson, A., Kanefsky, B., Morris, P., Rajan, K.R.A.K., Yglesias, J., Chafin, B.G., Dias, W.C., Maldague, P.F.: Mapgen: mixed-initiative planning and scheduling for the mars exploration rover mission. IEEE Intelligent Systems 19(1), 8–12 (2004)
5. Backes, P.G., Norris, J.S., Powell, M.W., Vona, M.A., Steinke, R., Wick, J.: The science activity planner for the mars exploration rover mission: Fido field test results. In: Proceedings of the IEEE Aerospace Conference, Big Sky, MT, USA (2003)
6. USARsim: Urban search and rescue simulator (2006), http://usarsim.sourceforge.net/
7. games, E.: Unreal engine (2003)
8. Karma: Mathengine karma user guide (2003)
9. Carpin, S., Lewis, M., Wang, J., Balakirsky, S., Scrapper, C.: Bridging the gap between simulation and reality in urban search and rescue. In: Lakemeyer, G., Sklar, E., Sorrenti, D.G., Takahashi, T. (eds.) RoboCup 2006: Robot Soccer World Cup X. LNCS (LNAI), vol. 4434, pp. 1–12. Springer, Heidelberg (2007)
10. Carpin, S., Lewis, M., Wang, J., Balarkirsky, S., Scrapper, C.: USARSim: a robot simulator for research and education. Proc. of the 2007 IEEE Intl. Conf. on Robotics and Automation (ICRA) (2007)
11. Carpin, S., Stoyanov, T., Nevatia, Y., Lewis, M., Wang, J.: Quantitative assessments of usarsim accuracy. In: Proceedings of PerMIS (2006)
12. Carpin, S., Birk, A., Lewis, M., Jacoff, A.: High fidelity tools for rescue robotics: results and perspectives. In: Bredenfeld, A., Jacoff, A., Noda, I., Takahashi, Y. (eds.) RoboCup 2005. LNCS (LNAI), vol. 4020, pp. 301–311. Springer, Heidelberg (2006)
13. Birk, A., Pathak, K., Schwertfeger, S., Chonnaparamutt, W.: The IUB Rugbot: an intelligent, rugged mobile robot for search and rescue operations. In: IEEE International Workshop on Safety, Security, and Rescue Robotics (SSRR). IEEE Press, Los Alamitos (2006)
14. Chonnaparamutt, W., Birk, A.: A new mechatronic component for adjusting the footprint of tracked rescue robots. In: Lakemeyer, G., Sklar, E., Sorrenti, D.G., Takahashi, T. (eds.) RoboCup 2006: Robot Soccer World Cup X. LNCS (LNAI), vol. 4434, pp. 450–457. Springer, Heidelberg (2007)
15. Birk, A., Carpin, S.: Rescue robotics - a crucial milestone on the road to autonomous systems. Advanced Robotics Journal 20(5), 595–695 (2006)
16. Birk, A., Markov, S., Delchev, I., Pathak, K.: Autonomous rescue operations on the IUB Rugbot. In: IEEE International Workshop on Safety, Security, and Rescue Robotics (SSRR). IEEE Press, Los Alamitos (2006)

17. Birk, A., Kenn, H.: A control architecture for a rescue robot ensuring safe semi-autonomous operation. In: Kaminka, G.A., Lima, P.U., Rojas, R. (eds.) RoboCup 2002. LNCS (LNAI), vol. 2752, pp. 254–262. Springer, Heidelberg (2003)
18. Iagnemma, K., Brooks, C., Dubowsky, S.: Visual, tactile, and vibration-based terrain analysis for planetary rovers. In: IEEE Aerospace Conference, vol. 2, pp. 841–848 (2004)
19. Iagnemma, K., Shibly, H., Dubowsky, S.: On-line terrain parameter estimation for planetary rovers. In: IEEE International Conference on Robotics and Automation (ICRA), vol. 3, pp. 3142–3147 (2002)
20. Lacroix, S., Mallet, A., Bonnafous, D., Bauzil, G., Fleury, S., Herrb, M., Chatila, R.: Autonomous rover navigation on unknown terrains: Functions and integration. International Journal of Robotics Research 21(10-11), 917–942 (2002)
21. Lacroix, S., Mallet, A., Bonnafous, D., Bauzil, G., Fleury, S., Herrb, M., Chatila, R.: Autonomous rover navigation on unknown terrains functions and integration. In: Experimental Robotics Vii. Lecture Notes in Control and Information Sciences, vol. 271, pp. 501–510 (2001)
22. Gennery, D.B.: Traversability analysis and path planning for a planetary rover. Autonomous Robots 6(2), 131–146 (1999)
23. Poppinga, J., Birk, A., Pathak, K.: Hough based terrain classification for realtime detection of drivable ground. Journal of Field Robotics 25(1-2), 67–88 (2008)
24. MER-Science-Team: Mars exploration rover (MER) mission data archives (2007), http://anserver1.eprsl.wustl.edu/anteam/merb/merb_main2.htm

Face Recognition for Human-Robot Interaction Applications: A Comparative Study[*]

Mauricio Correa, Javier Ruiz-del-Solar, and Fernando Bernuy

Department of Electrical Engineering, Universidad de Chile
jruizd@ing.uchile.cl

Abstract. The aim of this work is to carry out a comparative study of face-recognition methods for Human-Robot Interaction (HRI) applications. The analyzed methods are selected by considering their suitability for HRI use, and their performance in former comparative studies. The methods are compared using standard databases and a new database for HRI applications. The comparative study includes aspects such as variable illumination, facial expression variations, face occlusions, and variable eye detection accuracy, which directly influence face alignment precision. The results of this comparative study are intended to be a guide for developers of face recognition systems for HRI, and they have direct application in the RoboCup@Home league.

Keywords: Face Recognition, Human-Robot Interaction, RoboCup @Home.

1 Introduction

Face analysis plays an important role in building HRI (Human-Robot Interaction) interfaces that allow humans to interact with robot systems in a natural way. Face information is by far the most used visual cue employed by humans. There is evidence of specialized processing units for face analysis in our visual system [1]. Face analysis allows localization and identification of other humans, as well as interaction and visual communication with them. Therefore, if human-robot interaction could achieve the same efficiency, diversity, and complexity that human-human interaction has, face analysis could be extensively employed in the construction of HRI interfaces.

Currently, computational face analysis is a very lively and expanding research field. Face recognition, i.e. the specific process for determining the identity of an individual contained in an image area which has been already identified as containing a face (by a face detection system) and already aligned (by a face alignment process which usually includes eye detection), is a functional key for personalizing robot services and for determining robot behaviors, usually depending on the identity of the human interlocutor. Many different face recognition approaches have been developed in the last few years [2][3][4], ranging from classical Eigenspace-based methods (e.g. eigenfaces [5]), to sophisticated systems based on thermal information, high-resolution images or 3D models. Many of these methods are well suited to specific

---

[*] This research was partially supported by FONDECYT (Chile) under Project Number 1061158.

L. Iocchi et al. (Eds.): RoboCup 2008, LNAI 5399, pp. 473–484, 2009.

requirements of applications such as biometry, surveillance, or security. HRI applications have their own requirements, and therefore some approaches are better suited for them. It would be useful for developers of face recognition systems for HRI to have some guidelines about the advantages of some methodologies over others.

In this general context, the aim of this paper is to carry out a comparative study of face-recognition methods. This research is motivated by the lack of direct and detailed comparisons of these kinds of methods under the same conditions. The result of this comparative study is a guide for the developers of face recognition systems for HRI. We concentrated on methods that fulfill the following requirements of standard HRI applications: (i) *Full online operation*: No training or offline enrollment stages. All processes must be run online. The robot has to be able to build the face database from scratch incrementally; (ii) *Real-time operation* for achieving user interaction with low delays: The whole face analysis process, which includes detection, alignment and recognition, should run at least at 3fps, with 4-5fps recommended; and (iii) *One single image per person problem*: One good-quality face image of an individual should be enough for his/her later identification. Databases containing just one face image per person should be considered. The main reasons are savings in storage and computational costs, and the impossibility of obtaining more than one good-quality face image from a given individual in certain situations. In this context, good-quality means a frontal image with uniform illumination, adequate image size, and no blurring or artifacts due to the acquisition process (e.g. interlaced video), open eyes, and no extreme facial expression.

This study analyzes four face-recognition methods (Generalized PCA, LBP Histograms, Gabor Jet Descriptors and SIFT Descriptors), which were selected by considering their fulfillment the requirements mentioned above, and their performance in former comparative studies of face recognition methods [7][8]. Those comparative studies evaluate face recognition methods from a general-purpose point of view; [7] addresses eigenspace-based methods (holistic), while [8] addresses local matching approaches. The study analyzed aspects such as variable illumination, facial expression variations, face occlusions, and variable eye detection accuracy. Aspects such as scale, pose, and in-plane rotations of the face were not included in this study because it was assumed that the robot is mobile, and the active vision allows it to obtain faces with appropriate scales and rotations. The study is carried out in two stages. In the first stage, the FERET [14] *fa* and *fc* sets are considered. The original images are altered by including occlusions of some face areas, and by adding noise in the annotations of eye positions. In the second stage the new database UCHFaceHRI, especially designed for HRI applications, is employed.

As in [7][8] we consider illumination compensation as an independent preprocessing step. Illumination compensation will help every face-recognition method. Nevertheless, it is important to note that the analyzed face-recognition methods are illumination invariant to some degree, due to the use of quasi-illumination invariant features such as LBP (Local Binary Pattern), Gabor jets and SIFT (Scale-Invariant Feature Transform). This study has direct application in the RoboCup@Home league, where face recognition is an important tool.

This paper is structured as follows. The methods under analysis are described in section 2. In sections 3 and 4 the comparative analysis of these methods is presented. Finally, conclusions of this work are given in section 4.

2 Methods under Comparison

As mentioned above, the algorithm selection criteria are their fulfillment of the defined HRI requirements, and their performance in former comparative studies of face-recognition methods [3][7][8]. The first issue is that most of the holistic methods, which are normally based on eigenspace-decompositions, fail when just one image per person is available, mainly because they have difficulties building the required representation models. However, this difficulty can be overcome if a generalized face representation is built, for instance using a generalized PCA model. Thus, as a first selection, we decided to analyze a face-recognition method based on a generalized PCA model. In general terms, local-matching methods behave well when just one image per person is available, and some of them have obtained very good results in standard databases such as FERET. Taking into account the results of [8], and our requirements of high-speed operation, we selected two methods to be analyzed. The first one is based on the use of histograms of LBP features, and the second one is based on the use of Gabor filters and Borda account classifiers. Finally, we realized that local interest points and descriptors (e.g. SIFT) have been used successfully for solving some similar wide baseline-matching problems as fingerprint verification [12], and as a first stage of a complex face-recognition system [13]. Therefore, we decided to test the suitability of a SIFT-based face recognition system.

Thus, the methods under comparison are:

- *Generalized PCA*. We have implemented a face-recognition method that uses PCA as a projection algorithm, the Euclidian distance as a similarity measure, and modified LBP features [10] for achieving some degree of illumination invariance. In particular we used a generalized PCA approach, which consists on building a PCA representation in which the model does not depend on the individuals to be included in the database, i.e. on their face images, because the PCA projection model is built using a set of images that belongs to different persons. This allows applying this method in a case when just one single image per person is available. Our PCA model was built using 2,152 face images obtained from different face databases and the Internet. For compatibility with the results presented in [7], the model was built using face images scaled and cropped to 100x185 pixels, and aligned using eye information. We analyzed the validity of this generalized PCA representation by verifying that the main part of the eigenspectrum, i.e. the spectrum of the ordered eigenvalues, is approximately linear between the 10th and 1,500th components (we carried out a similar analysis to the one described in [9]). The RMSE [7] was used as a criterion to select the appropriate number of components to be used. To achieve a RMSE between 0.9 and 0.5, the number of employed PCA components had to be in the range of 200 to 1,050. Taking into account these results, we choose to implement two flavors of our system, one with 200 components and one with 500.

- *LBP Histograms*. A local-appearance-based approach with a single, spatially enhanced feature histogram for global information representation is described in [11]. In that approach, three different levels of locality are defined: pixel level, regional level and holistic level. The first two levels of locality are realized by dividing the face image into small regions from which LBP features are extracted for efficient texture information representation. The holistic level of locality, i.e. the global description of the face, is obtained by concatenating the regional LBP extracted features. The recognition is performed using a nearest neighbor classifier in the computed feature space using one of the three following similarity measures: histogram intersection, log-likelihood statistic and Chi square. We implemented that recognition system, without considering preprocessing (cropping, using an elliptical mask and histogram equalization are used in [11]), and by choosing the following parameters: (i) face images scaled and cropped to 100x185 pixels and 203x251 instead of 130x150 pixels, for having compatibility with other studies [7][8], (ii) images divided in 10, 40 or 80 regions, instead of using the original divisions which range from 16 (4x4) to 256 (16x16), and (iii) the mean square error as a similarity measure, instead of the log-likelihood statistic. We also carried out preliminary experiments for replacing the LBP features by modified LBP features, but better results were always obtained by using the original LBP features. Thus, considering the 3 different image divisions and the 3 different similarity measures, we get 9 flavors of this face-recognition method.

- *Gabor Jets Descriptors*. Different local-matching approaches for face recognition are compared in [8]. The study analyzes several local feature representations, classification methods, and combinations of classifier alternatives. Taking into account the results of that study, the authors implemented a system that integrates the best choice in each step. That system uses Gabor jets as local features, which are uniformly distributed over the images, one wave-length apart. In each grid position of the test and gallery image and at each scale (multiscale analysis) the Gabor jets are compared using normalized inner products, and these results are combined using the Borda account method. In the Gabor feature representation, only Gabor magnitudes are used, and 5 scales and 8 orientations of the Gabor filters are adopted. Face images are scaled and cropped to 203x251 pixels. We implemented this system using all parameters described in [8] (filter frequencies and orientations, grid positions, face image size). However, after some preliminary experiments we realized that the performance of the system depends on the background, that is, on the results of the Gabor filters applied out of the face area. For this reason we implemented 3 flavors of this system: a first one (*global*) that uses the original grid, and two others (*local1* and *local2*) that use reduced local grids. Considering that the methods are applied with face images cropped to 100x185 and to 203x251 pixels, the corresponding local grids are seen in Figure 1. (Figure 1 (a)-(c) shows the grids for the case of faces of 203x251 pixels):

- 100x185: *local1*/*local2* $0 \le x \le 100 \wedge 60 \le y \le 160$/ $0 \le x \le 100 \wedge 50 \le y \le 170$
- 203x251: *local1*/*local2* $50 \le x \le 150 \wedge 100 \le y \le 210$/ $45 \le x \le 155 \wedge 90 \le y \le 210$

- *SIFT descriptors*. Wide baseline-matching approaches based on local interest points and descriptors have become increasingly popular and have experienced an impressive development in recent years. Typically, local interest points are extracted independently from both a test and a reference image, and then characterized by invariant descriptors, and finally the descriptors are matched until a given transformation between the two images is obtained. Lowe's system [16] using SIFT descriptors and a probabilistic hypothesis rejection-stage is a popular choice for implementing object-recognition systems, given its recognition capabilities, and near real-time operation. However, Lowe's system's main drawback is the large number of false positive detections. This drawback can be overcome by the use of several hypothesis

Fig. 1. Face image of 2003x251 pixels. (a) *global*, (b) *local1*, and (c) *local2* grids used in Gabor-based methods. (d) Image with eye position (red dot) and square showing a 10% error in the eye position, (e) Image with partial occlusion.

rejection stages, as for example in the L&R system [12]. This system has already been used in the construction of robust fingerprint verification systems [12]. Here, we have used the same method for building a face-recognition system, with two different flavors. In the first one, *Full*, all verification stages defined in [12] are used, while in a second one, *Simple*, just the probabilistic hypothesis rejection stages are employed.

3 Comparative Study Using Standard Databases

We use the following notation to refer to the methods and their variations: A-B-C, where A describes the name of the face-recognition algorithm (H: Histogram of LBP features, PCA: generalized PCA with modified LBP features, GJD: Gabor Jets Descriptors, SD: L&R system with SIFT descriptors); B denotes the similarity measure (HI: Histogram Intersection, MSE: Mean square error, XS: Chi square, BC: Borda Count, EUC: Euclidian Distance); and C describes additional parameters of each algorithm (H: Number of image divisions; PCA: Number of principal components, GJD: filters grid, G - *Global*, L1 - *local1* or L2-*local2*; SD: verification procedure, *full* or *simple*).

Face images are scaled and cropped to 100x185 pixels and 203x251, except for the case of the PCA method in which, for simplicity, just one image size (100x185) was employed (the generalized PCA model depends on the image cropping). In all cases faces are aligned by centering the eyes in the same relative positions, at a fixed distance between the eyes, which was 62 pixels for the 100x185 size images, and 68 pixels for the 203x251 size images.To compare the methods we used the FERET evaluation procedure [14], which established a common data set and a common testing protocol for evaluating semi-automated and automated face recognition algorithms. We used the following sets: (i) *fa* set (1,196 images), used as gallery set (contains frontal images of 1,196 people); (ii) *fb* set (1,195 images), used as test set 1 (in *fb* subjects were asked for a different facial expression than in *fa*); and (iii) *fc* set (194 images,) used as test set 2 (in *fc* pictures were taken under different lighting conditions).

3.1 Experiments Using Annotated Eyes

Original fa-fb test. Table 1 shows the results of all methods under comparison in the original *fa-fb* test, which corresponds to a test with few variations in the acquisition process (uniform illumination, no occlusions). We also use the information of the annotated eyes, without adding any noise. It can be observed that: (i) The results obtained with our own implementation of the methods are consistent with those of other studies. The best H-X-X flavors were achieved in the 203x251 face images case, a similar performance (97%) to that in the original work [11]. Best GJD-X-X flavors achieved a slightly lower performance (97.6% vs. 99.5%) than in the original work [8]. There are no reports of the use of the generalized PCA or SIFT methods in these datasets; (ii) The performance of the GJD-X-X and SD-X methods depends largely on the size of the cropped images. Probably because the methods use information about face shape and contour, which do not appear in the 100x185 images; and (iii) The best results (~97.5%) were obtained by the H-XS-80, GJD-BC-G and SD-S methods (203x251 size). Nevertheless, other H-X-X, GJD-X-X and SD-X variants also got very good results. Interestingly, some H-X-X variants got ~97% even using 100x185 size images.

Eye Detection Accuracy. Most of the face-recognition methods are very sensitive to face alignment, which depends directly on the accuracy of the eye detection process; eye position is usually the primary, and sometimes the only, source of information for face alignment. For analyzing the sensitivity of the different methods to eye position accuracy, we added white noise to the position of the annotated eyes in the *fb* images (see example in fig. 1 (d)). The noise was added independently to the *x* and *y* eye positions. Table 1 shows the *top-1* RR (Recognition Rate) achieved by the different methods. Our main conclusions are: (i) SD-X methods are almost invariant to the position of the eyes in the case of using 203x251 face images. With 10% error in the position of the eyes, the top-1 RR decreases less than 2%. The invariance is due to the fact that this method aligns test and gallery images by itself; and (ii) In all other cases the performance of the methods decreases largely with the error in the eye position, probably because they are based on the match between holistic or feature-based representation of the images. However, if the eye position error is bounded to 5%, results obtained are still acceptable (~92%) by some H-X-X variants using 100x185 face images.

Partial Face Occlusions. For analyzing the behavior of the different methods in response to partial occlusions of the face area, *fb* face images were divided into 10 different areas (2 columns and 5 rows). One of these areas was randomly selected and its pixels set to 0 (black). See example in Figure 1 (e). Thus, in this test each face image of *fb* has one tenth of its area occluded. Table 1 shows the *top-1* RR achieved by the different methods. The main conclusions are: (i) Some H-X-X variants are very robust to face occlusions (e.g. H-HI-10, H-MSE-10, H-X-80) independent of using face images of 100x185 or 203x251 pixels. The same happens with SD-X methods in the case of 203x251 size images; (ii) H-X-80 methods achieve a top-1 RR of about 95% using images of either 100x185 or 203x251 pixels; and (iii) The performance of all GJD-X-X variants decreases largely (in more than 20% of the RR) with face occlusions. The performance of PCA-based methods decreases in about 10% of the RR.

Variable Illumination. Variable illumination is one of the factors with strong influence in the performance of face-recognition methods. Although there are some specialized face databases for testing algorithm invariance against variable illumination (e.g. PIE, YaleB), we choose to use the *fa-fc* test set, because (i) it considers a large number of individuals (394 versus 10 in Yale B and 68 in PIE), and (ii) the illumination conditions are more natural in the *fc* images. Table 1 shows the *top-1* RR achieved by the different methods in this test. The main conclusions are: (i) The results obtained with our own implementation of the methods are consistent with those of other studies. The best H-X-X flavors achieved in the 203x251 face images case, show a higher performance (89.7% vs. 79%) than the original work [11]. The best GJD-X-X flavors achieved a slightly lower performance (94.1% vs. 99.5%) than the original work [8]. There are no reports of the use of the generalized PCA or SIFT methods in the same database; (ii) Best performance is achieved by the GJD-BC-G and H-XS-80 methods, both of them using 203x251 images; (iii) In all cases much better results were obtained using larger face images (203x251); (iv) PCA-X-X and SD-X methods show a lower performance in this dataset; and (v) H-X-X methods with a large number of partitions show better performance than variants with a low number of partitions (~89.7% vs. 49.5%) in the case of using 203x251 images.

Table 1. FERET *fa-fb* and *fa-fc* test. Top 1 recognition rate. Noise in eye positions and face occlusion is tested in some experiments. O: Original. OC: Original + Occlusion.

Method	100x185						203x251					
	fb				OC	*fc*	*fb*				OC	*fc*
	O	Noise in eye positions					O	Noise in eye positions				
		2.5%	5%	10%				2.5%	5%	10%		
H-HI-10	95.7	**95.1**	**91.4**	**81.8**	**93.7**	12.9	95.1	20.1	18.7	14.7	**93.1**	49.5
H-MSE-10	95.7	**95.1**	**91.4**	**81.8**	**93.7**	12.9	95.1	20.1	18.7	14.7	**93.1**	49.5
H-XS-10	95.7	94.7	**92.7**	**82.3**	79.8	13.9	95.5	36.2	33.9	28.4	82.9	61.9
H-HI-40	92.3	88.8	78.5	55.1	75.0	22.7	87.0	19.0	16.0	9.0	69.6	59.8
H-MSE-40	92.3	88.8	78.5	55.1	75.0	22.7	87.0	19.0	16.0	9.0	69.6	59.8
H-XS-40	92.4	86.9	76.3	54.0	74.1	21.1	90.3	38.7	32.2	20.9	71.6	60.8
H-HI-80	**96.9**	96.1	91.3	74.9	95.9	49.5	97.0	32.6	30.0	21.4	**94.2**	83.0
H-MSE-80	**96.9**	96.1	91.3	74.9	95.9	49.5	97.0	32.6	30.0	21.4	**94.2**	83.0
H-XS-80	**96.7**	**95.1**	90.0	71.2	**94.8**	**50.0**	97.5	67.1	62.5	46.8	**96.2**	89.7
PCA-MSE-200	73.1	55.9	40.7	16.2	63.6	**52.1**	---	---	---	---	---	----
PCA-MSE-500	76.1	60.3	42.9	16.0	64.9	**57.2**	---	---	---	---	---	---
GJD-BC-G	69.3	65.4	57.9	33.1	47.5	32.5	**97.6**	86.7	80.2	51.1	73.4	**94.1**
GJD-BC-L1	67.1	62.8	54.6	30.3	46.2	36.6	93.2	70.6	59.6	31.2	77.7	82.0
GJD-BC-L2	69.4	65.6	57.2	32.5	47.4	36.6	95.1	75.4	65.3	37.5	82.8	86.1
SD-FULL	74.3	75.7	73.5	71.5	67.3	7.7	**97.1**	**96.2**	**95.7**	**95.3**	**95.6**	67.5
SD-SIMPLE	73.1	75.3	73.1	71.0	68.6	5.7	**97.5**	**96.7**	**96.4**	**96.2**	**95.3**	63.9

3.2 Experiments Using Automated Face and Eye Detection

For testing the methods under more real conditions, experiments using faces and eyes automatically detected by a state of the art detector [15] were carried out. As in the former section, experiments were performed using the *fa-fb* and *fa-fc* test sets. In the *fa* and *fb* sets 1,183 of 1,195 face images that appear simultaneously in both sets were detected. In the *fa* and *fc* sets 193 of 194 face images that appear simultaneously in both sets were detected. That means that the maximal top-1 RR that the methods can achieve is 99% and 99.5% in *fa-fb* and *fa-fc*, respectively. Table 2 shows the obtained top-1 RR, for two different face image sizes. From these results we can conclude: (i) In the *fa-fb* experiments, the best results were obtained by the H-X-80 and SD-X methods, using 203x251 size images (~95%). Second best results were obtained by

Table 2. FERET *fa-fb* and *fa-fc* tests. Top 1 recognition rate. Experiments were performed with automatically detected eyes.

Method	100 x 185		203 x 251	
	fa-fb	*fa-fc*	*fa-fb*	*fa-fc*
H-HI-10	88.9	8.8	91.9	29.0
H-MSE-10	88.9	8.8	91.9	29.0
H-XS-10	89.0	9.9	93.0	28.0
H-HI-40	85.2	20.7	82.2	39.4
H-MSE-40	85.2	20.7	82.2	39.4
H-XS-40	84.1	18.7	84.5	43.0
H-HI-80	**92.7**	30.1	**94.4**	52.3
H-MSE-80	**92.7**	30.1	**94.4**	52.3
H-XS-80	**92.6**	32.1	**95.5**	65.8
PCA-MSE-200	60.6	47.7	---	---
PCA-MSE-500	64.0	**51.8**	---	---
GJD-BC-G	73.8	40.4	87.6	**74.6**
GJD-BC-L1	72.9	42.0	62.3	36.3
GJD-BC-L2	74.1	43.0	66.0	37.3
SD-FULL	73.4	9.8	**95.4**	64.5
SD-SIMPLE	73.0	8.3	**95.9**	62.9

H-X-10 methods (~93%). In third place came GJD-X-X methods (~87%); (ii) In the *fa-fc* experiments, best results were obtained by the GJD-BC-G method (74.6%) using 203x251 image sizes. Second best results were obtained by H-XS-80 and SD-X methods; (iii) For most of the methods, better results were obtained when using 203x251 images. Larger variability in the results with the size of the images was observed in *fa-fc* experiments; (iv) Interestingly, in the *fa-fc* experiments, when using 100x185 images, the best results were obtained by the PCA methods.

3.3 Computational Performance

As already mentioned, in HRI applications it is very important to achieve real-time operation. In addition, the memory required by the different methods is very important in some mobile robotics applications where memory is an expensive resource. Table 3 shows the computational and memory costs of the different methods under comparison when images of 100x185 are considered. For the case of measuring the computational costs, we considered the feature-extraction time (FET) and the matching time (MT). In the case of measuring memory costs, we considered the database memory (DM), which is the required amount of memory for having the whole database (features) in memory, and the model memory (MM), which is the required amount of memory for having the method model, if any, in memory (PCA matrices

Table 3. Computational and memory costs. FET: Feature Extraction Time. MT: Matching Time. PT: Processing Time. DM: Database Memory. MM: Model Memory. TM: Total Memory. Time measures are in milliseconds, memory measures are in Kbytes. DB sizes of 1, 10, 100 and 1,000 faces are considered. An image size of 100x185 pixels is considered.

Method	FET	MT	PT (FET+MT)				DM	MM	TM (DM+MM)			
			1	10	100	1000			1	10	100	1000
H-X-10	15	0.11	15	16	26	120	11	0	11	110	1100	11000
H-X-40	15	0.29	15	18	44	305	41	0	41	410	4100	41000
H-X-80	15	0.42	15	19	57	435	80	0	80	800	8000	80000
PCA-MSE-200	170	0.02	170	170	172	190	0,8	137800	137801	137808	137878	138585
PCA-MSE-500	360	0.02	360	360	362	380	2	137800	137802	137820	137996	139757
GJD-BC-G	50	0.25	50	53	75	300	33	1240	1273	1572	4559	34427
GJD-BC-L1	35	0.18	35	37	53	215	22	1240	1262	1462	3465	23490
GJD-BC-L2	40	0.20	40	42	60	240	26	1240	1266	1498	3818	27021
SD-X	4.7	1.03	6	15	108	1036	428	0	428	4284	42845	428451

for the PCA case, and filter bank for the Gabor methods). We show the results for databases of 1, 10, 100 and 1000 individuals (face images).

If we consider that in typical HRI applications the database size is in the range 10-100 persons, the fastest methods are the H-X-X ones. The second fastest methods are the GJD-X-X ones. To achieve real-time operation with a database of 100 or fewer elements, all methods are suitable, except PCA-based methods. In databases of 10-100 individuals, H-X-X and GJD-X-X are suitable since they require less than 8 MByte of memory (they do not need to keep a model in memory). In the case of H-X-X methods, the required memory increases linearly with the number of partitions.

4 Comparative Study Using a Real HRI Database

We built the UCHFaceHRI database for comparing face analysis methods in tasks such as face detection, face recognition, and relative pose determination using face information. The database contains images from 30 individuals which were taken under several relative camera-individual poses (see Figure 2), in outdoor and indoor environments. Five different face expressions were considered for the case of the frontal face. For the experiments reported here the following tests were carried out: D-I: Distance Indoor and D-O: Distance Outdoor, using images acquired in points P10, P11 and P12 of the distance tests (fig 2.a). D-I* and D-O*, same as D-I and D-O, but using only images acquired in P12. R-I: Rotation-Indoor and R-O: Rotation-Outdoor, using images acquired in points P4, P5 and P6 of the out-of-plane rotation tests (Figure 2.b). E-I: Expression-Indoor and E-O: Expression-Indoor, using images containing 5 different face expressions, acquired at point P12 of the distance tests (Figure 2.a). Table 4 shows the top-1 recognition rates obtained in these tests. Main conclusions are: (i) In the D-I tests most of the methods show a similar performance,

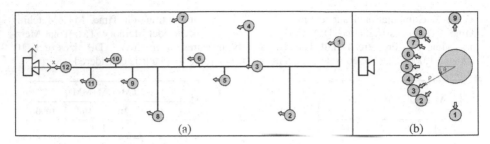

Fig. 2. Experimental setup for image acquisition at different distances (a) and angles (b). Arrows indicate the angular pose of the subjects. Distance/angle is indicated in centimeter/degrees. (a) Cartesian coordinates of acquisition points: P1 (1088,90), P2 (906,180), P3 (785,0), P4 (755,151), P5 (665,-51), P6 (574,30), P7 (514,181), P8 (423,-181), P9 (332,-61), P10 (272,30), P11 (181,-61), and P12 (90,0). (b) Polar coordinates of acquisition points: P1 (90,90°), P2 (90,45°), P3 (90,30°), P4 (90,15°), P5 (90,0°), P6 (90,-15°), P7(90,-30°), P8(90,-45°), and P9 (90,-90°).

Table 4. UCHFaceHRI tests. Top 1 recognition rate. Experiments are performed with detected eyes.

Method	100 x 185							
	D-I	**D-I***	**D-O**	**D-O***	**E-I**	**E-O**	**R-I**	**R-O**
H-HI-10	65.6	93.3	33.3	56.7	**95.0**	52.0	79.3	60.9
H-MSE-10	65.6	93.3	33.3	56.7	**95.0**	52.0	79.3	60.9
H-XS-10	**72.2**	93.3	34.4	46.7	92.5	50.0	79.3	56.3
H-HI-40	63.3	90.0	32.2	53.3	77.5	40.7	49.4	29.9
H-MSE-40	63.3	90.0	32.2	53.3	77.5	40.7	49.4	29.9
H-XS-40	61.1	90.0	33.3	53.3	76.7	40.7	55.2	33.3
H-HI-80	68.9	**96.7**	**43.3**	66.7	**93.3**	**64.0**	**81.6**	**72.4**
H-MSE-80	68.9	**96.7**	**43.3**	66.7	**93.3**	**64.0**	**81.6**	**72.4**
H-XS-80	**74.4**	**96.7**	42.2	60.0	90.0	45.3	77.0	66.7
PCA-MSE-200	67.8	93.3	27.8	36.7	58.3	26.0	51.7	33.3
PCA-MSE-500	68.9	93.3	33.3	50.0	62.5	30.0	51.7	35.6
GJD-BC-F	53.3	**100.0**	41.1	**76.7**	76.7	50.7	65.5	46.0
GJD-BC-L1	50.0	93.3	36.7	**70.0**	72.5	41.3	55.2	40.2
GJD-BC-L2	53.3	93.3	38.9	**73.3**	74.2	48.7	63.2	43.7
SD-FULL	66.7	83.3	7.8	10.0	75.0	4.7	67.8	4.6
SD-SIMPLE	66.7	83.3	8.9	10.0	75.8	6.0	66.7	9.2

except for the case of GJD-X-X. The reason seems to be the small size of the images obtained in P10 and P11. When these two locations are not considered (D-I* tests), the performance was increased largely by all methods. The best performance was obtained by GJD-BC-G and H-X-80 methods; (ii) When the same experiments are carried out in outdoor conditions, with variable sun light (D-O and D-O*), the performance decreases greatly, showing that all the methods are sensitive to irregular illumination. Under these conditions, the best results are obtained again by GJD-X-X (70%-76.7%) and H-X-80 (60%-66.7%) methods. Interestingly, the worse performance is shown by SD-X methods, showing that SIFT methods are very sensitive to outdoor illumination conditions; (iii) In the expression tests, the methods show a similar behavior to those in the occlusions tests (Table 1), because in both cases the face structure is altered. Best performance, in both indoor and outdoor conditions, is achieved by H-X-10 and H-X-80. In indoor conditions, both methods are almost insensitive to face expressions; and (iv) In the out-of-plane rotations tests (R-I and R-O), all methods show sensitivity to the rotation angle. The performance decreases in all cases. Acceptable results are only obtained by H-X-10 and H-X-80 methods (~80%).

5 Conclusions

In this article, a comparative study among face-recognition methods for HRI applications was presented. The analyzed methods were selected by considering their suitability for HRI use, and their performance in former comparative studies. The comparative study includes aspects such as variable illumination, facial expression variations, face occlusions, and variable eye detection accuracy, which directly influences face alignment precision.

It is difficult to select the best method, because the same method did not always obtain the best results in each test. However, we can say that H-X-80 methods show a high performance in most of the tests, being robust to alignment errors, face occlusions, expression variations and out-of-plane rotations, and showing an acceptable behavior in variable illumination conditions. In addition, their computational and memory requirements are adequate for HRI use. A second interesting family of methods is the GJD-X-X, which shows excellent performance under variable illumination.

As future work, we would like to extend this study by considering other methods and performing a deeper analysis of the results.

References

1. Sinha, P., Balas, B., Ostrovsky, Y., Russell, R.: Face Recognition by Humans: 19 Results All Computer Vision Researchers Should Know About. Proc. of the IEEE 94(11), 1948–1962 (2006)
2. Zhao, W., Chellappa, R., Rosenfeld, A., Phillips, P.J.: Face Recognition: A Literature Survey. ACM Computing Surveys, 399–458 (2003)
3. Tan, X., Chen, S., Zhou, Z.-H., Zhang, F.: Face recognition from a single image per person: A survey. Pattern Recognition 39, 1725–1745 (2006)
4. Chellappa, R., Wilson, C.L., Sirohey, S.: Human and Machine Recognition of Faces: A Survey. Proceedings of the IEEE 83(5), 705–740 (1995)
5. Turk, M., Pentland, A.: Eigenfaces for Recognition. Journal of Cognitive Neurosicence 3(1), 71–86 (1991)
6. Face Recognition Homepage (January 2008), http://www.face-rec.org/

7. Ruiz-del-Solar, J., Navarrete, P.: Eigenspace-based face recognition: a comparative study of different approaches. IEEE Transactions on Systems, Man and Cybernetics, Part C 35(3), 315–325 (2005)

8. Zou, J., Ji, Q., Nagy, G.: A Comparative Study of Local Matching Approach for Face Recognition. IEEE Transactions on Image Processing 16(10), 2617–2628 (2007)

9. Phillips, P., Flynn, P., Scruggs, T., Bowyer, K., Chang, J., Hoffman, K., Marques, J., Min, J., Worek, W.: Overview of the Face Recognition Grand Challenge. In: Proc. of the IEEE Conf. Computer Vision and Pattern Recognition – CVPR 2005, vol. 1, pp. 947–954 (2005)

10. Fröba, B., Ernst, A.: Face detection with the modified census transform. In: 6th Int. Conf. on Face and Gesture Recognition - FG 2004, Seoul, Korea, pp. 91–96 (2004)

11. Ahonen, T., Hadid, A., Pietikainen, M.: Face recognition with local binary patterns. In: Pajdla, T., Matas, J(G.) (eds.) ECCV 2004. LNCS, vol. 3021, pp. 469–481. Springer, Heidelberg (2004)

12. Ruiz-del-Solar, J., Loncomilla, P., Devia, C.: A New Approach for Fingerprint Verification based on Wide Baseline Matching using Local Interest Points and Descriptors. In: Mery, D., Rueda, L. (eds.) PSIVT 2007. LNCS, vol. 4872, pp. 586–599. Springer, Heidelberg (2007)

13. Mian, A.S., Bennamoun, M., Owens, R.: An Efficient Multimodal 2D-3D Hybrid Approach to Automatic Face Recognition. IEEE Trans. on Patt. Analysis and Machine Intell. 29(11), 1927–1943 (2007)

14. Phillips, P.J., Wechsler, H., Huang, J., Rauss, P.: The FERET database and evaluation procedure for face recognition algorithms. Image and Vision Computing J. 16(5), 295–306 (1998)

15. Verschae, R., Ruiz-del-Solar, J., Correa, M.: A Unified Learning Framework for object Detection and Classification using Nested Cascades of Boosted Classifiers. Machine Vision and Applications (in press)

16. Lowe, D.: Distinctive Image Features from Scale-Invariant Keypoints. Int. Journal of Computer Vision 60(2), 91–110 (2004)

17. RoboCup @Home Official Website, http://www.robocupathome.org/

xROB-S and iCon-X: Flexible Hardware, Visual Programming and Software Component Reuse

Stefan Enderle[1], Wolfgang Guenther[2], Hans-Juergen Hilscher[3], and Holger Kenn[4]

[1] KTB Mechatronics
enderle@qfix.de
[2] Hochschule Anhalt (FH)
w.guenther@emw.hs-anhalt.de
[3] Hilscher GmbH
hansjuergen@hilscher.com
[4] Europäisches Microsoft Innovation Center
Holger.Kenn@microsoft.com

Abstract. The following article describes an optimized system solution for Ro-boCup-applications, which reaches from RoboCup-Junior to humanoid robots.

It describes the versatile multiprocessor-hardware xROB-CUx. The control unit contains the communication processor netX®, based on an arm9-Core. A variety of communication- and hardware-interfaces as well as the simple scaling enables the use as a communication junction, e.g. for sub-layered DSP or FPGA-axis-controllers in humanoid robots or as a control unit for construction systems (e.g. fischertechnik®, Lego®, qfix®).

The paper also contains a description of the Windows® based software-system xROB-L , which contains the open visual programming-system iCon-L® as an essential part. The creation of user-programs, PC-simulation, data-archiving and the download to different target-systems is illustrated as well.

While the up to now available developer-environment iCon-MFB (Make Function Blocks) only supported the design-process of advanced function blocks in Visual C/C++, it is now possible to integrate ANSI-C written function blocks via the GNU-Compiler with the tool xROB-C. The main advantage to other Solutions is the Vision system, the relation of cost to performance and the wide area of possible applications. Because of the easy to use visual programming technology it can be used by all age groups.

1 Introduction

In the last years, Robocup, the Robot World Cup Soccer initiative has established itself as a standard benchmark for research in artificial intelligence and robotics. With the extension towards Robocup Rescue, the structured environment of the robot soccer games has been extended towards unstructured environments. At the same time, Robocup Junior has become a major event that enables young people to join the research community. What all robot leagues have in common is that the design of the actual robots and their control hard- and software are critical for team success. In some leagues such as the 4-legged-league, a common standard platform is used, but in

L. Iocchi et al. (Eds.): RoboCup 2008, LNAI 5399, pp. 485–494, 2009.

most leagues, robots and their control system are based on individual design. In some cases, the same control hardware can be used for different leagues such as the IUB 2002 Rescue and Small-Size league team [1,2] that used the CubeSystem [3] or RoboCup Junior teams who use qfix [19] for soccer, dance and rescue robots. However, one drawback of this particular hardware is its specific focus on mobile robots using wheels for locomotion. The control of more complex robot systems like humanoid robots is beyond the capabilities of this system.

In recent years, the introduction of new leagues such as the humanoid league has led to an increase in complexity of the robot systems used, both in hardware and software. One way to cope with this complexity is the use of well-tested software and hardware components that speed up the design process. Based on the CubeSystem, it has also been shown that between these different leagues, code reuse based on software components is possible [4], even by non-experts like participants of Robocup Junior.

The presented robot control system xROB-CU1 can especially be used for the RoboCup leagues Junior, ELeague, Small-Size, Rescue and Humanoid e.g. by dealing as a control unit for construction systems like. fischertechnik®, Lego® or qfix®.

In practice, the design of the mobile robot is connected with the development of mechanical structures, actuators, control systems (hardware) and control algorithm (software) etc. These subsystems often are exchangeable during the development process. To speedup this development process it will be helpfully to use automation tools, which simplify the design flow. xROB-L is based on a modified version of the open visual programming system iCon-L® [12]. xROB-C allows the easy integration of function block written in C language for the robot control system xROB-CU1.

A lot of applications for this flexible robot control system are within the education and hobby market. In this case standard interface adapter cable (i.e. for fischertechnik) can be used as much as possible. As a xROB-CU1 features several serial interfaces it can be used on medium size robots in combination with a notebook and/or in a network of xROB-CU1 devices.

The paper is structured as follows. Section two gives an overview of the features of the xROB-CU1 Hardware. In section three, the visual programming system xROB-L is introduced and its component structure is illustrated. Section four gives some preliminary results obtained by using the Hardware xROB-CU1 and the Software xROB-L and the C-Interface tool xROB-C for programming a soccer Robot. A more advance control system xROB-CU8 is used to control the 6-legged Robot ANTON. Section five discusses the current results and indicates future work.

2 xROB-CU1 Hardware Structure

The main design goal of the xROB-CU1 Hardware was to create a system which combines the need of efficient and flexible communication capabilities and a high amount processing power within limited power budget. The xROB-CU1 Hardware is based on a special communication ASIC called netX® [5].

The central element in the netX® is a Data Switch, which links the 32 Bit ARM CPU and the other Bus Masters with the memory and the internal peripherals. Unlike in other, simpler controller designs, it allows simultaneous access to various slave

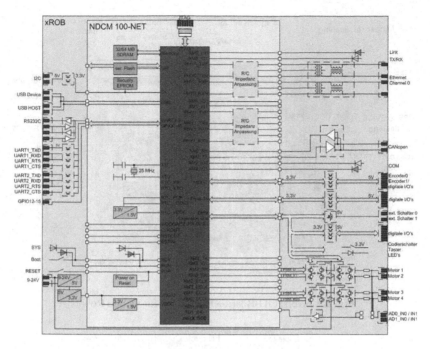

Fig. 1. xROB-CU1 Hardware Structure [15]

ports by the bus masters. The memory controller is a component of the Switch and ensures fast access to the external SDRAM. The internal structure of the communication processor is presented in detail in [5]. In order to free the central CPU from time-consuming IO operations, the ASIC contains a number of additional on-chip devices, the so-called xMAC (extended media access controller) and xPEC (extended protocol execution controller) devices. xMAC and xPEC form a realtime protocol engine that can be programmed with micro code to execute high-level tasks such as sensor/actuator signal processing and transmission without any intervention by the main CPU.

For instance, commands that are received via Realtime-Ethernet can be forwarded to a PWM control register while the current position of an encoder register can be transmitted.

In the in figure 1 shown xROB-CU1 Hardware structure the CPU is mounted on a netX DIMM module that additionally contains RAM and ROM. The hardware has a number of IO interfaces such as RS232, USB and Industrial Ethernet, four PWM outputs that can be used to control RC servos, 4 analog inputs and 24 digital I/O pins. Disadvantageously only 4 from the 8 internal ADU-Channels are connected to the plug of the NDCM 100-NET module.

As peripherals there are all the required functions such as Interrupt controller, Timer, Realtime-Clock, SPI and I^2C-Interfaces available. Three serial and one USB 1.1 interfaces, which can be driven as Host or Device are available for communication. Further function units are the ETM cell for setting Breakpoints and Traces of executed commands as well as a JTAG interface for debugging and for hardware testing.

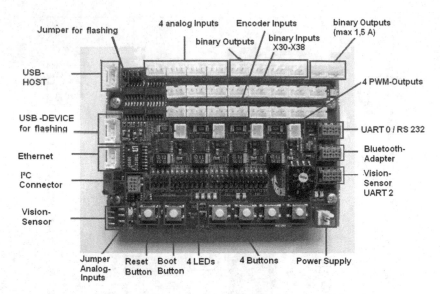

Fig. 2. Universal IO-Board XROB-CU1 for mobile Robots

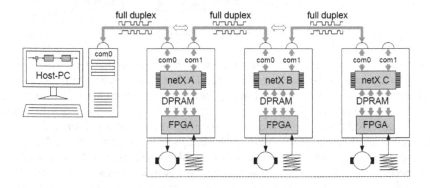

Fig. 3. xROB-CU8 Control Structure for 24 Joints via PC [15]

The xROB-CU8 Hardware structure is described in [15]. It uses the netX-50, a downsized version of the netX-500 and FPGAs to compute the control task for 8 joints within a cycle time less then 50 μs. Actual the xROB-CU8 is in prototype state. Existing documentation is insufficient for reuse of this sophisticated 8-joint-control unit. The communication between the PC and the tree xROB-CU8 is realized via EtherCAT with a cycle time less than 1 ms and a jitter less than 200 ns.

3 Software Structure

As the CPU core is based on the ARM9 CPU core, it can run system software for the ARM platform such as Linux and Windows CE. However, for realtime applications such as robotics, the use of a general purpose operating system has drawbacks such as the limited predictability of memory, network and scheduling. Therefore, a minimal realtime kernel has been implemented that is well adapted to the hardware of the communication processor. The so-called "realtime communication system for netX", short "RcX" [5] is a modularized preemptive multitasking operating system that implements a rich set of IPC mechanisms. Its core is quite small, using 24k of ROM and 6k of RAM, therefore the RcX core could be integrated into the boot ROM of the netX.

3.1 The Runtime System

The user software is a pointerlist, consisting of the functionblock calls and the parameters (see figure 4). To generate this pointerlist, the IDE has to read the device description file (see figure 5).

For the use of the visual programming system it is a precondition, that the runtime environment is ported to the actual target system [12]. First step is to develop a serial interface between the visual programming system and the target system. In this case the prototype solution was to implement the upload of the device description file and the download of the user software (pointer list) via RS232. The porting of the runtime environment to netX and RcX has been described in [7].

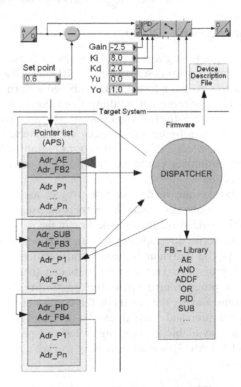

Fig. 4. The virtual pointer machine [8]

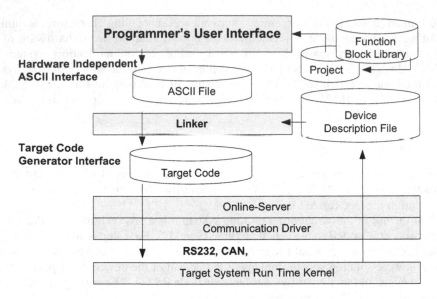

Fig. 5. Communication structure between Programmers User Interface und Target System

3.2 Visual Programming with xROB-L

iCon-L® is a visual programming and control system. The advantages and disadvantages of visual Programming are described in [9, 10, 11]. Graphical function blocks are placed in the different program levels and combined to form the user program.

A function block (FB) can represent anything from simple I/O to a complex control algorithm or even a graphical display. Function blocks are placed in the program as the user desires. As shown in figure 4 the blocks are connected together with 'wires' to form the program. Knowledge of classical programming techniques is not needed. Existing programs or subroutines can be easily tailored to meet changed specifications or be put into new functions.

Function block libraries are loaded with a project as needed. Programming of Function blocks may be done separately from the end user program in the development system iCon-MFB (Make function blocks) or in the new designed tool xROB-C. The overview of the powerful development system iCon-MFB is available on [12].

Function blocks may be individually tested and verified before use in the actual control or simulation programs. This division between function block programming and the actual control programming opens the possibility to anyone being able to make a control program, leaving the details of the control algorithm to the specialists. Function blocks, once written and verified, can be used and reused in any program.

An OPC-Interface is available to allowing Windows programs to have complete control over an target system. This is an easy way to give a windows program graphical control over a system.

3.3 Development of new Libraries for the xROB-CU1

The visual programming system provides a variety of visual function blocks for programming, which are included in libraries.

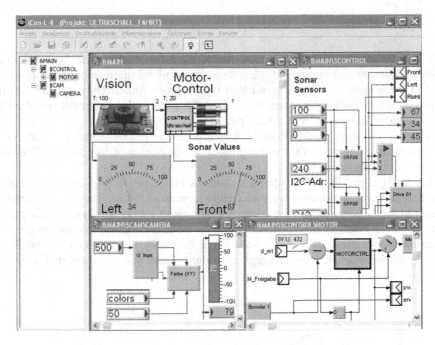

Fig. 6. Visual programming and debugging of a user program [15]

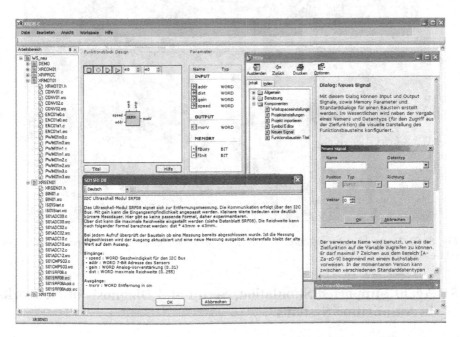

Fig. 7. Design of the Interface for a function Block, written in C

4 Applications

The robot control system xROB-CUx can be used for the RoboCup leagues Junior, Eleague, Rescue and Humanoid. First steps by using of the results of the described project were done in [6]. The team Red Devils from Anhalt University of Applied Science (Junior-Socker 1 against 1) has used the xROB-CU1 controller board for the qualification tournament in Magdeburg (2007) for the first time.

The figure 8 shows, that xROB-CU8 Hardware can be connected in EtherCAT topology, to realize control of the walking robot, which consist of 24 actuators, distributed within the robot's body.

Currently the additional requirements such as vision processing, environment sensing, and adaptive locomotion are preferred in humanoid robots. Thus, the software for sensor signal processing and for control of humanoid robots will become a nontrivial issue as more functionality is expected. Therefore it is desirable that the programs or function blocks created for the robots, could be easily interconnected, modified or configured for example within a graphical programming system and ported to the new generations of hardware.

Fig. 8. Six-Legged Walking Robot of University of Magdeburg with advanced walking capabilities [13, 15]

5 Agent Supported Simulation, Setup and Diagnosis of Networks in Automation Systems

This project is currently in the planning-stage (Fig. 9). The tool xROB-C, which has been developed in the xROB-project, will be used to simplify the development process for Plant Simulation libraries.

First step is a solution to train students of electrical engineering on an HIL-simulation system. With the netX-Starterkit it is possible to simulate the ET200 of the S7-System and a virtual plant (Fig. 8). When an application program is thoroughly verified on the target system, the PC may be removed and the system can run alone. The same program that has been developed and simulated can be downloaded to the target system and debugged [16].

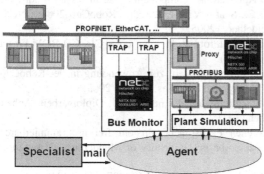
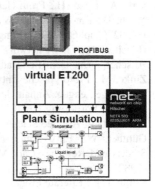

Fig. 9. Data Mining and Plant Simulation   **Fig. 10.** Plant Simulation

6 Conclusions

The robot control system xROB-CUx is scalable and powerful. Visual programming technology in xROB-L offers a graphical, object oriented development and control System to different robots. Actual the main use of visual programming is the entire automation and process control industry. Benefits include ease of use, high software reliability, cost savings, and hardware independence.

xROB-L bridges the gap between simulating and actual control of a real system for mobile robots. The same program that has been developed and simulated can be downloaded to the target System and debugged. When a program is thoroughly verified on the target System, the PC may be removed and the system can run alone. If actual test data is needed, simply connect the PC again and read archived data.

In the field of robotics there exist some limitations of visual programming. With the described very easy to use tool xROB-C it is possible to combine visual and textual programming.

The team Red Devils (Junior-Socker 1 against 1) has used the xROB-CU1 controller board for qualification tournament in 2007 in Magdeburg for the first time.

References

1. Birk, A., Kenn, H., Rooker, M., Agrawal, A., Balan, V., Burger, N., Burger-Scheidlin, C., Devanathan, V., Erhan, D., Hepes, I., Jain, A., Jain, P., Liebald, B., Luksys, G., Marisano, J., Pfeil, A., Pfingsthorn, M., Sojakova, K., Suwanketnikom, J., Wucherpfennig, J.: The IUB 2002 Rescue League Team. In: RoboCup 2002: Robot Soccer World Cup VI (2002)

2. Birk, A., Kenn, H., Rooker, M., Agrawal, A., Balan, V., Burger, N., Burger-Scheidlin, C., Devanathan, V., Erhan, D., Hepes, I., Jain, A., Jain, P., Liebald, B., Luksys, G., Marisano, J., Pfeil, A., Pfingsthorn, M., Sojakova, K., Suwanketnikom, J., Wucherpfennig, J.: The IUB 2002 Smallsize League Team. In: Proceedings of the RoboCup 2002: Robot Soccer World Cup VI (2002)

3. Birk, A., Kenn, H.: Walle, Th. On-board Control in the RoboCup Small Robots League. Advanced Robotics Journal 14(1), 27–36 (2000)

4. Birk, A., Kenn, H.: From Games to Applications: component reuse in rescue robots. In: Nardi, D., Riedmiller, M., Sammut, C., Santos-Victor, J. (eds.) RoboCup 2004. LNCS, vol. 3276, pp. 669–676. Springer, Heidelberg (2005)

5. Hilscher, H.-J.: netX networX on chip. Alle Informationen auf einer CD. CD-ROM Firma Hilscher, Firmenschriften Hattersheim (2008)

6. Zinke, M.: Prototypische Entwicklung einer Hardware zur Anwendung in der RoboCup Small Size League. Anhalt University of Applied Sciences Köthen (2005)

7. John, U.: Portierung des iCon-L Laufzeitkerns auf die netX CPU. Diplomarbeit. Anhalt University of Applied Sciences. Köthen (2005)

8. Günther, W.: Entwicklung einer offenen Fachsprache für ein frei programmierbares Steuer- und Regelgerät auf Mikroprozessorbasis. University of Magdeburg, Dissertation (1988)

9. Knuth, D.E.: Literate Programming. The Computer Journal 27(2), 97–111 (1984)

10. Takayuki Dan Kimura, Juli W. Choi, Jane M. Mack: Show and Tell: A Visual programming Language, in Glinert 90-P&S, 397–404

11. Schiffer, S.: Visuelle Programmierung: Grundlagen und Einsatzmöglichkeiten. Addison-Wesley/ Longman (1998)

12. http://www.pro-sign.de

13. Palis, Dzhantimirov, Schmucker, Zavgorodniy, Telesh. HIL/SIL by development of six-legged robot SLAIR 2. In: 10th Int. Conference on CLAWAR, July 16-18, Singapore (2007)

14. Hilscher, H.J., Günther, W.: Verwendungsnachweis – Teil Sachbericht zum Proinno-II-Teilprojekt; Hard und Firmware zur xROB-Produktpalette, iCon-L-Programmiersystem für xROB. Hattersheim, Dez (2007)

15. RobotsLab at University of Magdeburg,
    http://www.uni-magdeburg.de/ieat/robotslab

16. Damm, M.: Entwicklung eines auf dem netX-Starterkit lauffähigen und über PRFIBUS-DP an die SPS koppelbaren Anlagensimulators für die Ausbildung. Diplomarbeit. Anhalt University of Applied Sciences. Köthen (2008)

Multi-level Network Analysis of Multi-agent Systems

Pejman Iravani

Department of Mechanical Engineering
University of Bath
Bath, BA2 7AY
p.iravani@bath.ac.uk

Abstract. This paper presents a multi-level network-based approach to study complex systems formed by multiple autonomous agents. The fundamental idea behind this approach is that elements of a system (represented by network vertices) and their interactions (represented by edges) can be assembled to form structures. Structures are considered to be at one hierarchical level above the elements and interactions that form them, leading to a multi-level organisation.

Analysing complex systems represented by multi-level networks make possible the study of the relationships between network topology and dynamics to the system's global outcome. The framework proposed in this paper is exemplified using data from the RoboCup Football Simulation League.

1 Introduction

Increasingly, engineers are faced with having to understand and design complex systems. For example, systems such as traffic in order to optimise time and reduce CO_2 emissions, urban transport systems to improve their efficiency, holistic manufacturing systems, health systems, electric power grid, large companies, etc. Traditional systems engineering approaches can not deal with some of the characteristics of complex systems such as emergent behaviour, evolution, co-evolution and adaptation [1]. In order to start understanding complex systems, it is of relevance to develop frameworks which can be used to analyse this phenomena in a systematic manner.

There is no real consensus about what are the essential characteristics that make systems complex. One of the most important characteristics and the focus of this paper is that the behaviour of complex systems is the result of the interactions of its parts. For example, traffic is the result of the interactions among the various vehicles that share the road network, traffic signals, roads, etc. The behaviour of the electric power grid depends on the number of generators connected, the loads, the transmission network, the protection elements, the weather conditions, etc. This is known as *upward causality*, from the interactions of the parts moving upwards to the system's behaviour as a whole.

L. Iocchi et al. (Eds.): RoboCup 2008, LNAI 5399, pp. 495–506, 2009.

Similarly, the behaviour of the whole will have effects on the behaviour of the parts, for example, traffic jams will force drivers to take alternative routes. This is known as *downward causality*. Thus, in complex systems, there is a constant upward-downward causality that drives the dynamics of the system.

Generally the engineer is interested in the system's *macro-dynamics* (dynamics of the system as a whole) rather than on the *micro-dynamics* (dynamics of the parts). For example, engineers will be interested in avoiding traffic jams rather than controlling the speeds and distances among individual vehicles. Unfortunately, the engineer can not directly influence the macro-dynamics of the system, and to make any changes, actuation must be at the micro-level. Therefore, a main research challenge is to understand the interplay between upward and downward causality.

Multi-robot football is an excellent test-bed to study this phenomena as the local interactions of players produce the global outcomes of the game (e.g. attacking, defending), and these global outcomes drive the players' behaviours (e.g. covering, breaking-away from a cover). This paper concentrates on the study of the *multi-level dynamics* of complex systems, and proposes to use a framework based on a multi-level networks to analyse it. The paper is organised in two large sections, in the first half it discusses how complex network can be used to represent and study multi-agent systems, in the second part it sketches the multi-level approach.

2   Network Representation of Robot Football Interactions

Networks can be used to represent a wide variety of systems in which vertices represent elements and edges represent their interactions. For example a network could use vertices to represent web-sites and edges to represent the existence of hyper-links from one to another. They could also be used to represent people as vertices and edges as their friendship relation. In this paper, vertices are used to represent players and edges interactions.

At all times during the simulated football game there are various types interactions occurring. Some interactions involve only players others also involve the environment (positions in the field, the ball and the goals), for example:

- *ClosestTeammate*: two players of the same team are closest to each other.
- *Supporting*: two players of the same team share controlled areas.
- *Cluster*: players from either team are at close distance.

These type of interactions can be defined using Boolean rules and they are considered to exist if the rule evaluates to *true*. For example, the *ClosestTeammate* rule will be true if two players belong to the same team and if the distance between them is the minimum of all the pairwise distances to the other teammate players.

Figure 1 illustrates some of the interactions occurring during the 208th timestep of the 2007 RoboCup final game. It also illustrates their network representation. This paper focuses only on Boolean (existing or non-existing) interactions, thus two vertices will be connected through and edge if the defining rule is *true*. Interaction rules could be extended to have continuous values or directed links, but this is out of the scope of this paper.

Fig.1(a) displays the *ClosestTeammate* interaction as it happens during the game while Fig.1(b) illustrates its representation as a network. Fig.1(c) illustrates the voronoi diagram formed by the players' positions and the limits of the field. This diagram illustrates the areas that are closest to each player, in some sense these can be seen as the areas controlled by the players. Fig.1(d) illustrates the network of *Supporting* interactions which have been defined over the voronoi diagram as follows. If two players of the same team share two or more vertices of the voronoi diagram (their controlled areas are in contact) then the interaction is considered active. For example, the player w_{11} shares two edges of its voronoi area with w_1, w_2, w_3, w_4 and thus is defined to be supporting them (we will use b_i and w_i to refer to black and white player i respectively).

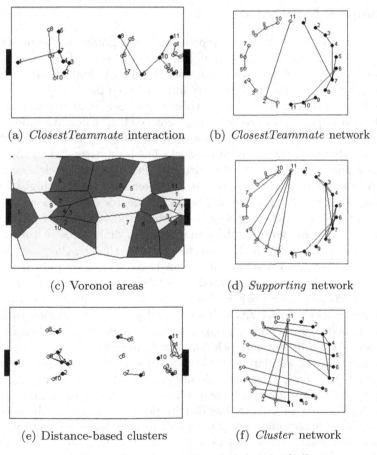

(a) *ClosestTeammate* interaction   (b) *ClosestTeammate* network

(c) Voronoi areas   (d) *Supporting* network

(e) Distance-based clusters   (f) *Cluster* network

Fig. 1. Some interactions in robot football

Figure 1(e) illustrates the *Cluster* interaction which is based on clustering players according to their distances, Fig.1(f) illustrates the network representation. In this particular network, edges between players from different teams can be seen as covering interactions, while edges within the same teams can be seen as support ones. These are just some of the interactions that take place during a football game and can be represented by networks.

The first questions that spring into mind after viewing multi-robot interactions represented by networks are the following:

- Do these networks have any structural patterns or are they random?
- If networks present structural patterns, which ones are beneficial for the team?

In order to address these questions, the next section reviews some of the theory of complex networks specially focusing on how topology affects network functionality.

3   Complex Networks

Mathematically a network can be represented as a graph that consists of two sets, a set of *vertices* $V = \{v_1, v_2, ..., v_n\}$ and a set of *edges* $E = \{e_1, e_2, ..., e_m\}$, such that $V \neq 0$ and E is a set of unordered pairs of elements of V. In other words, vertices are points and edges are links between pairs of points.

Since its inception in 1736, graph theory has been studying the properties of graphs and applied them to solve problems such as calculating flow in pipe networks, distinct colouring of neighbouring areas using minimum number of colours, etc. Recently, the study of networks has been extended to the analysis of complex networks, *i.e.* networks with complex topology which evolve over time [2,5]. Complex topology refers to characteristics such as high clustering coefficient, hierarchical structure and heavy-tail degree distribution.

Network topology refers to the way in which the vertices are connected to each other. There are several values that indicate differences in topology. For example, vertices have different *connectivity degree* that is, they have different number of incident edges, some will be lowly-connected by only a few edges while others may be highly-connected by a large number of edges. The *degree distribution* $P(k)$ of a network is defined as the probability that a randomly selected vertex has connectivity degree k, and gives an indication of how the connectivity is distributed among the vertices of a network. For example, Figure 2 illustrates four different network topologies and their degree distributions. Fig.2(a) illustrates a regular network which has $P(k_i) = 1$ meaning that all vertices have exactly k_i edges. Randomly connected networks Fig.2(c) have a gaussian degree distribution with the peak being the mean value of k. Scale free networks [3] Fig.2(d) have power law distributions meaning that the majority of vertices have few edges and only a few of them have large connectivity degree, this means that some vertices act as *hubs* connecting to many other vertices.

Another important measure that relates to topology is the *average shortest path length*, L, that is the average of the minimum number of edges that separate

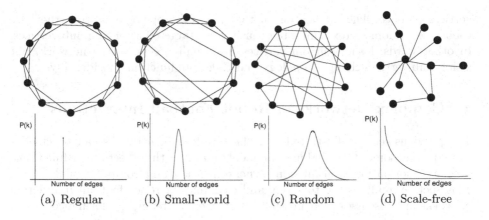

(a) Regular   (b) Small-world   (c) Random   (d) Scale-free

Fig. 2. Different network topologies and their degree distribution

any two vertices, in other words, the average of the shortest paths connecting any two vertices. Equation 1 defines L, where n is the number of vertices in V and d_{ij} is the Euclidean distance between vertices i and j. Networks with small path lengths indicate that is some-how 'faster' to go from one vertex to any other vertex when compared to networks with larger path lengths. This is an important characteristic in networks as it relates to the efficiency of information transmission. A shortcoming with this equation is that L diverges with disconnected vertices, as their distance is infinite. There are two ways to solve this, one is to measure L of the largest connected group, and the other is to use the *efficiency* E measure as defined in Equation 1 where disconnected vertices have a zero effect over the E value.

$$L = \frac{1}{n(n-1)} \sum_{i,j \in V,\ i \neq j} d_{ij} \qquad E = \frac{1}{n(n-1)} \sum_{i,j \in V,\ i \neq j} \frac{1}{d_{ij}} \qquad (1)$$

Watts and Strogatz [4] studied the effects of taking a regular network (Fig.2(a)) and randomly eliminating edges and re-wiring them into random vertices. This led into what is known as the *small-world* topology (Fig.2(b)) which is characterised by having a short L when compared with the same type of network (same number of vertices and edges) with a regular topology. This decrease in L is attributed to some edges that act as 'short-cuts' to otherwise distant parts of the network, allowing to travel long distances within the network in only a few steps. The authors presented various different networks with the small-world topology, including the US power grid and the neural network of the *Caenorhabditis* worm. Some of the implications of the small-world topology over the network's function is that there is an enhanced signal-propagation speed and computational power. From the dynamics viewpoint, small-world networks have enhanced synchronisation capabilities [4].

Scale-free networks [3] (Fig.2(d)) have even shorter average path lengths than small-world networks, this is due to the existence of hubs which connect many vertices at once. The implication of the scale-free topologies is that they are robust to random vertex failure but sensitive to targeted attack on hub vertices. In other words, losing random vertices has small effects on the network, but losing hub vertices changes topology and therefore function considerably.

4   Complex Networks of Robot Football Interactions

The previous section illustrated how the topology of networks can be charac-terised using some statistical measurements and how this affects their functions, such as information transmission, synchronisation and computational power. This section analyses the topology and dynamics of robot football interaction networks and discusses their possible effect over the teams' performance.

4.1   Statistical Analysis of Network Topology

These experiments are conduced over the 2001 to 2007 RoboCup log-files Sim-ulation League finals. Networks are analysed for each time-step for the 7 finals, as each game contains approximately 6000 time-steps, this study is comparing about 42000 networks. For each final, and in order to study the implications of network topology in team performance, the teams have been divided into *winners* (champions of the simulation league) and *losers* (second best team). It should be noted that both finalist teams are high performing, thus only small network differences are expected.

This section studies the *ClosestTeammate*, *Supporting* and *Cluster* interac-tions (explained in Section 2) by calculating the previous topological measurements.

Figure 3(a) illustrates the degree distribution $P(k)$ of the *ClosestTeammate* network. This shows that although the majority of vertices (over 60%) have only $k = 1$ there are few cases in which players have up to $k = 4$ closest neighbours. This means that during the game there are players that act as hubs connecting to many others. This measurement does not show any significant differences be-tween winner and loser teams, and may only be caused by the spatial distribution of players in the field.

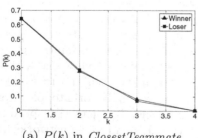
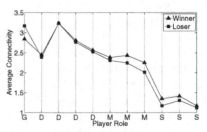

(a) $P(k)$ in *ClosestTeammate*   (b) Average connectivity in *Supporting*

Fig. 3. Degree distributions of two different interaction networks

Table 1. Winner vs Loser statistics

Year	2001		2002		2003		2004		2005		2006		2007	
Result	W	L	W	L	W	L	W	L	W	L	W	L	W	L
E Supporting	0.21	0.19	0.19	0.22	0.19	0.18	0.23	0.20	0.23	0.18	0.19	0.21	0.22	0.19
%FC Supporting	36.2	21.4	18.3	35.9	23.1	21.2	40.9	24.6	45.9	14.6	22.7	39.7	41.3	17.6
#Clusters	12.49		11.92		11.82		10.88		9.66		9.85		9.72	

Figure 3(b) illustrates the average connectivity of the *Supporting* network (Fig.1(d)) with respect to each of the player's role, one goalie (G), four defenders (D), four mid-fielders (M) and three strikers (S). In other words, the graph illustrates the average number of supporting team mates in relation to the player's role. It shows that defenders are the group with highest connectivity while strikers are the ones with lowest, thus, teams tend to play with cohesive defenders and distributed strikers. The graph also shows connectivity differences between winner and loser teams, specially for their mid-fielders and strikers. The average of differences are 7.07% and 8.48% for mid-fielders and strikers respectively. A significance test (ttest) on the data showed over a 99% confidence on the difference of the mean values. This differences indicate that mid-field and striker players in winner teams are better supporting each other, in other words, they better position themselves to make and receive passes (in previous analysis it was shown that the *Supporting* interaction allows for higher probability of successful passes [8]).

Table 1 illustrates some more measurements related to network topology for winner (W) and loser (L) teams. The *E Supporting* row is the average *Efficiency* (Eq.1) of the *Supporting* network. Larger values indicate that the network has improved transmission capability, in the football scenario this implies teams which could pass the ball faster from one part of the field to another. It seems that winning teams have higher E value (5 out of 7 finals). This is not the case for the 2002 and 2006 finals, which indicates that having a good team-connectivity or high E is not sufficient to win or lose a game.

The *%FC Supporting* row indicates the percentage of time that the *Supporting* network is fully connected, or in other words, the times that there is a path in the network that connects all vertices. During the game this shows a path for fast passing the ball from defenders to strikers. The pattern for this measurement is the same as the discussed above and has the same explanation.

The *#Clusters* row indicates the average number of distance-based clusters that appear during each final, it is interesting to observe that this number has been decreasing during the evolution of the League from 2001 to 2007 indicating that teams are playing closer to each other, or in other words, they are involving more players into the active areas of the game.

4.2 Dynamics of Complex Networks

As games progress the interactions among players change and thus do the networks they define. Figure 4 illustrates a sequence from the 2007 final in which

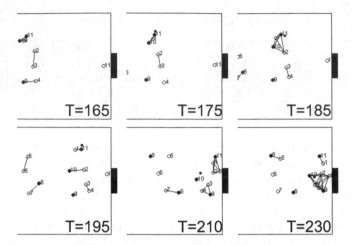

Fig. 4. Goal sequence in 2007 log-file

the winning team scores a goal. In summary, what occurs is that b_{11} dribbles from the mid-field up to the opponent's left-corner (from T=165 to T=195), at this point b_{10} has moved into a good position (being uncovered and in front of the goal). At T=210 player b_{11} passes to b_{10} which then dribbles and scores.

The edges between players in Figure 4 represent the *Cluster* interaction, an edge exists if players are close to each other (distance < 10m). Between players of different teams, this interaction resembles a covering relation, whereas among players of the same team it resembles a support relation. It is important to observe what happens to this interaction between T=185 and T=195, in this period the b_{10} has moved away from its team mate b_{11} but also from the defender w_1 (the interaction with these players at T=195 is inactive). By T=210 b_{10} has broken the interaction with its cover w_2 and its now well positioned to receive a pass from b_{11} and score; this is exactly what happens.

Is it possible to abstract this type of information by studying how networks evolve in time? Figure 5 seems to illustrate some patterns in the dynamics of networks and the game's outcome. It displays the evolution of the *Cluster* interaction during a fraction of the game, between the kick-off (T=0) to the T=500 time-step. The continuous line indicates the number of clusters, the horizontal line indicates the mid-field position and the discontinuous line indicates the ball position. High number of clusters means that players are far from each other in the field, low number of clusters means that players are in close proximity. In this period of time, the winner team scores two goals, at T=247 and T=492.

The figure shows that when the winner team attacks (climbing ball position) the total number of clusters increases. Specially on the build-up towards the goal from T=120 to T=220 and T=330 to T=460. The opposite happens when the loser team is attacking (descending ball position), and the total number of clusters decreases. In other words, when the winner team attacks its players move in such a way that they separate from their covers (as seen in the sequence in Fig.4) increasing the probability of receiving a pass and scoring a goal. This

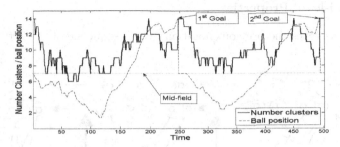

Fig. 5. Evolution of the *Cluster* network

exemplifies that different network dynamics may have different outcomes in the game, obviously one example is not enough to generalise to all games, but it is a good starting point to show the influence of network dynamics.

5 Multi-level Network Analysis Framework

A fundamental idea behind our work is that elements of sets can be assembled under network relations to form structures. Structures exist at a higher level than its elements in a multi-level hierarchy. The framework used to represent and study multi-level structures is defined below, these resembles the work presented in [6] but has some clear differences in the way structures are assembled. In their work, a special process is defined to assemble structures. In our work, structures are defined by the self-organising networks of interactions and thus only interaction rules need to be defined.

Let us assume that certain structures exist at the base level of the hierarchy (Level N), in our example the structures at this level correspond to players and they can be represented in the following manner:

$$\sigma^l_{i,t} = \langle s, r, p \rangle$$

where i is structure number, t is the time at which the structure is observed, l is the level, s is the state of the structure *e.g.* position, orientation, ball-possession; r is a set of interaction rules *e.g.* the *Cluster* interaction rule; and p is a set of properties, *e.g.* the probability of scoring a goal.

At the next level (Level N+1) structures $\sigma^{l+1}_{i,t}$ are formed by the assembly of Level N structures defined by their interaction network. Thus, if a number of Level N structures are interacting, then they assemble into a Level N+1 structure. Level N+1 structures have state, interaction rules and properties but these may be different from the ones defined at the lower level. Properties at Level N+1 can be considered to be *emergent* if they do not exist at Level N, *e.g.* if single players are structures at Level N they can not have the property of passing the ball as this property requires more than one player. The next sections present some of the multi-level analysis conducted over the 2007 RoboCup final game which was won by the black team.

5.1   Multi-level Properties

Figure 6 illustrates the previous notation over a sequence of the 2007 game.

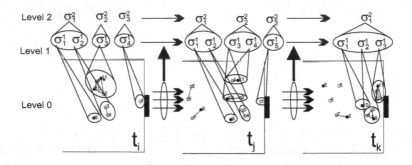

Fig. 6. Multi-level assembly and its dynamics

Let us refer to the base level as Level 0 (L0). Base level structures are defined as: i is the player number; t is the simulator time-step; s is the player's position in the field; r is the *Cluster* interaction rule (active when in close proximity of other players) and p is a property that indicates to what team the structure belongs to. For example, at time t_i player b_{10} generates the following structure:

$$\sigma^0_{b_{10},t_i} = \langle s = position(b_{10}), r = Cluster, p = black \rangle$$

The interaction of L0 structures results in Level 1 (L1) structures, where i is a number from $1, 2, ...m$ with m being the total number of networks[1] defined by the structures at L0. The state s is defined as the centroid of the positions of the L0 structures; r continues being a *Cluster* interaction, this time with threshold (distance < 20m). The property p indicates which team has more players in the structure, $p = W$ more white players, $p = B$ more black players, and $p = X$ same number of players. A $p = W$ structure is beneficial for white and a $p = B$ for black. For example, at time t_i the following is a L1 structure with $p = X$ as it contains as many white as black players:

$$\sigma^1_{3,t_i} = \langle s = centroid(\sigma^0_{b_{10},t_i}, \sigma^0_{b_{11},t_i}, \sigma^0_{w_1,t_i}, \sigma^0_{w_2,t_i}), r = Cluster, p = X \rangle$$

The interaction of L1 structures will generate new structures at Level 2 (L2). L2 structures have s defined as the centroid of L1 structures and p as before. This time, no further interaction rules are defined ($r = \phi$), thus L2 will be the highest on the hierarchy as seen in Figure 6. The following is a L2 structure:

$$\sigma^2_{1,t_i} = \langle s = centroid(\sigma^1_{1,t_i}, \sigma^1_{2,t_i}), r = \phi, p = X \rangle$$

---

[1] We also consider single vertices as networks.

Table 2 shows the structures' properties at different levels in the hierarchy for the 2007 final. At L0 there is the same proportion of structures with the white (W) and black (B) properties, and none with the X. This is because both teams have the same number of players. At L1, structures start to show more interesting properties, with X being the largest followed by B and W. This indicates that most structures are equilibrated (43.4%) but with a slight advantage for the black team. L2 reveals a stronger presence of black structures over white ones (38.4% vs 34.7%).

Table 2. Multi-level p

	W	B	X
Level 0	50%	50%	0%
Level 1	27.7%	28.8%	43.4%
Level 2	34.7%	38.4%	26.7%

This analysis shows how properties have different meanings at different levels, at L0 the property p indicates that both teams have the same number of players. At L1 most structures are in equilibrium $(p = X)$ this is because at this level structures mainly represent local interactions between close-by players, usually by 1-vs-1 interactions. At L2 the proportion of structures with team advantage $(p = W$ or $p = B)$ is larger than the equilibrated ones. This indicates that L2 is more strategic, and in this case, favourable to the black team.

6 Conclusion and Further Work

This paper has presented a framework to study complex multi-agent systems based on representing the system using multi-level networks. The main idea is that agents can be represented as nodes and interaction as edges of a network. Different interactions over the same agents define different networks, for example this paper showed how three different interactions (*ClosestTeammate*, *Supporting* and *Cluster*) result on different networks over the same set of agents.

The first part of the paper reviewed some of the complex network literature, the main purpose was to show that network topology affects network functionality. Some evidence of this was then provided in relation to the RoboCup Simulation League in which slightly better teams display different network topology during the game in comparison with the slightly worse teams. The main evidence was that better teams have slightly higher connectivity and efficiency in the *Supporting* network, specially in the mid-field and attacking positions.

The second part of the paper introduced the multi-level analysis framework. One of the main ideas in the paper was that (sub)networks formed by players and their interactions can be considered to be a structure that exists at a higher hierarchical level. It was argued that structures have different properties at different levels. For example the majority of Level 1 structures have the same number of players from both teams. Level 2 structures start to be more strategic and show how some teams position better their players to gain structural advantage by having one or more players free of cover. This analysis is interesting as many properties of complex system are not visible at low-levels but are the result of the interaction of their parts.

One of the main aims of the RoboCup Coach competition was to have an agent that observes the opposition team and is capable of modelling their behaviour. The basic principle exploited here was to define behaviours as functions know *a-priory* and using statistics to calculate the parameters of the function. For example, the positioning behaviour could be a function of the home-position and ball-position [9], therefore this type of pattern recognition was based on statistical function fitting. We believe that approach presented in this paper could be used to a similar extent in this league with the difference that patterns would not be defined as parametric functions but as relational functions and that multi-level information could be used. The results presented in this paper are based on past final games in which the teams have similar performance (being the best two of that year), the following experiments will be conducted on controlled teams with different capabilities.

References

1. Ottino, J.M.: Engineering Complex Systems. Nature 427, 398–399 (2004)
2. Boccaletti, S., et al.: Complex Networks: Structure and dynamics. Physics Reports 424, 175–308 (2006)
3. Barabasi, A., Bonabeau, E.: Scale-Free Networks. Scientific American, 50–59 (2003)
4. Watts, D.J., Strogatz, S.H.: Collective dyanamics of 'small-world' networks. Nature 393, 440–442 (1998)
5. Albert, R., Barabasi, A.L.: Statistical mechanics of complex networks. Reviews of Modern Physics 74, 47–97 (2002)
6. Rasmussen, S., et al.: Ansatz for Dynamical Hierarchies. Artificial Life 7, 329–353 (2001)
7. Johnson, J.H., Iravani, P.: The Multilevel Hypernetwork Dynamics of Complex Systems of Robot Soccer Agents. ACM Transactions on Autonomous Systems 2, 1–23 (2007)
8. Iravani, P.: Discovering relevant sensor data using Q-analysis. In: Bredenfeld, A., Jacoff, A., Noda, I., Takahashi, Y. (eds.) RoboCup 2005. LNCS, vol. 4020, pp. 81–92. Springer, Heidelberg (2006)
9. Kuhlmann, G., Knox, W.B., Stone, P.: Know Thine Enemy: A Champion RoboCup Coach Agent. In: Proceedings of the Twenty-First National Conference on Artificial Intelligence, pp. 1463–1468 (2006)

A Decision-Theoretic Active Loop Closing Approach to Autonomous Robot Exploration and Mapping

Xiucai Ji, Hui Zhang, Dan Hai, and Zhiqiang Zheng

College of Mechatronics Engineering and Automation,
National University of Defense Technology, 410073 Changsha, China
{jxc_nudt,zhanghui_nudt,haidan,zqzheng}@nudt.edu.cn

Abstract. One of the challenges of SLAM (Simultaneous Localization and Mapping) for autonomous robots is the loop closing problem. In this paper, a decision-theoretic active loop closing approach is presented, which integrates the exploration planning with loop closing. In our approach, the active loop closing process is modeled as a multi-stage decision problem, and a frontier-based auxiliary topological map is build to assist the decision process. The autonomous robot chooses its actions according to the sequential decision results. The unknown range most likely to close a loop is selected to explored, and a particle-filter-based localization and smoothing method applied to partial maps is used in the loop validating and loop constraints building process. Experiments have shown that our approach can practically implement loop closure and obviously improve the mapping precision compared to passive exploration strategy.

Keywords: SLAM, active loop closing, decision theory, particle filter.

1 Introduction

Simultaneous Localization and Mapping (SLAM) is the problem of learning maps of unknowns environment for a robot using the information obtained by sensors mounted on it, and it the key problem for the robot to be really autonomous [1]. A SLAM solution can be divided into at least the following three parts [2]:

- *Incremental SLAM* - The process of building an incremental map from sensor data, and simultaneously localizing itself.
- *Loop Closing* - The process of detecting if and where the robot entering an previously explored area.
- *State Optimization* – An optimization process that calibrates the map and robot localization (all the whole path) according to all the constraints from incremental mapping and loop closing.

It has been shown that SLAM is essential an incremental self-localization process, and the robot position estimation error will increase with the exploration of unknown area [3]. So it's difficult to detect loops by the incremental localization, and the loop closing problem is very critical to an autonomous robot to explore and map a large

L. Iocchi et al. (Eds.): RoboCup 2008, LNAI 5399, pp. 507–518, 2009.
© Springer-Verlag Berlin Heidelberg 2009

environment. If the robot hasn't the ability to close loops in its exploration, an area in the physical environment will correspond to multiple ones in its incremental map, and the map will be topologically inconsistent. On the contrary, if the robot has the loop closing ability, not only the consistent map will be obtained, but also the robot will know which area is truly unknown and explore the whole environment efficiently and reliably.

Some on-line loop closing approaches have been proposed [4-7], and they differ by the estimation algorithms (scan matching, Monte Carlo filter, etc.). Most of them are independent from the exploration strategy, and only when the robot occasionally meets a loop, the loop closing is activated; and some approach even assuming that the loop closing point (where the robot entering an previously explored area) is known. So these methods are *passive*.

In this paper, aiming at detecting and closing loops as soon as possible to reduce the mapping error, an active loop closing approach is presented, which combines the exploration strategy with the loop closing process. In our approach, the loop closing process is modeled as a multi-stage decision problem, and the robot selects its action according to sequential decisions. The robot actively explores the unknown areas more likely to close potential loops. When a potential loop is detected, a naval particle-filter-based globally robot localization algorithm in partial map is used in loop validation and then a path soothing approach is used to build multiple loop constraint. Experiments have shown that our approach can practically implement loop closure, and at the same time obviously improve the mapping precision compared to passive loop closing methods.

2   Decision-Theoretic Architecture for Active Loop Closing

Our active loop closing process is a combination of some sub-processes including exploration planning, loop detecting, validating and constraint building, and so on. We divide the exploration of the robot into two modes: *regular exploration* and *loop closing exploration*. In regular exploration mode, the robot explores the unknown area that most likely to close a potential loop, and at the same time it detects if it's possible that the robot has returned to a previously explored area, which is called *loop detection*. If it has, the robot enters the loop closing exploration mode to validate the potential loop. During the loop closing exploration, the robot needs to validate whether or not there is really a loop by a series of sensing perception, and this process is called *loop validation*. If the robot makes certain that there is a loop, it should then identify where the loop closure is, and create the loop constraints, and this process is *constraint formation*. If it's proven that there isn't a loop at the moment, then the robot returns to regular exploration.

During the loop closing process, the autonomous robot should make the following decisions to determine its actions:

1) *Loop Proposal Decision*

   During the regular exploring, the loop detection should decide whether or not to start a loop closing exploration, and this is the loop proposal decision process, which tests the following tow mutually exclusive hypotheses:

H_0^1 : There is a potential loop at the moment;

H_1^1 : There are no potential loops at all;

If H_0^1 is accepted, a new loop closing exploration is started. Otherwise, H_1^1 is accepted and the robot keeps the regular exploration.

2) *Unknown Area Exploration Decision*

During the regular exploration, there may be many unknown areas available for the robot to explore. In our approach, the unknown area exploration decision needs determine which one to explore allowing for the exploration cost and utility of every area.

3) *Loop Validation Decision*

It's possible that the loop detect process puts forward a spurious loop proposal, so every loop proposal must be checked carefully. The loop validation decision should test the following two tow hypotheses:

H_0^2 : There is a loop at the moment;

H_1^2 : There isn't a loop at all;

It must be very carefully to make this decision, for if it's wrong, the robot will get an inconsistent map. If the robot couldn't determine anyone of the above hypotheses, it maintains the loop exploration for more sensing evidences.

Fig. 1 shows the basic scheme of the active loop closing process. The active loop closing process is driven by the results of sequential decision processes. At the beginning of every time step of the regular exploration, the loop proposal decision process determines whether or not to propose a potential loop closure. If H_1^1 is accepted, after loop closing constraints are constructed, the regular exploration will continue, and the unknown area exploration decision process will select the best area to explore. If H_0^1 is accepted, the loop closing exploration will be activated, and the loop validation

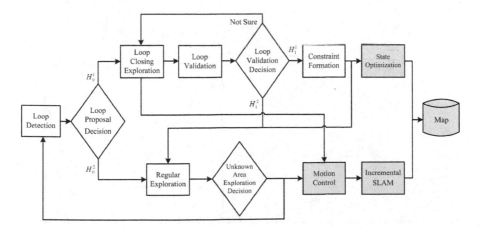

Fig. 1. The flow chart of our active loop closing process. The diamond frames are decision processes, and the shadowed frames are other processes not for loop closing.

process will begin to execute. At every time step of the loop closing exploration, the loop validation decision process determine to accept the loop closure proposal, reject it, or be not sure about it. If it's not sure, the loop closing exploration continues. If the loop closure proposal is accepted, the loop constraint formation process will be activated, and then the robot reenters into the regular exploration mode. When loop closure proposal is rejected, the robot will reenter into the regular exploration mode directly.

3 Active Loop Closing Decision Processes

In this section, we will describe the detailed implements of every decision process in our active loop closing architecture.

3.1 An Auxiliary Topological Map for Decisions

In order to facilitate the decisions in our active closing process, we develop an auxiliary topological map G, which is composed by discrete points of robot's path and the entrances of unknown areas. The entrance of an unknown area is represented by the midpoint of the continuous border between explored free-space and the unexplored area. The entrance are called *frontiers* [8], and denoted by f_i, $i=1,...,N$, where N is the number of frontiers. The set of all frontiers is denoted as $F=\{f_i \mid_{i=1,...,N}\}$. The direction of a frontier is defined as the normal direction of the border from the explored area to the unexplored. To construct G, we initialize it with using the starting location of the robot. The present robot location x_t will be added to G if the distance between x_t and all other path nodes of G exceeds a certain threshold c_1, or if all other path nodes of G are invisible to x_t :

$$G^t =\{x_t \cup G^{t-1}\} \quad \text{if} \quad \forall x_i \in G^{t-1} : \quad [dist(x_t, x_i) > c_1 \quad \vee \quad invisible(x_t, x_i)]. \tag{1}$$

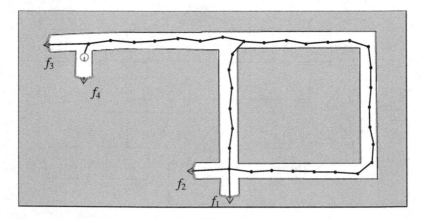

Fig. 2. A sketch of a topological map and frontiers. Three frontiers (*green lines*) are added to the topological map (*black and red points and lines between them*). Frontier f_4 is being explored, so deleted from the topological map.

After x_t is added, an edge between x_t and the nearest and local reachable point x_i is attached to G. Any frontier will be added to G at the moment it was founded, at the same time an edge between the frontier and the nearest path node in G which is visible to the frontier is also attached to G.

Fig. 2 is a sketch map of the topological map and frontiers. In this figure, white area is explored free-space, and gray area is unknown. The black lines between them are obstacle borders. The black points are path nodes, and the lines between them are edges of the topological map. There are four frontiers f_1, f_2, f_3, f_4, indicated by green lines and red points. Because f_4 is being explored, it isn't added to the topological map.

3.2 Loop Proposal Decision

If the robot has returned to a previously explored area along a loop path, it must enter the area from a frontier in the topological map G. Supposing that the present robot position (location and facing) estimation is $\hat{x}_t$, and the error covariance matrix is Σ_t, which can be obtained by most incremental SLAM algorithms. Then $(x-\hat{x}_t)^T \Sigma_t^{-1}(x-\hat{x}_t)$ follows the χ_d^2 distribution of dimension $d = \dim x$. The following set

$$R_x = \left\{ x \mid (x-\hat{x}_t)^T \Sigma_t^{-1}(x-\hat{x}_t) < \delta_{\sigma_1} \right\} \tag{2}$$

is a confidential interval around $\hat{x}_t$ with the confidential level $1-\sigma_1$. We define a set of frontiers in G as

$$I = \left\{ f_i \in G \mid \overline{f}_i \in R_x, dist_G(f_i, \hat{x}_t) > c_2, i = 1,...,N \right\}, \tag{3}$$

where $dist_G(x, y)$ is the distance between the node x and y in G, c_2 is a threshold insuring that f_i is enough far away from $\hat{x}_t$, and $\overline{f}_i$ is the point with the same position and contrary direction with f_i. If $I \neq \varnothing$, there are some frontiers around $\hat{x}_t$, so the robot possibly has entered a previously explored area. Then the hypothesis H_0^1 is accepted, a potential loop is proposed, and the loop closing exploration is activated. Otherwise, the robot continues the regular exploration.

3.3 Unknown Area Exploration Decision

During the regular exploration, there may be many frontiers available for the robots to explore. The method to decide which one is to explore is called *exploration strategy*. Usual methods choose the nearest or biggest frontiers [8]. In our active loop closing approach, the exploration strategy is simple: trying to select the frontier most likely to close a loop, at the same time the exploration cost is also taken into account.

Let $C(i)$ denotes the exploration cost of frontier f_i, and $U(i)$ denotes its loop closing utility. Then our unknown area exploration decision can be expressed as

$$i^* = \arg\max_i [U(i) - C(i)]. \tag{4}$$

The cost and utility of each frontier can be calculated as follows.

Cost: The exploration cost of frontier f_i is given by the minimum path from the robot location to the frontier in the map built by far:

$$C(i) = dist(\hat{x}_t, f_i). \tag{5}$$

The minimum path can be computed efficiently by graph searching methods such as the A^* algorithm or Dijkstra's algorithm.

Utility: In our approach, the loop closing utility of each frontier is evaluated by the possibility to close a loop when the robot explore at that frontier. In our approach, the possibility of a frontier is evaluated by the distance and direction from it to others:

$$U(i) = \max\left\{-k_{ij} \times mdist(f_i, f_j) \mid dist_G(f_i, f_j) > c_3, j = 1, 2, \ldots, N \text{ and } i \neq j\right\}, \tag{6}$$

where, $mdist(f_i, f_j)$ is the Mahalanobis distance of f_i and f_j, which is calculated in a new coordinate system with the direction of f_i as an axis. In order to avoid robots' influence to the local exploration, distance of f_i and f_j in the topological map G must larger than a certain threshold c_3. k_{ij} is a punishment coefficient for the directions of f_i and f_j. If f_i and f_j are in the direction of being closer, then $k_{ij} = 1$. Otherwise, k_{ij} is calculated by the following formula:

$$k_{ij} = \Delta\theta_{ij} / 90°, \tag{7}$$

where, $\Delta\theta_{ij}$ is the minimum angle about which the robot must turn around if it explores ahead through f_i to close a loop at f_j. For example, in Fig.2 $k_{42}=1.0$, $k_{32}\approx2.0$ and $k_{31}\approx3.0$.

As shown in Fig. 2, according to the design of the exploration cost and utility, the unknown area exploration decision process selects the frontier f_4 to explore.

3.4 Particle-Filter-Based Loop Validation Decision

The task of the loop validation decision is to decide whether or not the robot has truly entered a previously explored area using sensor readings. The straightforward approach would be to compare the present local map with the previously built map. However, such an approach could be extremely inefficient since the searching space is too huge. An alternative approach is to localize the robot in the previously built map [5, 6]. If the robot could localize itself well in the previously built map, we believe that the robot has entered into it, and make sure that there is a loop.

The localization problem for mobile robot comes in two flavors: *global localization* and *position tracking*. Since the initial robot position in the previously built map, which denoted as M_e, is not known, the loop validation is a global localization. Recently the *Monte Carlo localization*, also called *Particle Filter*, has been proven to be a practical, efficient and robust global localization approach and been used widely [9]. The loop validation through the global localization in M_e is different from the classical global localization problem. Since at the beginning of the validation, the robot may be not in M_e, it's essential a global localization problem in a partial map and the classical particle filter is incompetent for the loop validation decision.

We have modified the classical particle filter, and put forward a global localization algorithm in a particle map based on the work of Fox [11]. Similar to the implementation of classical particle filter, our algorithm represents the posterior distribution of the robot position at each time step by a set of particles with importance weights: $S_t = \{\langle x_t^{[i]}, w_t^{[i]} \rangle \mid i = 1, \ldots, N_t\}$, where N_t is the number of particles. The algorithm updates the particle set according to the robot movement and sensor readings.

Table 1 outlines the algorithm. Suppose that the loop validation starts at time step 0, and N_0 particles with uniform importance weights are initialized in M_e following the distribution of the robot position estimation by far $p(x_0)$. We don't know when and where the robot enters into M_e, but it must enter M_e through the frontier. At each iteration of the particle filter, a little of εN_t new particles starting nearby the frontier (Step 7) are generate following the distribution of the robot position estimation by far $p(x_t)$. Because there are new particles added at every time step, the algorithm resamples particles at every time step. In case of deleting the particles near the enter point of the robot in the resampling step, the algorithm will keep these new particles unresampled for a fixed time Δt and just adjust their importance weight. So at every time step t the particles can be divided into two parts: new particles generated in time interval $[t - \Delta t, t]$, which are call *frontier particles*, the set of them is denoted as S_t^f and their number is N_t^f. Others are called *inner particles*, their set is denoted as S_t^i and their number is N_t^i. So

$$S_t = S_t^f \cup S_t^i. \tag{8}$$

At time step $t+1$, those frontier particles generated before $t - \Delta t$ in S_{t+1}^f will become inner particles and are put into S_{t+1}^i (Step 5). The algorithm resamples the particles in S_t^i (Step 11), and just adjusts the importance weights of the frontier particles (Step 9).

If a wrong loop validation decision is made, an inconsistent map will be generated, the robot may get lost in the inconsistent map and the exploration task will fail. So the loop validation decision is critical to exploration and mapping of the autonomous robot, and it must be greatly careful to make a decision. When there are no enough sensing evidences to support H_0^2 and H_1^2, the robot should to continue the loop closing exploration. Yet, the robot couldn't take the loop closing exploration ceaselessly, and there should be some means to end such a ceaseless exploration. Supposing the present robot position estimation at the beginning of the loop closing exploration is $\hat{x}_0$, and the estimation error covariance matrix is Σ_0, then the following set:

$$P = \left\{ x \mid (x - \hat{x}_0)^T \Sigma_0^1 (x - \hat{x}_0) \le \delta_{\sigma_2} \right\} \tag{9}$$

is a confidential interval around $\hat{x}_0$ with the confidential level $1 - \sigma_2$ ($\sigma_2 < \sigma_1$). If at time t' the robot position estimation $\hat{x}_{t'} \notin P$, then we believe that if there really is a loop, then the robot must have entered into M_e, no new frontier particles are generated any more, and the loop closing validation decision must be made in limited time.

Table 1. Outline of the particle filter based localization algorithm for loop validation decision

1. Inputs: $S_{t-1}^i = \left\{ \left\langle x_{t-1}^{[j]}, w_{t-1}^{[j]} \right\rangle \mid j = 1, \ldots, N_{t-1}^i \right\}, S_{t-1}^f = \left\{ \left\langle x_{t-1}^{[k]}, w_{t-1}^{[k]}, cnt_{t-1}^{[k]} \right\rangle \mid k = 1, \ldots, N_{t-1}^f \right\}$,

   control information u_t, observations z_t, particle map M_e

2. $S_t^i = \varnothing, S_t^f = \varnothing; \ \Omega_t^i = 0; \Omega_t^f = 0;$ //Initialize

   $N_t^i = N_{t-1}^i; N_t^f = N_{t-1}^f;$

3. for all samples in S_{t-1}^i do   //Generate S_t^i

        $x_t^{[j]} \sim p(x_t \mid x_{t-1}^{[j]}, u_t);$ //Predict

        $w_t^{[j]} = w_{t-1}^{[j]} p(z_t \mid x_t^{[i]});$ //Update importance weights

        $S_t^i := S_t^i \cup \{x_t^{[j]}, w_t^{[j]}\};$

        $\Omega_t^i = \Omega_t^i + w_t^{[j]};$ //Compute the sum of importance weights

4. for all samples in S_{t-1}^f do   // Generate S_t^f

        $x_t^{[k]} \sim p(x_t \mid x_{t-1}^{[k]}, u_t);$ //Predict

        $w_t^{[k]} = w_{t-1}^{[k]} p(z_t \mid x_t^{[i]});$ // Update importance weights

        $cnt_t^{[k]} ++;$ //Count the time step of particle k

5.   if $(cnt_t^{[k]} > \Delta t)$ then   // Change frontier particles generated

            $S_t^i := S_t^i \cup \{x_t^{[k]}, w_t^{[k]}\};$ // before $t - \Delta t$ into inner particles

            $N_t^i = N_t^i + 1; \ \ N_t^f = N_t^f - 1;$ // Update the size of S_t^i and S_t^f

            $\Omega_t^i = \Omega_t^i + w_t^{[k]};$

6.   else

            $S_t^f := S_t^f \cup \{x_t^{[k]}, w_t^{[k]}, cnt_t^{[k]}\};$

            $\Omega_t^f = \Omega_t^f + w_t^{[k]};$

7. for $i = 1 : \varepsilon N_t$ do   // Generate new frontier particle

        $x_t^{[i]} = sample\_near\_frontiers(M_e);$ // Sample near the frontier

        $w_t^{[i]} = p(z_t \mid x_t^{[i]}) / N_t; \ cnt_t^{[i]} = 1;$ // Update importance weights

        $S_t^f := S_t^f \cup \{x_t^{[i]}, w_t^{[i]}, cnt_t^{[i]}\};$

        $\Omega_t^f = \Omega_t^f + w_t^{[k]};$

8. $N_t^f = N_t^f + \varepsilon N_t;$ //Compute the number of particles in S_t^f

9. for all samples in S_t^f do   //Normalize importance weights in S_t^f

        $w_t^{[k]} = w_{t-1}^{[k]} / (\Omega_t^i + \Omega_t^f);$

10. $N_t^i = N_t - N_t^f;$ // Update the size of S_t^i

11. $w_t^i = \Omega_t^i / [(\Omega_t^i + \Omega_t^f) N_t^i];$ //Compute importance weight of every inner

                                   // particle

12. $resample(S_t^i, N_t^i, w_t^i);$ //Resample S_t^i

In the loop validate decision, all inner particles are used to test H_0^2 and H_1^2. The center of all the inner particles is

$$\hat{x}_t = \frac{1}{N_t^i} \sum_{j=1}^{N_t^i} x_t^{[j]} w_t^{\prime [j]}. \tag{10}$$

$w_t^{\prime [i]}$ is an importance weight before resampling. Let Q is a set around $\bar{x}_t$ such that

$$Q = \left\{ x \mid: \parallel x - \bar{x}_t \parallel < \delta_q \right\}, \tag{11}$$

where δ_q is a threshold to control the volume of Q. We make the loop validation decision by using the mean of the importance weights of all the inner particles before resampling and the sum of them after resampling respectively:

$$\bar{w} = \frac{1}{N_Q} \sum_{x_t^{[i]} \in Q} w_t^{\prime [i]} , \quad \Omega_q = \sum_{x_t^{[i]} \in Q} w_t^{[i]} , \tag{12}$$

Where N_Q is number of the inner particles in Q, and $w_t^{[i]}$ is an importance weight after resampling. When $\bar{w}$ and Ω_q are all big enough:

$$\bar{w} > \eta_1 \quad \vee \quad \Omega_q > (1 - \sigma_3) N_t^i / N_t , \tag{13}$$

we will accept H_0^2 and make sure that there really is a loop. After the time step $t' + \Delta t_1$, if H_0^2 is still not accepted and $\bar{w}$ and Ω_q are all small enough:

$$\bar{w} < \eta_2 \quad \vee \quad \Omega_q < (1 - \sigma_4) N_t^i / N_t , \tag{14}$$

we will accept H_1^2 and reject the loop proposal. Otherwise, we believe that there is no enough sensing evidence to make a decision and continue the loop closing exploration. After the time step $t' + \Delta t_2$ ($\Delta t_2 > \Delta t_1$), if neither H_0^2 or H_1^2 is accepted, we will accept H_1^2 and return to regular exploration in case of a ceaseless loop closing exploration. How to set δ_q, η_1, σ_3, η_2, σ_4, Δt_1 and Δt_2 is related to the perception models of sensors and the environments, and can be learned from experiments [9].

Once we have accepted H_0^2, we will simultaneously get a robot position estimation in M_e, and a loop closure constraint is established. We also can use the particle filter as a path smoother to localize the path of the robot M_e and establish multiple constraints for more precise state optimization.

4   Experiments

We used a simulation environment just like the open experiment dataset: *ut_austin_aces* (*http://cres.usc.edu/radishrepository/outgoing/ut_austin_aces3/aces3_publicb.rtl*) described to test our active loop closing method. Because we use a different exploration strategy which results in a different exploration path, the real experiment dataset couldn't be used. Our simulation environment is about 90m×90m, has four adjoining loops, as shown in Fig.4.a, and the robot needs to run about 540m to explore of the whole environment. In the simulation experiment, a simulated laser range finder is

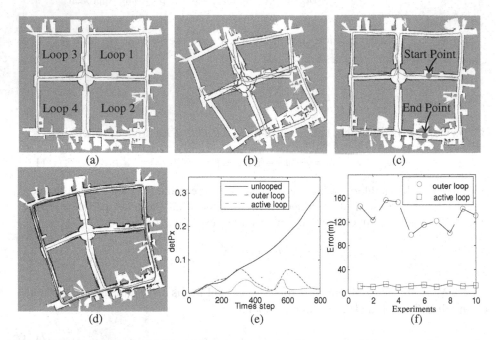

Fig. 3. Experiment result. (a) The map and robot path without error. (b) Incremental map and path. (c) The map and path after active loop closing. (d) The map and path when first explore the outer loop (*read line*). (e) The changing of the determinant of the robot position error co-variance matrix with time. (f) The sums of the absolute robot position error of the whole path in ten experiments.

used, *Histogram Correlation* is used to incrementally build grid maps [11], and a local map is maintained and registered to a global map at intervals.

Fig. 3 shows our experiments results. Fig. 3(a) depicts the map and robot path (*blue line*) without error. According our active loop closing exploration strategy, the robot explores Loop 1 firstly, then Loop 2 and Loop 3, at last Loop 4 is explored. Fig. 3(b) is the incremental map and path without loop closure optimization. Obviously, the map is inconsistent, and can't be used for robot navigation. Fig. 3(c) is the map and path obtained through our active loop closing approach and optimized by the method put forward by Lu etc [4], and Fig. 3(d) is the optimized map and path when the robot explored the big outer loop (*red line*) first and then the middle part. They show that the map obtained by our active loop closing approach is much more precise. Fig. 3(e) compares the changing of the determinant of the robot position error covari-ance matrix with time in deferent situations: incremental mapping, our active loop closing and firstly exploring the outer loop. And Fig. 3 (f) depicts the sums of the absolute robot position error of the whole path in ten experiments by two different exploration strategies: our active loop closing and firstly exploring the outer loop. It shows that the average sum of the absolute robot position error of the later strategy is about 10 times larger than our approach. So our active loop closing approach can improve the mapping precision greatly and efficiently.

In the loop proposal decision, we set $\sigma_1 = 0.05$ to ensure most potential loops can be proposed. In the implementation of the particle filter for the loop validation decision,

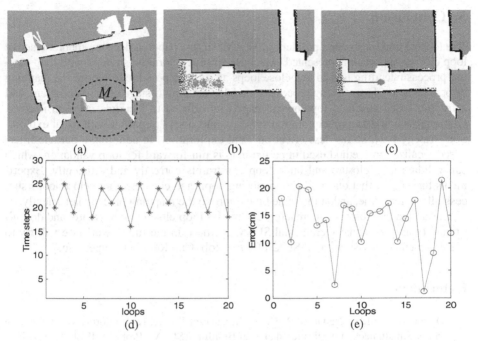

Fig. 4. Experiment results of particle-filter-based loop validation. (a) Particles (*green points*) at the initial time step. (b) Particles (*red points are frontier particles and green ones are inner particles*) when the robot enter M_e. (c) Particles when loop closure was validated. (d) Time steps needed to converge for 20 loops. (e) Position error at the converged points for 20 loops.

1200 particles are used and each iteration step takes less than 0.1s on a notebook PC with 1.73 CPU and 512 Mb memories, so the algorithm can be used in real time. The width of a frontier is set as 0.5m. The ratio of new frontier particles is set as $\varepsilon = 8\%$, and they are kept for two time steps to be not resampled. So there are at most 192 frontier particles. We have found that when ε is more than 4% and less than 12%, the algorithm all could validate an existing loop successfully. When the particles near the center of inner particles is more than 90%, the loop validation will accept H_0^2 to believe that for a certainty a loop is closed. Fig. 4(a) shows the initiation of particles following the distribution of robot position estimation at the beginning of the loop validation. Fig. 4(b) shows the particles at the time when the robot entered into M_e, the red points are frontier particles and the green ones are inner particles. Fig. 4(c) depicts distribution of particles at the time when H_0^2 was accepted.

Because the range of M_e is not too big and we have a prior distribution of the robot localization, especially have a prior knowledge about the facing of the robots, so the particle filter can converge very fast once the robot enter into M_e. Fig. 4(d) shows the time step at which a loop is validated in 20 loop closing processes of 5 experiments. Fig. 4(e) depicts the absolute localization error of at the point where loops were validated, and the error is some large, and this is the common faults of particle-filter-based localization approaches. In order to improve the precision of state optimization, we have saved all the sensor readings during the loop validation process, and used the particle filter as a path smoother to gain multiple constraints.

5 Conclusion

The loop closing is an open problem of SLAM. In this paper, a decision-theoretic active loop closing approach is presented, which integrate exploration strategy with loop closing process intending to find and close loops as soon as possible to reduce the mapping error. The active loop closing process is modeled as a multi-stage decision problem, and a frontier-based auxiliary topological map is introduced to facilitate decision making. Our approach is a macro exploration strategy, and doesn't conflict with usual exploration strategy [12], so has less effect on local exploration. In addition, a particle-filter-based localization method used in partial map is put forward for loop validation, which can validate loop closure and build loop constraints correctly and efficiently. Experiments has shown that our active loop closing approach can detect and close loops successfully in real time, and at the same time improve the mapping precision efficiently.

Our active loop closing approach is essential a loop closing architecture, and doesn't limited to any particular incremental SLAM method. In the future work, we will try to use this method in vision-based SLAM for the RoboCup Rescue Competition.

References

1. Dissanayake, M.W., Newman, P., Clark, S., Durrant-Whyte, H.F., Csorba, M.: A Solution to the Simultaneous Localization and Map Building (SLAM) Problem. IEEE Transactions on Robotics and Automation 17, 229–241 (2001)
2. Andrea, C., Gian, D.T.: Lazy Localization Using a Frozen-Time Filter, http://purl.org/censi/research/2008-icra-fts.pdf
3. Ji, X.-C., Zheng, Z.-Q., Hui, Z.: Robot Position Convergency in Simultaneous Localizaton and Mapping. In: IEEE Intl. Conf. on Mechatronics and Automation, pp. 320–325. IEEE, Harbin (2007)
4. Lu, F., Milios, E.: Globally Consistent Range Scan Alignment for Environment Mapping. Autonomous Robots 4, 333–349 (1997)
5. Gutmann, J.S., Konolige, K.: Incremental Mapping of Large Cyclic Environments. In: Proceedings of the IEEE International Symposium on Computational Intelligence in Robotics and Automation, Monterey, CA, USA, pp. 318–325 (1999)
6. Neira, J., Tardos, J.D., Castellanos, J.A.: Linear Time Vehicle Relocation in SLAM. In: IEEE Intl. Conf. on Robotics and Automation, Taipei, Taiwan, pp. 427–433 (2003)
7. Stachniss, C., Hahnel, D., Burgard, W., Grisetti, G.: On Actively Closing Loops in Grid-Based FastSLAM. Advanced Robotics 19, 1059–1079 (2005)
8. Yamauchi, B.: Frontier-Based Exploration Using Multiple Robots. In: Proceedings of the 2nd International Conference on Autonomous Agents, pp. 47–53. ACM, New York (1998)
9. Thrun, S., Fox, D., Burgard, W., Dellaert, F.: Robust Monte Carlo Localization for Mobile Robots. Artificial Intelligence 128, 99–141 (2001)
10. Fox, D.: Distributed Multi-Robot Exploration and Mapping. In: Proceedings of the 2nd Canadian Conference on Computer and Robot Vision, pp. 15–25. IEEE Press, Los Alamitos (2005)
11. Thomas, R.: Using Histogram Correlation to Create Consistent Laser Scan Maps. In: IEEE Intl. Conf. on Robotics Systems, pp. 625–630. IEEE, EPFL, Lausanne (2002)
12. Gonzalez-Banos, H.H., Latombe, J.: Navigation Strategies for Exploring Indoor Environments. International Journal of Robotics Research 21, 829–848 (2002)

Domestic Interaction on a Segway Base

W. Bradley Knox, Juhyun Lee, and Peter Stone

Department of Computer Sciences,
University of Texas at Austin
{bradknox,impjdi,pstone}@cs.utexas.edu
http://www.cs.utexas.edu/~AustinVilla

Abstract. To be useful in a home environment, an assistive robot needs
to be capable of a broad range of interactive activities such as locat-
ing objects, following specific people, and distinguishing among different people. This paper presents a Segway-based robot that successfully
performed all of these tasks en route to a second place finish in the
RoboCup@Home 2007 competition. The main contribution is a complete description and analysis of the robot system and its implemented
algorithms that enabled the robot's successful human-robot interaction
in this broad and challenging forum. We describe in detail a novel person recognition algorithm, a key component of our overall success, that
included two co-trained classifiers, each focusing on different aspects of
the person (face and shirt color).

1 Introduction

The population distribution of the world's industrialized countries is becoming
denser in the age range that may require assisted living. The U.S. Census Bureau
estimates that from 2005 to 2030, the percentage of the country's population
that is above 85 years of age will increase by 50% [1]. The cost of providing
such human care will be massive. The mean cost of assisted living in the U.S.,
according to a recent assessment, is $21,600 per year[1]. To give adequate care, we
must find ways to decrease the economic cost per retired person. One solution
for the not-so-distant future may be replacing the human home assistants with
robots.

An assistive robot needs to be capable of a broad range of interactions. Among
other capabilities, it must be able to locate and identify common objects; it must
be able to follow people or guide people to places of interest; and it must be able
to distinguish the set of people with whom it commonly interacts, while also
successfully identifying strangers.

RoboCup@Home is an international competition designed to foster research
on such interactive robots, with a particular focus on domestic environments.
In 2007, its second year of existence, RoboCup@Home attracted eleven custom-
built robots from ten different countries and five different continents.

---

[1] www.eldercare.gov/eldercare/Public/resources/fact_sheets/assisted_living.asp

L. Iocchi et al. (Eds.): RoboCup 2008, LNAI 5399, pp. 519–531, 2009.
© Springer-Verlag Berlin Heidelberg 2009

This paper presents the UT Austin Villa RoboCup@Home 2007 entry, a Segway-based robot and the second-place finisher in the competition. The robot demonstrated its ability to complete versions of all three of the tasks mentioned above. The main contribution of this paper is a complete description of the robot system and its implemented algorithms which enabled the robot's successful human-robot interaction in this broad, challenging, and relevant event. A key component of our overall success was a novel person recognition algorithm that included two, co-trained classifiers, each focusing on different aspects of the person (face and shirt color).

The remainder of the paper is organized as follows. Section 2 briefly describes the RoboCup@Home competition. Section 3 introduces the UT Austin Villa robot. Section 5 describes our specific solutions and performance for each task. The final competition results are described in Section 6. Section 7 discusses related work and Section 8 concludes the paper.

Fig. 1. The RoboCup@Home domestic setting

2   RoboCup@Home

RoboCup@Home is an international research initiative that aims "to foster the development of useful robotic applications that can assist humans in everyday life"[2]. In The 2007 Competition, robots in a living room and kitchen environment (see Figure 1) had to complete up to four of six specified tasks. These tasks can be considered fundamental building blocks toward the complex behavior and capabilities that would be required of a fully functional home assistant robot. The specific tasks are described in Table 1.

Table 1. 2007 RoboCup@Home tasks

Task	Description
Navigate	navigate to a commanded location
Manipulate	manipulate one of three chosen objects
Follow and Guide a Human	follow a human around the room
Lost and Found	search for and locate previously seen objects
Who Is Who	differentiate previously seen and unseen humans
Copycat	copy a human's block movement in a game-like setting

Within each task, there were two levels of difficulty. The easier level, called the first phase, existed as a proof of concept and often abstracted away part of the problem (e.g. object recognition or mapping and navigation). The second, more difficult phase of each task was structured similarly to how the task would need to be performed in a real domestic setting. During each phase, there was a ten minute time limit to complete the task objectives.

---

[2] www.ai.rug.nl/robocupathome/documents/rulebook.eps

After the specific tasks, all teams performed a free-form demonstration in what was called the *Open Challenge,* during which they showed off their most impressive technical achievements to a panel of other team leaders. Each event was scored and five teams advanced to the Finals. In the Finals, the five finalists performed demonstrations for trustees of the RoboCup organization, who determined the final standings.

Our robot attempted three of the six possible @Home tasks. These tasks were *Lost and Found, Follow and Guide a Human,* and *Who Is Who.* Each task is described in the following subsections. Our specific approaches to the three tasks are detailed in Section 5.

2.1   Lost and Found

This task tested a robot's ability to find an object that had been "lost" in the home environment. We competed in only the first phase of the *Lost and Found* task. In that phase, a team would hide a chosen object somewhere in the living environment at least five meters from their robot and out of its view. Then the task began. The task ended successfully when the robot had moved within 50 cm of the item and had announced that it found it.

2.2   Follow and Guide a Human

In *Follow and Guide a Human,* a robot followed a designated human as he or she walked throughout the home and then, optionally, returned to the starting position (thus "guiding" the human).

First Phase. In the first phase, a team member led his or her robot across a path determined by the competition referees. The leader was permitted to wear any clothing or markers he chose.

Second Phase. The rules were the same except that the human leader was a volunteer chosen from the audience. Therefore the algorithm needed to robustly identify a person without markers or pre-planned clothing.

2.3   Who Is Who

The *Who Is Who* task tested person-recognition capabilities on a mobile robot. Both phases of the task involved the robot learning to recognize four people, the referees rearranging the people and adding one new person (a "stranger"), and the robot subsequently identifying the four known people and the stranger accurately.

First Phase. In the first phase of the *Who Is Who* task, the four people lined up side-to-side while a robot moved among them and learned their appearances and names. Once the robot finished training, the four people and a stranger were arranged into a new order by the referees. Then, the robot again moved among the people, announcing their names as each was identified. One mistake was allowed.

Second Phase. The second phase was much like the first, but after the robot finishes training, the four known people and the stranger were placed by the referees in various locations around the entire living room and kitchen environment. The robot then had to search them out *and* correctly identify them.

3   The Segway Robot

This section introduces the hardware and software systems of the UT Austin Villa RoboCup@Home 2007 entry, shown in Figure 2.

The robot consists of a Segway Robotic Mobility Platform (RMP) 100[3], supporting an on-board computer and various sensors. The Segway provides controlled power in a relatively small package. This suits a domestic environment well, for it is small enough to maneuver a living environment built for humans and powerful enough to reliably traverse varying indoor terrain including rugs, power cords, tile, and other uneven surfaces. The large wheels easily navigate small bumps that challenged other indoor robots during the competition. The two-wheeled, self-balancing robot reaches speeds up to six mph, exerts two horsepower, and has a zero turning radius. . The Segway moves with two degrees of freedom, receiving motion commands in the form of forward velocity (m/sec) and angular velocity (radians/sec). It provides proprioceptive feedback in the form of measurements of odometry and pitch. With a payload capacity of 100–150 lbs., the Segway could easily carry several times the weight of its current load.

Fig. 2. Our Segway home assistant robot

A 1GHz Fujitsu tablet PC sits atop the Segway platform, performing all sensory processing, behavior generation, and the generation of motor commands on-board. It interfaces with the Segway via USB at 20 Hz.

Two cameras and one laser range finder are available to sense the robot's environment. The Videre Design STOC camera[4] provides depth information, but is not used for the tasks and experiments described in this paper. Higher picture quality is obtained by the second camera, an inexpensive Intel webcam which sends 30 frames per second. The attached Hokuyo URG-04LX[5] is a short range, high resolution laser range finder that is well-suited for indoor environments. It collects 769 readings across 270 degrees at 10 Hz. Also, a Logitech microphone and USB speakers are attached.

The Segway RMP 100 is based on the p-Series Segway line for human transport. Despite its power, the robot is quite safe, featuring safety mechanisms such as automatic shut-off, an emergency kill rope, and speed caps at both the hardware and software levels.

---

[3] www.segway.com

[4] www.videredesign.com/vision/stereo_products.htm

[5] www.hokuyo-aut.jp/02sensor/07scanner/urg.html

A multi-threaded program, written from scratch, operates the robot. The program's structure can be divided into six modules: the camera input processing, the laser range finder input processing, the motion input/output, speech output, the high-level behavior unit, and the GUI.

4 Dual-Classifier Person Recognition

We use a dual-classifier system for person recognition. Face recognition is used as a starting point, but it is augmented by tracking the more frequently visible, but perhaps less uniquely identifying, characteristic of shirt color. Primary, uniquely identifying facial characteristics are dynamically associated with secondary, more ambiguous, possibly transient, but more easily computable characteristics (shirt colors). When primary characteristics are identifiable, they can be used to provide additional training data for the secondary characteristics visible on the person. The secondary characteristics can then be used to track the person, even when the primary characteristics are undetectable.

In this section, we summarize the full version of our novel person-recognition algorithm that includes two co-trained classifiers. Since the main contribution of this paper is the description of our overall, integrated system, we refer the reader to [13,14] for the full details and related experiments. We used this algorithm in the finals of the competition (Section 5.4) and the second phase of Follow and Guide a Human (Section 5.2). A modification of the algorithm was used for the more constrained, earlier task of Who Is Who (Section 5.3). We note the changes in that section as appropriate.

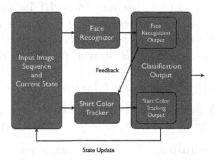

Fig. 3. A conceptual diagram of our dual-classifier person-recognition algorithm

The algorithm described in this section learns a single person's characteristics and differentiates that person from unknown people. In Section 5.3, we extend it to differentiate among multiple known people.

4.1 Primary Classifier (Face)

We use the face as the input for our primary classifier. Challenging in its own right, face-recognition becomes even more difficult when performed by a mobile robot. Perspectives change as both robots and humans move. Illumination changes as humans move through different lighting conditions. When the mobile robot is in motion the image quality from its camera(s) is decreased. Also, computational limitations are acute. Faces must be recognized in real time with what computational resources are connected to the robot. A number of successful face-recognition algorithms exist (e.g. [2,12]), but we found none that fit the needs of a home assistant robot.

Our algorithm is aided by two previously published algorithms. We detect the person's face in the camera image using Viola and Jones' real-time face-detection algorithm [19], and facial features are extracted as scale-invariant feature transforms (SIFT) features [16], shown in Figure 4).

(a) same person   (b) different people

Fig. 4. Matched SIFT features are connected with a line

Training. Using the face detection algorithm, we collect 50 image samples of the person being learned. SIFT features are extracted from these samples. The collected features are stored for use during testing.

Testing. After training, the task of the face classifier is to determine whether the known person is visible in the camera image. To do this, it first identifies a face in the picture using Viola and Jones' face-detection algorithm. If one is found, it then extracts SIFT features from the face image. It then attempts to find matches between these SIFT features and the known identity's bag of SIFT features. If the number of SIFT feature matches are above a certain threshold, the face image is positively classified as that of the known person. Otherwise, the face image is deemed to be that of a stranger.

4.2 Secondary Classifier (Shirt)

A person's is more often visible than his or her face, so we support face recognition by using the shirt as a secondary classifier.

Training. To sample the person's shirt during training, first the algorithm scans incoming video images for a face, again using Viola and Jones' face detection algorithm. If it detects a face, the bounding box of the face (as given by face detection) is used to choose three other bounding boxes that provide positive and negative shirt samples. One box is drawn directly below the face to provide the positive sample pixels from the shirt. For negative samples, two other bounding boxes are drawn to the left and to the right of the face. Figure 5 illustrates the training process. Training data consists of 50 samples drawn from face-containing frames, taken simultaneously with the face classifier samples.

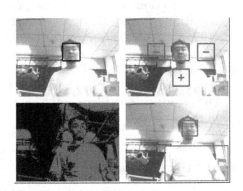

Fig. 5. Training and testing of the shirt-classifying algorithm. Top-left: face-detection; top-right: positive and negative sampling of the shirt; bottom-left: Boolean mapping of shirt colors onto image; bottom-right: shirt detection.

At this point the collected samples are analyzed. Both the positive and negative sample pixels are inserted into respective histograms, each of which is a 16x16x16 RGB cube. Each histogram is normalized and a final histogram is created by subtracting the negative histogram from the positive histogram. Values in this final histogram that are above a threshold are considered positive shirt colors. From this final histogram, an RGB color cube was created in which each RGB value contained a Boolean value indicating whether or not the RGB value was associated with the shirt color.

Testing. Once the training is over, our robot is ready to track the person. The classifier maps a 320 × 240 webcam image to a same-sized image with Boolean values replacing RGB values at each pixel location. After the mapping, the classifier looks for blobs of Boolean true pixels in the mapped image. Among many possible shirt candidate blobs, the blob representing the shirt is chosen by its height and its proximity to recent shirt blobs.

4.3   Inter-classifier Feedback

In this system, the secondary classifier acts as a backup to the primary classifier. When the robot does not detect a face, the secondary classifier is acted upon. If a face is detected, the primary classifier's output determines the output of the person classifier. Additionally, the shirt pixels below the detected face were used to check the accuracy of the shirt classifier. If the two classifiers disagreed (e.g. the face is classified as negative and the shirt is positive), the secondary classifier was retrained using newly taken samples. A conceptual diagram of the algorithm is shown in Figure 3.

5   UT Austin Villa Approach and Performance

This section describes the strategies and algorithms the Segway used in the tasks described in Section 2. All tasks were performed in the same home environment, shown differently in Figure 1 and Figure 6.

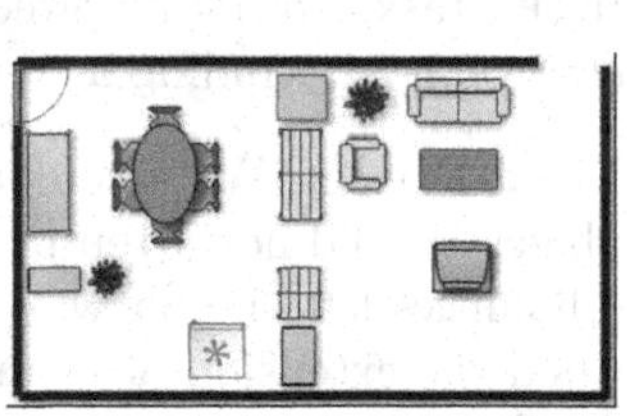

Fig. 6. An approximate recreation of the RoboCup@Home floor plan

5.1   Lost and Found

In *Lost And Found*, a robot searched for a known object that had been placed in an unknown location in the home environment. The task setup is described in Section 2.1. Our robot competed in the first phase of *Lost and Found*.

First Phase. We used an ARTag marker as the target object[8]. ARTag is a system of 2D fiducial markers and vision-based detection. The markers are robustly detected from impressive distances (more than 5 meters at 320 × 240

resolution in our lab with a $20cm \times 20cm$ marker) with varying light and even partial occlusion. Each marker is mapped to an integer by the provided software library. We did not observe any false positives from our ARTag system.

For the *Lost and Found* task, our robot searched the environment using a reflexive, model-free algorithm that relied on a fusion of range data and camera input. The Segway moved forward until its laser range finder detected an obstacle in its path. It would then look for free space, defined as an unoccupied rectangular section of the laser plane 0.75 m deep and a few centimeters wider than the Segway, to the left and right and turned until facing the free space. If both sides were free, the robot randomly chose a direction. If neither side was free, it turned to the right until it found free space. Algorithmically, free space was determined by a robustly tuned set of pie pieces in the laser data which overlapped to approximate a rectangle (see Figure 7).

We placed the object on a table at the opposite end from where the Segway began. A straight line between the two would have passed through a television, shelves, and a kitchen table. The robot had neither prior knowledge of the object's location nor any model of the environment. The Segway successfully completed its search with more than three minutes to spare. Of the six teams that attempted *Lost and Found*, only three teams, including our team, completed it.

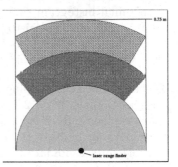

Fig. 7. The laser range finder data is checked for occupancy at three different ranges and angles to approximate a rectangle that is a bit wider and deeper than the Segway

5.2   Follow and Guide a Human

In this task, a robot followed behind a human as he or she walked within the home environment, winding around the furniture. Its setup is described in Section 2.2.

First Phase. We attempted only the following (not guiding) portion of this phase. We did not attempt the extension because time constraints and technical difficulties left the Segway without functional mapping software. (No team finished the extension of returning back to the starting point.) Again, we used an ARTag marker on the shirt of the leading human. The robot flawlessly followed the human leader, without touching furniture or the human. Six of eight teams that attempted the first phase of *Follow and Guide a Human* completed this portion of the task.

Second Phase. Without the ARTags of the first phase, the robot instead trained and used a shirt classifier as described in Section 4.2. Since we anticipated following a human with his back turned, the face recognition component of our person recognition algorithm was not used.

In the competition, the referees chose a volunteer wearing a white shirt. Much of the background was also white, so the negative samples collected during training were not distinguishable by our algorithm from the samples of his shirt. Instead of tracking the volunteer's shirt as intended, the robot classified a large portion of the wall as the person and was unable to follow the volunteer. The choice of volunteer revealed a weakness in our algorithm, but in later rounds (described in Section 5.4) we showed that, given a shirt color that is distinguishable from the background colors, the robot can follow a person for whom it has no a priori data.

5.3   Who Is Who

The *Who Is Who* task tested a mobile robot's ability to meet and later recognize humans. We used a modification of the Section 4 dual-classifier person-recognition algorithm for both phases of the *Who Is Who* task, so we explain it here before describing the specific phases.

To learn the faces of multiple people, we train a face classifier for each person as described in Section 4.1. For *Who Is Who*, the output of the multiple-face classifier is the set of identities which had a number of SIFT feature matches above an empirically determined threshold. If the output set is empty, then the threshold is lowered and the classifier is rerun.

Given the set of candidate identities, a shirt classifier takes over. This classifier gathers samples as described in Section 4.2, but otherwise the shirt classifier is different, having been modified to eliminate blob selection. Since the face is easily detectable in this task, the shirt pixels are simply taken from below the face as in training. For each candidate identity, the Euclidean distance between the average R, G, and B values of the pixels on the persons shirt (a 3-tuple) and the average R, G, and B values of the specific identity's shirt samples is calculated. If at least one candidate's shirt distance is above a shirt threshold, then the candidate with the shortest distance is chosen as the identity of the person. If none are above the shirt threshold, the person is announced as a stranger.

First Phase. In the first phase, we chose the four people and their shirts. We gave them strongly distinguishable shirt colors – red, green, blue, and yellow. Our robot correctly identified four of the five people. The stranger was misidentified as one of the known people.

We believe this error occurred specifically on the stranger for two reasons. First, the volunteer's SIFT features matched many features of at least one of the known people. Second, the volunteer's shirt was colored similarly to the person whose SIFT features were similar. With both the primary characteristic (the face) and the secondary characteristic (the shirt) testing as false positives, the person tracker did not correctly classify the stranger.

Of seven teams that attempted this task, some of which used commercial software packages, only one other received points by identifying at least four of the five people.

Second Phase. The training of the second phase is the same as in the first, except the persons were chosen randomly by the committee. The testing is especially more challenging in the second phase. The five people (four known and one stranger) are not standing in a line anymore, but are instead randomly distributed throughout the home environment.

As in the *Lost and Found* task, we used a stochastic search to look for candidate people as recognized by positive identification from the face detection module. During the allotted time, the robot found one of the people and correctly identified him. No other team identified a single person during the second phase.

5.4   Open Challenge and Finals

Once all teams had attempted their specific tasks, each competed in what was called the *Open Challenge*. This consisted of a presentation and free-form demonstration. Going into this event, after receiving scores from the specific tasks, UT Austin Villa ranked third of eleven. A jury of the other team's leaders ranked us second for the *Open Challenge*. The robot's demonstration was a simplified version of the one performed in the Finals, so it will not be described.

Finals. The top five teams competed in the Finals. Having ranked third in the specific tasks and second in the open challenge, UT Austin Villa advanced, along with Pumas from UNAM in Mexico, AllemaniACs from RWTH Aachen in Germany, RH2-Y from iAi in Austria, and Robot Cognition Lab from NCRM in France. The Finals were judged by a panel of trustees of the RoboCup organization, all well-known robotics researchers.

Before describing the demonstration itself, we begin with some motivation for the scenario we focused on. Accurate person-recognition will be essential in any fully functional home assistant robot. Rigidly learning a person's exact appearance at one moment will likely not be sufficient to identify him or her after a significant change in appearance (e.g. haircut, aging, etc.). A home assistant robot will need to be flexible, adapting to such changes.

Our scenario was designed to display our algorithm's robustness and adaptability. Specifically, it shows person identification using shirt color as a secondary classifier in the absence of the primary classifier, the face. It also mimics the frequent occurrence of a human changing clothes, showing the robot adapt to this change in the secondary classifier's input. Lastly, it shows that the Segway robot can effectively follow a recently learned person without markers, as we unfortunately were unable to show during the second phase of the *Follow and Guide a Human* task. The difference was that we used shirt colors which stood out from the colors of the background (as a white shirt did not).

Before the demonstration, we again presented a short talk about the robot and our algorithms. A video of the presentation and demonstration can be found at our team webpage, `http://www.cs.utexas.edu/ AustinVilla/?p=athome`.

The demonstration involved two people, one with whom the robot intended to interact and another who was unrelated to the robot's primary task (stranger). The person-recognition component of the demonstration algorithm is described in Section 4. At the beginning, the robot trains classifiers for the intended person's face and shirt. It then follows the learned person based on only shirt color when his face is not visible, first with a green shirt and later with a red shirt.

The Segway twice gets "passed" to a stranger, whose back is turned (i.e. face invisible) and is wearing the same shirt color. Each time, it follows the stranger until it can see his face. At that point, the face classifier returns a negative classification and supersedes the shirt classifier, and the robot announces that it has lost the learned person and turns away to look for him. Upon finding the original person based on a positive facial classification, it retrains the person's shirt, subsequently stating whether the shirt color has changed.

The demonstration went smoothly, with only one noticeable flaw. At one point, the robot turned away from the red-shirted human and towards a mahogany set of shelves, again revealing the same limitation in our shirt classifier that hurt us in the second phase of *Follow and Guide a Human*.

6   @Home Final Results

The panel of judges scored the presentations and demonstrations of each finalist, determining each team's final standing in RoboCup@Home 2007. We finished in second place (full results in Figure 8). Of the top three teams, we had a couple of unique characteristics. Our at-RoboCup@Home team size of three people was half that of the next smallest team. We were also the only team

Team	Final Score
AllemaniACs	256
UT - Austin	238
Pumas	217
RH2-Y	199
Robot Cognition Lab	190

Fig. 8. 2007 RoboCup@Home final results

in the top three that was competing for the first time. We were very successful as well in the specific tasks in which we competed. We received more points than any other team in the person-recognition task of *Who Is Who* and accomplished all tasks that we attempted in the first phases of *Lost and Found* and *Follow and Guide a Human*.

7   Related Work

A variety of home assistant robots have been created in the past decade. Many exhibited impressive specific capabilities. Care-O-bot II [10] brought items to a human user and took them away in a domestic setting. It also functioned as a walking aid, with handles and an interactive motion system that could be controlled directly or given a destination. Earlier systems include HERMES [4] and MOVAID [5].

Various studies have been conducted on face recognition [2,12,19]. In contrast with these recognition methods, which rely on careful face alignment, we extract SIFT features [16] from faces, similar to work proposed in [3], and classify faces by counting the matching SIFT features in near real-time. To track faces through changing perspectives and inconsistent lighting, we augment a face classifier with another classifier (the shirt classifier). Previous work on integrating multiple classifiers has shown that integrating multiple weak learners (ensemble methods) can improve classification accuracy [17]. In [15], multiple visual detectors are co-trained to improve classification performance. This method merges classifiers that attempt to classify the same target function, possibly using different input features. In contrast, the classifiers we merge are trained on different concepts (faces and shirts) and integrated primarily by associating their target classes with one another in order to provide redundant recognition and dynamically revised training labels to one another. Tracking faces and shirts is a known technique [20], but previous work did not utilize inter-classifier feedback.

Person-following specifically has received much attention from researchers. A recent laser-based person-tracking method was developed by Gockley et al. [9]. Their robot Grace combined effective following with social interaction. Asoh et al. [6] couple face recognition with a sound detection module, using the sound's direction to indicate where the face is likely to be. A vision-based approach similar to our own was created by Schlegel et al. [18]. In their system, the robot also tracked shirts using color blobs, but the shirts had to be manually labeled in the training images. Some more recent approaches have used stereo vision and color-based methods to track humans [7,11].

8   Conclusion and Future Work

RoboCup@Home will continue to move towards higher robotic autonomy and tasks that demand more effective navigation, object and person-recognition, and object manipulation. In order to keep pace with these rising challenges, our system will need to move towards general object recognition instead of relying an the ARTag system, and it will need to include mapping capabilities. Meanwhile, although person recognition is the current strength of our system, it can be improved by strengthening the face classifier's robustness to low-contrast conditions, improving its accuracy at identifying unknown people, and adding the ability to learn patterns of different colors on shirts.

The main contribution of this paper was the complete description of our Segway-based platform that performed successfully in the 2007 RoboCup@Home competition. Leveraging our main technical innovation of using co-training classifiers for different characteristics of a person (face and shirt), it was able to follow a person, distinguish different people, identify them by name, and ultimately combine these abilities into a single robust behavior, adapting to a person changing his or her clothes.

Acknowledgments

This research is supported in part by NSF CAREER award IIS-0237699 and ONR YIP award N00014-04-1-0545.

References

1. Interim state population projections. US Administration on Aging within the Population Division, U.S. Census Bureau (2005)
2. Belhumeur, P., Hesphana, J., Kriegman, D.: Eigenfaces vs. fisherfaces: Recognition using class specific linear projection. PAMI (1997)

3. Bicego, M., Logorio, A., Grosso, E., Tistarelli, M.: On the use of sift features for face authentication. In: CVPR Workshop (2006)
4. Bischoff, R.: Design concept and realization of the humanoid service robot hermes (1998)
5. Dario, P., Guglielmelli, E., Laschi, C., Teti, G.: Movaid: a mobile robotic system for residential care to disabled and elderly people. In: MobiNet Symposium (1997)
6. Asoh, H., et al.: Jijo-2: An office robot that communicates and learns. IEEE Intelligent Systems 16(5), 46–55 (2001)
7. Enescu, V., et al.: Active stereo vision-based mobile robot navigation for person tracking. Integr. Comput.-Aided Eng. (2006)
8. Fiala, M.: Artag, a fiducial marker system using digital techniques. In: CVPR (2005)
9. Gockley, R., Forlizzi, J., Simmons, R.: Natural person-following behavior for social robots. In: Proceedings of Human-Robot Interaction, pp. 17–24 (2007)
10. Graf, B., Hans, M., Schraft, R.D.: Care-o-bot ii – development of a next generation robotic home assistant. Autonomous Robots (2004)
11. Kwon, H., Yoon, Y., Park, J.B., Kak, A.: Human-following mobile robot in a distributed intelligent sensor network. ICRA (2005)
12. Lee, D., Seung, H.: Learning the parts of objects by non-negative matrix factorization. Nature 401(6755), 788–791 (1999)
13. Lee, J., Knox, W.B., Stone, P.: Inter-classifier feedback for human-robot interaction in a domestic setting. Journal of Physical Agents 2(2), 41–50 (2008); Special Issue on Human Interaction with Domestic Robots
14. Lee, J., Stone, P.: Person tracking on a mobile robot with heterogeneous inter-characteristic feedback. ICRA (2008)
15. Levin, A., Viola, P., Freund, Y.: Unsupervised improvement of visual detectors using co-training. In: ICCV (2003)
16. Lowe, D.: Distinctive image features from scale-invariant keypoints. IJCV (2004)
17. Schapire, R., Singer, Y.: Improved boosting algorithms using confidence-rated predictors. Machine Learning (1999)
18. Schlegel, C., Illmann, J., Jaberg, H., Schuster, M., Worz, R.: Vision based person tracking with a mobile robot. In: BMVC (1998)
19. Viola, P., Jones, M.: Robust real-time face detection. IJCV (2004)
20. Waldherr, S., Romero, R., Thrun, S.: A gesture based interface for human-robot interaction. Autonomous Robots 9(2), 151–173 (2000)

Combining Policy Search with Planning in Multi-agent Cooperation

Jie Ma and Stephen Cameron

Oxford University Computing Laboratory,
Wolfson Building, Parks Road, Oxford, OX1 3QD, UK
{jie.ma,stephen.cameron}@comlab.ox.ac.uk
http://www.comlab.ox.ac.uk

Abstract. It is cooperation that essentially differentiates multi-agent systems (MASs) from single-agent intelligence. In realistic MAS applications such as RoboCup, repeated work has shown that traditional machine learning (ML) approaches have difficulty mapping directly from cooperative behaviours to actuator outputs. To overcome this problem, vertical layered architectures are commonly used to break cooperation down into behavioural layers; ML has then been used to generate different low-level skills, and a planning mechanism added to create high-level cooperation. We propose a novel method called *Policy Search Planning* (PSP), in which Policy Search is used to find an optimal policy for selecting plans from a *plan pool*. PSP extends an existing gradient-search method (GPOMDP) to a MAS domain. We demonstrate how PSP can be used in RoboCup Simulation, and our experimental results reveal robustness, adaptivity, and outperformance over other methods.

Keywords: Policy Search, Planning, Machine Learning, Multi-agent Systems.

1   Introduction

Cooperation is one of the most significant characteristics of multi-agent systems (MASs). Compared with a single agent, cooperating agents may gain over autonomous agents in *efficiency, robustness, extensibility* and *cost*. Sometimes cooperation is necessary to achieve goals due to the observation or action limitations of a single agent.

In order to reduce the learning space of cooperative skills, most of today's MASs tend to adopt *vertical layered architectures* [1, 2]. Such structures can arguably balance decision accuracy and speed, and simplify the learning process for high-level (deliberative) skills — such as cooperation — by decomposing them into lower-level skills. Those lower-level skills can be further decomposed until the lowest-layer reaction skills.

Although machine learning approaches have shown advantages in solving low-level skills [3, 4, 5], there still remain two difficulties in learning cooperation. Firstly, when the number of agents increases, state and action spaces become too

L. Iocchi et al. (Eds.): RoboCup 2008, LNAI 5399, pp. 532–543, 2009.

large for current machine learning approaches to converge. In the MAS domain, previous machine learning methods showed very limited cooperation and most of that cooperation was demonstrated in a stationary environment with only 2 players. Secondly, in an adversarial MAS application such as RoboCup and Combat Simulation, cooperation is usually very complex and highly related to its opponent, and repeated work has shown that it is difficult to yield such cooperation directly through machine learning [2, 6]. Although planning methods have arguably represented such cooperation, few of these methods have shown successful generalised adaption.

We propose a novel method called Policy Search Planning (PSP) which combines machine learning with a symbolic planner. We claim it can increase cooperation quality in POMDPs (Partially Observable Markov Decision Processes). Plenty of human knowledge on complex cooperation can be presented in a symbolic planner that allocates subtasks to appropriate agents; policy search is then used to find an optimal cooperation pattern in an unknown environment.

To evaluate our method, we employed RoboCup Soccer 2D Simulation as our test-bed, which is essentially a dynamic adversarial MAS environment. In this domain, a number of low-level individual skills such as *Positioning* [4], *Interception* [3], and *Dribbling* [5] were addressed and solved by machine learning approaches. Under high-level cooperation however, although some attempts have been made to present and decompose cooperation using planning, few of them have demonstrated how to learn to select plans (cooperative tactics) to maximise overall performance in an unknown environment. Our experimental results show adaptive cooperation among 11 agents and a significant improvement in performance over pure planning methods.

In §2 we discuss previous work, and we then present our PSP method in §3, where we consider a MAS as a generalised POMDP. The details of PSP in our test-bed RoboCup 2D Simulation are provided in §4, and we conclude in §5.

2   Related Work

Accompanied by the booming research on MAS, increasing interest has been shown in extending machine learning (ML) to the multi-agent domain. There are mainly three kinds of ML: *supervised, unsupervised* and *reinforcement learning*. In terms of learning cooperation, reinforcement learning is the most appropriate one among the above three because the mappings from cooperative actions to a global goal are usually obscure thus supervised and unsupervised learning is not suitable for yielding cooperative strategies.

Under reinforcement learning (RL), Q-Leaning is the most common learning approach in MAS scenarios. Recently however, there has been an increasing interest in another method of reinforcement learning, namely policy search. As a reasonable alternative to Q-learning, it can arguably promote performance in POMDP (as shown below).

In this section, we review how policy search was used in single-agent and multi-agent applications. Moreover, how reinforcement learning can be used to promote or simplify cooperative *planning* is also reviewed.

2.1 Policy Search

In a large POMDP problem, such as RoboCup, traditional Q-Learning sometimes has difficulty in approximating the Q-functions [7]. Especially in the MAS
domain, since agents have to broadcast their local Q-function to the other agents,
Q-Learning is significantly restricted by storage space and communication bandwidth. In recent years, Policy Search that directly finds the optimal policy is
stepping into the spotlight.

Since the policy is usually parameterised, the optimal policy can be found by
searching the parameter space θ, and using gradient-based search (also known as
Policy Gradient) can substantially increase the search speed. Under policy gradient, the Boltzmann[1] distribution is widely used for computational simplicity.

In a single-agent POMDP system, as the state space cannot always be completely observed, *biased estimation* using an eligibility trace has been proposed
by Kimura et al [9]. Baxter and Bartlett further suggested a GPOMDP (Gradient
of Partially Observable Markov Decision Process) method, which was proved to
converge under some assumptions [10, 11]. GPOMDP is essentially a biased and
estimated gradient using the *eligibility trace* method [12]. The learning equations
are given as follows:

$$
\begin{aligned}
z_{t+1}\left(\gamma\right) &\leftarrow \gamma z_t\left(\gamma\right) + \frac{\nabla\mu_{U_t}(\theta, Y_t)}{\mu_{U_t}(\theta, Y_t)} \\
\Delta_{t+1} &= \Delta_t + \tfrac{1}{t+1}[R(U_t)z_{t+1} - \Delta_t]
\end{aligned}
\tag{1}
$$

where z_t is an eligibility trace and Δ_t is a gradient estimate. For each observation
Y_t, control U_t and its reward $R(U_t)$, $\mu_{U_t}(\theta, Y_t)$ represents the probability distribution with parameter θ at time t. $\gamma \in [0, 1)$ is a discount factor, where γ close
to 1 yields a smaller bias but a larger variance. The contribution of GPOMDP
is that the action transition probabilities and the probability distribution over
the observation space are not necessarily required. Due to the space limitation
here, the details of generalised GPOMDP algorithm can be found in [10, 11].

In the MAS domain, on the other hand, only a little work has been done
using policy search. Tao et al suggested a possible way to adopt GPOMDP in
network routing [13]. Routers are regarded as agents, each of which has a set of
local parameters $\theta_i = \left(w_1^i, w_2^i, \cdots, w_m^i\right)$, and each parameter controls an action.
Essentially, every agent learns its local parameters θ_i from a local perspective
with the global rewards $R\left(\overrightarrow{U_t}\right)$. This algorithm has also been employed in a simple cube-pushing game in recent research [14]. Experimental results supported
the performance of GPOMDP. Although this method has shown strong robust
performance in some circumstances, it has two particular limitations: the action
space in previous experiments was small, and the cooperation among agents was
not complex.

Another relatively more complex MAS application has been demonstrated
by Peshkin et al in [6]. Their algorithm essentially extends the REINFORCE
algorithm [15] to an MAS domain. It can guarantee convergence to local optimality in a parameterised policy space. This algorithm has been adopted in a

---

[1] Essentially the same as Softmax and Gibbs distributions [8].

simple football game, and experiments have demonstrated outperformance over Q-learning in a partially observable environment. However, when the agent population grows the state space will become too large to be practical.

2.2 Combining Reinforcement Learning with Planning

Planning enables an agent to automatically achieve a goal by searching a set of actions. It is a significant way to undertake deliberative reasoning. In single-agent systems, there exist some mature planning methods, such as STRIPS (Stanford Research Institute Problem Solver), ADL (Action Description Language), HTN (Hierarchical Task Network) and PDDL (Planning Domain Definition Language).

The usages of planning in MASs are different from those in single-agent systems. In a single-agent system, planning is mainly used to find an action sequence to directly achieve a goal. In the MAS domain, however, planning tends to generate advanced cooperation. According to the taxonomy of Marinova [16], there are mainly three types of multi-agent planning: *centralised planning for distributed actions*, *distributed planning for centralised actions* and *distributed planning for distributed actions*. Today, most planning approaches in MASs are of the type centralised for distributed actions.

Pecora and Cesta proposed a hierarchical structure to apply PDDL planning [17] to MASs. The HTN method has been employed in MASs by Obst and Boedecker [18, 19]. Their method can arguably promote expert knowledge in dynamic POMDPs and speed up the planning process due to its hierarchical planning structure, though the role mappings in this method are rather stationary. To extend the flexibility of role mapping, Fraser and Wotawa presented a possible way to apply traditional STRIPS to MAS domains [20]. Before a plan starts, an agent can select its role, and broadcast it to others.

In the planning process, RL has been adopted in two ways. Firstly, RL is used to learn advanced individual skills, and so planners are able to search these skills instead of the actuators at the lowest-level. Secondly, some attempts have been made to directly promote planning decisions by using RL.

Using the first combination, recent work of Grounds and Kudenko [21] supports this approach using an example problem of an agent navigating through a grid. In a single agent grid square, Q-learning is used to generate low-level behaviours (choosing the direction of motion), while a STRIPS-based planner encodes high-level knowledge. Experimental results reveal that their PLANQ method performs better than pure Q-learning in a small domain, but on the other hand it will lose its strength when the state space grows.

Strens and Windelinckx employed Q-learning in a multi-robot task allocation problem [22]. Experimental result showed significant energy saving compared with a greedy method. Their main contribution is extending the action space to plan space. However, there are two limitations of this algorithm. Firstly, at any state, the active plans have to be the same for all the agents. Secondly, this algorithm requires a pure planning decision structure (pure deliberation architecture). Therefore, the algorithm is difficult to apply to generalised MASs.

Recently, Buffet and Aberdeen proposed have a planner called Factored Policy Gradient (FPG) [8, 23, 24]. Their algorithm combines a single-agent planner (PDDL) with the aforementioned policy search method, GPOMDP. Although the authors did not mention this, its potential advantage is that a pure deliberative architecture is not always needed. But the drawback is that all the plans have to share the same action space.

3   Policy Search Planning (PSP)

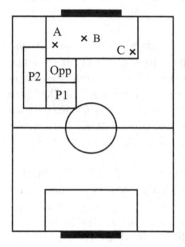

Fig. 1. A Planning scenario in RoboCup

In complex MASs, particularly in a system with hybrid individual architectures, planning plays a different role compared with that in traditional domains. In a simplified single-agent system, planning is used to directly find a goal. In dynamic MASs, however, the goal is usually difficult to achieve, or sometimes it is difficult to describe the goal. In addition, the traditional action effects will lose their original meaning: environmental state can also be changed by other agents at the same time, or sometimes it continually varies without any actions. For example, consider a scenario from RoboCup as shown in Figure 1: $P1$ and $P2$ are two team members with $P1$ controlling ball, Opp is an opponent, and they are all located in different areas. A traditional planner might construct a plan in which $P2$ dashes to point A and then $P1$ passes the ball. However, in this situation, points B and C are also potential target points for $P2$. Even from a human's perspective it is difficult to say which plan is better before fully knowing the opponent's strategies. Therefore, in multi-agent systems planning tends to be regarded as a "tutor" to increase cooperative behaviours so as to improve overall system performance. Expert knowledge can be embodied in such planning, without which agents mainly execute individual skills.

We propose a novel method called Policy Search Planning (PSP) for POMDPs, which is essentially a centralised planner for distributed actions. In the example of Figure 1, PSP can try to find the most appropriate policy for selecting a plan even without the opponent's model. Specifically, it can represent a number of complex cooperative tactics in the form of plans. Plans are shared by all the agents in advance, and policy search is used to find the optimal policy in choosing these plans. As a plan is not designed to find the goal directly but to define cooperative knowledge, the style of it is not very critical. One possible presentation, a PDDL-like planner, is shown in Figure 2.

Compared with original PDDL, *:goal* will not be included as PSP aims not to achieve it directly, and *:effect* is not needed unless it is used for parameters in policy search. The concept of *stage* is introduced, which makes complex

```
(define (PLAN_NAME)
   (:plan_precondition CONDITION_FORMULA)
   ((:agentnumber INTEGER(N))
   (ROLE_MAPING_FORMULA(1))
   (ROLE_MAPING_FORMULA(2))
    ...
   (ROLE_MAPING_FORMULA(N)))
   ((:stagenumber INTEGER(M))
    ((:stage_1_precondition    CONDITION_FORMULA)
     (:stage_1_success         CONDITION_FORMULA)
     (:stage_1_failure         CONDITION_FORMULA)
     (:stage_1_else            CONDITION_FORMULA)
     (:action1                 ACTION_FORMULA)
     (:action2                 ACTION_FORMULA)...)
    ((:stage_2_precondition    CONDITION_FORMULA)
     ...)   ...
    ((:stage_M_precondition    CONDITION_FORMULA)
     ...))
   [(:effect EFFECT_FORMULA)])
```

Fig. 2. A PDDL-like Plan Structure in PSP

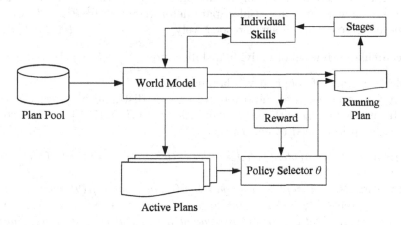

Fig. 3. Learning Process in PSP algorithm

cooperation possible, whereby if and only if the success condition of the current stage is met a planner moves to the next stage; and *role mapping* formulae are introduced to find the most appropriate agents to implement actions.

In the PSP algorithm, a plan is actually a cooperative strategy. We can define plenty of offline plans by hand to establish a *plan pool*, which is essentially an expert knowledge database. If the external state satisfies the precondition of a plan, the plan will be called an *active plan*. At time t, if there is only one active plan, it will be marked as the *running plan* and actions will be executed stage by stage. However, along with the growth of the plan pool, multiple active plans may appear at the same time.

Previous solutions [18, 19] chose a plan randomly, which is clearly a decision without intelligence. Q-learning is apparently a wiser approach, but unfortunately Q-learning is difficult to adopt in generalised decision architectures because all the plans cannot guarantee activation.

In this paper we employ another reinforcement learning method, policy search, to overcome this difficulty. The learning framework is illustrated in Figure 3. Say there exist n agents, which are organised in a hybrid architecture. From a global perspective, although all the actions are executed by the lowest level, decisions come from two different directions: individual skills and cooperative planning. Then the global policy value that can be evaluated by the accumulated reward can be represented as:

$$\rho\left(\theta, \theta^1, \theta^2 \cdots \theta^n\right) = \sum_{s \in S, \phi \in \varphi} R_{\phi, \vec{a}} P\left(\theta, \phi\right)$$

$$= \sum_{s \in S, a^i \in A^i, i \in [1..n]} R\left(\vec{a}\right) P\left(\theta^i, a^i\right) + \sum_{s \in S, \phi \in \varphi} R\left(\phi\right) P\left(\theta, \phi\right)$$

(2)

where: φ is the action plan pool in state s; $P\left(\theta, \phi\right)$ is the probability distribution of selecting plan ϕ under plan parameter vector θ; $\theta^1, \theta^2 \cdots \theta^n$ are policy parameter vectors of each agent respectively; $\sum\limits_{s \in S, a^i \in A^i, i \in [1..n]} R\left(\vec{a}\right) P\left(\theta^i, a^i\right)$ is the accumulated reward of individual skills; and $\sum\limits_{s \in S, \phi \in \varphi} R\left(\phi\right) P\left(\theta, \phi\right)$ is the accumulated reward of cooperative planning.

In hybrid architectures, however, we cannot distinguish where the reward come from, thus along with three assumptions of GPOMDP [10, 11], two additional assumptions need to be satisfied:

Assumption 1. *Individual policies $\theta^1, \theta^2 \cdots \theta^n$ are independent of planning policy θ.*

Assumption 2. *During the observation of policy value $\rho\left(\theta\right)$ over active plan pool φ, probability distributions over individual actions under local policies $P\left(\theta^i, a^i\right)$ are stationary and individual actions yield a stationary accumulated reward:* $\sum\limits_{s \in S, a^i \in A^i, i \in [1..n]} R\left(\vec{a}\right) P\left(\theta^i, a^i\right) = C$

Under the above two assumptions, the global policy value (Equation 2) can be rewriten as:

$$\rho\left(\theta, \theta^1, \theta^2 \cdots \theta^n\right) = \sum_{s \in S, \phi \in \varphi} R_{\phi, \vec{a}} P\left(\theta, \phi\right) = C + \sum_{s \in S, \phi \in \varphi} R\left(\phi\right) P\left(\theta, \phi\right) \quad (3)$$

From the equation above, we find that under individual policies $\theta^1, \theta^2 \cdots \theta^n$ the accumulated reward of the individual skills is independent of the global policy value $\rho\left(\theta, \theta^1, \theta^2 \cdots \theta^n\right)$. In other words, under our assumptions the global policy value is only determined by the planning policy value, and thus we can directly use the global policy value to evaluate the planning policy value:

$$\rho\left(\theta\right)' = \rho\left(\theta, \theta^1, \theta^2 \cdots \theta^n\right) = \sum_{s \in S, \phi \in \varphi} R_{\phi, \overrightarrow{a}} P\left(\theta, \phi\right) \tag{4}$$

Therefore, we can extend the GPOMDP (Equation 1) to a MAS planning domain. Agents adjust the planning parameters independently without any explicit communication. The adjustment is based on the local observation and a shared planning control. For an agent i, the PSP learning equations are as follows:

$$z_{t+1}^i\left(\gamma\right) \leftarrow \gamma z_t^i\left(\gamma\right) + \frac{\nabla \mu_{\overrightarrow{U_t}}\left(\theta, Y_t^i\right)}{\mu_{\overrightarrow{U_t}}\left(\theta, Y_t^i\right)}$$
$$\Delta_{t+1}^i = \Delta_t^i + \frac{1}{t+1}[R(\overrightarrow{U_t})z_{t+1}^i - \Delta_t^i] \tag{5}$$

where θ is a planning parameter; and z_t^i and Δ_t^i are the eligibility trace and the gradient estimate for the agent i respectively at time t. For each local observation Y_t^i, global planning control $\overrightarrow{U_t}$ and its global reward $R(\overrightarrow{U_t})$, $\mu_{\overrightarrow{U_t}}\left(\theta, Y_t^i\right)$ represents the probability distribution with the planning parameter at time t.

4   Application in RoboCup

RoboCup simulation is a suitable application to evaluate the PSP algorithm because of the following three reasons: firstly, humans have knowledge about football and need to apply it to robots; secondly, intelligent cooperation is useful; and thirdly, there exist no universal optimal policy and so adaptive and adversarial strategies are required. In order to show universality, our plans were constructed in a RoboCup coach language, *CLang* [25]; however, due to space limitations, we are unable to show the *CLang* form of the plan here.

We created two opponent teams *OppA* and *OppB* which tend to defend from side and centre respectively. All the plans have three features, which are the extents to which a plan will change the ball or player positions towards the left sideline, right sideline, and the opponent goal respectively. They are calculated by vector operations, and selected under a policy π. The policy π is parameterised by a vector of weights $\theta = (w_1, w_2, w_3)$. The probability for selecting plan ϕ with parameter vector θ is taken from a Boltzmann distribution.

Our experiments consist of three parts. In our first experiment, 30 plans were defined in the plan pool, and agents play against *OppA*. In order to verify the robustness of PSP, an additional 15 plans were defined in our second experiment under the legacy policy from the first experiment. *OppB* is also used to test adaptivity in our third experiment.

During the PSP learning, we used a discount factor of $\gamma = 0.95$; when agents score a goal $R_{\phi, \overrightarrow{a}} = 1$; and the average number of goals per 100 seconds were calculated every 15 minutes. The PSP learning processes lasted for 20 hours, and was compared with a non-planning method and a planning without learning method. Experimental results are shown in Figures 4–7.

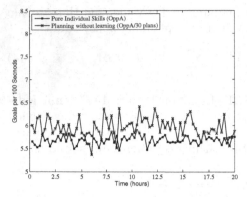

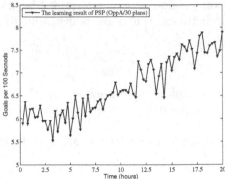

Fig. 4. Performance of pure individual skills and planning without learning

Fig. 5. The learning result of PSP (OppA/30 plans)

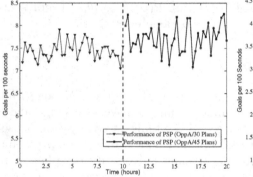

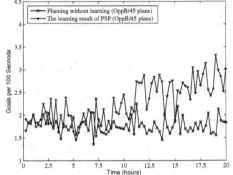

Fig. 6. Performance of PSP (OppA/30 and 45 plans)

Fig. 7. The learning result of PSP (OppB/45 plans)

Figures 4 and 5 show that multi-agent planning causes a larger fluctuation in the average reward, but can promote overall performance. It supports previous research on MAS planning [18, 19, 20, 26]. However, randomly selecting active plans without learning leads to only a slight performance improvement in our first experiment. In the first 4 hours, learning speed was very slow; the performance difference between learning and non-learning is not very clear. It is followed by a steady reward increase from around 6 to more than 7.5 goals per 100 seconds. The figures illustrate that PSP leads to a notable increase in system performance compared with non-learning planning and non-planning methods.

The result of our second experiment (Figure 6) shows slightly better performance with 15 additional plans using the legacy planning policy. It suggests that PSP is reasonably robust — with the same opponent, legacy policy is also compatible with new plans that can lead to further improvement of the decision quality.

Although the legacy policy may not be useful when playing against a completely new opponent, Figure 7 shows that through reinforcement learning agents will finally find a policy to beat their new opponent, namely *OppB* which has a stronger defence strategy. Without learning, pure planning can only make fewer than 2 goals per 100 seconds; after about 8 hours' learning, PSP increases the average goal rate to just below 3. Therefore, adaptive cooperations are established without knowing the opponent model, which supports our theoretical analysis.

5   Conclusion and Future Work

Cooperative behaviours has long been regarded as one of the most important features of MASs. The main difficulty in establishing cooperation using machine learning is the large learning space. Today, layered learning frameworks are widely used in realistic MAS applications. Under this architecture, ML is used to establish low-level skills while planning is used to define high-level tactics.

Under a layered decision architecture, we proposed a novel method called PSP in a generalised POMDP scenario, in which a large selection of cooperative skills can be presented in a plan pool; and policy search is used to find the optimal policy to select among these plans. The innovations of our PSP method include:

1. enabling agents to learn to find the most efficient cooperation pattern in an unknown environment;
2. the design of a learning framework to yield robust and adaptive cooperation among multiple agents; and
3. extending GPOMDP to a MAS domain where complex cooperation is useful.

We demonstrated why and how PSP can be used in RoboCup 2D Simulation, and experimental results show explicit robustness, adaptivity and outperformance over non-learning planning and non-planning methods. In our more general (unquantified) experience with PSP it appears able to find solutions for problems that cannot be solved in a sensible timescale using earlier methods.

PSP is our first attempt to learn the optimal cooperation pattern amongst multiple agents. Our future directions are two-fold. Under RoboCup we are planning to define more plans and more features for PSP in our OxBlue 2D and 3D teams to further verify the robustness of our method. Meanwhile, third party developed opponents will be used to evaluate the performance of PSP. As to a generalised POMDP, we are keen to explore how the different architectures of a planner can effect learning quality, and to verify its generality in other MAS applications.

References

[1] Perraju, T.S.: Multi agent architectures for high assurance systems. In: American Control Conference, San Diego, CA, USA, vol. 5, pp. 3154–3157 (1999)
[2] Stone, P., Veloso, M.: Layered learning and flexible teamwork in roboCup simulation agents. In: Veloso, M.M., Pagello, E., Kitano, H. (eds.) RoboCup 1999. LNCS, vol. 1856, pp. 495–508. Springer, Heidelberg (2000)

[3] Nakashima, T., Udo, M., Ishibuchi, H.: A fuzzy reinforcement learning for a ball interception problem. In: Polani, D., Browning, B., Bonarini, A., Yoshida, K. (eds.) RoboCup 2003. LNCS, vol. 3020, pp. 559–567. Springer, Heidelberg (2004)

[4] Bulka, B., Gaston, M., desJardins, M.: Local strategy learning in networked multi-agent team formation. Autonomous Agents and Multi-Agent Systems 15(1), 29–45 (2007)

[5] Ma, J., Li, M., Qiu, G., Zhang, Z.: Q-learning in robocup individual skills. In: China National Symposium on RoboCup (2005)

[6] Peshkin, I., Kim, K.E., Meuleau, N., Kaelbling, L.P.: Learning to cooperate via policy search. In: Sixteenth Conference on Uncertainty in Artificial Intelligence, pp. 307–314. Morgan Kaufmann, San Francisco (2000)

[7] Kok, J.R., Vlassis, N.: Collaborative multiagent reinforcement learning by payoff propagation. J. Mach. Learn. Res. 7, 1789–1828 (2006)

[8] Buffet, O., Aberdeen, D.: FF+FPG: Guiding a policy-gradient planner. In: The International Conference on Automated Planning and Scheduling (2007)

[9] Kimura, H., Yamamura, M., Kobayashi, S.: Reinforcement learning by stochastic hill climbing on discounted reward. In: ICML, pp. 295–303 (1995)

[10] Baxter, J., Bartlett, P.: Direct gradient-based reinforcement learning. Technical report, Research School of Information Sciences and Engineering, Australian National University (1999)

[11] Baxter, J., Bartlett, P.: Infinite-horizon policy-gradient estimation. Journal of Artificial Intelligence Research 15, 319–350 (2001)

[12] Barto, A.G., Sutton, R.S., Anderson, C.W.: Neuronlike adaptive elements that can solve difficult learning control problems, 81–93 (1990)

[13] Tao, N., Baxter, J., Weaver, L.: A multi-agent policy-gradient approach to network routing. In: ICML, pp. 553–560 (2001)

[14] Buffet, O., Dutech, A., Charpillet, F.: Shaping multi-agent systems with gradient reinforcement learning. Autonomous Agents and Multi-Agent Systems 15(2), 197–220 (2007)

[15] Williams, R.J.: Simple statistical gradient-following algorithms for connectionist reinforcement learning. Machine Learning 8(3), 229–256 (1992)

[16] Marinova, Z.: Planning in multiagent systems. Master's thesis, Department of Information Technologies, Sofia University (2002)

[17] Micalizio, R., Torasso, P., Torta, G.: Synthesizing diagnostic explanations from monitoring data in multi-robot systems. In: AIA 2006, IASTED International Conference on Applied Artificial Intelligence, Anaheim, CA, USA, pp. 279–286. ACTA Press (2006)

[18] Obst, O.: Using a planner for coordination of multiagent team behavior. In: Bordini, R.H., Dastani, M., Dix, J., El Fallah Seghrouchni, A. (eds.) PROMAS 2005. LNCS, vol. 3862, pp. 90–100. Springer, Heidelberg (2006)

[19] Obst, O., Boedecker, J.: Flexible coordination of multiagent team behavior using HTN planning. In: Bredenfeld, A., Jacoff, A., Noda, I., Takahashi, Y. (eds.) RoboCup 2005. LNCS, vol. 4020, pp. 521–528. Springer, Heidelberg (2006)

[20] Fraser, G., Wotawa, F.: Cooperative planning and plan execution in partially observable dynamic domains. In: Nardi, D., Riedmiller, M., Sammut, C., Santos-Victor, J. (eds.) RoboCup 2004. LNCS, vol. 3276, pp. 524–531. Springer, Heidelberg (2005)

[21] Grounds, M., Kudenko, D.: Combining reinforcement learning with symbolic planning. In: Tuyls, K., Nowe, A., Guessoum, Z., Kudenko, D. (eds.) ALAMAS 2005. LNCS, vol. 4865, pp. 75–86. Springer, Heidelberg (2008)

[22] Strens, M.J.A., Windelinckx, N.: Combining planning with reinforcement learning for multi-robot task allocation. In: Kudenko, D., Kazakov, D., Alonso, E. (eds.) AAMAS 2004. LNCS, vol. 3394, pp. 260–274. Springer, Heidelberg (2005)

[23] Aberdeen, D.: Policy-gradient methods for planning. In: Neural Information Processing Systems (2005)

[24] Buffet, O., Aberdeen, D.: The factored policy gradient planner (IPC 2006 version). In: Fifth International Planning Competition (2006)

[25] Chen, M., Dorer, K., Foroughi, E., Heintz, F., Huang, Z., Kapetanakis, S., Kostiadis, K., Kummeneje, J., Murray, J., Noda, I., Obst, O., Riley, P., Steffens, T., Wang, Y., Yin, X.: Robocup soccer server for soccer server version 7.07 and later (August 2002)

[26] Pecora, F., Cesta, A.: Planning and scheduling ingredients for a multi-agent system. In: UK PLANSIG Workshop, pp. 135–148 (November 2002)

Model-Free Active Balancing for Humanoid Robots

Sara McGrath[1], John Anderson[2], and Jacky Baltes[2]

[1] Cogmation, Winnipeg, Canada
ummcgrath@cs.umanitoba.ca
[2] Autonomous Agents Lab, University of Manitoba, Winnipeg, MB, Canada,
R3T 2N2
{andersj,jacky}@cs.umanitoba.ca
http://aalab.cs.umanitoba.ca

Abstract. To be practical, humanoid robots must be able to manoeuvre over a variety of flat and uneven terrains, at different speeds and with varying gaits and motions. This paper describes three balancing-reflex algorithms (threshold control, PID control, and hybrid control) that were implemented on a real 8 DOF small humanoid robot equipped with a two-axis accelerometer sensor to study the capabilities and limitations of various balancing algorithms when combined with a single sensor. We term this approach a *model-free* approach, since it does not require a mathematical model of the underlying robot. Instead the controller attempts to recreate successful previous motions (so-called baseline motions). In our extensive tests, the basic threshold algorithm proves the most effective overall. All algorithms are able to balance for simple tasks, but as the balancing required becomes more complex (e.g. controlling multiple joints over uneven terrain), the need for more sophisticated algorithms becomes apparent.

1 Introduction

For humanoid robots to move from fantasy to reality and become practical, they must be able to move over a variety of different terrain with different speeds and gaits without falling over and without the need to manually tune the gait. At present few small humanoid robots competing at the large international robotics competitions FIRA and RoboCup use any type of feedback control, and do little in the way of active balancing. For example, the technical challenge at RoboCup 2005 involved a section of uneven terrain, but this challenge was removed because most teams were unable to complete it. Of the few that did, some effectively cheated by adding skis to the feet of their robot.

Integrating dynamic balancing into robots will allow them to not only deal with changing surfaces, but also allow them to compensate for sudden changes in their equilibrium. Further, balancing will also allow robots to move to new gaits and tasks, such as crawling or load-bearing with greater ease and robustness.

Humans themselves use multiple sensors to balance: vision, position/force feedback (muscle feedback) and tilt/acceleration sensors (inner ear organs). Based on

L. Iocchi et al. (Eds.): RoboCup 2008, LNAI 5399, pp. 544–555, 2009.

an analysis of the dynamics of a robot, force feedback and motion based sensors have been used by many researchers in combination. A lot of these approaches have only been implemented in simulation and require an accurate mathematical model of the robot.

Most human balancing occurs subconsciously using balancing reflexes that can be demonstrated (by, for example tapping a human on the shin). These reflexes are present in any human motion, allowing people to preform basic balancing whatever their actions may be. Little research has been done on simple balancing reflexes for humanoid robots that employ a simple algorithm and a single sensor. However, to evaluate the capabilities and limitations of various algorithm and sensor combinations, this type of experimentation is extremely important. Multiple sensors potentially mask or ignore important sensor data.

To develop balancing reflexes, we decided to investigate each sensor individually and determine the extent of its usefulness.

A two-axis accelerometer was mounted on *Lillian*, a humanoid robot, as the sole balancing sensor. The choice of an accelerometer was based on the fact that these are small, cheap, and easily available. We often use mobile phones for on-board computation in our robots, and many modern mobile phones come equipped with accelerometers, but not other sensors. Another reason was that accelerometers are similar to the human inner ear and can thus be used to study and exploit human balancing reflexes.

Lillian, a robot from the University of Manitoba's Autonomous Agents Lab, has 8 degrees of freedom (DOF) (actuated with servo motors), a Memsic 2125 two-axis +/- 1.5g accelerometer and an Eyebot controller board. Figure 1 shows a model of the actuator setup for Lillian. Lillian uses 8 DOF, 4 for each leg.

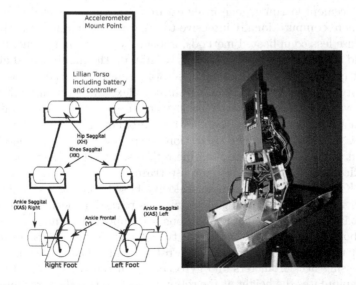

Fig. 1. Lillian, an 8 DOF robot. Kinematics are shown on the left, the actual robot during a tilting test with mounted accelerometers on the right.

The ankle can be moved in the sagittal plane using joints labeled XAS as well as the frontal plane using the joints labeled Y. One degree of freedom in the knee allows movement in the sagittal plane using joints XK. The hip can also be moved in the sagittal plane using joints XH.

The main design principles in building the robot were simplicity and frugality, as this forces the resulting algorithms to be more robust, versatile solutions.

2   Related Work

Several special-purpose algorithms to adjust a robot's motions are described in the literature. They can broadly be classified into two types: (a) Center Of Mass (COM) based algorithms keep the robot's COM in the supporting polygon of the robot's feet, (b) Zero Moment Point (ZMP) algorithms calculate the point in the horizontal plane at which all the moments are zero, and keep the ZMP in the supporting polygon. On the other hand, the versatile PID (Proportional Integral Derivative) controller is a basic control strategy that adjusts the error of a feedback output to a desired reading (baseline), making corrections based on a percentage of the error, and the integral and derivative components of the error. Similarly threshold control, with minimal corrections applied to readings outside a given threshold, was developed in previous research by [1].

Control methods can also be divided into simple balancing reflexes, with tight control loops between sensors and simple algorithms, and more specialized algorithms such as AutoBalancer [2]. Exploration into new sensors tends to begin with simpler control methods and then more complicated algorithms. While these intensive algorithms can use more complicated methods to be fine-tuned (ie, reinforcement learning, genetic algorithms), they apply to the specific robot, and are more computationally intensive to create, often requiring preprocessing. Reflexes use less complicated methods, and are simpler, more general functions that could ideally transfer more effectively than do the highly-tuned algorithms.

Huang et al. [3] investigated using sensory reflexes on their robot BHR-01, incorporating a ZMP reflex, a landing phase reflex, and a posture reflex into the dynamic walking pattern. These reflexes were triggered by sensory information, and when active, would compensate for any imbalances in the walk by adjusting the ankles, hips or knees. The corrections were used to adjust the offline pre-calculated walk pattern. These realtime reflexes, added to the walking pattern, proved effective in walking over uneven terrain.

Team KMUTT [4] used a simpler velocity based control to dynamically balance their robot. This balancing mechanism is part of a specific walk, chosen by the robot if its sensors indicate conditions are appropriate, not a standard part of the robot's behaviour. The robot has two PD-controlled walks: a slower static walk that uses the force sensors on the robot, and a faster dynamic walk that balances using accelerometers and gyroscopes. In the static walk, the PD controller manipulates the height at the robot's hip based on the force sensors in the foot. The dynamic walk controls the velocity at the hip with its PD controller. Team KMUTT competed at RoboCup 2006 using this code.

The University of Manitoba's Tao-Pie-Pie is the sole robot to use only gyroscope readings for active balancing [1]. The readings are processed and run through a Threshold controller, compensating for perturbations in the gait. The Threshold controller simply applies a minimal correction when sensor readings break a predefined boundary. Balance is explicitly added-on as corrections made to the pre-calculated walk gait. These corrections were used in competition to compensate for the poor surface, and were found to be better than the previous gait. Further, the Threshold method is extremely simple to implement and tune.

3 Determination of the Baseline Walk

Most approaches to humanoid balancing are model-based. In a model-based approach, the researcher creates a mathematical model of the robot to work on. Then a control algorithm is implemented and tested on the model (often in simulation). Finally, the controller is moved to the physical robot. Often, significant readjustments are required at this point to transfer the controller. One of the major disadvantages of this method is the fact that it is often difficult to develop a mathematical model for an existing robot that is accurate enough to help develop balancing algorithms.

Model-based control approaches attempt to overcome this problem by adding sensors to implement a feedback loop to overcome limitations in the model or external disturbances. This is problematic as in many cases the control variable of interest (e.g., center of mass or ZMP) can not be measured directly. So a controller will only be able to provide feedback via indirect measurements (e.g., joint positions). Unfortunately, the mapping of joint positions to ZMP is dependent on the model and the robot will fall over even if small changes to the weight of arms or torso are being made.

In contrast, our approach is implementation-based and does away with the need for a model of the robot. The robot itself is used as a test platform. This removes the necessity of adjusting the control strategy for the robot or unforeseen physical factors, as the testbed is perfectly accurate. The drawback is in maintaining the robot, as well as the length of time required for each test. Our methodology is to modify a pre-existing gait to improve it. Thus, our control algorithms are applied to the current gait, allowing the results of each strategy to be directly compared with each other and the uncorrected walk on the real robot.

Our approach relies on being able to record baseline motions (previously successful motions and their associated sensor readings) and on a common sense analysis of the joint motions of the robot (e.g., the position of the robot in the frontal plane is controlled by the ankle and hip servo, which has to compensate for gravity as well as the weight of the torso).

Figure 2 shows the X and Y readings for several dozen steps of a successful walk. As can be seen, most readings in a successful walk fall in a small band.

After analyzing the baseline walk, an upper and lower threshold limit was introduced as shown in Fig. 2. These threshold limits were used in the threshold controller as described in Section 4.

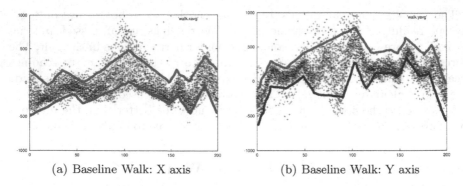

(a) Baseline Walk: X axis   (b) Baseline Walk: Y axis

Fig. 2. Sensor reading for the baseline walk in the X and Y axis

4 Methodology

Three popular control algorithms were used here to convert sensory readings into motion corrections and thus implement a balancing reflex: a standard PID controller, a threshold based controller, and a hybrid version of the two. For an accelerometer-based PID control, a baseline is created by either taking a sample from previously programmed good motions (e.g., a walk gait), or setting the baseline to be unmoving (stand). The closer the baseline conforms to the actual readings, the better the corrections. Threshold balancing first determines a threshold area, where no corrections will be made. This threshold is currently centered on the PID baseline, but with a broader range to avoids corrections from causing oscillation when the robot is standing still. Corrections are only applied once they pass a certain error value. This allows for a simpler means of adjusting settings, and eventual comparison or combination with PID settings. Both methods listed above have their own faults: PID, a quick reaction but a tendency to overcorrect, and Thresholds less of a tendency to overcorrect, but also less able to react quickly to larger errors. Thus a hybrid method was proposed and implemented to combine the best of both methods. It uses Thresholds for smaller corrections, but PID-based corrections for larger errors.

The Sum of Absolute Error (SAE) is a quantitative measurement used here to determine the relative goodness of varying walks. As all the correction methods have a baseline (or a set of thresholds), the deviations can be measured to directly compare trials. Summing the absolute errors measures the total deviations from the baseline. The greater the total deviation, the less the walk conforms to the baseline, and the less the corrections are helping.

As with most balancing approaches, Lillian was tuned for one plane before complicating matters with multiple planes, inclines, or uneven terrain. Further, as differing gait disturbances produce oscillation at differing points in the tuning, small increments of complexity are necessary to allow for the robot to be properly tuned. As many researchers use only one plane, or do not specify a method by which to tune for increasingly complicated balancing, one is given here.

Tuning begins with standing still, before moving to tilting, walking, and then multiple joints and planes. First, the corrective methods are tuned not to oscillate during stands. This gives a minimal value to use as a base for the PID settings and threshold bounds. The next step is for the robot to stand still on a surface that tilts in either the frontal or sagittal plane, thus forcing corrections for deviations from its desired pattern. In order to minimize possible complicating factors, the robot will correct in one axis at a time, attempting to maintain a sensor reading of zero. Once Lillian can remain stable while tilting, a more complicated sensor pattern can be used, and the tuning adjusted. Following a walk baseline on a flat surface instead of a steady line adds another level of complexity to the balancing, exposing previously hidden oscillation, as the balancing must compensate at many levels and speeds. Balancing in two axes is much more difficult than any of the previous tasks, as any oscillation (or even a too quick correction) in one axis can produce a rebound and perhaps oscillation in the other axis. Thus, two axes are not tested until balancing is working effectively in one axis. Again, two axes balancing starts with a simple tilting platform before moving to a more complicated pattern — walking on a even surface, then uneven terrain. Tuning is limited here to walking on a flat surface. Further tests are used to evaluate the algorithms.

While this methodology only uses a walk gait for its most complicated sensor pattern, any gait will have a repetitive pattern that can be used to calibrate a baseline. These gaits include crawling and running, to begin with. Once Lillian has been tuned for multiple planes of balancing, it is possible to replace the walk baseline with one created from a crawling gait, for instance. Corrections would then be made based on the new gait, but use the same tuning as the previous gaits. This allows the balancing reflexes implemented here to transfer quickly and easily to new gait patterns.

Tuning configuration began with the PID and threshold methods on the tilting platform. The test results were used to refine the PID and Threshold walking tests. The best results were used to test the hybrid method. A side-by-side comparison of all the best results was used to choose settings for the final tests with the perturbed walks, and the stepping field tests.

Testing on the tilting platform began with the PID and Threshold correction methods applied to individual joints: ankle sagittal (XAS), knee sagittal (XK), and ankle lateral (Y), and then the best results were used to create further walking tests. The platform was tilted from -30° to +30°, from a starting position of 0°, with an angular velocity of 240° per minute. The tests were coarse grained, running a trial with controller gains of 150, 450, 750 and 1050, with delays of 10, 40, and 70 milliseconds required between corrections.

After tuning the P settings, the best setting from each joint was taken and tested with a range of differing D settings, by setting a baseline that stayed at +1000 for the first half of the gait, and then moved to -1000 for the second half. The only movement was provided by the corrections adjusting the accelerometer readings. It was thus possible to look for the overshoot caused (or avoided) by varying the D parameter, running each trial once for an initial exploration.

Results from the single joint tests were used to tune for correcting multiple joints simultaneously. Results that improved on the baseline where possible, and the best results available otherwise, were selected for further tuning and divided into two (best and good). Each of the best joint settings was paired with each of the best and good settings for another joint, to allow reasonably thorough testing without factorial explosion. Each of these tests were run three times, to reduce the noise in the data. Further, X joints were first tested and combined with each other (XAS and XK) before adding in a second plane with the Y ankle joint. This follows the ideology of the prior tests in adding in as little complexity as possible to each test for a clear picture of the effects of each new factor.

5   Evaluation

A basic walking gait was first used to evaluate the best of the tuning results side by side. Any result that improved on the gait (in the plane it was correcting) was used for the final evaluation. Tests include randomly perturbing the walk to varying degrees, and running Lillian over a stepping field, with and without balancing reflexes, for approximately twenty seconds. This stepping field (Figure 3) was constructed of layered pieces of cardboard, always providing a height difference of 3 mm between neighbouring pieces, but possibly more than one piece over the length or width of Lillian's foot.

Gaits were perturbed by randomly varying the control points of one good gait over a spread of 5 or 10 set points ([-2 .. 2], or [-5 .. 5]), at multiple points throughout the gait. The disturbances were applied to both joints, as the balancing control assumes that the movements of the joints are coupled.

The sheer number of tests carried out to tune the controllers on the robot make it difficult to determine what settings best improves the walk. Therefore, the best settings in each method, PID and Threshold, were directly compared against each other, though the SAE is normally calculated differently for both. As the thresholds are currently set equidistantly from the PID baseline, the SAE was calculated using the PID baseline on all non-tuning trials.

Fig. 3. Stepping field used in this evaluation

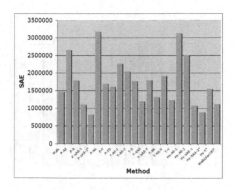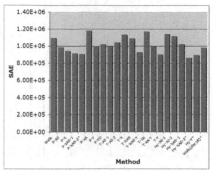

Fig. 4. Random walking test results, by method and settings. Y axis shows SAE of raw sensor data.

The PID controller settings chosen were the best for each joint, the best combination of settings for the X joints, the best for all the joints, and the best P and PD controllers for Y. No D settings were chosen for any of the X joints as they had not improved the controllers. found to improve on the basic walk, no further combinations of settings involving them were chosen.

The Threshold settings were similarly chosen as the best setting for each joint, and the best combinations of X and XY joints. Two XY settings were chosen, as one improved on the basic walk for Y, and the other was the best set of corrections for X. Two more settings were chosen for testing: XAS + Y and XK + Y, as one of the XY settings was the same as the three best individual joint settings. Further, each of the single joint settings (unlike the PID controller) actually improved on the basic walk, suggesting that a single X joint setting with a Y setting could prove effective.

As the hybrid controller depends on combining the Threshold and PID controllers, it was only tested after the best settings for the simpler controllers were determined. The hybrid controller is tested on settings that improve the walk: on the top two XAS settings for both PID and Threshold; the combination of the Y controllers; and for comparison's sake, the two best Threshold XY controllers with the best PID XY controller.

Figure 4 shows the results of the direct comparison. Less than half actually improve upon the uncorrected walk, in the plane(s) that they are correcting for. This criteria is used to select settings for the final evaluations, giving P-XAS-1, P-XAS-2, Hy-XAS-1, Hy-XAS-2, T-XAS, T-XK, T-Y, P-Y, and Hy-Y.

5.1 Random Walks

We tested our algorithm by perturbing the walk (simulating control error in the servo motors) by varying degrees (5%,10%, and 15%). Even with small perturbations of 5%, the need for corrections becomes apparent. The SAE of the methods correcting in the X plane is a third to a quarter of the uncorrected methods. Solely Y corrections are not as impressive, leaving the walk unimproved.

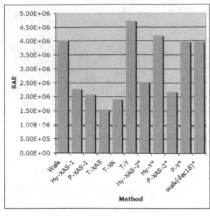
(a) Random Walk 10%: X

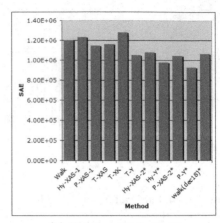
(b) Random Walk 10: Y

Fig. 5. Random walking test results with a perturbation spread of 10%, by method and settings. Y axis shows SAE of raw sensor data.

The differences between correction methods in this test are much less than the difference between the corrected and uncorrected tests.

Increasing the perturbations to up to a spread of 10% caused the differences between the methods to become yet more pronounced, as shown in Figure 5. The Threshold results are the best in the X plane, but with a greater difference between them and the other methods. Any correction method still shows a marked improvement over the uncorrected walk.

5.2   Stepping Field

Results from the stepping field (see Figure 6) show that the balancing degrades here. For the X plane, only the PID corrections improve on the uncorrected walk. While the hybrid method actually worsens the SAE readings, Threshold corrections merely leave the walk mostly unchanged. Unlike previous tests, Threshold corrections on the XK joints are actually slightly better than those of the XAS joint. Compared to the differences between the corrected and uncorrected walks in the perturbation test, however, the differences between the results are relatively minor, indicating the X corrections are not making a large difference to the walk. As expected, the lack of corrections to the Y plane by these methods is clearly shown, as the Y readings for all the X correction methods are worse than the uncorrected Y walk. The threshold corrections for Y improve on the uncorrected walk for both X and Y, not just Y. Therefore, unlike the perturbation test, where the main corrections to be made were to the X plane, here the most effective corrections are to the Y plane, and correcting in X without correcting Y will not be very effective.

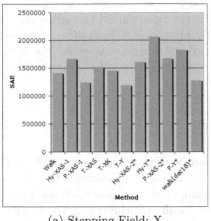
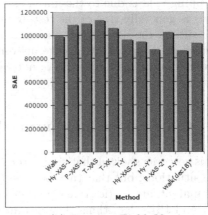

(a) Stepping Field: X   (b) Stepping Field: Y

Fig. 6. Stepping field test results, by method and settings. Y axis shows SAE of raw sensor data.

5.3 Analysis

Directly comparing all results with each other led to some interesting observations. The more joints a controller attempted to control, the worse the balancing became. Similarly, more complicated controllers, such as PD instead of P, did not improve the balancing. The simplest ideas, such as the threshold controller, were just as effective in controlling the walk, and maintaining the desired accelerometer readings. The simpler threshold controller also responded better to adding more joints, not overreacting as much as the PID controller.

This conclusion of complexity not being handled well was further born out by the random walks. Different joints were also more effective in controlling the balancing. The XAS joint is much more effective at controlling the robot's balance than any of the other tested joints. Further, as the perturbations increased, Threshold showed as more effective than PID. The hybrid controller is still better than no controller, but not as good as either of the simpler controllers.

The stepping field upsets the previous trends with PID actually outperforming Thresholds. This suggests the threshold controller is unable to react quickly and strongly enough for the corrections required. The random walks had a change of perhaps 5 or 10 servo settings over 1 second; the threshold controller allows approximately 10-12 servo setting changes per second. Thus the threshold controller could compensate for the changes in the walk, while the PID controller tended to overreact. Once on the stepping field however, the changes would occur instantly, and require compensation of 5-10 servo settings (or perhaps more). The PID controller could react instantly, while the threshold control would move more slowly to correct. This slower reaction time would allow the robot enough time to accumulate inertia in the wrong direction, making it even more difficult for the robot to compensate. A further explanation of the poorer results on the stepping field is due to the initial assumption that all joint actions are coupled. This is generally true of the testing, as with the randomized perturbations and the tilting, the robot's feet remain relatively aligned with each other. The stepping field, however, due to the unevenness of the terrain, allows for the feet to become misaligned, and this is not easily corrected by the balancing reflex, as currently implemented.

6   Conclusions and Future Work

This research showed that simple reflexes can be used to balance a robot in simple situations, but that they become unable to handle the complexities of more normal situations (ie, walking while controlling multiple joints). Both Threshold and PID algorithms showed impressive results on the tilting tests, with a broad range of settings providing beneficial corrections to the robot. Moving those corrections to a walk demonstrated the shortcomings of the reflex algorithms, as the speed and complexity of balancing a walking robot started to overcome the balancing capabilities of the controllers. This was most noticeable with PID, as Thresholds still had several useful settings. Adding multiple joints to be controlled, or moving the robot to an uneven surface further demonstrated the inability of the balancing reflexes to compensate for the amount of variability in the walk. While the reflexes were not able to fully compensate for any surface, they did improve the walk noticeably against smaller, more regular changes, as shown by the random perturbation tests.

No one algorithm was consistently best; rather, the most effective algorithm depended on the circumstances in which the robot was placed. Overall, Thresholds appeared best for slower, steadier changes, while PID responded better to occasional larger changes. The Hybrid method was never the best method, but was almost always in between the two other algorithms in terms of goodness; never the best, but regularly the runner-up. In every test, however, at least one of the correction methods matched or outperformed the uncorrected walk, showing that a tuned correction method is better than no correction.

Overall, due to its ease of tuning and general performance, the threshold method is the easiest and most useful choice for future balancing. While there are differences between the methods, for a single joint they all do improve upon the uncorrected walk, and thus any method is better than none.

This research has focused on a walk gait; this should be expanded to include more complex motions such as crawling or load-bearing in future. The simple reflexes used here can apply directly to a different gait (such as crawling) by simply calibrating the crawl to give Lillian a new baseline, and applying the same corrections used for the walk on a crawl. Load-bearing is even simpler, as it relates to a previous motion. Changes should not have to be made to the robot, but simply weights added to the robot, and the balancing reflexes should immediately begin to compensate for the extra weight.

Future work should address the initial assumption was that all joint movement would be coupled, investigating the possibilities and difficulties involved with allowing each foot or leg to be corrected separately. Differing threshold bounds were used in prior work [1], but have not been investigated here. They may improve the corrections made by the threshold algorithm, as it allows for corrections to be more or less sensitive without extra tuning. Finally, physical modifications to the robot could make it more difficult for the robot to balance, such as by adding weight at the head for extra sway.

This research provides an initial foundation for work looking into balancing reflexes with accelerometers, as it shows that it is possible to balance with only an accelerometer, and a simple control method. However, it also shows that these methods only work for reasonably simple balancing. More complex adjustments are not implemented well with these methods. Terrain such as the stepping field will require in addition more complicated (or at least more effective) means of balancing than either a PID or a threshold controller.

References

1. McGrath, S., Baltes, J., Anderson, J.: Active balancing using gyroscopes for a small humanoid robot. In: Mukhopadhyay, S.C., Gupta, G.S. (eds.) Second International Conference on Autonomous Robots and Agents (ICARA), pp. 470–475. Massey University (2004)
2. Kagami, S., Kanehiro, F., Tamiya, Y., Inaga, M., Inoue, H.: Autobalancer: An online dynamic balance compensation scheme for humanoid robots. In: Proceedings of the Fourth International Workshop on Algorithmic Foundations on Robotics (2000)
3. Huang, Q., ming Zhang, W., Li, K.: Sensory reflex for biped humanoid walking. In: Proceedings of the 2004 International Conference on Intelligent Mechatronics and Automation (2004)
4. Kulvanit, P., Srisuwan, B., Siramee, K., Boonprakob, A., Laowattana, D.: Team kmutt: Team description paper. In: Proceedings of the RoboCup 2006, December 12 (2006), www.humanoidsoccer.org/qualification/KMUTT_TDP.pdf

Stereo-Vision Based Control of a Car Using Fast Line-Segment Extraction

Brian McKinnon, Jacky Baltes, and John Anderson

Autonomous Agents Laboratory, University of Manitoba, Winnipeg,
MB, Canada, R3T 2N2
{jacky,andersj}@cs.umanitoba.ca
http://aalab.cs.umanitoba.ca

Abstract. This paper describes our work on applying stereo vision to the control of a car or car-like mobile robot, using cheap, low-quality cameras. Our approach is based on line segments, since those provide significant information about the environment, provide more depth information than point features, and are robust to image noise and colour variation. However, stereo matching with line segments is a difficult problem, due to poorly localized end points and perspective distortion. Our algorithm uses integral images and Haar features for line segment extraction. Dynamic programming is used in the line segment matching phase. The resulting line segments track accurately from one frame to the next, even in the presence of noise.

1   Introduction

This paper describes our approach to stereo vision processing for the control of a car-like mobile robot using low-quality cameras. The end goal of our work is a mobile platform that can be viewed as expendable for applications such as Urban Search and Rescue (USAR), necessitating inexpensive equipment. We are working on making stereo vision an inexpensive and robust alternative to the more expensive forms of sensing that are currently employed for robotic mapping and exploration (e.g. high-resolution laser scanners). A focus on inexpensive equipment also leads to the development of more broadly applicable intelligent software solutions, compared to relying on expensive hardware to deal with difficulties such as image noise.

The goal of all stereo vision processing is to produce dense depth maps that can be used to accurately reconstruct a 3-D environment. However, our target application imposes two additional constraints that make this problem more difficult. First, since the stereo system is intended for a mobile robot, the approach must support real-time depth map extraction. Second, since we are using inexpensive equipment, the approach must be robust to errors such as noise in the acquired images and the calibration of the cameras.

Most of the focus in stereo vision processing to date has been on point feature-based matching. The most notable feature extraction methods are corner detection [1,2], SIFT [3], and SURF [4]. Most point feature based stereo matching

L. Iocchi et al. (Eds.): RoboCup 2008, LNAI 5399, pp. 556–567, 2009.

relies on the epipolar constraint [5] to reduce the search space of the feature matcher to a 1-D line. The epipolar constraint can be applied using the Essential or Fundamental matrix, which can be determined using several methods [5].

The problem encountered in attempting to employ these mechanisms under conditions typical of USAR with inexpensive equipment is that cameras tend to shake as the robot moves. Unlike the heavy equipment used in events such as the Grand Challenge, inexpensive robots designed to be deployed in large groups in expendable settings must be based on far less expensive materials. In our USAR work, for example, we employ very cheap ($10) webcams mounted on 3mm aluminum sheet-metal tie, connected to a servo so that the stereo vision mechanism can be panned. Under these conditions, even small changes in the position of one camera relative to the second require the recalculation of the epipolar lines, which can be an expensive and error-prone task during the operation of the robot. While a more elaborate hardware setup would be another solution, forcing the work to be performed on simple hardware not only makes the solution more widely deployable, it demands a more robust and generally applicable solution that serves to advance the state of the art. In this case, solving this problem using an inexpensive vision system demands a more robust feature set that would be trackable over a larger 2-D search space.

These weaknesses led us to explore region-based stereo matching [6] to provide a more robust feature set. However, region segmentation proved to be a difficult task in complex scenes under real-time constraints. In addition, regions would not always form in a similar manner, making the hull signature matching approach we employed prone to errors. The most important piece of information gleaned from this work was that most of the information in a region was contained on the boundary between two regions. This intuitively led to a shift in our approach towards the extraction and matching of boundaries - in particular, line segments. Though several authors have proposed methods for line segment matching [7,8], none could achieve the desired speed and accuracy needed for our mobile robotics applications.

The line segment-based approach taken in this paper emphasizes the dual requirements of real-time processing and robustness to camera jitter due to robot movement. The latter requirement means that knowledge of the epipolar lines in particular cannot be assumed, nor can we assume rectified stereo frames. In addition, we allow the off-the-shelf webcams in our implementation to freely modify the brightness and contrast of an image (typical of low-priced hardware), and so we also cannot assume colour calibration.

We describe two algorithms designed for these conditions, involving fast line segment extraction and two-frame line segment matching. The fast line segment extraction makes use of integral images to improve the performance of gradient extraction, as well as provide an edge representation suitable for binary image segmentation. Line segment matching is achieved using a dynamic programming algorithm, with integral images providing fast feature vector extraction.

We begin by introducing existing methods for line segment based stereo processing, and then describe our algorithms in detail. This is followed by some initial results that have been generated using the described line segment based method.

2   Related Work

This section describes existing line segment extraction and matching algorithms. The term *Line segment extraction* refers to the process of identifying boundaries between image regions, and grouping the points into a line segment feature. *Matching* is the process of identifying the new location of a previously observed feature given some change in the position of the cameras.

The extraction of line segments can be approached using either global or local methods [9]. Both methods rely on identifying boundary pixels using one of several methods of edge detection (for more information refer to [10]).

Most global line segment extraction algorithms are extensions of the generalized Hough transform [11]. In the generalized Hough transform each boundary pixel votes for candidate lines in the Hough space accumulator. Once all boundary pixels are processed, a search is done in the accumulator to find peaks that exceed a threshold number of votes. This Hough transform is used to extract infinite lines, but many methods have been presented [11,12] to deal with the extraction of line segments with start and end points. The primary difficulty involved with global line segment extraction algorithms is that they are computationally expensive, and can generate imaginary lines formed by textured surfaces.

The simplest form of local line segment extraction uses chain coding [13]. Chain code values represent the location of the next pixel in the chain using either a 4 or 8 connected neighbourhood. The boundary is followed starting from an initial edge point (generally the top-leftmost point in the chain) and followed until the chain returns to the start. Noise can be filtered from the chain code using a median filter over the connection direction. The final chain code is then examined for line segments by finding runs of equal value neighbour connections. Local line segment methods are more sensitive to noise in the image, and so most recent work focuses on joining small segments into larger segments [9,11].

Line segment matching is generally considered to be a more difficult problem than interest point matching. Although it seems simple to view line segments as simply a connected start and end point, the problem is that end points on line segments tend to be very unstable due to noise in the image [8]. The two primary features of a line segment that are used for matching are the colour information around the line segment, and the topology of line segments. Bay et al. [8] used colour information from points three pixels to the left and right of a line segment to generate histograms. The histograms from the two candidate lines were normalized, and the distance between them was calculated using the Euclidean distance between histogram bins in the defined colour space. The colour information produced soft matches [8], reducing the number of potential matches. By applying the sidedness constraint [8], the incorrect matches were filtered out to produce the final set of matched lines. The use of colour information was limited to situations where the capture devices produced very consistent colour output. In the more common situation, colour information varies greatly between the images [14] due to automatic brightness and contrast corrections differing between the capture devices.

If colour information is ignored, then matching must be based on the topology of the images. Hollinghurst [14] defined a set of rules for topological matching that includes initial matching and refinement operations. Using geometric constraints, initial matches are generated based on overlap, length ratio, orientation difference, and disparity limit. The initial matches are refined by applying constraints that either support or suppress a potential match. Matches are suppressed if they match multiple line segments, and if they violate the sidedness constraint. Support is generated by collinearity and connectivity, including parallelism, junctions at endpoints, and T-junctions. These constraints are applied to the initial matches until only supported matches remain.

To deal with the problem of unstable endpoints, Zhang [7] measured the amount of overlapping area in a given calibration. In his experiments, the extrinsic calibration for the two cameras was unknown, and instead the essential matrix was estimated by using rotated points on an icosahedron. He found that given a set of matched line segments, the essential matrix could be estimated as accurately as a calibrated trinocular system. The system did show that matched line segments could be used to generate a depth map of the image. One problem is that there is no discussion of the line segment matching problem, and it appears that pre-matched (possibly by a human operator) line segments were used. As a result, it is difficult to predict how line segment-based stereo vision will work in the presence of noise found in real-world images.

3   Implementation

This section describes our algorithms for line segment extraction and matching. For simplicity and grounding, the method described in this paper will be applied to 8-bit gray-scale stereo images of dimensions 320x240. The sample images used to describe our work are shown in Fig. 1.

3.1   Line Segment Extraction

Our algorithm for line segment extraction is divided into five steps:

1. Integral Image Generation
2. Haar Feature Extraction
3. Gradient Thinning
4. Binary Segmentation
5. Overlap Clean-up

The integral image is an efficient data structure for extracting large amounts of data from images in constant time. In an integral image, the value stored in each pixel is the sum of all intensities in the area above and to the left of that pixel.

Fig. 1. The left and right input images used in the description of the algorithm

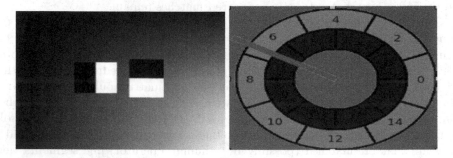

Fig. 2. The two main data structures used by the algorithm. Left, two Haar features overlayed over the integral image from Fig. 1. Right, the encoding of the gradient of a line using an edge cell.

This means that the sum of any rectangular subregion inside the integral image can be extracted with only four table look-ups and three addition/subtraction operations.

The most practical application of integral images is to extract arbitrarily sized Haar features from an image [15]. While there are several Haar features that can be extracted from an image, this algorithm focuses on the vertical and horizontal gradient features. During the Haar feature extraction stage of the algorithm, the two gradient features are extracted at a kernel size of 4x4, 8x8, and 12x12 pixels. The values are cached into a look-up table to improve performance during the matching stage. These features then provide the input for both gradient thinning and dynamic program matching.

The gradient thinning algorithm is based on the Canny method, with the addition of the Haar feature gradients and an efficient data structure. The goal of this method is to extract the peak gradients by keeping only the strongest values along the gradient orientation. From the cached Haar features, one kernel size is selected for thinning. A threshold based on the magnitude of the gradient is initially applied to the pixels. Pixels exceeding the defined threshold are used to activate an edge cell (EC).

The EC is an efficient data structure for storing gradient orientations, since it allows binary comparisons of the gradient orientation. Each EC is a dual layer ring discretized into 8 cells, allowing a resolution of $\pi/4$ per ring. The inner layer is offset from the outer layer by a distance of $\pi/8$, and the combined layers produce a $\pi/8$ resolution for the EC. The gradient orientation of each pixel activates one cell in each layer of the pixel's EC. Using this activation, the EC can be quickly tested for a horizontal, vertical, or diagonal orientation with a binary logic operation. ECs can be compared against each other using either an equal operator, which returns true if both cell layers overlap, or a similar operator, which returns true if at least one cell layer overlaps. The thinning direction is determined by the EC activation, and only the strongest gradient pixels along the thinning orientation are processed by the binary segmentation.

Binary segmentation is the core of the line segment extraction, and makes use of the EC data structure described above. For each active pixel, the EC is separated into two Split ECs (SEC) with the first containing the inner ring activation, and second containing the outer ring activation. The 8-neighbour binary segmentation then tests each of the two SECs separately against its neighbours. The binary test for the neighbour returns true if that neighbour EC is similar (at least one cell overlaps) to the current SEC. The final result of the segmentation is that each pixel is a member of two thinned binary regions, or line segments as they will be referred to from now on. The overlapped line segments are useful, but we generally prefer that each pixel only belong to a single line segment.

There are two cases of overlap that need to be cleaned up to produce the final result. The first clean-up step is the removal of any line segments contained largely within another longer line segment. This is achieved by simply having each pixel cast a vote for the longer of the two line segments for which it is a member. Any line segment with a number a votes below a threshold is discarded. The second clean-up step involves finding an optimal cut point for any partially overlapping line segments. This is achieved by finding the point in the overlapping area that maximizes the area of a triangle formed between the start of the first line segment, the cut point, and the end point of the second line segment. The overlapping line segments are then trimmed back to the cut point to produce

Fig. 3. The final line segments extracted by the described algorithm

the final line segment. Any line segments smaller than a defined threshold are discarded at the end of this stage.

Fig. 3 shows the output of our line extraction algorithm. The result are good line segments for stereo vision. However, more post processing could be performed to improve the line segments and further reduce noise. A few examples include merging line segments separated by noise, or extending end points to junctions. These features await future implementation.

3.2   Dynamic Matching

The matching of line segments is a difficult task, since the end points are generally poorly localized and the segments can be broken into multiple pieces due to noise in the image. In addition, some line segments may not be completely visible in all images due to occlusion. To address these problems, a dynamic programming solution is applied to the line segment matching problem. The dynamic programming table consists of points from the source line segment, $ls1$, making up the row header, and the points from the matching line segment, $ls2$, making up the column header. The goal is to find the overlapping area that maximizes the match value between the points of the two line segments. To compare the points of two line segments, we return to the Haar features extracted earlier. The match value for two point feature vectors, $v1$ and $v2$, is calculated as:

$$M = \sum (sign(v1_i) == sign(v2_i)) * \min(v1_i, v2_i) / \max(v1_i, v2_i)$$

Each insertion into the table involves selecting a previous match value from the table, adding the current match value, and incrementing a counter for the number of points contributing to the final result.

The insertion of the match values into the dynamic programming table requires certain assumptions to prevent the table from becoming degenerative. The assumptions are defined algorithmically using the variables, x for the current column, y for the current row, Dp for the dynamic programming table, St for the match sum table, Ct for the point count table, and M for the current match value. The Dp value is generated from St/Ct and for simplicity any assignment $Dp[x][y] = Dp[x-1][y-1]$ is actually $St[x][y] = St[x-1][y-1] + M$ and $Ct[x][y] = Ct[x-1][y-1] + 1$. With this in mind, the assumptions used to generate the Dp table are:

1. If a line segment diverges from the center line, it cannot converge back to the center line. This prevents oscillation in the line match, and is enforced using the rules:
   - if $x = y$ then
     $Dp[x][y] = Dp[x-1][y-1]$
   - if $x > y$ then
     $Dp[x][y] = best(Dp[x-1][y-1], Dp[x-1][y])$
   - if $x < y$ then
     $Dp[x][y] = best(Dp[x-1][y-1], Dp[x][y-1])$

2. No point in the first line can match more than two points in the second line. This prevents the table from repeatedly using one good feature to compute all matches, and is enforced by defining the behaviour of the *best* function as:
   - if $Ct[x1][y1] > Ct[x2][y2]$ then
     $best(Dp[x1][y1], Dp[x2][y2]) = Dp[x2][y2]$
   - else if $Ct[x1][y1] < Ct[x2][y2]$ then
     $best(Dp[x1][y1], Dp[x2][y2]) = Dp[x1][y1]$
   - else
     $best(Dp[x1][y1], Dp[x2][y2]) = \max(Dp[x1][y1], Dp[x2][y2])$

The best match value is checked in the last column for all rows with an index greater than or equal to the length of $ls2$, or for any entries in Ct with a count equal to the size of $ls1$. Once a best match value is found, a back trace is done through the table to find the point matched that produced the best match. The position difference between the matched points in $ls1$ and $ls2$ is used to determine the disparity of the matched pixels. Linear regression is used to find a slope for the disparity for all points in $ls1$. The best match value is then recalculated using the disparity generated by the linear regression line, since the best match value should be based on the final disparity match. The matching process in repeated for all $ls2$ segments in the potential match set.

To reduce the number of line segments that are compared, a few simple optimizations are performed. The primary filtering method is to only compare line segments that have similar edge cell activations. A secondary filtering method is to apply two different thresholds to the maximum search space. The first threshold involves placing a bounding rectangle extended by the size of the search space around $ls1$, with a second rectangle placed around $ls2$. If the two bounding rectangles overlap, then the comparison continues. The second threshold is done on a point-by-point basis, with points outside the search space receiving a match value of zero. These simple optimizations greatly increase the speed of the algorithm, making real-time performance achievable.

4   Experiments

To evaluate these algorithms, we performed intial testing using the well-known Middlebury 2005 and 2006 datasets [16].

Using the 27 image pairs from the 2005 and 2006 Middlebury test set, the algorithm described above has a matching accuracy of $71.8\% + / - 11.5\%$ for matches within one pixel of error (See Fig. 4. If only unoccluded pixels are considered (assuming that some form of occlusion detection has been added to the algorithm) the accuracy rises to $78.8\% + / - 10.0\%$. In addition, the line segments cover an average of $7.0\% + / - 2.4\%$ of the image, providing far more density than point matches.

The most important results, however, are observed in the environments in which we intend to apply this work: unstructured robot control domains using low quality webcams. In such environments, the minimum and maximum

Fig. 4. Left, depth map generated by the line segment matching algorithm. Right, ground truth depth map from the Middlebury set. Depth maps have a small dilation applied to improve visibility.

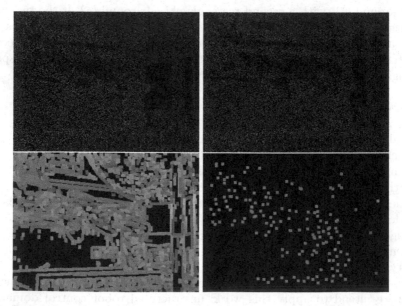

Fig. 5. Top, left and right stereo web cam images. Bottom, line segment matching disparity (left), SURF feature matching (right).

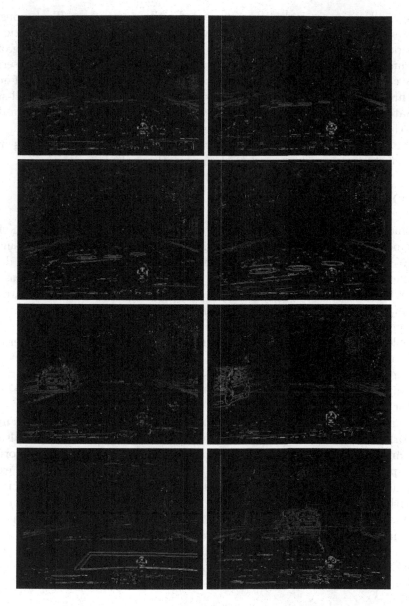

Fig. 6. Output of our algorithm on a sequence from an vehicle equipped with stereo vision cameras

disparity in the horizontal and vertical directions are unknown. For the sake of execution time, a $+/-$ 80 pixel horizontal and $+/-$ 20 pixel vertical disparity limit are applied to 320x240 webcam images. The results are compared qualitatively against a SURF [4] implementation, by examining the disparities generated at neighbouring points.

Fig. 5 illustrates a single stereo image under these conditions. To demonstrate performance on a sequence of images, we equipped an inexpensive robotic plat-form (a large-scale plastic RC car) with a stereo-vision system consisting of two webcams mounted on an aluminum bar, which was itself mounted on a servo to allow camera panning. The image sequence in Fig. 6 shows the output of our al-gorithm. The extracted line segments are robust and are particularly well-suited as matching features for stereo vision in this domain, since apart from a depth map these also incorporate the first stage of object recognition.

The image sequence shows that the extracted line information includes crucial areas of the image such as road markings or other vehicles.

5   Discussion

The method of line segment extraction described in this paper has been shown in demonstration to produce good quality matches, and has proven its suitabil-ity to real-world image sequences such as the ones from autonomous driving vehicles. By combining integral images and dynamic programming it has been shown that real-time processing rates are achievable using commodity hardware. The matching algorithm also shows that good quality matching can be achieved without the epipolar constraint. If exact calibrations are not required then it is possible to build stereo camera rigs with the ability to converge the cameras and focus on specific distances.

The immediate step in applying this work is to integrate it with our previ-ous implementations of stereo vision-based USAR [17] to form a more robust platform for this application.

We are also currently performing a larger empirical comparison using a mobile platform under the real-world conditions depicted in Fig. 6. There are also sev-eral improvements to both the line segment extraction and matching algorithms being proposed for future development. These include improved gradient extrac-tion, using one of the methods evaluated by Heath et al [10]. The extraction algorithm is very robust, and therefore future development will focus on merg-ing line segments split by noise, and extending end points to meet at junctions. The matching method must be improved to address inaccuracies when dealing with repeated patterns, and occlusion boundaries. Repeated patterns will require more of a global optimization over the matching values, rather than the greedy method currently used.

References

1. Shi, J., Tomasi, C.: Good features to track. In: IEEE Conference on Computer Vision and Pattern Recognition (1994)
2. Tomasi, C., Kanade, T.: Detection and tracking of point features. Carnegie Mellon University Technical Report CMU-CS-91-132 (1991)
3. Lowe, D.G.: Object recognition from local scale-invariant features. In: Proceedings of the International Conference on Computer Vision ICCV, Corfu., pp. 1150–1157 (1999)

4. Bay, H., Tuytelaars, T., Gool, L.V.: Surf: Speeded up robust features. In: Leonardis, A., Bischof, H., Pinz, A. (eds.) ECCV 2006. LNCS, vol. 3951, pp. 404–417. Springer, Heidelberg (2006)
5. Hartley, R.I., Zisserman, A.: Multiple View Geometry in Computer Vision. Cambridge University Press, Cambridge (2004)
6. McKinnon, B., Baltes, J., Anderson, J.: A region-based approach to stereo matching for USAR. In: Bredenfeld, A., Jacoff, A., Noda, I., Takahashi, Y. (eds.) RoboCup 2005. LNCS, vol. 4020, pp. 452–463. Springer, Heidelberg (2006)
7. Zhang, Z.: Estimating motion and structure from correspondences of line segments between two perspective images. IEEE Trans. Pattern Analysis and Machine Intelligence 17, 1129–1139 (1995)
8. Bay, H., Ferrari, V., Van Gool, L.: Wide-baseline stereo matching with line segments. In: IEEE Conference on Computer Vision and Pattern Recognition, vol. 1, pp. 329–336 (2005)
9. Jang, J.H., Hong, K.S.: Fast line segment grouping method for finding globally more favorable line segments. Pattern Recognition 35, 2235–2247 (2002)
10. Heath, M., Sarkar, S., Sanocki, T., Bowyer, K.: A robust visual method for assessing the relative performance of edge-detection algorithms. IEEE Transactions on Pattern Analysis and Machine Intelligence 19, 1338–1359 (1997)
11. Kim, E., Haseyama, M., Kitajima, H.: Fast line extraction from digital images using line segments. Systems and Computers in Japan 34 (2003)
12. Mirmehdi, M., Palmer, P.L., Kittler, J.: Robust line-segment extraction using genetic algorithms. In: 6th IEEE International Conference on Image Processing and Its Applications, pp. 141–145 (1997)
13. Gonzalez, R.C., Woods, R.E.: Digital Image Processing. Addison-Wesley Longman, Boston (2001)
14. Hollinghurst, N.J.: Uncalibrated stereo hand-eye coordination. PhD thesis, University of Cambridge (1997)
15. Viola, P., Jones, M.: Rapid object detection using a boosted cascade of simple features. In: Proceedings of the IEEE Conference on Computer Vision and Pattern Recognition, vol. 1 (2001)
16. Scharstein, D., Szeliski, R.: A taxonomy and evaluation of dense two-frame stereo correspondence algorithms. International Journal of Computer Vision 47 (2002)
17. Baltes, J., Anderson, J., McKinnon, B., Schaerer, S.: The keystone fire brigade 2005. In: Bredenfeld, A., Jacoff, A., Noda, I., Takahashi, Y. (eds.) RoboCup 2005. LNCS, vol. 4020. Springer, Heidelberg (2006)

A Layered Metric Definition and Evaluation Framework for Multirobot Systems

Çetin Meriçli and H. Levent Akın

Department of Computer Engineering
Boğaziçi University
İstanbul, Turkey
{cetin.mericli,akin}@boun.edu.tr

Abstract. In order to accomplish it successfully, the top-level goal of a multi-robot team should be decomposed into a sequence of sub-goals and proper sequences of actions for achieving these subgoals should be selected and refined through execution. Selecting the proper actions at any given time requires the ability to evaluate the current state of the environment, which can be achieved by using metrics that give quantitative information about the environment. Defining appropriate metrics is already a challenging problem; however, it is even harder to assess the performance of individual metrics. This work proposes a layered evaluation scheme for robot soccer where the environment is represented in different time resolutions at each layer. A set of metrics defined on these layers together with a novel metric validation method for assessing the performance of the defined metrics are proposed.

1 Introduction

In a multi-robot system, both individual robots and the entire team are confronted with a set of decisions for achieving both short-term and long term goals. In order to make a decision, one needs to evaluate the current situation of the environment. Evaluation of the current situation requires statistically consistent quantitative metrics but both defining appropriate metrics and validating them are challenging processes.

Robot soccer is a good platform to test and develop multi-agent applications because it has some physical limitations such as limited and noisy sensorial information and limited moving capability as in the real life and it also has a highly dynamic, real-time environment.

Definition of metrics for performance evaluation of multirobot systems is is not a deeply investigated research topic so far. VerDuin *et al.* drew the attention on the importance of evaluating the performance evaluation metrics by investigating the efficiencies of different criteria used in model evaluation in machine learning [1]. Yavnai proposed a set of metrics for classification of system autonomy [2]. Horst applied previously published Naive Intelligence Metric on autonomous

L. Iocchi et al. (Eds.): RoboCup 2008, LNAI 5399, pp. 568–579, 2009.
© Springer-Verlag Berlin Heidelberg 2009

driving problem [3]. Olsen and Goodrich proposed six metrics for evaluating human computer interactions [4]. Balch has proposed a metric called Social Entropy based on Shannon's Information Entropy for measuring the behavioral diversity of a team of homogeneous robots [5]. On evaluating real soccer games, Rue and Salvesen applied Markov Chain Monte Carlo method on the final results of the games played during a limited time period for predicting the possible results of the forthcoming games [6]. Yanco,discussed the methods of defining metrics for robot competitions in order to be able to judge the participants efficiently [7].

In RoboCup domain, Kok *et al.* used the distance and the orientation of the ball with respect to the opponent goal, and the position of the opponent goalkeeper for determining the optimal scoring policy in RoboCup 2D Simulation environment [8]. Dylla *et al.* have initiated a qualitative soccer formalism for robot soccer [9]. They proposed a top-down approach to the soccer knowledge, following the classical soccer theory. Quantitative information like the distance and orientation to the ball, distance and orientation to the opponent goal and distance to the nearest teammate are widely used in role assignment or individual behavior selection [10,11]. Quinlan *et al.* proposed to use more high-level measurements like the goal difference and the remaining time to the end of the game in determining team aggression level [12].

In this work, we propose a three-layered decomposition of a soccer game in which each layer deals with the system at a different time resolution. A set of metrics built on top of position information of players and the ball in three different time resolutions are also presented. Finally, a novel, contingency table based validation method for metric consistency is given. Main contributions of this work are:

- Metric validation problem is stated as a challenging problem where autonomous decision making systems are in use.
- A novel statistical method is proposed for addressing the metric validation problem.
- A three-layered decomposition of the soccer game is given and a set of metrics are defined on different time resolutions.

Organization of the rest of the paper as follows: In Section 2, proposed approach is explained in detail. Section 3 contains explanation of metric validation process and finally, we conclude and point out some future works in Section 4.

2   Proposed Approach

The most primitive information we can estimate in the environment is the position information of players and the ball so we defined a set of metrics calculated from the position information for different time resolutions. It is assumed that the positions of opponent players are also known with a degree of error.

Since there are both team-level long-term goals and individual-level short-term goals, we need a game decomposition in different time resolutions. We propose three layers defined on different time resolutions in a game:

- Instantaneous Level
- Play Level
- Game Level

2.1   Instantaneous Metrics

Instantaneous metrics are calculated from one time-step position information. Since getting the control of the ball is the most important sub-task, most of the metrics are proposed for evaluating the chance of getting control of the ball.

Convex Hull Metrics

Convex Hull of a set of points is defined as the smallest convex polygon in which all of the points in the set lies. In an analogous manner, by subsituting points with the players in the soccer, we obtain a new concept: *Convex Hull of a Team*. We propose two metrics involving the convex hull of a team:

- *The Area of Convex Hull* tends to measure the degree of spread of the team over the field. The value of this metric increases as the team members are scattered across the field.
- *Density of Convex Hull* is applied only if the ball falls within the convex hull. The formal definition of the density is given in Equation 1.

$$Density = \frac{\sum_{i=1}^{N} \sqrt{(X_{player(i)} - X_{ball})^2 + (Y_{player(i)} - Y_{ball})^2}}{N} \qquad (1)$$

where, N is the number of players on the corners of the convex hull. If the ball is in the own half of the field, the goalkeeper is included in the convex hull calculation. If the ball is in the opponent field, the goalkeeper is excluded.

The density value is calculated only if the ball falls within the convex hull. If the ball falls outside of the convex hull, then the value of the metric is 0. The probability of the ball falling within the convex hull increases as the area of convex hull increases and it is expected that if the ball falls within the convex hull, the probability of getting the control of the ball increases. On the other hand, it is expected that the probability of getting the control of the ball increases as the density of the convex hull increases.

Vicinity Occupancy

Vicinity Occupancy measures the ratio of the teams players to the opponent players within a vicinity of the object of interest. The formal definition of the vicinity occupancy is given in Equation 2.

$$Occupancy = \frac{P_{own} - P_{opp}}{P_{own} + P_{opp}} \qquad (2)$$

where, P_{own} is the number of own players in the vicinity of the object of interest, P_{opp} is the number of opponent players in the vicinity, and P is the total number of players in the vicinity. The result is a real number in the interval $[-1, 1]$ where

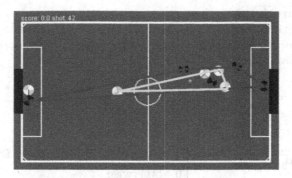

Fig. 1. Convex Hulls of two teams at a time point

−1 means that the vicinity is dominated by the opponent players, 0 means that there is no dominance and finally, 1 means that the vicinity is dominated by our players. Vicinity Occupancy is calculated for three objects of interest:

- Ball
- Own Goal Area
- Opponent Goal Area

The ball is the most important object in the game. Dominating the vicinity of the ball can be interpreted as having the control of the ball since the probability of controlling the ball increases as the number of own players in the vicinity increases and decreases as the number of opponents in the vicinity increases.

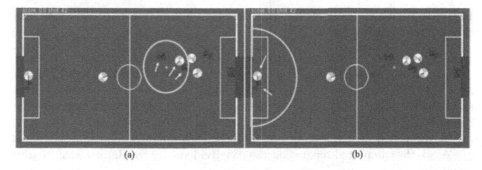

Fig. 2. a) Occupancy in the vicinity of Ball. b) Occupancy in the vicinity of Own Goal.

Dominating the vicinity of own goal is desired since it can be interpreted as a good defensive tactic. Dominating the vicinity of opponent goal is basically the opposite situation of occupancy of own goal case. As a result, in the ideal case, it is desired to dominate vicinities of both ball and goals but dominating the ball vicinity is the most important issue.

Pairwise separation Metrics

Pairwise separation is aimed to measure the degree of separation of an object of interest with opponent team. Equation for calculation of pairwise separation is given in Equation 3.

$$S_{Object} = \frac{\sum_{i=1}^{n} \sum_{j=1}^{n} \sum_{k=1}^{m} separates(P_{own}^{i}, P_{own}^{j}, P_{opp}^{k}, Object)}{2}$$

(3)

$$separates(P_1, P_2, P_3, Object) = \begin{cases} 1 & \text{if } Line(P_1, P_2) \text{ intersects } Line(P_3, Object), \\ 0 & \text{otherwise.} \end{cases}$$

(4)

where, n is the number of own players, m is the number of opponent players, and P_{own} and P_{opp} are the sets of own and opponent players, respectively.

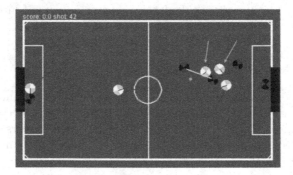

Fig. 3. Pairwise separation of Ball from Opponent Team: Robots pointed with light arrows are separated from the ball

Pairwise separation depends on the assumption that if an opponent player is *separated* from the object of interest, it is more likely for us to prevent it from accessing the object of interest. For example, if the pairwise separation value for ball is high, our chance to control the ball will also be high. Since separation test is performed for each player and with each teammate, each tuple is counted twice. So, the calculated separation value is divided by 2 to eliminate this double count.

Clearance of the path between two points

Clearance metric measures the accessibility of one point from another point. Clearance depends on the existence and positions of players and their movement capability. It is assumed that a player has control over an area called *Area of Impact*. The size and shape of area of impact depends on locomotional abilities of the robot. For a robot with omnidirectional movement and shooting ability with any side (for example Teambots robots or MIROSOT robots), shape of the area of impact will be a circle.

The area of impact depends on the speed of the robot. A fast robot would have a larger area of impact than a slower robot. The area of impact of a robot is considered as a physical obstacle along with the body of the robot when calculating the clearance. If the path between two points is occluded by the area of impact of at least one opponent player, it is considered that the way between the two points is not clear.

We calculate three clearance metrics:

- Clearance to the ball
- Clearance of ball to the opponent goal
- Clearance of ball to the teammates

Once a player reaches to the ball, there are three actions it can take:

- Shooting to the goal
- Dribbling with the ball
- Passing the ball to a teammate

Since it is assumed that the player must reach the ball before shooting or passing, only the clearance of the ball to the opponent goal and to the teammates are important. Sample clearance situations are shown in Figure 4.

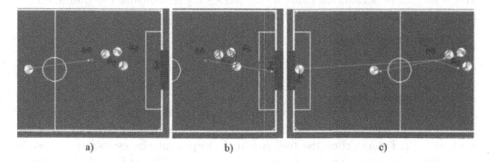

Fig. 4. a) Clearance to the Ball. b) Clearance to the Opponent Goal for the Ball. c) Clearance of the Ball to the Teammates.

2.2 Play Level Metrics

Play level metrics tend to measure the two important issues in the soccer game: Reachability of a position from another position and ball possession. The proposed predicates $isReachable(Position_{from}, Position_{to})$ and $hasBall(Player)$ are calculated by using instantaneous metrics over a time period. Both $isReachable$ and $hasBall$ are boolean metrics so we need to map the output of the metric combination from a real number to a boolean value.

isReachable Predicate

The function $isReachable(Position_{from}, Position_{to})$ returns *True* if the path between points $Position_{from}$ and $Position_{to}$ is clear from obstacles (Sec. 2.1). Since clearance metrics are instantaneous metrics and can be quite noisy, clearance is calculated by examining the consecutive values of the clearance metrics. If the path between two positions is *Clear* for consecutive N time-steps, *isReachable* is set to *True*. Contrarily, if the path between two positions is *Occluded* for consecutive N time-steps, *isReachable* is set to *False*. Determining the number N is another optimization problem. Since such an optimization is beyond the scope of this work, we arbitrarily select $N = 10$ and leave finding the optimal value of N as a future work.

hasBall Predicate

hasBall is used to check whether a certain player has the ball possession or not. *hasBall(Player)* returns *True* if the *Player* has the ball possession or not. As in the *isReachable* predicate, output is calculated from the values of the metrics developed for measuring the ball possession over a number of consecutive time-steps. We used the same value of 10 for the window size variable N for calculating the value of the *hasBall*.

2.3   Game Level Metrics

Game level metrics are proposed for measuring the statistics about the game over a long time period. All game level metrics try to measure the dominance of the game. Three metrics are calculated in game level:

- Attack/Defense Ratio
- Ball Possession
- Score Difference

Attack/Defense Ratio

Attack/Defense Ratio (ADR) tends to measure the dominance of the game by comparing the longest time the ball spends in our possession area in the game field with the longest time the ball spends in opponent possession area. Possession areas are defined as goal-centered semi-circles. The radius of the circle is a hyper-parameter and it needs some machine-learning and optimization techniques for finding the optimal value of the radius, but we simply select the radius of the circle as half of the field height.

The Attack/Defense Ratio is the difference of largest consecutive time-steps that the ball is in opponent possession area and the largest consecutive number of time-steps that the ball is in our own possession area divided by the sum of them. Formula for Attack/Defense Ratio is given in Equation 5.

$$ADR = \frac{Pos_{own} - Pos_{opp}}{Pos_{own} + Pos_{opp}} \tag{5}$$

This value is a real number in the interval $[-1, 1]$. A positive value of this metric indicates that the ball is spending more time in the opponent possession field than it spends in own possession field meaning that our team is more aggressive and dominating the game.

Ball Possession

Ball Possession (BP) measures the dominance of the game by comparing the longest time our team has the ball possession with the longest time opponent team has the ball possession. Play Level predicate *hasBall* is used in the calculation of ball possession. Formula for Ball Possession is given in Equation 6.

$$BP = \frac{Ball_{own} - Ball_{opp}}{Ball_{own} + Ball_{opp}} \tag{6}$$

where $Ball_{own}$ is the number of consecutive time steps that $hasBall(Player_{own})$ is *True* for one of our players. $Ball_{opp}$ is the number of consecutive time steps that $hasBall(Player_{opp})$ is *True* for one of the opponent players. A positive value of the metric indicates that our team has the control of the ball more than the opponent team.

Score Difference

Score difference (SD) is probably the most popular and trivial metric which is calculated from the scores of the teams. The equation for calculating score difference is given in (7).

$$SD = Score_{own} - Score_{opp} \tag{7}$$

The result is an integer indicating the dominion over game so far.

3   Evaluating Metrics

For the evaluation of defined metrics, a total of 200 games were played with our team against four different opponents in Teambots simulation environment [13]. In order to reveal the performance of the opponent teams in all aspects and to eliminate ceiling and floor effects in evaluating the performance of our own team, we have tried to use stratification in selecting the opponent teams so we choose both weak, moderate and powerful teams as opponents.

After the games are played and the position data for the players and the ball are recorded, each game is divided into *episodes* which starts with a kick-off and ends with either a score or end of half or end of game whistle. Episodes ending with own scores are marked as positive examples and episodes ending with opponent scores are marked as negative examples. Episodes ending with end of half or end of game whistle are ignored. At the end of 200 games, 81 negative and 1016 positive episodes were recorded. Each episode is then divided into smaller sequences of time-steps that are separated by a touch (or kick) to the ball. These sub-episodes are also marked as positive/negative examples depending on which team has touched the ball at the end of the sub-episode. If the ball is kicked by own team and the previous kick was performed by the opponent team, that sub-episode is marked as a *Positive* example. If the ball is kicked by opponent players and the previous kick was made by home players, that sub-episode is marked as a *Negative* example. The sub-episodes that are started and ended with the kicks of same team are ignored. Then, the marked sub-episodes are used to evaluate metrics related to the ball possession.

3.1 Metric Validation

Proposing metrics is a challenging task but it is even harder to evaluate the performance of a metric. We use metrics to obtain quantitative information about the environment but how can we be sure that the metric we proposed really *measures* the property it is supposed to measure. So we are confronted with another challenging problem: Metric validation. In order to consider a metric as *informative*, the metric should show the same trends in the same situations. For example, we can propose the *distance to the ball* metric for assessing the probability of getting the control of the ball. However, distance might not be the right indicator. So we should check whether the distance metric has the same trends in positions having the same ending (our team got the control of the ball, or opponent team got the control of the ball). Due to noise and sudden changes in positions of ball and other players, recorded metric data contain noise making the observation of trends in metric data difficult. In order to extract trends in recorded noisy data, some smoothing algorithms are applied to the recorded data. We have tried two smoothing algorithms on the recorded metrics:

– 4253h,Twice Smoothing
– Hodrick-Prescott Filter

3.2 4253h, Twice Smoothing

In 4253h, Twice algorithm, running median smoothers with window sizes 4, 2, 5 and 3 are applied consecutively. Then *Hanning* operator is applied. Hanning operator replaces each data point P_i with $\frac{P_{i-1}}{4} + \frac{P_i}{2} + \frac{P_{i+1}}{4}$. Then the entire operation is repeated [14]. Performing two or three consecutive 4253h, Twice resulted in great reduce in noise but the trend extraction is still hard in resultant smoothed data.

3.3 Hodrick-Prescott Filter

Hodrick-Prescott filter is proposed for extracting underlying trend in macroeconomic time series [15]. In the Hodrick-Prescott (HP) Filter approach, the observable time series y_t is decomposed as:

$$y_t = g_t + c_t \tag{8}$$

where g_t is a non-stationary time trend and c_t is a stationary residual. Both g_t and c_t are unobservable. We think y_t as a noisy signal for the g_t. Hence, the problem is to extract g_t from y_t.

HP Filter solves the following optimization problem:

$$\underset{\{g_t\}_{t=1}^{T}}{Min} \sum_{t=1}^{T}(y_t - g_t)^2 + \lambda \sum_{t=2}^{T}[(g_{t+1} - g_t) - (g_t - g_{t-1})]^2 \tag{9}$$

where λ is a weight for a signal against a linear time trend. $\lambda = 0$ means that there is no noise and $y_t = g_t$. As λ gets larger, more weight is allocated for the linear trend. So as $\lambda \to \infty$, g_t approaches to the least squares estimate of y_t's linear time trend. Selecting the value of λ is another design problem. In our work, we used 14400 as the value of the λ which is used to smooth monthly data in original implementation.

Figure 6.a shows the kicks in which the team with the ball possession is changed. In Figure 6, bold spikes denotes the kicks that are performed by our team and preceding by an opponent kick and, narrow spikes denotes the kicks that are performed by the opponent team and preceding by an own kick. In order to test the correlation among the sub-episodes with the same mark (positive or negative), a straight line is fitted on metric data in the sub-episode by using Least Squares Fitting. Then, the possible correlation between the mark of the

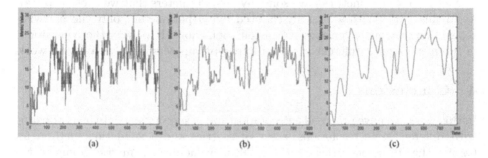

(a)   (b)   (c)

Fig. 5. Smoothing: a) Raw data, b) 4253h,Twice, c) Hodrick-Prescott Filter

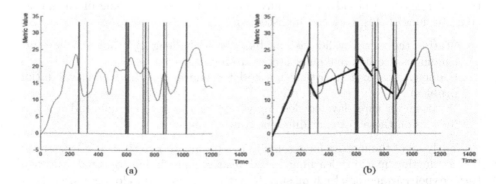

(a)   (b)

Fig. 6. a) an example Pairwise Separation of the Ball Metric with positive and negative kicks. b) after fitting a Least-Squares Line to the metric.

Table 1. The Kick-Slope distribution for Pairwise Separation of the Ball

	Own Kick	Opponent Kick
Positive Trend	94	27
Negative Trend	19	50

sub-episode and the sign of the first derivative of the fitted line (i.e. slope of the line) is investigated. It is expected that the signs of the slopes of fitted lines on the metric data in sub-episodes with the same mark are the same.

In Figure 6.b, fitted lines on the pairwise separation of the ball metric data between two kicks can be seen. It is seen in the figure that the fitted lines to the positive sub-episodes have positive slopes where the fitted lines to the negative sub-episodes have negative slopes.

Table 1 shows that the pairwise separation of the ball metric has a positive correlation with the sub-episode mark. Whenever the metric shows an increasing trend, our own team performs a kick and since performing a kick requires the ball possession, it can be said that if the pairwise separation of the ball metric shows an increasing trend, our own team has the ball possession.

Nearly all of the metrics have some hyper-parameters that we chose arbitrarily in this work. With arbitrarily selected hyper-parameters, only the pairwise-separation metrics have shown consistent behaviors. Exploring the consistency of the metrics with different values of hyper-parameters is left as a future work.

4   Conclusions

In this work, we have proposed a decomposition of soccer game into layers dealing with different time resolutions, a set of metrics and a validation method for testing the consistency (hence, the informativeness) of a metric. Some of the metrics are novel and a metric validation method is proposed for the first time. Although the proposed decomposition is applied on robot soccer, it is not limited to soccer and can be adapted to any multi-robot system. Some of the major contributions of this work can be listed as:

- Stating the metric validation problem as a challenging problem where autonomous decision making systems are in use.
- Proposing a novel statistical method for addressing the metric validation problem.
- Proposing a three-layered decomposition of the soccer game and a set of metrics on these layers at different time resolutions.

Nearly all of the metrics have some hyper-parameters so there is a large room for conducting machine learning based research on finding optimal values of these hyper-parameters. Finding such hyper-parameters, developing methods for dealing with uncertainty in real life, investigating the possibility of a spatial decomposition of the soccer and combining metrics defined on different layers of both spatial and temporal decompositions are left as future work.

Acknowledgments

This work is supported by TUBITAK Project 106E172.

References

1. VerDuin, W.H., Kothamasu, R., Huang, S.H.: Analysis of performance evaluation metrics to combat the model selection problem. In: PERMISA Workshop (2003)
2. Yavnai, A.: An information-based approach for system autonomy metrics part i: Metrics definition. In: PERMISA Workshop (2003)
3. Horst, J.A.: Exercising a native intelligence metric on an autonomous on-road driving system. In: PERMISA Workshop (2003)
4. Olsen, D.R., Goodrich, M.A.: Metrics for evaluating human-robot interactions. In: PERMISA Workshop (2003)
5. Balch, T.: Social entropy: a new metric for learning multi-robot teams. In: 10th International FLAIRS Conference (FLAIRS 1997) (1997)
6. Rue, H., Salvesen, Ø.: Predicting and retrospective analysis of soccer matches in a league
7. Yanco, H.A.: Designing metrics for comparing the performance of robotic systems in robot competitions. In: Workshop on Measuring Performance and Intelligence of Intelligent Systems (PERMIS) (2001)
8. Kok, J., de Boer, R., Vlassis, N.: Towards an optimal scoring policy for simulated soccer agents. Technical report (2001)
9. Dylla, F., Ferrein, A., Lakemeyer, G., Murray, J., Obst, O., Röfer, T., Stolzenburg, F., Visser, U., Wagner, T.: Towards a league-independent qualitative soccer theory for robocup. In: Nardi, D., Riedmiller, M., Sammut, C., Santos-Victor, J. (eds.) RoboCup 2004. LNCS, vol. 3276, pp. 611–618. Springer, Heidelberg (2005)
10. Veloso, M., Lenser, S., Vail, D., Roth, M., Stroupe, A., Chernova, S.: Cmpack 2002: Cmu's legged robot soccer team. Technical report (2002)
11. Röfer, T., et al.: Germanteam 2006. Technical report, The GermanTeam (2007)
12. Quinlan, M.J., Henderson, N., Middleton, R.H.: The 2006 nubots team report. Technical report, Newcastle Robotics Laboratory (2007)
13. Balch, T.: Teambots (2000), http://www.teambots.org
14. Cohen, P.R.: Empirical Methods for Artificial Intelligence. MIT Press, Massachusetts (1995)
15. Hodrick, R.J., Prescott, E.C.: Postwar u.s. business cycles: An empirical investigation. Journal of Money, Credit and Banking 29 (1997)

RobotStadium: Online Humanoid Robot Soccer Simulation Competition

Olivier Michel[1], Yvan Bourquin[1], and Jean-Christophe Baillie[2]

[1] Cyberbotics Ltd., PSE C - EPFL, 1015 Lausanne, Switzerland
Olivier.Michel@cyberbotics.com, Yvan.Bourquin@cyberbotics.com
http://www.cyberbotics.com
[2] Gostai SAS, 15 rue Vergniaud 75013 Paris, France
baillie@gostai.com
http://www.gostai.com

Abstract. This paper describes robotstadium: an online simulation contest based on the new RoboCup Nao Standard League. The simulation features two teams with four Nao robots each team, a ball and a soccer field corresponding the specifications of the real setup used for the new RoboCup Standard League using the Nao robot. Participation to the contest is free of charge and open to anyone. Competitors can simply register on the web site and download a free software package to start programming their team of soccer-playing Nao robots. This package is based on the Webots simulation software, the URBI middleware and the Java programming language. Once they have programmed their team of robots, competitors can upload their program on the web site and see how their team behaves in the competition. Matches are run every day and the ranking is updated accordingly in the "hall of fame". New simulation movies are made available on a daily basis so that anyone can watch them and enjoy the competition on the web. The contest is running online for a given period of time after which the best ranked competitors will be selected for a on-site final during the next RoboCup event. This contest is sponsored by The RoboCup federation, Aldebaran Robotics, Cyberbotics and Gostai. Prizes includes a Nao robot, a Webots package and a URBI package.

1 Why We Need Another Simulation Competition

1.1 Introduction

The RoboCup is an excellent robotics benchmark [4] allowing different researchers to address a common challenge and to compare their results with each other. Moreover, RoboCup allows for a measurement of the progress of research over years as the benchmark specifications are fairly stable over time. This turns out to be a very valuable tool to evaluate the performance of mobile robots in the real world.

The RoboCup is divided into a number of robot soccer leagues, including the small size league, the middle size league, the standard platform league, the

L. Iocchi et al. (Eds.): RoboCup 2008, LNAI 5399, pp. 580–590, 2009.
© Springer-Verlag Berlin Heidelberg 2009

humanoid league and the simulation league [5]. There are also other leagues (rescue, at home, junior, etc.) which are out of the scope of this paper.

The Nao robot was recently chosen to become the official platform for the standard league competition of the RoboCup, replacing the discontinued Aibo robot from Sony. During the 2008 edition of the RoboCup, the first Nao-based standard league competition will be held in parallel with the last Aibo-based standard league.

However, it is very costly for a research lab to invest in RoboCup research as it requires to purchase, build and maintain a large number of mobile robots as well as a soccer field which occupies a large space with a good control on the lighting conditions. An alternative to these investments is to compete in the simulation league. However, although there are currently a number of simulation league, including 2D and 3D simulation, none of these simulations correspond to a real RoboCup setup. For example the 2D simulation is not a realistic simulation software in the sense that it doesn't try to simulate the physics of objects, sensors and actuators of the robots. Hence it makes it difficult to reuse research developed within a simulation league and transfer it to real robots. Moreover, the 3D simulation is currently in a very early stage as a new simulator was introduced only recently (2007).

This matter of fact is very disappointing because it has been observed that the strategies and level of intelligence developed in the simulation league were very interesting from a scientific point of view (very advanced AI, learning, multi-agent coordination, etc.). This can probably be explained by the fact that the researcher working in simulation can spend more time focusing on real control and AI problems rather than on mechanical, electronics or low level software problems. Moreovoer, simulations are often easier to reproduce, analyse, compare, etc. It would be really nice if such simulations could be easily transfered to the real world. But this is unfortunately not the case so far.

Making simulations closer to real RoboCup setups would help a lot researchers to develop new ideas in simulation and transfer them more easily onto the real robots. That would also allow for interesting comparisons between simulations and real setups. Finally, it would benefit in a straightforward way to the overall RoboCup competition.

2   Going Further with RoboCup simulation

2.1   Realistic Simulation

Accuracy is a very important aspect of a robot simulation. The environment, robots and rules should be modelled very carefully to match the features of the real robots. This requires a calibration phase of the simulation in which a series of control algorithms are tested on both the simulation and the real setup. The results are then compared and the simulation models are tuned to better fit the real setup. This operation is repeated several times until a satisfactory level of reliability is achieved. Such calibrations methods have been conducted several times on the Webots simulation software [6] to refine complex robot model and allow an easy transfer from simulation to the real robot. They involve several levels of calibration, including sensor calibration (i.e, lookup tables), actuator calibration (physical parameters, like the maximum torque of a servo or the precise range of movement), physics calibration (mass distribution, friction parameters, etc.). A Fujitsu HOAP-2 humanoid robot was successfully calibrated in Webots using such a method [3]. Moreover, the utilization of the URBI middleware [2] for programming both the simulation and the real robots can dramatically facilitate the transfer onto the real robots.

This way, the simulation can be transfered into a real robot fairly accurately, hence making the use of a simulation software even more interesting.

2.2   Free and Easy to Use

Unlike real robots, simulation software can be made available very easily to a large community of users. Although our Webots simulation environment is a commercial package, including the URBI middleware, a special simulation of the RoboCup Nao Standard League was released free of charge, so that anyone can download it, install it and use it to develop robot controller programs within the framework of the RoboCup Nao Standard League. Our Webots-based simulation environment runs on all major platform, including Windows, Linux and Mac OS X. The utilization of the URBI middleware in combination with Java makes it very simple and powerful to program the Nao robots. Examples provided in the package help users to get started with programming their simulated Nao robots within minutes. Moreover, a number of graphical tools facilitating the generation of servo trajectories are included in the package (see figure 1).

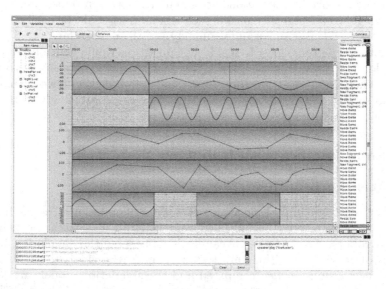

Fig. 1. The URBI motion editor

2.3   Online Competition

In order to be available to a wider audience, a simulation contest can be held on the Internet. This allow everyone to enter the competition regardless of his/her geographic location or ability to travel abroad to an international conference. Having the competition available on the web is also very attrative for potential competitors who can learn about the competition, register and get started within minutes.

Another very attractive aspect is the fact that competition match are held on a regular basis very often (like once a day). This way, competitors can view the results in real time, improve their robot controllers and submit new versions until the end of the online competition. The main consequence of this is that the developed robot controllers are continuously tested in real condition during the development phase. This undoubtably yields to better results at the end. The Rat's Life contest is an example of successful online competition [7,8].

2.4   Durable Benchmarking

Since the RoboCup is recognized as a reference benchmark in robotics, our simulation environment should also be used a useful benchmark, contributing to scientific progress in robotics research. A benchmark is useful if users can compare their own results to others and thus try to improve the state of the art. Hence a benchmark should keep a data base of the solutions contributed by different researchers, including binary and source code of the robot controller programs. These different solutions should be ranked using a common performance metrics, so that we can compare them to each other and evaluate the general progress over time. Keeping an archive of the contributed robot controllers can help the RoboCup community to measure the quality of their results.

3   Competition Organization

This first edition of the simulated Nao RoboCup online contest is sponsored by the RoboCup federation along with three private companies: Aldebaran robotics, the makers of the Nao robot, Gostai, the developers of URBI and Cyberbotics, the developers of Webots.

3.1   Nao

Bruno Maisonnier, founder of Aldebaran Robotics [10], has been convinced for 25 years that the era of personal robotics is coming. During these years, he has developed prototypes, evaluated technologies, met with research teams and analyzed the markets, in addition to serving as the CEO of several companies in multicultural contexts. With the rise of mobile technologies and the coming together of key collaborators, the potentials have now become the possibilities: In 2005 he launched Aldebaran Robotics, the first French company dealing with humanoid robotics. Aldebaran Robotics team, which currently consists of 44

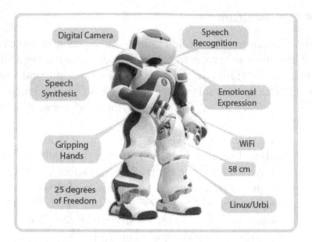

Fig. 2. The Nao robot

members, entirely dedicated to the development of its first robots, will continue to grow throughout 2008.

The Nao project, launched in early 2005, aims to make available to the public, at an affordable price, a humanoid robot with mechanical, electronic, and cognitive features (see figure 2).

Initially, delivered with basic behaviors, the robot will be, at its market introduction, the ideal introduction to robots. Eventually, with many improved behaviors, it will become an autonomous family companion. Finally, with more sophisticated functions, it will adopt a new role, assisting with daily tasks (monitoring, etc.).

Featured with an intuitive programming interface, the entire family will be able to enjoy the robot experience. Yet, full of new technologies, the robot will also satisfy the demanding techno-addicts expectations.

Designed for entertainment purposes, the robot will be able to interact with its owner, with evolving behaviors and functionalities. Additionally, the user will be able to teach the robot new behaviors using a computer with Wi-Fi connectivity. The behaviours creating software is designed to fit with any users levels: from graphical blocs editing for beginners to code for more skilled users. The possible behaviors are limited only by the imagination of the users!

With a conviction that design is key to successful adoption in a home environment, Aldebaran Robotics has partnered with a Parisian design school, particularly with designers Thomas Knoll and Erik Arlen. The robots hull will include customizable features, allowing each to have a unique appearance.

The robot is based on a Linux platform and scripted with URBI, an easy-to-learn programming language, with the option of a graphic interface for beginners or code commands for experts.

Currently in the final phase of development, the first Nao has been presented early 2007. The first units, dedicated to laboratories and universities were sold early 2008 for the RoboCup competition, and the general public release is planned for the end of 2008.

3.2 URBI

Created in 2006 after 4 years of R&D in an academic research lab (CogRob, ENSTA), Gostai [12] has a 6 years of experience in complex AI-driven robotics applications and development tools. The core of its technology is the Urbi middleware for robotics, initially developed for the Aibo by J.-C. Baillie, a former researcher at Sony Computer Science Lab.

Urbi [1,2] is a middleware for robotics, which includes dedicated abstractions to handle parallelism and event-based programming from within C++, Java or Matlab, together with a distributed component architecture called UObject which can be interfaced with Microsoft Robotics Studio and CORBA. The main focus is on simplicity, flexibility and code re-use, while providing convenient abstractions needed in the development of complex robotics applications.

Urbi is based on a client/server architecture and a dynamic language called Urbiscript that can be used to coordinate the UObject components. The novelty of Urbiscript as a programming language is that it brings new abstractions to handle parallelism and event-based programming, directly integrated into the language semantics. Urbiscript programs routinely run hundreds of parallel threads and react to several events at the same time.

The idea is to separate the logic of the program on one side (Urbiscript) and the fast algorithms on the other side (UObjects in C++), and use the dynamic capabilities of script languages to build the glue (see figure 3). This is already a widely used approach in videogames, with script languages like LUA or Python. Urbi brings this same approach to robotics and improves the script language side with parallelism and event-based programming.

UObject is a C++ based component architecture. Any C++ class can inherit UObject and become visible inside Urbiscript as a regular object that can interact with other objects already plugged. UObjects can be either linked to the Urbi

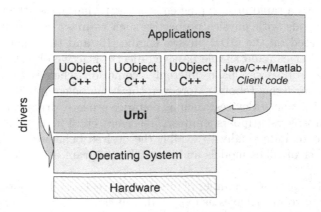

Fig. 3. URBI

Engine (the interpretor of the Urbiscript language) and share its memory and other UObjects memory, or it can also be run remotely as an autonomous separate process. In that case, the interface between Urbi and the UObject is done transparently through the network via TCP/IP and the C++ object is reflected inside the Urbiscript language in exactly the same way as in the "linked" mode. This allows to have large distributed network of objects interconnected through scripts coordinating them in a parallel and event-driven way, which contrasts with the traditional one-to-one component interactions of CORBA or similar component architectures.

CORBA and other component architectures can be bridged through UObject using the dynamic feature of the Urbiscript language, which can dynamically create a proxy object within the language to access the remote component. Using this feature, Urbi aims at providing a universal bridge between heterogeneous component architectures.

Urbi also includes a set of graphical programming tools called Urbi Studio, which comprises a hierarchical finite state machine editor, an animation editor and a universal remote controller for Urbi equipped applications. These tools are graphical front-ends which generate Urbiscript code that can be reused in various contexts.

Urbi is now compatible with 12 different robots and simulators, including Webots, and is used by industrial companies and more than 30 universities in the world.

3.3   Webots

Webots [6] is a commercial software for fast prototyping and simulation of mobile robots. It was originally developed by Olivier Michel at the Swiss Federal Institute of Technology in Lausanne (EPFL) from 1996 and has been continuously developed, documented and supported since 1998 by Cyberbotics Ltd. [11]. Over 500 universities and industrial research centers worldwide are using this software for research and educational purposes. Webots has already been used to organize robot programming contests (Rat's Life, ALife contest and Roboka contest).

Webots offers a rapid prototyping environment, that allows the user to create 3D virtual worlds with physics properties such as mass, joints, friction coefficients, etc. based on the ODE physics engine [13] and the OpenGL rendering engine. The user can add simple passive objects or active objects called mobile robots. These robots can have different locomotion schemes (wheeled robots, legged robots, or flying robots). Moreover, they may be equipped with a number of sensor and actuator devices, such as distance sensors, drive wheels, cameras, servos, touch sensors, grippers, emitters, receivers, etc. Finally, the user can program each robot individually to exhibit the desired behavior. Webots contains a large number of robot models and controller program examples to help users get started.

Webots also contains a number of interfaces to real mobile robots, so that once your simulated robot behaves as expected, you can transfer its control program to a real robot like Khepera, Hemisson, LEGO Mindstorms, Aibo, etc.

Fig. 4. Webots

Although Webots is a commercial software, a demo version is freely available from Cyberbotics's web site. This demo version includes the complete Nao RoboCup simulation with a URBI programming interface. So, anyone can download, install and practice the simulation of the Nao RoboCup competition at no cost (see figure 4).

4   Competition Description

This paper doesn't claim to be a technical reference for the simulated Nao RoboCup competition. Such a technical reference is available on the Robotstadium web site [9].

4.1   Rules

In order to get started rapidly and to take advantage of the feedback provided by early competitors, no strict rule will defined from the beginning. The idea is to invite constants to respect the standard basic soccer rules (like do not catch the ball with the hands) and to develop fair robot controllers (like robots should not attempt to block the game by locking the ball, or they should not try to attack other players, etc.).

A standard Nao RoboCup soccer environment is provided to the competitors (see figure 5) which includes a supervisor process responsible for counting the time (a match has two half time of 10 minutes each), counting the score, resetting the ball and the player at initial position once a goal is scored, etc.

The supervisor process is actually an automatic referee which saves the contest organization from the human supervisor. Hence matches can be run automatically by a script program and results can be published automatically as well.

This supervisor process will be continuously developed to refine the rules (to possibly introduce some penalties if some illegal actions are taken by some

Fig. 5. Screenshot of the robotstadium simulation model

robots) during the competition. This will allow the organizers to detect problems in the rules and correct them while the contest is running. However, we will try to introduce only slight modifications of the rules in respect of the principles of the standard basic soccer rules and fair play. Moreover, the rules will be fixed about one month before the end of the contest. This will allow contestant to rely on a fair and stable version of the rules for the last mile of competition.

4.2   Online Contest System

Web site: The Robotstadium online contest (see figure 6) will allow contestant to learn about the contest, view simulation movies of contest matches, register, download the necessary software, learn how to develop their own robot controllers, upload their robot controller, see their ranking in the "hall of fame" updated every day. In order to provide all these services, the web site will contains a number of standard web tools, including a registration system, a forum, a file management system, a shout box, and a number of sections: movies, documentation, rules, FAQ, etc.

Participation to the Contest: In order to participate in the online contest, the competitors can download the free version of Webots from Cyberbotics' web site [11]. They can program the simulated Nao robots to play soccer on the simulated soccer field. Then, they have to register a contestant account on the contest's web site [9]. Once open, this account allows the competitors to upload the controller programs they developed with the free version of Webots. Participation to the contest is totally free of charge.

Fig. 6. Banner of the robotstadium web site

Ranking System: Every day at 12 PM (GMT) a competition round is started in simulation and can be watched online from the Robotstadium web site [9]. A hall of fame displays a table of all the competitors registered in the data base and who submitted a robot controller program. If there are N competitors in the hall of fame, then $N - 1$ matches are played. The first match of a round opposes the last entry, i.e., number N at the bottom of the hall of fame, to the last but one entry, i.e., number $N - 1$. If the robot number N wins, then the position of these two robots in the hall of fame are switched. Otherwise no change occurs in the hall of fame. This procedure is repeated with the new robot number $N - 1$ (which may have recently changed due to the result of the match) and robot number $N - 2$. If robot number $N - 1$ wins, then it switches its position with robot number $N - 2$, otherwise nothing occurs. This is repeated with robots number $N - 3$, $N - 4$, etc. until robots number 2 and 1, thus totaling a number of $N - 1$ matches.

This ranking algorithm is similar to the bubble sort. It makes it possible for a newcomer appearing initially at the bottom of the ranking, to progress until the top of the ranking in one round. However, any existing entry cannot loose more than one position in the ranking during one round. This prevents a rapid elimination of a good competitor (which could have been caused by a buggy update of the controller program for example).

And The Winner Is... The contest is open for a fixed period of time. During this period of time, new contestants can register and enter the contest. The contestants can submit new versions of their controller program any time until the closing date. Once the closing date is reached. The top five entries of the hall of fame will be selected for the finals to be held during the next RoboCup event.

The finals will take place at RoboCup event to ensure that a large number of people, including a scientific committee, attends the event and can check that nobody is cheating the contest.

The contest will run continuously over years so that we can measure the progress and performances of the robot controllers over a fairly long period of robotics and AI research.

5   Expected Outcomes and Conclusions

Thanks to the Robotstadium contest, it will become possible to evaluate the performance of various approaches to the RoboCup benchmark with humanoid robots. The performance evaluation will allow us to make a ranking between the different control programs submitted, but also to compare the progresses achieved over several years of research on this problem. For example, we could compare the top 5 controller programs developed in 2008 to the top 5 controller programs developed in 2012 and evaluate how much the state of the art progressed.

The control programs resulting from the best robot controllers could be transfered to the real world robotics setup of the RoboCup, especially the new

Nao-based standard league. This will help the teams involved in the standard league to accelerate their developments. Hopefully, this will allow researchers to go further in the development of advanced intelligent control architectures.

Moreover, because of the easy availability of the robotstadium simulation setup, a large number of researchers is expected to start investing their time in developments for the simulated Nao standard league. In turn these people will be keen to try to transfer their results in the real Nao standard league, thus bringing even more success to the real league.

Finally, by making available a large number of simulation movies on the Internet, this simulation competition may attract the interest of a wider audience, including general public and the media.

We hope that this initiative is a step towards a more general usage of realistic simulation and advanced programming languages in robotics research, as we are convinced that roboticists need more high quality software tools in order to focus more efficiently on their research and achieve significant breakthoughts in robotics research.

References

1. Baillie, J.C.: Design Principles for a Universal Robotic Software Platform and Application to Urbi. In: Proceedings of ICRA 2007 Workshop on Software Development and Integration in Robotics (2007)
2. Baillie, J.C.: Urbi: Towards a Universal Robotic Low-Level Programming Language. In: Proceedings of IROS 2005, pp. 820–825 (2005)
3. Cominoli, P.: Development of a physical simulation of a real humanoid robot, Master's thesis, EPFL, Swiss Federal Institute of Technology in Lausanne (2005)
4. Dillmann, R.: Benchmarks for Robotics Research, EURON (April 2004), http://www.cas.kth.se/euron/eurondeliverables/ka1-10-benchmarking.pdf
5. Kitano, H., Asada, M., Kuniyoshi, Y., et al.: RoboCup: the robot world cup initiative. In: IJCAI 1995 Workshop on entertainment and AI/ALife (1995)
6. Michel, O.: Webots: Professional Mobile Robot Simulation. Journal of Advanced Robotics Systems 1(1), 39–42 (2004), http://www.ars-journal.com/International-Journal-of-Advanced-Robotic-Systems/Volume-1/39-42.pdf
7. Michel, O., Rohrer, F., Bourquin, Y.: Rats Life: A Cognitive Robotics Benchmark. In: European Robotics Symposium 2008 (EUROS 2008), vol. 44, pp. 223–232. Springer, Heidelberg (2008)
8. Rat's Life contest, http://www.ratslife.org
9. Robotstadium contest, http://www.robotstadium.org
10. Aldebaran Robotics SAS, http://www.aldebaran-robotics.com
11. Cyberbotics Ltd., http://www.cyberbotics.com
12. Gostai SAS, http://www.gostai.com
13. ODE: Open Dynamics Engine, http://www.ode.org

Real-Time Simulation of
Motion-Based Camera Disturbances

Dennis Pachur[1], Tim Laue[2], and Thomas Röfer[2]

[1] Fachbereich 3 - Mathematik und Informatik, Universität Bremen,
Postfach 330 440, 28334 Bremen, Germany
pachur@informatik.uni-bremen.de
[2] Deutsches Forschungszentrum für Künstliche Intelligenz GmbH,
Sichere Kognitive Systeme, Enrique-Schmidt-Str. 5, 28359 Bremen, Germany
{Tim.Laue,Thomas.Roefer}@dfki.de

Abstract. In the RoboCup domain, many robot systems use low-cost
image sensors to perceive the robot's environment and to locate the robot
in its environment. The image processing typically has to handle image
distortions such as motion blur, noise, and the properties of the shutter
mechanism. If a simulator is used in the development of the control software, the simulation has to take account for these artifacts. Otherwise,
the performance of the image processing system in the simulation may
not correspond to its performance on the real robot; it may even perform
worse.

The effect of motion blur has been widely used for special effects both
for movies and for computer games. While real-time algorithms using
modern graphics hardware came up in recent years, the image distortion
resulting from a so-called rolling shutter has not been in focus so far.
In fact, this effect is not relevant for gaming, but it is for simulating
low-cost cameras of robots.

In this paper, we present an efficient way to simulate the rolling shutter effect using per-pixel velocities. In addition, we improve the velocity
buffer method for creating motion blur using the current speed of each
pixel in real-time. The application of our approach is shown exemplarily
for the head-mounted camera of a humanoid soccer robot.

1 Introduction

The quality of a robot simulation is inter alia determined by its ability to calculate realistic measurements for the sensors simulated. Each simulation can only
approximate reality up to a certain level of detail. For many applications, the
missing details are not relevant and the interaction between the robot software
and the simulated world is not affected. Nevertheless, the characteristics of some
sensors matter to an extent that leads to a significant mismatch between simulation and reality. Such a mismatch is relevant in two cases: the first and naïve
case is that the robot control software is initially developed only in a simulator,
and it does not handle sensor distortions that are not present in the simulation,
but surely will be in reality. In the other case, the more experienced developer

L. Iocchi et al. (Eds.): RoboCup 2008, LNAI 5399, pp. 591–601, 2009.
© Springer-Verlag Berlin Heidelberg 2009

Fig. 1. Examples of image distortion caused by the rotation of a CMOS image sensor with a rolling shutter

would develop the software on the real robot and in simulation side by side. However, software tuned for real sensor readings may perform worse or even fail if confronted with unrealistically simple sensor readings coming from a simulator, rendering the simulator as a tool rather useless. Hence, it is important that a simulation generates all the sensor distortions that are actively handled by the robot control software developed.

Figure 1 shows examples of typical distortions, using images from a CMOS camera that is part of the head of a humanoid robot. While the blurry impression is caused by motion blur, the distortion that bends and squeezes the yellow goal is produced by the so-called rolling shutter. Instead of taking images at a certain point in time, a rolling shutter takes an image pixel by pixel, row by row. Thus the last pixel of an image is taken significantly later than the first one. If the camera is moved, this results in image distortions as shown in figure 1. Both artifacts, motion blur and rolling shutter effect, can be generalized as temporal aliasing effects caused by the light integration of the image sensor. While the blurriness of a pixel results from the movement in the scene while the pixel is exposed, the rolling shutter causes the pixels of an array to be exposed in a serialized way.

Compensating this effect becomes significantly important when working with robots which move their cameras fast or operate in a rapidly changing environment. In the RoboCup domain, this particularly affects robots with pan-tilt heads, e. g. in the Four-legged and the Humanoid league. To overcome problems regarding world modeling, which obviously arise from these disturbances, different compensation methods have been developed, e. g. by Nicklin et al. [1] or by Röfer [2]. A necessity of simulating these disturbances can be derived from this explicit handling by the software of different teams.

Within the RoboCup community, a variety of robot simulators is currently used. Among the most advanced and established ones are, for example, Webots [3], Microsoft Robotics Studio [4], and the USARSim [5]. In general, they are able to simulate camera images at a high level of detail, e. g., by simulating lights and shadows. But so far, none of them addresses the problem of image disturbances.

The contribution of this paper is an efficient approach for simulating common image disturbances, i. e. the rolling shutter effect and motion blur, in real-time. This approach has been implemented for an existing robot simulator; a movable camera of a simulated humanoid robot is used as an example application.

The paper is organized as follows: Section 2 provides an overview of related work about generating image disturbances. In Section 3, our approach is presented, follow by the experimental results in Section 4.

2 Related Work

This section gives a short overview of existing models and methods for image distortion. Considering performance as well as their capabilities, design choices for our approach are derived.

2.1 Image Distortion Model

In [6], the image distortion caused by a rolling shutter within a CMOS image sensor is given by the pixel position x at a certain time t. x_t^u is the undistorted position of a pixel at time t and u_t is its motion. By taking into account the motion since the start of the acquisition of the current image, the position is transformed by the motion b_t within the interval $[0, t_x]$. The time of the pixel exposure at t_a can be calculated by

$$t_x \approx \frac{x_{x,t}^u}{WH} + \frac{x_{y,t}^u}{H} \tag{1}$$

with W and H being the width and the height of the pixel array, and $x_{x,t}^u$ and $x_{y,t}^u$ being the coordinates of the pixel in the undistorted image. The distorted pixel position x_t^d can then be calculated by equation 2:

$$x_t^d = x_t^u + \int_{t=0}^{t_x} u_t dt \tag{2}$$

The motion during image acquisition is assumed to be constant.

For using this equation for distorting an image, the velocity has to be calculated. Whereas camera rotations cause a homogeneous velocity vector field for the complete image area, camera translations and object movements cause local velocities. To take these movements into account, per-pixel velocities have to be determined. This is common practice for a class of algorithms generating motion blur.

2.2 Integrating Motion Information into Rendered Images

Computer-generated images can be produced either in consideration of the image quality or of the time needed for the generation. The algorithms applied to calculate real-time graphics are based on complex formulas containing approximating parts. For the field of motion information (e. g. motion blur), real-time algorithms commonly approximate the visibility function. This results in generating artifacts in image areas with little spatial information. A complete equation without approximation is given in [7] but can only be used by a ray-tracer which calculates one pixel at once. Real-time rendering engines, such as OpenGL, use a rasterizer to approximate the spatial information by constructing surfaces between points and projecting the rasterized surface points to the image plane by interpolating the information along the edges.

For real-time graphics, a few classes of algorithms have made their way into everyday use: the accumulation of images from different points in time within the light integration interval [8], the extension of the object geometries in direction of the object's motion direction [9,10], and the post processing of the image using velocities calculated previously for either the objects [11] or the image pixels [12,13]. Due to the scene-independent performance of post processing, algorithms arose for iterative applications [14] that can create motion-blurred images with almost no artifacts.

Such methods are also used in the work presented here since the applied image distortion model is based on post-processing the image using per-pixel velocities. Hence motion-blur and the rolling shutter effect can be integrated without any additional pre-calculations.

2.3 Calculating Per-Pixel Velocities

The velocity v_x of a pixel x is given by the difference of the projections dP of a point at two time steps $t, t-1$ and can be calculated as in equation 3.

$$v_x = \frac{dP}{dt} \qquad \text{with } dP = P_t - P_{t-1} \qquad (3)$$

Therefore, the matrix stack from modelview to projection for both time steps must be known. For image generation, this is implicitly known for each vertex sent through the rendering pipeline. To generate velocities for each pixel, the difference calculated for each vertex can be interpolated over the adjacent polygons by varying an attribute while rasterizing the surface. By using up-to-date hardware, this velocity vector field can be rendered into an image using multiple render targets, requiring a single rendering pass only. The result can be copied to a texture and be stored for later use if more than a single velocity image is needed within the motion blur appliance and the render target is not chosen as a texture anyway.

A more recent method to calculate the per-pixel velocity is described in [15]. The velocities are determined by taking the matrices of both time steps and reconstructing a three-dimensional image point through employing the depth image commonly used by the rasterizer to solve the visibility function.

Once the per-pixel velocities have been calculated, the image can be distorted by using a per-pixel operation. Nowadays, this operation can be accelerated using an image-aligned plane, a feature of a modern fragment shader, and a variable shading language such as CG, GLSL or HLSL.

3   Simulating Image Disturbances

Our simulator SimRobot [16] uses OpenGL to visualize the scene and to generate camera images. Hence, we use frame buffer objects to render the image data and the per-pixel velocities in only a single pass. The rendering system supports antialiasing and render-target-switching to obtain high visual quality and to store up to six velocity images for later use. The motion blur approach of [13] was chosen to blur the image using multiple velocity buffers.

3.1   Calculating Per-pixel Velocities

For a single image pixel $\boldsymbol{x}$, the fragment shader determines the pixel position offset by interpolating the velocities of the last two velocity buffer images $\boldsymbol{v}_{\boldsymbol{x}}^{curr}$ and $\boldsymbol{v}_{\boldsymbol{x}}^{prev}$. If the image resolution is known, the interpolated velocity $\boldsymbol{v}_{\boldsymbol{x}}$ of a pixel $\boldsymbol{x}$ can be calculated using

$$\boldsymbol{v}_{\boldsymbol{x}} = f\boldsymbol{v}_{\boldsymbol{x}}^{curr} + (1 - f)\boldsymbol{v}_{\boldsymbol{x}}^{prev} \qquad (4)$$

with

$$f = \frac{\boldsymbol{x}_x}{WH} + \frac{\boldsymbol{x}_y}{H}. \qquad (5)$$

For simplifying the image distortion with negligible imprecision, the temporal offset caused by the horizontal position $\boldsymbol{x}_x$ in the image can be discarded. Thus a fragment shader can use texture coordinates to obtain generalized image coordinates in the range of $[0\ldots1]$ that can be applied to any image resolution. In this case, the position of the pixel in the image can be simply calculated using

$$f = \boldsymbol{x}_y \qquad y \in [0\ldots1] \qquad (6)$$

The interpolated velocity adds information about the change of speed. If the camera starts rotating within the time of exposure, the bottom left pixel will

Fig. 2. Examples of image distortion for a rolling shutter simulation based on per-pixel velocity. The left image has been generated using a constant velocity, whereas interpolated velocities have been applied to the right image.

receive most of the speed of the velocity calculated previously while the upper right pixel will almost use the current velocity. This facilitates not only creating the image distortion but also adjusting the motion blur itself.

To adjust the pixel position x in the local image, equation 2 is used. The coordinates of the current pixel x have to be distorted by the calculated velocity.

$$x^d = x - f v_x \tag{7}$$

Figure 2 shows two images generated by this method. While the right image uses interpolated velocities and identifies the change of rotation speed of the camera, the left image only uses the current velocity v_x^{curr} as distortion base.

3.2 Reducing Image Artifacts

The images in Figure 2 show distortion artifacts at the top left corner of the goal. They result from calculating per-pixel velocities for a rasterized image as mentioned before.

To minimize image artifacts, the represented time integral can be adjusted and be shifted slightly into the future. This can be done by determining the new pixel position x^d with $f' = f - 0.5$. The center of the distortion is shifted from the top of the image to its center resulting in a negative distortion direction in the bottom half and a positive distortion direction in the top half of the image. The shorter distortion vectors result in fewer artifacts. If the change of speed is too fast within two time steps, the bottom half starts to seem swinging as the prediction of the upcoming motion is failing.

When the camera is moving, a subset of the distorted image points is located outside the image area. To avoid leaving the image space (and thus leaving points outside the plane clear or repeating the image at its borders) the render system has to provide more than the usual visual image field. By adjusting the field of view before rendering the scene and correcting the texture coordinates afterwards, the area of the working image can be extended (cf. Fig. 3).

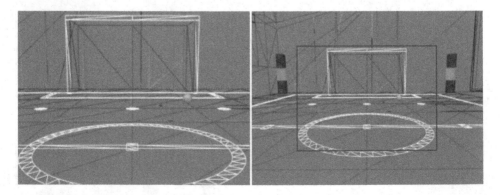

Fig. 3. The left image shows the part of the model that should be rendered. The right image shows the extended field of view used to be able to determine the pixel velocities correctly close to the image border.

Fig. 4. Combining interpolated velocities used for the rolling shutter distortion with a commonly used motion blur method

3.3 Applying Motion Blur

For the new pixel location x^d, another velocity has to be calculated to determine the locally distorted velocity for applying the motion blur itself. Again, using interpolated velocities will result into more image information. It has to be mentioned that the interpolation method for the velocity buffers determines the change of motion in the final image. Hence, using more buffers will result in a better approximation.

As known from other pixel motion blur methods, colors are accumulated and divided by the number of accumulated samples along the vector of the velocity image. The number of samples can be determined by the length of this vector resulting in a higher rendering speed when there is little motion. If interpolated velocities are used, the blur scales correctly with the change of speed. Figure 4 shows various images of the combined usage of interpolated velocities for image distortion and blurring. In the upper left image, nearly no movement is present at the bottom line, resulting in less distortion and blur distance. At the upper end of the image, both distortion and blur are present. The upper right image shows a camera movement towards the goal. The camera speed is decreased and stops within the integration interval.

3.4 Noise

To simulate CMOS image sensors, our simulator uses the previously presented techniques in conjunction with dynamically generated noise. The noise model is based on variances for each color channel taken by difference images of an image sensor chosen. The model tries to adopt the static, color, and edge noise but fails on the dynamic temporal behavior of the noise. Therefore only the noise

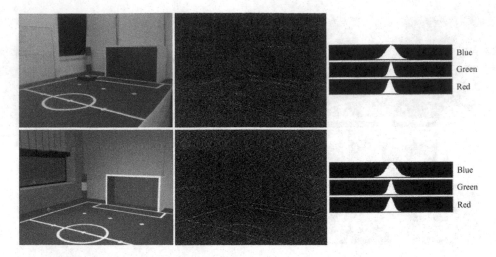

Fig. 5. Comparison of a real camera image at the upper left and its simulated equivalent image at the bottom right. The middle column shows the corresponding contrast enhanced difference images of the noise. The third column shows normalized histograms per color channel.

generated for a single image shows nearly correct statistics and the corresponding visual effect. This calculation is part of a post process of the rendering and it only marginally affects the rendering time. Figure 5 shows a real camera image in the top left corner and a contrast enhanced difference image to another image of the same scene in the middle. The histograms in the upper right corner show the color channel variances of the unenhanced difference image. In comparison, the lower row shows a simulated static scene with the image noise generated. The noise helps to create inhomogeneous color areas in the image, which also appear in real camera images.

4   Results

The generic distortion model allows a variety of different image effects. In addition, it can simply extend an integrated temporal antialiasing algorithm based on per-pixel velocities with nearly no noticeable performance change on contemporary hardware. For comparison, we implemented the relevant state-of-the-art motion blur methods and measured the rendering times in milliseconds. The benchmark scene contains 73 different objects from a simple box up to a complete room. Per frame, about 12000 triangles are rendered. In addition, the scene uses five light sources. The object lighting is calculated per fragment. Every image is rendered using 16 samples for smooth edges. Table 1 shows averaged times taken on an up-to-date desktop PC (AMD Athlon 64 X2 Dual Core, 2.4 GHz, NVIDIA GeForce 8800 GTX).

Table 1. Average rendering times. Times not measured are blanked

	1 Sample	6 Samples	24 Samples	240 Samples
Accumulation Buffer	6.632ms	39.114ms	151.424ms	-
Velocity Buffer	-	13.436ms	14.372ms	15.153ms
Generic Distortion Model	-	-	-	16.266ms

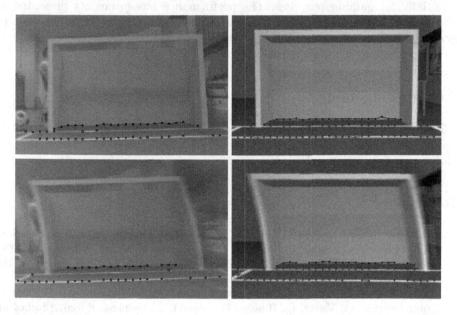

Fig. 6. Comparison of real images on the left and equivalent simulated images on the right. The dots and lines represent percepts of the robot's vision system.

While the accumulation buffer method only handles the motion blur of a scene, the default velocity buffer method using two buffers shows better performance. It allows the integration of other post processes without loosing performance, as it has been done for the work presented here. The time needed for an accumulated image increases linearly with the number of images rendered. The time for the double buffered velocity method depends on the performance of the graphics hardware. The additional calculations performed by the fragment shader units have a low impact on the rendering time. Hence, our extensions made to realize image distortions and noise do not significantly impede the performance.

Finally, Figure 6 compares real images on the left with their simulated counterparts on the right. The top row shows a static scene while the bottom row shows an image from a rotating camera.

5 Conclusion

In this paper, we presented an approach that is able to simulate common image disturbances, i. e. motion blur and the rolling shutter effect. Especially the latter might influence the robot's perception in a way that it has to be handled explicitly within the vision software. Thus its simulation is necessary to allow a more realistic testing within a simulation environment. The quality of our approach is depicted in Figure 6.

Contemporary graphics hardware already provides mechanisms that allow an efficient implementation of this simulation, although originally being developed especially for gaming purposes. The performance measurements presented in Table 1 show that the approach is capable of operating in real-time, i.e. to generate distorted images at a frame rate of up to 60 Hz. This could be assumed to be satisfactory for most applications.

Acknowledgements

This work has been partially funded by the Deutsche Forschungsgemeinschaft in the context of the Schwerpunktprogramm 1125 (*Kooperierende Teams mobiler Roboter in dynamischen Umgebungen*).

References

1. Nicklin, S.P., Fisher, R.D., Middleton, R.H.: Rolling shutter image compensation. In: Lakemeyer, G., Sklar, E., Sorrenti, D.G., Takahashi, T. (eds.) RoboCup 2006: Robot Soccer World Cup X. LNCS (LNAI), vol. 4434, pp. 402–409. Springer, Heidelberg (2007)
2. Röfer, T.: Region-based segmentation with ambiguous color classes and 2-D motion compensation. In: Visser, U., Ribeiro, F., Ohashi, T., Dellaert, F. (eds.) RoboCup 2007: Robot Soccer World Cup XI. LNCS (LNAI), vol. 5001, pp. 369–376. Springer, Heidelberg (2008)
3. Michel, O.: Cyberbotics Ltd. - WebotsTM: Professional mobile robot simulation 1(1), 39–42 (2004)
4. Jackson, J.: Microsoft Robotics Studio: A technical introduction. Robotics and Automation Magazine 14(4), 82–87 (2007)
5. Wang, J., Lewis, M., Gennari, J.: USAR: A game based simulation for teleoperation. In: Proceedings of the 47th Annual Meeting of the Human Factors and Ergonomics Society (2003)
6. Cho, W.h., Kim, D.W., Hong, K.S.: CMOS digital image stabilization. IEEE Transactions on Consumer Electronics (August 2007)
7. Sung, K., Pearce, A., Wang, C.: Spatial-temporal antialiasing. IEEE Transactions on Visualization and Computer Graphics 8(2), 144–153 (2002)
8. Haeberli, P., Akeley, K.: The accumulation buffer: Hardware support for high-quality rendering. ACM SIGGRAPH Computer Graphics 24(4), 309–318 (1990)
9. Tatarchuk, N., Brennan, C., Isidoro, J.: Motion blur using geometry and shading distortion. In: ShaderX2, Shader Programming Tips and Tricks with DirectX 9, pp. 299–308. Wordware Publishing, Inc. (October 2003)
10. Jones, N.E.: Real-time geometric motion-blur for a deforming polygonal mesh. Master's thesis, Texas A&M University, Heatherdale, Houston, Texas (May 2004)
11. Nelson, M.L., Lerner, D.M.: A two-and-a-half-D motion-blur algorithm. ACM SIGGRAPH Computer Graphics 19(3), 85–93 (1985)

12. Green, S.: Stupid OpenGL shader tricks. In: Game Development Conference, pp. 3–16 (2003)
13. Microsoft Corporation: Pixelmotionblur sample, MSDN Microsoft Developer Network DirectX SDK Sample (visited 15.07.07) (2005), http://msdn2.microsoft.com/en-us/library/bb147267.aspx
14. Shimizu, C., Shesh, A., Chen, B.: Hardware accelerated motion blur generation. Technical Report 2003-01, University of Minnesota Computer Science Department, Twin Cities, Minnesota (2003)
15. Rosado, G.: Motion blur as a post-processing effect. In: Nguyen, H. (ed.) GPU Gems 3, pp. 575–581. Pearson Education, Inc., Bosten (2007)
16. Laue, T., Spiess, K., Röfer, T.: SimRobot – A General Physical Robot Simulator and Its Application in RoboCup. In: Bredenfeld, A., Jacoff, A., Noda, I., Takahashi, Y. (eds.) RoboCup 2005. LNCS (LNAI), vol. 4020, pp. 173–183. Springer, Heidelberg (2006), http://www.springer.de/

Database Driven RoboCup Rescue Server

Rahul Sarika, Harith Siddhartha, and Kamalakar Karlapalem

International Institute of Information Technology, Hyderabad,
Gachibowli, Hyderabad-500032, India
{rahul_sarika,harith}@students.iiit.ac.in
kamal@iiit.ac.in
http://cde.iiit.ac.in

Abstract. The RoboCup Rescue Simulation, a generic urban disaster simulation environment constructed on a network of computers, has been in existence for many years. The server used in the simulation league has problems of scaling up. Further, it requires considerable effort to understand the server code to make any additional changes. Therefore, it is difficult for newcomers to quickly enhance the server. The architecture and the functional design of the current server are excellent. This helps us provide a database driven architecture that can scale up the current server to 10-15 times the number of agents that can be simulated. Moreover, it is now easy for others to implement many other subsystems that can provide additional functionality. We have also shown in this paper a new scoring strategy for agent teams which can be customized to emphasize, test and evaluate different concepts and strategies employed by the agent teams.

Keywords: RoboCup Rescue Simulation, multi-agent systems, database.

1 Introduction

Disasters like earthquakes and floods cause heavy damage to a city's infrastructure such as buildings, roads, etc. and endanger the lives of civilians. During such critical emergencies, efficient co-ordination among various rescue teams of paramedics, police and fire fighters is very crucial to minimize the overall loss. The major objective is to save as many civilians as possible along with control over damage to the city. Response time is a major factor in rescue operations, as every second counts.

In order to take the most favourable decisions in real time with minimal error, computer aided simulations are of extensive help. The experience gained in disaster tactics and management through such simulations will be of immense use during an actual disaster [1].

RoboCup Rescue Simulation[10] is one such generic urban disaster simulation environment executed on a network of computers. Its main purpose is to provide emergency decision support through the integration of disaster information, prediction, planning, and human interface. Any simulated environment must provide percepts that enable agents to take appropriate actions in the stipulated

L. Iocchi et al. (Eds.): RoboCup 2008, LNAI 5399, pp. 602–613, 2009.

time. The complexity of the entire rescue and disaster management procedure increases exponentially as the size of the city and the population increases. This provides a substantial challenge for multi-agent systems researchers to enhance the simulator's performance.

1.1 Related Work

The original server [2], based on the modular simulator design [3], underwent many changes since the first competition which was held at RoboCup, in 2001. The changes include the use of a GIS file of a virtual city map and a stable traffic simulator, as well as a considerable amount of code reorganization and cleanup [4]. A new team performance evaluation rule was also proposed which is still being used in the current version, v0.49.9. Takahashi [4] summarized the competitions held in 2002 and mentioned the communication models used by the winner and the runner-up in RoboCup Rescue in 2002. The winner, Arian, thought that PDAs or cellular phones should be used at disaster areas, while YowAI2002 believed that communication lines would be damaged by earthquakes and hence cannot be relied on. However, the scoring system did not consider this aspect during the evaluation of the teams. In real life scenarios, YowAI2002's assumption is a valid one, and the communication model is said to be added as a parameter for future competitions.

A considerable amount of work has been done in the field of Massive Multi Agent Systems Simulators to enhance their performance. SPADES [6] is a programming language independent, distributed MAS simulation environment. JADE [7] is another popular software, a middleware for developing and deploying MAS applications. DMASF [8] is a new Python toolkit that can be used to simulate a large number of agents over a distributed computer network. ZASE [9] is another approach to simulate massive multi-agent systems. It is essential for the RoboCup Rescue Simulation to scale up the numbers in order to make the simulation more realistic and these new toolkits require building of the RoboCup Rescue server from scratch. The new architecture presented in this paper addresses this.

1.2 Challenges in Scaling Up the Current RoboCup Rescue Server

The performance of an agent depends on the space and time complexity for gathering environmental percepts, communication with other agents and computing its next action based on the percepts received and goal(s) to achieve. The whole simulation will slow down if the server takes a long time in simulating the world and generating percepts, regardless of the speed of the agents. It is apparent that the main memory alone cannot be depended on for simulating bigger environments with more number of agents. The introduction of a secondary persistent storage like a DBMS paves way to simulate thousands if not billions [8] of agents. However, introducing such a module addresses only a part of the problem. The communication speed between the slow secondary storage and the fast main memory will be the new performance bottleneck. This is a major issue because data in the secondary storage needs to be loaded into the main memory and then operated upon. Hence, this calls for the use of an appropriate external memory algorithm which will optimize the use of main memory resources.

1.3 Organization of the Paper

This paper is organized as follows. In Section 2, we show that the current RoboCup Rescue server is incapable of running scaled up simulations. In Section 3, we propose a new architecture that overcomes this problem. We also try to rectify the current scoring policy by introducing a new Score Vector. In Section 4, we show the results and discuss the same. Finally, we conclude the paper in Section 5.

2 Background

The architecture (see figure 1) of the current server [5] is wrapped around the system's main memory.

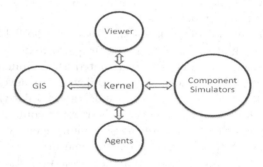

Fig. 1. Architecture of the current server

The kernel and various simulators like *misc simulator, fire simulator, traffic simulator*, et cetera, that connect to it maintain a copy of the pool of objects comprising the world model which is a collection of buildings, roads, nodes and agents. The kernel is the center for all communication. It is responsible for sending agents their percepts, receiving their commands, sending simulators valid agent commands, and receiving updates from simulators (that is, all the steps in a cycle).

The main component of the current RoboCup Rescue server is *LibRescue*. The server was written in a very good object oriented fashion and LibRescue is the library containing the definitions of all objects (for example, Humanoid, Building, Node) used on the server side. By looking at a small snapshot of the Civilian object (see figure 2), many levels of inheritances and high degree encapsulations can be observed.

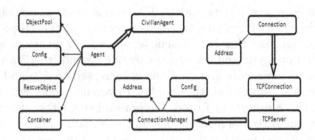

Fig. 2. Snapshot of the Civilian object. A bold arrow and a normal arrow from object A to object B indicate that B inherits A and B is a member of A respectively.

This way of object oriented programming is not suitable for scaling up simulations because the main memory fills up and is unable to handle the large number of instantiations which limits the number of agents that can be created. The main memory cannot handle maps of large cities which contain thousands of buildings (Building object) and tens of thousands of nodes (Node object). In the case of a DBMS, with a proper database schema, all the objects can be treated as tables. Members of the objects will be columns in the respective tables (see figure 3).

Buildings

Id	X	Y	floors	attributes	ignition	fieryness	brokenness entrances
apexes	ground area	total area	building code	importance	temperature	last updated time	

Nodes

Id	X	Y	No Of edges	Signals	last updated time

Fig. 3. The Buildings and Nodes tables contain the information of all buildings and nodes in the map respectively

The object oriented structure was too difficult to work with and reuse in the new architecture. Hence, modules like *gis, kernel, civilian simulator, misc simulator*, et cetera had to be re-written. Nevertheless, the conversion of bigger and more complex modules like traffic and fire simulators was too big a task. Therefore, we developed temporary interfaces which convert the format of the data (MySQL data —> Objects and vice versa) to support communication between the new server and the current component simulators.

In order to find out the upper bound on the number of agents the current server can handle, we conducted a few experiments. The experiment was conducted on the Kobe map starting with 30 rescue agents and 100 civilians and going up to 2000 rescue agents and 6000 civilians in an incremental fashion, maintaining the ratio of civilians to rescue agents as 3. The time taken for a regular Kobe simulation, that is 24 rescue agents and 72 civilians, is a little less than 1.5 seconds per cycle and it increases by one second (to 2.5 seconds) when the number of rescue agents and civilians is increased to 150 and 100 respectively. For the next simulation, the number of rescue agents was increased to 175. However, *the simulation failed*, the reason being either the simulators or the agent program ran out of memory during the initialization process. All these experiments were conducted on the default system configuration as specified in [11]. It is evident from these experiments that only main memory is not sufficient for scaling up the simulation.

Another point to be noted here is that in every cycle, the kernel sends the same updated information to every simulator regardless of its type. For example, suppose a fire fighting agent submits a fire extinguish command. The traffic simulator only simulates the motion of the agents. The traffic simulator does not need the information of the fire fighter which has submitted an extinguish request. Similar redundancies can be observed during a kernel-simulator conversation.

An aspect of the current server which needs revision is its scoring policy which remains unchanged since 2002. The formula used to calculate the score V as mentioned in the rules for the competition [11] is given by,

$$V = (P + \frac{S}{S_{int}}) * \sqrt{\frac{B}{B_{int}}} \tag{1}$$

where, P: Number of living agents, S: Remaining HP of all agents, S_{int}: Total HP of all agents at start, B: Area of houses that are undestroyed, B_{int}: Total area at start.

In equation (1), factor P overshadows (S/S_{int}) because (S/S_{int}) is always less than 1 whereas P is very likely to be far greater than 1. No penalties are issued to the teams for any negligent behavior (for example, wasting resources). The performance of a team cannot be judged by considering the above two parameters only. The end result is definitely to see the maximum number of civilians saved and the minimum amount of the city damaged however, optimal utilization of resources must also be considered. A lot of teamwork is involved in achieving this end result and must be given importance. By increasing the number of parameters, new scenarios can be set to challenge the teams and weaknesses in the current scoring method can be addressed.

3   Database Driven RoboCup Rescue Server

It is evident that massive simulations cannot be run on main memory alone. A secondary persistent storage is necessary. To overcome this problem, we propose a new architecture integrated with a DBMS, namely Database Driven RoboCup Rescue Server (DDRRS). We studied the modelling of the RoboCup Rescue environment, agents and the percepts provided to the agents by the kernel.

With this perspective, we designed a database using all the relevant information to ensure that existing teams work with DDRRS seamlessly. The current server maintains a pool of objects in the simulation. Every agent, building, node, road is an object with its attributes such as coordinates, health, stamina as the members of the object. In DDRRS, these objects are translated to rows in specific tables in the database. Their attributes are columns in the table (see figure 3). As is the case with the current server, additional simulators can be simply plugged in and run. The architecture of DDRRS is shown in figure 4.

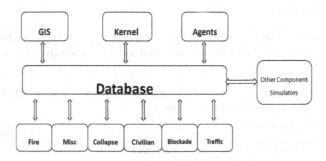

Fig. 4. Architecture of DDRRS

The primary bottleneck in the current server is the main memory which is replaced by a combination of main memory and a database in the new server. In a cycle, most of the time is spent by the kernel on retrieving the percepts for all agents. In the current server, using standard system configurations, the maximum number of rescue agents that can be simulated is just above 150. Assuming that the main memory is capable of holding scaled up simulations, a new problem arises. The kernel in the current server needs to query the object pool for every agent to retrieve the AK_SENSE percept. The time taken to do so for 150 rescue agents is approximately 2 seconds. This time will increase drastically with increase in the number of rescue agents. However, in DDRRS, this problem is solved (see Section 4) which helps support large scale simulations. The algorithms used by the current server concentrate on the kernel side. Currently, work is going on to improve the utilization of the main memory with the help of external memory algorithms.

The problem mentioned above can be divided into two parts, namely the Kernel Side and the Agent Side. As mentioned earlier, the kernel's functionality has been optimized to run the simulation with minimal response time. However, it is observed that as the number of agents increases, the time taken to connect and initialize their states increases exponentially. This problem is on the agent's side. The API written for agents does not support scaling up of simulations. In the current server, the simulation fails in the preliminary stages. However, in DDRRS, there exists no such issue except for slow agent connectivity. In order to substantiate the capability of DDRRS, several scaled up pseudo simulations were conducted, the results of which are very satisfactory. In the pseudo simulation, virtual agents connected to the kernel and submitted random commands. With the standard RoboCup Rescue Simulation Server configuration [11], the results of the new solution to the core problem are quite satisfactory. For the second part of the aforementioned problem, namely the Agent Side, there exist two possible solutions. Either the current API on the agent side can be re-written in a distributed fashion and be light weight, or we can increase the number of machines to run the agents with no change in the API. The better of these two solutions will be implemented in the next version of DDRRS.

3.1 Score Vector

Factors which we thought need to be considered for better evaluation of the teams so as to enhance the level of competition are given in table 1.

Table 1. Factors influencing the performance of a rescue team, the type of influence on the score and the objectives for the teams

No	Factor	Influence	Objective
A	Agents in the following categories:		
	A.1. Dead (0<=HP<=10)	Negative	Minimize
	A.2. Critical (11<=HP<=40)	Negative	Minimize
	A.3. Average (41<=HP<=70)	Positive	Maximize
	A.4. Healthy (71<=HP<=100)	Positive	Maximize
B	Time spent by a rescue agent travelling in the city	Negative	Minimize
C	Average number of messages passed amongst rescue agents	Negative	Optimize
D	Ratio of civilians in refuge	Positive	Maximize
E	Ratio of civilians rescued	Positive	Maximize
F	Percentage of building area destroyed	Negative	Minimize
G	Ratio of fires extinguished	Positive	Maximize
H	Average time taken to		
	H.1. Rescue a civilian	Negative	Minimize
	H.2. Extinguish a fire	Negative	Minimize
	H.3. Transport a civilian to a refuge	Negative	Minimize

The factors involved are easy to calculate by utilizing the database of parameters collected. For example, the traffic simulator updates the column "timeTravelled" in the "Agent" table whenever it simulates the motion of an agent. At the end of each cycle, the viewer calculates the score as is the case with the current server. The viewer in this example calculates the sum of "timeTravelled" column for all agents and obtains the average. The remaining factors can be calculated in a similar fashion.

In the current server, a configuration file is given to the server that defines the game for the next 300 cycles. Nevertheless, competition organizers can change a few things in the scenario like setting more buildings on fire or adding more rescue agents, based on the current performance of the agent teams, to set additional challenges. This is possible with DDRRS. There are tables which contain the configuration data of all simulators, environment and agents which can be modified by using specific SQL queries.

4   Performance of DDRRS

We have designed and implemented version 1.0 of DDRRS whose performance results are given in the following subsections.

4.1   Scale Up Results

Using the same parameters fixed for the scale up experiment conducted to test the limit of the current server, pseudo simulations were conducted on DDRRS, the results of which are shown in figure 5. The current version of DDRRS is capable of simulating 500 rescue agents and 1500 civilians on the Random_Large map provided with the current server v0.49.9 in under 4 seconds per cycle. This sums up to 20 minutes of simulation (300 cycles) which is reasonable. This shows that we are able to scale upto 16 times the current number of rescue agents, not to mention, 15 times the current number of civilians. The current server takes less than 1.5 seconds to simulate a cycle with 30 agents in the Kobe scenario. DDRRS takes the same amount of time to simulate 240 rescue agents trying to help out 720 civilians in a city bigger than Kobe. From this, it can be seen that DDRRS is quite capable of running large scale simulations. With distributed simulation on a network of computers, we expect to scale up further.

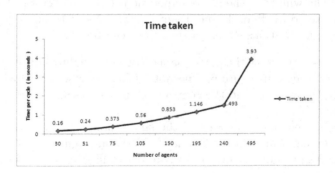

Fig. 5. Scale Up Experiment Results

4.2   Score Vector Analysis

We used the agent codes of six teams which participated in RoboCup Rescue 2006 [10] to test the new scoring method. Of the six, we could run only three as either there were some compatibility issues with the code or the code required some initial training to be done. Given below (see table 2) are the scores of those three teams evaluated using the Score Vector. The game was run on the

Kobe map. The formulae used for the calculation of the parameters are given in equations (2) and (3),

$$MovingAverageFactorUtilization = \frac{\frac{X}{N}}{Y} \tag{2}$$

$$FactorSuccessPercentage = \frac{W}{Z} \tag{3}$$

where, (X, N, Y, W, Z): Depend on the parameter in question

Table 2. Scores of Teams A, B and C at the end of 300 cycles on the Kobe map with reference to Table 1

Factor	Team A	Team B	Team C	Rankings (1, 2, 3)
A.1	11	35	6	C, A, B
A.2	8	4	7	B, C, A
A.3	27	31	18	B, A, C
A.4	50	26	65	C, A, B
B	2.76	2.30	2.12	C, B, A
C	10316	7999	9069	B, C, A
D	47	35	64	C, A, B
E	0.995	0.994	0.997	C, A, B

The ranks of every team on all parameters were summed up to obtain the final ranks. The winner in the above experiment (see table 2) was Team C with 12 points, followed by Team B and Team A with 17 and 19 points respectively.

The utility of equations (2) and (3) can be seen below.

Fire Fighters - We show the performance of the fire fighters of the three teams in extinguishing fires in figure 6. The dotted line represents "Moving Average *Fire Fighter* Utilization" and the continuous line represents "*Fire Fighter* Success Percentage".

Y: Time elapsed since the start of the simulation, X: Time spent by all the fire fighters extinguishing fires till time Y, N: Total number of fire fighters, W: Number of fires put out by time Y, Z: Number of fires spawned since time Y

Paramedics - Similarily (see figure 7), Moving Average *Paramedic* Utilization (dotted line) which is measured in terms of time to rescue a civilian and *Paramedic* Success Percentage (solid line), can be calculated in the same way as done for the fire fighters.

Civilians - The status of the civilians as time progresses can be seen in Fig. 8,

Vector of Parameters - The proposed scoring policy gives a vector composed of parameters which are believed to measure the true performance of the participating teams. This vector can also be mapped to a scalar, to give a quick view of the team standings.

4.3   Discussion

A subtle difference can be observed between the proposed scoring mechanism in table 2 and the current scoring function which just returns a number.

In figure 6, we see that Team C extinguished all the fires by the 30th cycle. However, Team B struggled till the end (solid lines). Also, the utilization of the fire fighters of Team C decreased gradually as there were not any more fires. We can say that Team C assigned high priority to fires so as to reduce any further damage done to the city after the earthquake.

In figure 7, Team B and Team C rescued civilians till the end of the game. However, Team A stopped the process at approximately the 170th cycle having rescued almost all civlians by that time (see table 2). This behaviour is reflected

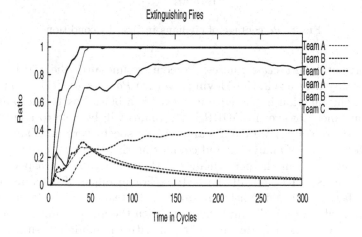

Fig. 6. Performance of the fire fighters

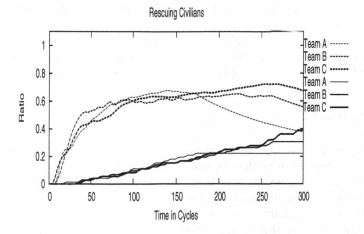

Fig. 7. Performance of the paramedics

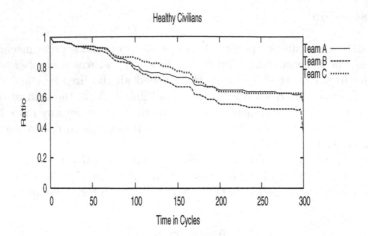

Fig. 8. Percentage of civilians in Healthy condition

in the horizontal *Paramedic* Success Percentage line after the 170th cycle and also the steady decrease in the Moving Average *Paramedic* Utilization.

The scoring function in the current server is a black box to the participants during run time. However in DDRRS, the teams will be able to identify which aspects of their strategy were good and which aspects were not. This gives the teams a better way of analyzing and comparing their performances with others and more details about the performance of the teams can be grasped.

Every team is ranked on all the factors. For example, if 4 teams are being ranked on factors A.1 and D listed in table 1, the team with the least number of dead agents will be ranked 1 and the team with the highest number of civilians in refuge will be ranked 1. Hence, the team with the minimum sum of all such ranks will be the winner as the sum reflects team effort. For example, if the police agents do not clear high priority blockades, more time will be spent by all agents to travel around the city which increases the risk of losing injured civilians as well as wastage of fuel.

The RoboCup Rescue Simulation Technical Committee can decide on the number and type of parameters to judge the teams. New parameters can be introduced accordingly which open doors to new scenario designs.

5   Conclusion

In this paper, we enhance the existing RoboCup Rescue Simulation server to support 10-15 times the number of agents it currently supports. We employ the database centric architecture. The main advantages of our approach are: (i) less dependence on the main memory (Java objects tend to use large amounts of main memory), (ii) enhanced ease in adding different subsystems to cater to additional functionalities that can augment the server, (iii) hassle free integration of legacy agent code with DDRRS, (iv) new scoring parameters that can enable the competition organizers to set simulations to test and evaluate multi-agent systems principles. We also rank three teams on the basis of new scoring parameters, thus showing the utility of the new server. In our on-going work, we are addressing the agent set up problem for agent teams, and extending this server to run on a network of computers.

References

1. Takahashi, T., et al.: Agent Based Approach in Disaster Rescue Simulation - From Test-Bed of Multiagent System to Practical Application. In: Birk, A., Coradeschi, S., Tadokoro, S. (eds.) RoboCup 2001. LNCS, vol. 2377. Springer, Heidelberg (2002)
2. Takahashi, T., et al.: RoboCup-rescue disaster simulator architecture. In: Stone, P., Balch, T., Kraetzschmar, G.K. (eds.) RoboCup 2000. LNCS, vol. 2019. Springer, Heidelberg (2001)
3. Noda, I.: Modular Simulator - Draft of New Simulator for RoboCup. In: WorkShop (ABS-4) Notes, IJCAI (1999)
4. Takahashi, T.: RoboCupRescue simulation league. In: Kaminka, G.A., Lima, P.U., Rojas, R. (eds.) RoboCup 2002. LNCS, vol. 2752, pp. 477–481. Springer, Heidelberg (2003)
5. Kaneda, T., et al.: Simulator Complex for RoboCup Rescue Simulation Project - As Test-Bed for Multi-Agent Organizational Behavior in Emergency Case of Large-Scale Disaster. In: Stone, P., Balch, T., Kraetzschmar, G.K. (eds.) RoboCup 2000. LNCS, vol. 2019. Springer, Heidelberg (2001)
6. Riley, P.F., et al.: Spades - A Distributed Agent Simulation Environment with Software-in-the-Loop Execution. In: Winter Simulation Conference (2003)
7. Bellifemine, F., et al.: JADE - A FIPA compliant agent Framework. In: PAAM (1999)
8. Rao, I.V.A., et al.: Towards simulating billions of agents in thousands of seconds. In: AAMAS (2007)
9. Yamamoto, G., et al.: A Platform for Massive Agent-based Simulation and its Evaluation. In: AAMAS (2007)
10. RoboCup Rescue Homepage, http://www.robocupcuprescue.org
11. RoboCup 2006 Rescue Simulation League Rules (2006)

What Motion Patterns Tell Us about Soccer Teams

Jörn Sprado and Björn Gottfried

Centre for Computing Technologies, University of Bremen, Germany
{sprado,bg}@tzi.de

Abstract. A qualitative representation of motion patterns is presented that forms an interface between low-level concepts of behaviours and high-level concepts of reasoning. How the patterns can be employed for characterising interaction patterns in soccer is demonstrated using the simulation league; also, specific soccer scenes from real games prove their adequacy. The advantages of our approach are: it supports the limited abilities of robots in the different RoboCup leagues, i. e. it relies on coarse positional distinctions that are reliably obtainable and easily translated into action; the analysis is directly derived from raw data without the need for any preprocessing steps; both situations can be dealt with, egocentric viewpoints of individuals and the bird's eye view; the approach is independent on the domain, i. e. generalises to arbitrary spatiotemporal interaction patterns.

1 Introduction

This paper investigates motion patterns among soccer players for analysing team behaviours. The purpose of this analysis is to find patterns which are typical for a specific team [7,14], for their defenders or other subgroups, and to come to specific conclusions about successful and unsuccessful behaviours in order to find promising team interactions. More advanced topics include the automatic search for specific situations in databases of recorded games up to the automatic recognition or even realisation of strategies.

A more specific sub-goal we shall focus on consists in characterising adequately specific scenes. The adequacy derives from a description which considers the specific role of each player, namely his viewpoint which determines how he interacts with others. This egocentric viewpoint has to rely on as simple distinctions as whether another player is left of him, in front of him, towards his right, or back. More precise descriptions would not reflect the distinctions a player, or rather a robot, is capable to make reliably. It is therefore the question, how a behaviour model looks like that captures those distinctions and whether such a model sufficiently characterises scenes of soccer games.

The model of motion patterns we shall finally arrive at has the following characteristics:

(a) It forms an interface between low-level concepts of behaviours and high-level concepts of reasoning.

L. Iocchi et al. (Eds.): RoboCup 2008, LNAI 5399, pp. 614–625, 2009.

(b) Giving the limited abilities of robots in the different RoboCup leagues, the proposed coarse qualitative motion patterns are well suited for they do only require simple abilities regarding both sensors and effectors; the patterns rely on coarse positional distinctions that are reliably obtainable and that can be easily translated into action.

(c) The qualitative behaviour patterns can be directly derived without any pre-processing steps, making the algorithms defined on them particularly fast and applicable for realtime analysis (cf. [1]).

(d) The proposed motion pattern representation is suited for both cases to deal with the egocentric viewpoint of individuals and for analysing a scene from the bird's eye view.

(e) The approach generalises to arbitrary spatiotemporal behaviour investigations that can built upon coarse qualitative patterns.

This approach is employed for characterising behaviour patterns in soccer using the complete 2d-simulation league championships of 2007; in addition, specific soccer scenes from real games prove the adequacy of this approach.

1.1 Overview

After having stated fundamental problems that characterise more thoroughly the situation, the most similar related work on motion patterns is reviewed. A number of basic motion patterns are introduced afterwards. These patterns are used in the following in order to analyse how they behave when considering different temporal and spatial granularity levels; this is for the purpose of showing that the approach handles raw positional data without any preprocessing steps. A number of combined patterns are introduced in the next section, showing how simple patterns combine to more sophisticated interaction behaviours. Such interaction behaviours are employed afterwards, giving first insights into how the approach will manage more sophisticated issues in future work, for example the description and recognition of tactics.

2 Motion Analysis

The spatiotemporal behaviour of objects can be analysed in many different ways. The obvious thing to do would be to take the positions of all objects for all time points in every temporal window of interest. Then however many problems arise:

– Which temporal windows are of interest?
– What is an appropriate temporal granularity level for the time points?
– What do we have to do with those many positions of each object in each time window in order to obtain meaningful patterns?
– How to integrate the great number of positions of different objects in different time windows in order to obtain information about how objects interact?

2.1   Related Methods

Each method for the analysis of spatiotemporal behaviours has to deal with these problems. In particular the need to represent interactions is of great significance in order to evaluate team behaviour. However many of the methods restrict themselves to single trajectories: [12] consider only single trajectories instead of object interactions; [6] relate trajectories to the spatial context of a soccer pitch but they do also not consider object interactions; [9] provide a framework for intention recognition based on motion patterns, but again, do not consider interactions.

From those methods that have been applied in the soccer context [10] also considers interactions. They compare motion attributes of objects such as speed, change in speed, and motion azimuth; patterns are matched on these attributes in order to describe *constancy* (sequences of equal attributes for consecutive time steps), *concurrence* (objects showing the same attributes), and *change* (the change of attributes). Based on these patterns more complex patterns can be defined, such as *trend-setter*: an object anticipating motion attributes of other objects. This approach as well as others using such fundamental attributes have the disadvantage that there is the need to define interactions based on basic motion attributes before comprehensible patterns can be derived. An approach capturing directly significant object interactions would represent in a compact way complex behaviours of a group of objects.

Another approach applied to the soccer simulation league considers similar attributes such as motion azimuth, distances, and their changes for deriving also information about speed [11]. They calculate four types of time series from the raw positional information of soccer players: the motion direction and speed of individual players, as well as the directions and distances among pairs of objects. Then two steps of abstraction are performed, namely the temporal segmentation of the time series into time intervals of homogeneous motion and a mapping of the attribute values describing the intervals to qualitative classes; the latter being domain dependent concepts which can be further processed by a rule-based interpretation system based on predicate logic.

3   Basic Patterns

Looking for an appropriate repertoire of motion patterns for modelling interactions the relationships between objects should be directly taken into account. These relationships should be defined in a relative way in order to directly capture relative object motions which do not require to take into account absolute

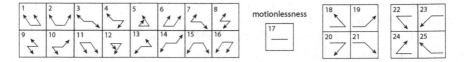

Fig. 1. Atomic motion patterns: 16 $\mathcal{TLT}$ and 4 $\mathcal{BLT}$ classes

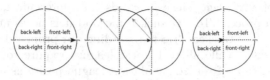

Fig. 2. How $\mathcal{TLT}(1)$ is defined

positions. In this way we model object interactions as defined in [5] which are shown in Fig. 1; since each pattern describes object interactions by using three line segments these patterns are referred to as *Tripartite Line Tracks*, $\mathcal{TLT}$ for short. In each pattern two objects as well as their positional relation among two time points is described; the middle line segment connects the positions of both objects at t_0 while the heads of the arrows show their positions at t_1. In this way, each $\mathcal{TLT}$ shows the spatial interaction between two objects, implying the consideration of relative positions and directions of movement.These patterns satisfy two important conditions: they can be generated efficiently and they can be easily comprehended; this enables their application in the context of game analysis. Even more important is that the patterns take into account the viewpoints of individual players: for this purpose the orientation grid of [15] is employed, there, motivated from the cognitive point of view. Fig. 2 shows how the orientation grid defines these patterns by taking the first pattern, $\mathcal{TLT}(1)$, as an example; it also shows its adequacy for the egocentric point of view, because it is simple to reliably make the distinctions between left, right, front, and back of one's own perspective; it is however also interesting for the coach in that the very same distinctions can be easily made from the bird's eye view.

While the motion patterns of the previous paragraph describe how two players move relative to each other between two time points, it might occur that one of the players does not move or that even both keep where they are. While the latter is the trivial pattern that connects two positions (referred to as motionlessness), in the former case four possibilities (referred to as semi-motionlessness) exist for each of the two players which are shown on the right hand side of Fig. 1. As a consequence there is a total of $16 + (2 \cdot 4) + 1 = 25$ basic motion interactions we shall distinguish.

4   Temporal and Spatial Granules

The aim is to provide efficient methods for game analysis, i. e. methods that directly process the raw data of tracked objects. That is, instead of accumulating time series [11] which are to be segmented into meaningful intervals, which in turn are to be categorised with respect to domain knowledge, it would be of great advantage to avoid the segmentation step. This is mainly because both threshold based and monotonicity based criteria for the segmentation of time series (both applied in [11]) are problematic since they apply general criteria to specific scenes that might significantly vary. A method that directly processes the raw positional data would be of advantage, since it avoids the representational vagueness of the segmentation process. To test such a method it is necessary to analyse the raw positional data of tracked objects regarding our proposed representation of interaction patterns. While all real world leagues require to track all objects of interest, in the case of the 2d-simulation league positions are directly given (by server logfiles). For a comprehensive investigation we have considered all 88 games of the RoboCup Championships 2007 in Atlanta. The results of the analysis have to reveal whether the raw positional data are in fact appropriate for game analysis taking into account the temporal and spatial dimensions. First hints would be that the motion pattern distributions do not significantly vary among different games which are analysed on the same temporal and spatial granularity levels. After that, for single specific scenes it is also to be shown that typical behaviour patterns are adequately characterised by our approach.

4.1 Temporal Granularity

The temporal granularity concerns the temporal distance between two successive states of a game. In a sense, for each real word league the temporal granularity is the same and from the point of view of the human observer continuous spatiotemporal patterns arise. For an appropriate analysis however the question arises as to how large the temporal distance between successive states should be. Regarding the simulation leagues the temporal dimension is partitioned into discrete slices: a number of 6000 snapshots define a game. These temporal slices can be directly taken for the analysis of such simulation games and for real word leagues similar slices are to be explicitly defined.

The first analysis looks for differences occurring when changing the temporal granularity. This analysis is of importance since an appropriate temporal granularity level is to be chosen for all other investigations. The result of this analysis shows that the number of $\mathcal{TLT}(1)$, $\mathcal{TLT}(6)$, $\mathcal{TLT}(11)$, and $\mathcal{TLT}(16)$ patterns increase when choosing coarser granules, while the number of those patterns decreases for which one or both objects do not change their positions. An explanation is that at finer temporal granules it occurs more often that players roughly keep where they are, while at coarser temporal granules the distance among successive positions increases. Of great interest is the dominance of $\mathcal{TLT}(1)$, $\mathcal{TLT}(6)$, $\mathcal{TLT}(11)$, and $\mathcal{TLT}(16)$ patterns at each temporal granularity level. It shows that objects almost ever move into similar directions, namely towards that direction where the main action of a game takes place; this is a clear tendency shown in most games.

The temporal granularity has been analysed by using a spatial granularity (see below) of 0.1. For the finest case, i. e. a granularity of 1 (each time point is taken into account) the standard deviation is 0.79, for a temporal granularity of 500 the standard deviation is 0.93; it shows that on each of the temporal granularity levels the games show similar motion pattern distributions. In other words the frequency distributions are very similar for all of the 88 games, even on quite coarse granularity levels. At a temporal granularity level of 500 however significant distinctions disappear, as shown by the black bars and the patterns

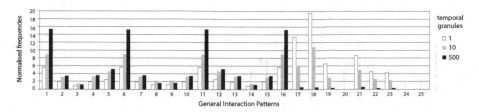

Fig. 3. Different temporal granules compared

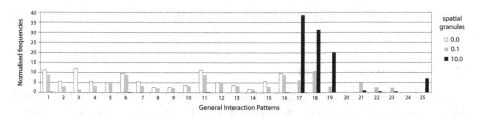

Fig. 4. Different spatial granules compared

on the right hand side of Fig. 3, i. e. patterns 17 to 25. Therefore, a granularity level should be chosen at which these patterns can be still distinguished; this is still the case at a temporal granularity level of 10. Note that at this granularity level we not only work on the raw positional data but even have to take into account only a tenth part of the original data.

4.2 Spatial Granularity

The positional data of the 2d-simulation league is very precise; small changes are not visible. As a consequence a parameter is used that can be adjusted to an appropriate spatial granularity level, that is a level at which changes in position are visible. This brings in the problem that many small successive positional changes would be regarded as positional idleness, although they together sum up to large positional changes. We treat this problem by choosing a temporal granularity level, t, that correlates with the spatial granularity level, s, as follows: the larger the spatial threshold, the more positional changes will be regarded as spatial idleness; the temporal threshold has to be accordingly high: $t = f(s)$; the function f assures that t is large enough in comparison to the spatial granularity. Conversely, the lower s, the finer the temporal granules can be chosen.

Fig. 4 shows that when the spatial granularity level is too large then the obtained motion patterns are quite inaccurate, for there are many idle motion patterns (mainly patterns 17, 18, and 19); on the other hand the interesting patterns between 1 and 16 disappear. The experiments indicate that a spatial granularity of 0.1 (in terms of the spatial unit of the 2d-simulation league) is appropriate together with a temporal granularity level of 10 (cycles), since at

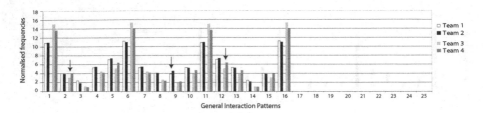

Fig. 5. Two games with four different teams

this granularity levels there is no significant loss in information compared with the raw data.

4.3   Reactivity Behaviour

Single games have been analysed as follows: the motion patterns between each pair of teammates have been determined; on the other hand, patterns among two players of different teams have not been considered. The purpose of this restriction is to measure the behaviour of teams as opposed to how a single player behaves in relation to all other opponents. Fig. 5 shows two games (the first one of loser round 1, the second is the final game), i. e. for altogether four different teams the frequency distributions of interactions are shown. Team 1 (AT Humboldt) played against team 2 (NCL), and team 3 (Brainstormers) against team 4 (WrightEagle). Clearly, in both games the frequency distributions are very similar. There are only a few outliers, such as those three indicated by arrows between team 1 and team 2. An explanation for the similarity of the frequency distribution of opposing teams is that the team behaviours are quite reactive (cf. [13]), entailing similar behaviours. This can in fact be comprehended very well when observing games in the simulation league. Note that this observation is not specific to these two games but shown in most of the 88 games of the 2007 championships. Furthermore, we have chosen deliberately one game of the loser round 1 and the final game; in other words this behaviour is shown in very different games. We assume that in human soccer such reactivity behaviours will not be as dominant as in simulation soccer.

Comparing two different games it shows however that the frequency distributions differ (Fig. 5), i. e. the frequencies of team 1 and team 2 are clearly different to those of team 3 and team 4. This simply indicates differences concerning the dominance of interaction patterns in different games.

5   Specific Scenes

Having analysed all 88 games from a holistic point of view, i. e. by determining histograms of motion patterns for the entire games, we will now turn our attention to specific scenes, in order to investigate whether the motion patterns are also appropriate for characterising specific interaction patterns in a game.

A specific scene is characterised by a number of motion patterns which define how the players behave in that scene. Several characteristic behaviour patterns are analysed in the following paragraphs. That is, basic patterns can be combined to arbitrarily complex patterns in order to model interaction patterns of a number of $k > 2$ players and along the temporal dimension.

Definition 1 (Motion Pattern). *A motion pattern is a 5-tuple (m, o_1, o_2, t_0, t_1) with two objects $o_1, o_2 \in \mathcal{O}$ among which a motion pattern $m \in \mathcal{M}$ holds between two time point $t_0, t_1 \in \mathbb{N}$ with $t_0 < t_1$.*

Definition 2 (Complex Motion Pattern). *A complex motion pattern is a conjunctive set of motion patterns.*

5.1   Complex Temporal Patterns

This category of patterns combines $\mathcal{TLT}$s along the temporal dimension: for two successive time intervals the basic patterns are identified, defining a more complex pattern.

Definition 3 (Complex Temporal Motion Pattern). *A complex temporal motion pattern is a conjunctive set of motion patterns for successive time intervals.*

Fig. 6 illustrates a complex temporal pattern. The meaning of such a passing manoeuvre depends on whether both players are teammates or not. In addition, the direction of such a passing manoeuvre regarding the pitch is of significance; this direction can be determined by defining motion patterns between players and specific landmarks, e. g. the corners of the pitch.

5.2   Complex Spatial Patterns

Fig. 7 shows a complex spatial pattern in the 73rd minute of the game Germany–Costa Rica of the world championship 2006: three players of the white team (Costa Rica) attack the black team (Germany); player 1 (left most) tries to run towards the front middle in order to receive the ball the attacker is going to pass through the German players. Table 1 shows the motion patterns for the scenario shown in Fig. 7 (Scene 1) and Fig. 9 (Scene 2). The motion patterns have been determined among all players within each team and among players of different teams.

Fig. 6. Complex temporal patterns: passing manoeuvre with different meanings depending on whether both players are teammates or not; differences occur since B might turn to the left, $\mathcal{TLT}(11)\ \mathcal{TLT}(12)$, or to the right, $\mathcal{TLT}(10)\ \mathcal{TLT}(11)$

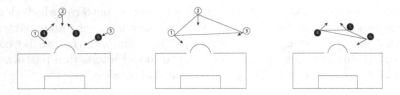

Fig. 7. White attacks black (left); motion patterns among teammates (middle & right)

Fig. 8. Motion patterns among opponents

Table 1. $\mathcal{TLT}$ motion patterns for two scenes

	White Team	Black Team	Both Teams
Scene 1	12, 12, 12	05, 08, 04	10, 09, 12, 05, 12, 12, 08, 07, 06
Scene 2	*04*, *04*, 12	05, 08, 04	*02*, *01*, *04*, 05, 12, 12, 08, 07, 06

Fig. 9. Scene 2: no. 1 moves towards another direction, entailing the change of patterns

Definition 4 (Complex Spatial Motion Pattern). *A complex spatial motion pattern is a conjunctive set of motion patterns of a number of $k > 2$ objects.*

The pattern among the teammates of Costa Rica (middle of Fig. 7) is an attack pattern in which all players move towards the goal. No. 1 runs in the penalty area, preparing for getting the pass from no. 2. The pattern among the German players (right of Fig. 7) is a typical defense pattern: two of the teammates show the tendency to meet in order to avoid a gap and to get the ball. That is Metzelder (no. 4) and Mertesacker (no. 5) move towards the attacker while Friedrich (no. 6) moves towards the middle expecting the ball in the middle, trying to beat Wanchope. Comparing these two patterns within the teams it clearly shows that the strategies of individual players (of the Costa Rica team) and their relations to their teammates follow one strategy, while the German defenders follow simultaneously two different strategies: no. 4 and no. 5 fight the attacker, while no. 6 prepares for dealing with no. 1 of the other team. A common strategy of the German team could have been an offside trap. For this purpose no. 6 would have had to avoid to run towards the goal too deeply. This alternative strategy can be easily communicated to human players. But even formally such an alternative strategy can be characterised using $\mathcal{TLT}$ motion patterns: no. 6 would have to avoid a motion pattern like $\mathcal{TLT}(12)$ with no. 1 as shown on the left hand side of Fig. 8 and should rather try to realise a pattern like $\mathcal{TLT}(9)$. Clearly, a finer argumentation would be that no. 6 could maintain pattern $\mathcal{TLT}(12)$, however, by simultaneously running not closer to the goal than no. 1. This strategy is much more difficult to realise for a robot with limited running abilities and sensors. By contrast, the realisation of $\mathcal{TLT}(9)$ instead of $\mathcal{TLT}(12)$ is much simpler since the distinctions of positional relations can be reliably made without sophisticated measurements.

After player 1 has changed his direction of movement, five of the motion patterns change (see *italic numbers* in Table 1). The situation is completely different in that the white attacker now has no one to pass the ball to. However the motion pattern search algorithm is capable of telling apart these situations. The discussed scene ended by Wanchope (no. 1 in Fig. 7) kicking the ball into the net, changing the score to 2:3 for Costa Rica. This scene and several other scenes show that in fact $\mathcal{TLT}$ motion patterns and their combinations make significant distinctions regarding team behaviours, as can be comprehended by further typical scenarios which are presented in [3].

5.3   Complex Spatiotemporal Patterns

In this section we explain how to describe more complex situations by combining complex spatial with temporal patterns.

Definition 5 (Complex Spatiotemporal Motion Pattern). *A complex spatiotemporal motion pattern is a conjunctive set of motion patterns of a number of $k > 2$ objects for successive time intervals.*

As an instance of a complex spatiotemporal problem, we exemplarily describe an offside situation which is defined by the Fédération Internationale de Football Association (FIFA) [4] as follows: A player is in an offside position if he is nearer to his opponents goal line than both the ball and the second last opponent. In addition to that there are some restrictions to the offside law. Thus, a player is not in an offside position if he is in his own half of the playfield or he is level with the second last opponent or he is level with the last two opponents. There is, furthermore, no offside offence if a player receives the ball directly from a goal kick or a throw-in or a corner kick which reduces the states we have to verify. Detecting offside situations we analyse motion patterns among $k \geq 4$ objects, at least two offensive players, an opponent and the ball. In general, to detect an offside position it is almost sufficient to regard two successive scenes. However, to identify an offside offence in which a striker is involved in active play or interfering with an opponent or gaining an advantage by being in an offside position, successive time intervals have to be taken into account. Fig. 10 shows

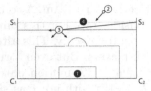

	S_1	S_2	C_2	C_1
$\leftarrow i$	21, 23, 24	21, 22, 24	22, 23, 24	21, 22, 23
$i \rightarrow$	22, 23, 24	21, 22, 23	21, 22, 24	21, 23, 24
$\leftarrow o$	21, 22, 23	22, 23, 24	22, 23, 24	21, 22, 23
$o \rightarrow$	21, 23, 24	21, 22, 24	21, 22, 24	21, 23, 24

Fig. 10. No. 2 passes the ball to his teammate no. 3 who holds an offside position (left); possible initial positions and directions for motion patterns (right)

a situation where a player no. 2 passes the ball to his teammate no. 3 who holds an offside position. Generally, the offside area can be defined by the orthogonal intersection outgoing from the second last opponent and the sidelines of the pitch. In this case, the offside area is limited by the vertices S_1, S_2, C_1, C_2. Thus, the challenge of identifying an offside position corresponds to the issue of determining the objects' position regarding a particular area. That is, we determine if a player is contained in the offside area A or not by analysing motion patterns with respect to the vertices of A. Fig. 10 shows general distinctions which can be made in this context outgoing from possible positions (inside or outside) and possible moving directions (left or right). For example, a combination of motion patterns for player no. 3 and a time interval [0,1] exists: $(3, S_1, 22, 0, 1)$, $(3, S_2, 22, 0, 1)$, $(3, C_2, 21, 0, 1)$ and $(3, C_1, 21, 0, 1)$. In this case, we can unambiguously infer the initial position (inside) and direction (right) of the player.

6 Summary

In this paper we propose motion patterns as a qualitative representation to describe spatial and temporal behaviours in the soccer domain. It shows that these motion patterns can be directly projected onto symbolic patterns. That this works has been demonstrated for all 88 games of the simulation Championships in 2007. We also discuss temporal and spatial granules to draw up motion patterns among objects. We learn that, when being combined, these result motion patterns will enable much more complex scenarios and will form the basis in future work for plan recognition [8] by means of argumentation frameworks [2].

References

1. André, E., Binsted, K., Tanaka-Ishii, K., Luke, S., Herzog, G., Rist, T.: Three robocup simulation league commentator systems. AI Magazine 22(1), 57–66 (2000)
2. Bench-Capon, T.J.M., Dunne, P.E.: Argumentation in artificial intelligence. Artif. Intell. 171(10–15), 619–641 (2007)
3. Bode, G.: WM Analyse. Fussball training 24, 39–47 (2006)
4. Fédération Internationale de Football Association (FIFA). Laws of the game 2007/2008 (2007)
5. Gottfried, B.: Representing short-term observations of moving objects by a simple visual language. Journal of Visual Languages and Computing (2007)

6. Gottfried, B., Witte, J. (Sprado). Representing spatial activities by spatially contextualised motion patterns. In: Lakemeyer, G., Sklar, E., Sorrenti, D.G., Takahashi, T. (eds.) RoboCup 2006: Robot Soccer World Cup X. LNCS (LNAI), vol. 4434, pp. 330–337. Springer, Heidelberg (2007)
7. Hirano, S., Tsumoto, S.: Finding interesting pass patterns from soccer game records. In: Boulicaut, J.-F., Esposito, F., Giannotti, F., Pedreschi, D. (eds.) PKDD 2004. LNCS, vol. 3202, pp. 209–218. Springer, Heidelberg (2004)
8. Kautz, H., Allen, J.: Generalized plan recognition. In: AAAI 1986, pp. 32–37. AAAI Press, Menlo Park (1986)
9. Kiefer, P., Schlieder, C.: Exploring context-sensitivity in spatial intention recognition. In: Gottfried, B. (ed.) 1st Workshop on Behaviour Monitoring and Interpretation (BMI 2007), vol. 296, pp. 102–116. CEURS Proceedings (2007)
10. Laube, P., Imfeld, S., Weibel, R.: Discovering relative motion patterns in groups of moving point objects. International Journal of Geographical Information Science 19(6), 639–668 (2005)
11. Miene, A., Visser, U., Herzog, O.: Recognition and prediction of motion situations based on a qualitative motion description. In: Polani, D., Browning, B., Bonarini, A., Yoshida, K. (eds.) RoboCup 2003. LNCS (LNAI), vol. 3020, pp. 77–88. Springer, Heidelberg (2004)
12. Musto, A., Stein, K., Eisenkolb, A., Röfer, T., Brauer, W., Schill, K.: From motion observation to qualitative motion representation. In: Habel, C., Brauer, W., Freksa, C., Wender, K.F. (eds.) Spatial Cognition 2000. LNCS, vol. 1849, pp. 115–126. Springer, Heidelberg (2000)
13. Riedmiller, M., Gabel, T.: Brainstormers 2D - team description 2007. In: Visser, U., Ribeiro, F., Ohashi, T., Dellaert, F. (eds.) RoboCup 2007: Robot Soccer World Cup XI. LNCS (LNAI), vol. 5001. Springer, Heidelberg (2008)
14. Visser, U., Weland, H.: Using online learning to analyze the opponent's behavior. In: Kaminka, G.A., Lima, P.U., Rojas, R. (eds.) RoboCup 2002. LNCS, vol. 2752, pp. 78–93. Springer, Heidelberg (2003)
15. Zimmermann, K., Freksa, C.: Qualitative Spatial Reasoning Using Orientation, Distance, and Path Knowledge. Applied Intelligence 6, 49–58 (1996)

Designing Grounded Agents:
From RoboCup to the Real-World

Christopher Stanton

Innovation and Technology Research Laboratory,
The University of Technology, Sydney
cstanton@it.uts.edu.au

Abstract. This paper discusses the nature and role of "grounding" in
designing programs for controlling autonomous mobile robots. Since its
inception, artificial intelligence has been plagued by problems of scaling
and brittleness. A fundamental problem impeding the development of
artificial intelligence is our dependence on grounding agents by design.
That is, currently agents tend to be grounded by their designer's understanding of the world, task, and robot. However, little (if any) of the
knowledge of "how to ground" is embedded in the artificial agent. Consequently, brittle, purpose-built systems result. This paper explores how
the intellectual burden of grounding can be shifted from the programmer
to the program by designing robots capable of grounding themselves. An
overview of a grounding oriented design methodology (Go-Design) is presented - an initial step towards the longer-term objective of developing
autonomous grounding capabilities.

1 Introduction

A common feature of research in the field of autonomous robotics involves the
development of systems capable of performing specific tasks in controlled, micro-
worlds - environments in which the complexity and unpredictability of the real-
world is simplified, either by the designer choosing to ignore large parts of it,
and/or by the designer controlling or modifying the environment. The driving
assumption of the micro-world paradigm is the artifacts produced for the micro-
world environment can serve not only as a proof-of-concept, but also as a starting
point for "scaling-up" to more sophisticated, "intelligent" and general-purpose
programs. The history of AI is littered with examples of micro-worlds, such as
blocks world, chess, and more recently robot soccer. However, the great promise
offered by many systems developed for micro-worlds has yet to translate to the
real-world, and thus we have the related problems of scaling [1,2] and brittleness [3].

In this paper we discuss the nature of *grounding* [4,5,6] and consider the
implications for both RoboCup and artificial intelligence more generally. The
paper begins (Section 2) with an interpretation of the grounding process, i.e.
what "grounding" exactly is, and how we currently approach the problem. In
Section 3, the role of grounding in RoboCup is discussed, and it is argued that our

L. Iocchi et al. (Eds.): RoboCup 2008, LNAI 5399, pp. 626–637, 2009.

ability to develop robust and scalable systems is restricted by our failure to treat the grounding problem as a problem *per se*. Section 4 discuss the longer-term goal of developing autonomous grounding capabilities. In Section 5, a (very) brief overview of a grounding oriented design methodology (Go-Design) is presented, together with examples of its application in the robot soccer domain. The paper concludes by discussing the benefits and limitations of Go-Design, together with future work.

2   Grounding

The term *grounding* in the context of artificial intelligence and cognitive science is related to the concepts of "meaning" and "understanding". For example, symbol grounding [5] concerns how the symbols of a symbol system can become meaningful to the symbol system, while the physical grounding hypothesis argues that representations should be "grounded in the physical world" [6]. In layman's terms, to say a person is "not grounded" could mean that their understanding or beliefs about the world (or a particular topic) are incorrect, irrelevant, or even delusional. In contrast, a grounded person is indeed the opposite - for example, "the mechanic has a solid grounding with regards to truck engines" implies the mechanic has either experience, or a thorough understanding, of the mechanics of truck engines. It is this meaning of grounding - loosely described as a perceptually-related "understanding" of the world - that we are concerned about in this paper.

While many researchers argue that for an artificial agent to be grounded it must autonomously learn the "meaning" of representations [5,8,7], the fact is, *all* artificial agents are embedded by design in their environments. That is, designers use their knowledge of the problem to design appropriate control mechanisms, regardless of the balance between knowledge acquired through learning versus innate knowledge endowed to the robot's control program *a priori*. So, is grounding the process of writing control programs? No, but it is involved in this process. More specifically, *we ground robots by understanding the world for them*. The process of embedding an artificial agent in an environment involves identifying regularity and structure within the world that can be used for decision-making. For the designers of robot control programs, it can be as simple as making decisions regarding "what" to perceive and represent. For example, the designer of a control program for an autonomous vehicle may decide to represent and perceive other cars, traffic lights, pedestrians and so forth. Likewise, with regard to implementing a representation design, developers may make decisions such as identifying a sensor state (e.g. a sensor value or pattern of values) which correlates to something in the real world, or in the case of a learning system, develop a (biased) learning algorithm to learn such a relationship. Unfortunately, we often encode little of this knowledge of how we understand the world into the program, and a "code-and-fix" iterative development paradigm results.

2.1 Groundedness

Robot soccer presents an excellent domain for comparing the *groundedness* of systems. Anyone who has ever had "hands-on" development experience in robotics knows a robot's model of the world will invariably have a degree of error, and the nature of that error can vary. For example, a robot soccer player may miscalculate the distance of the soccer ball from the robot, or alternatively fail to "see" the ball entirely. The term "groundedness" [9] refers to the quality of an agent's grounding - i.e. how *well* the agent is grounded. Systems from airline reservation databases to autonomous mobile robots rely on grounded representations. For example, an airline reservation system must manage information about flights and passengers in a way that corresponds to real flights and real passengers. Similarly, an autonomous mobile robot that navigates a physical space will be more effective in achieving its objectives if its internal representations of physical barriers correspond to real physical barriers in its environment. Despite the varied approaches to grounding, little attention has been paid to measuring and assessing the performance of different theoretical approaches and practical implementations (except for [9]). For example, consider a soccer playing robot and its representation of the location of the soccer ball. The representation of the ball's location may be in error by the smallest of distances (e.g. 1 millimetre), a large distance (e.g. 1 metre), or somewhere in between. Thus, when considering if a representation is grounded, our judgements must be made with respect to task requirements and task performance. That is, designers must specify and understand what constitutes a grounded representation for each grounding problem.

3 Grounding: From RoboCup to the Real-World

The domain of robot soccer provides a common benchmark for demonstrating, comparing, evaluating and (hopefully) sharing state-of-the-art techniques in robotics and artificial intelligence, with the aspirational goal of bettering the world's best human players by 2050. Examples of grounding in the RoboCup domain are abundant - sensor data is interpreted by designers/programmers as being "about something"; designers choose what to represent, modeling things as the soccer ball and the robot[1]; and designers build elaborate decision-making routines which rely on the state (and "correctness") of these models. The common aspect of all approaches to grounding in robot soccer is that systems are predominately grounded by their designers, with little or no knowledge of "how to ground" embedded in the resultant programs. As a consequence, the systems we build tend to be brittle in the face of unanticipated (from the perspective of the designer) changes in the environment or task, and as such, robotics development is (generally) a highly iterative code-and-fix paradigm[2], in which developing systems capable of scaling up to human levels of intelligence has (so far) proven unattainable.

---

[1] And anything else that is useful in winning soccer games.
[2] But not always - e.g. a Mars rover needs to work first time.

Due to the ease and effortlessness of which we "make sense" of the world, developers often tend to underestimate the complexities and difficulties in building autonomous agents. The grounding problem is a difficult problem, and has been compared in difficulty to the entire problem of artificial intelligence[5]. The embodied, real-time nature of robot soccer means the grounding problem can't be avoided - all robots must be grounded in someway. Rather, the pertinent questions are *"how* are they grounded?", and "how *well* are they grounded?" (i.e. their *groundedness*). Importantly, the autonomous systems developed for RoboCup are overwhelmingly *grounded by design* (that is, grounded by their designer's understanding of the world and task), with minimal *autonomous grounding* capabilities endowed to the autonomous system. Thus, such solutions tend to be handcrafted to the specific RoboCup environment of each competition, in which the soccer field configurations are (generally speaking) slowly transitioning towards more natural environments (e.g. through the gradual removal of distinctive landmarks and controlled lighting). Designers typically handcraft perception mechanisms, with the role of learning usually restricted to gray-box[10] parameter-optimization (rather than learning new "concepts" or entirely novel ways of doing things).

Despite the gradual and ongoing evolution of RoboCup soccer fields towards more natural environments, programmers have failed to to turn over the process of grounding to the programs they create. Today's implementations are successful because of a designer's insight into each particular problem. For example, consider the legged-league[3]. In the legged-league, over many years, distinctive[4] landmarks to assist with localization have gradually being reduced in number and in their "distinctiveness". However, as landmarks have been removed, rather than build systems which find their own landmarks, we (designers) have chosen new landmarks for the robots to localize off (e.g. instead of writing vision modules for perceiving beacons we write vision modules for perceiving field lines), and busily set about hand-crafting a solution for perceiving the new chosen landmark. Thus, the designers of the program (rather than the program we created to control the robot) are making the decisions about which entities in the environment are worth perceiving. Unfortunately, such an approach is not scalable - in the real-world designers can't make such decisions on behalf of so-called "autonomous" robots.

4   The Objective: Autonomous Grounding

Systems that are robust, adaptable, and scalable will need to be capable of performing their own grounding. Ideally, the process of adapting to a new or changing environment, task and/or embodiment should be autonomous. Grounding representations has two key problems - the problem of "relevance", which involves deciding "what" to represent; and the problem of "reference", which

---

[3] Sadly, 2008 is the Legged-League's final year of competition due to the discontinued Sony AIBO.

[4] i.e. easily perceived.

involves maintaining the correspondence between representations and their referents [13]. Ensuring the groundedness of agents' representations is imperative for successful autonomous decision-making, with the manual creation and maintenance of representation being one of the largest costs associated with deploying robots. Any artificial agent operating in a changing environment is required to respond appropriately to that environment, and thus must have a "solution" to its particular grounding problem. However, when grounding agents, decisions of relevance (i.e. what to represent) and reference (i.e. how to maintain that representation over time) are, on the whole, made manually by designers, rather than by programs - i.e. robotic agents are grounded by design. As such, current approaches to grounding robotic agents are ad-hoc - we make the decisions regarding relevance and reference on a case-by-case, system-by-system, task-by-task basis, but we embed little (if any) of the knowledge of how we find structure and meaning in the world. Therefore, such solutions are usually restricted to the particular domain for which the agent operates. As our robots can not autonomously ground themselves, robotic systems tend to brittle in the face of change, and a highly iterative "code-and-fix" development paradigm results. Thus, a general solution to the grounding problem is required.

4.1   Current Approaches

A large body of multi-disciplinary research has been generated by the grounding problem. While grounding-related research varies in implementation and application, most approaches have focused on ascribing meaning through categorical perception. These approaches assume grounding can be achieved by linking "symbolic" representations to sensorimotor, "subsymbolic" representations that are "invariantly" correlated (or are "causally" related) with the real-world phenomena being represented. In other words, I know what a pizza means because I know what it tastes like, smells like, looks like, feels like, and so forth - and therefore, if we could do the same for a robotic system (using cameras, taste sensors, etc), the robotic system's meaning of pizza would also be grounded [14].

In practice, this "robot functionalism"[15] is realized by most grounding approaches utilizing machine learning techniques to map associations between representations and sensor data. All are concerned with how to make representation meaningful, though what constitutes representation and meaning is different under each approach. Hybrid systems ascribe meaning to a high-level symbolic reasoning system by connecting the symbol system to machine learning algorithms capable of forming categorical (subsymbolic) representations of sensorimotor data, e.g [5]. In such systems, the process of symbolic theft can enable the construction of new concepts, including those that are abstract or imagined [11]. Behaviourist approaches treat the grounding problem as a practical problem of finding the right "function" to control a robot's behaviour, and thus the representation being grounded is the entire control program itself [6,12]. Lastly, developmental robotics embraces a life-long learning approach to robot development (e.g. [16]), and proponents of this approach argue that if a robot learns its own representations then such representations are grounded. Most grounding approaches focus on the problem of reference, and there is little (if any) grounding research devoted to the problem of relevance, with most decisions regarding relevance (and to a lesser extent reference) made by designers, not programs.

5   Go-Design

"Grounding Oriented Design" [13] (or *Go-Design* for short) is a methodology for designing and grounding the "minds" of robotic agents. Due to our interest in grounding, Go-Design focuses on understanding (both of the developer and robot) by providing a simple means for modeling knowledge and decision-making in robotic systems. The design process has two main components:

- Processes, guidelines and techniques for designing a robot's mind in terms of units of encapsulated abilities which we call *skills*.
- A modeling notation for representing a mind's design through *skill diagrams*.

In this section we present a (very) brief overview of Go-Design.

5.1   Motivation

The development of Go-Design was motivated by the need to solve the grounding problem. Go-Design, however, is not a solution to the grounding problem, but rather a small first step towards understanding the difficult problem of how we ground, with a longer-term view of automating this process. Thus, while Go-Design focuses on grounding by design (as opposed to autonomous grounding), Go-Design can offer developers assistance with the "here-and-now" problems related to the development of control programs for autonomous robots.

5.2   Objectives

When developing Go-Design, two main objectives were formulated:

1. To build a methodology which would improve the groundedness of systems built using the methodology; i.e. using the methodology should result in better grounded systems than if no methodology was used.
2. To build a methodology which, in the longer-term, can improve our understanding of the grounding process, and that in time can be extended to automate the grounding process.

Both objectives concern improving the groundedness of systems - they differ only in scope. While the methodology focuses on assisting developers with grounding robot minds by design, the methodology is intended to provide a base from which insights into how we ground can be gleaned, with the longer-term view of developing systems capable of autonomous grounding.

5.3   Design Considerations

How can we build a methodology to improve the groundedness of systems? Go-Design was created based upon the following assumptions about the grounding problem:

Grounded Designers. Designing grounded robots requires grounded designers. As grounding is task-specific, the first step of Go-Design is understanding the nature of the current problem - a process called "context-level analysis". This process is similar to the requirements elicitation and requirements specification processes that occur when building traditional software. However, Go-Design's context-level analysis is specifically tailored for the development of control programs for autonomous robots.

A Software Problem. Robot control programs will usually be implemented in software, and therefore Go-Design treats the grounding problem as a software development problem. The designs produced by Go-Design are sufficiently detailed (pseudo-code is required) that translation from a grounding design to a software implementation is a straightforward process. Go-Design provides a single diagramming notation which captures the key aspects of both software design (such as hierarchical, modular, layered designs, flow-of-control between modules, and pseudo-code) and grounding design (such as representation, referents, perception, behavior, and decision-making).

The Relevance Problem. One of the main grounding problems is the problem of relevance - i.e. choosing what to represent. Thus, Go-Design provides a set of structured steps to assist the designer in identifying relevant entities that should be represented. This process involves firstly understanding the requirements and nature of the task-at-hand, and then identifying the knowledge required to achieve the task.

Problem Decomposition and Decision-Making. Knowing what needs to be represented requires understanding the subtleties of what the agent must do. Therefore, to identify the relevant entities that should be represented the designer must identify the decisions the agent must make to respond appropriately to changes in the environment. Go-Design involves iteratively decomposing the problem task into subproblems until decisions are identified which map to actions and behaviors the agent is capable of performing.

The Reference Problem. Grounding involves maintaining representations with respect to a changing world. Therefore, Go-Design forces designers to identify and define (in terms of decision-making processes) the perceptual mechanisms responsible for maintaining the correspondence between representations and their referents, while also explicitly identifying the decision-making processes which *rely* upon those perceptions.

Groundedness. When is the quality of an agent's grounding "good enough"? How do we "debug" ungrounded robots? A designer should be able easily explain *why* a robot is behaving in a particular manner - building a grounded robot requires a grounded designer. To help ensure a high quality of grounding, Go-Design guides designers through a process of identifying groundedness requirements for representations based upon the consequences of decision-making processes which rely upon those representations. Also, Go-Design creates transparent, easily understood designs in which the decisions that are dependent upon particular representations can be easily traced and identified, as well as offering structured processes for testing robotic systems.

5.4 Skills

In Go-Design, the main unit of abstraction are *skills*. Skills are encapsulated, task-based abilities; they refer to "things" a robot mind can do; i.e. they are labels we attach to processes which do or achieve something. Skills can be anything - the ability to see, to throw and catch a ball, or to even "think". We could have called "skills" many other names, such as "abilities", "capabilities" or "things a robot can do". Skills can use other skills - for example, being able to "do the grocery shopping" requires the ability to "drive the car" coupled with the ability to "find the shopping center", and then after arriving at the shopping center, the ability to "find a car park", and to "find the grocery shop inside the shopping center", and then (once all the previous skills have been accomplished successfully) there is the art of "finding and controlling a shopping trolley", and so forth - intelligent behaviour requires a rich assortment of skillful skills.

5.5 Skill Diagrams

Interactions and collaborations between skills are be modeled through the use of *skill diagrams*. Skill diagrams can model three main types of interaction:

- *Skill sequences.* Skill sequences are skills that operate sequentially over time, one after the other, like a chain. For example, "1. Find the car; 2. Start the car; 3. Drive the car to the shops", and so forth. In a skill sequence the term *skill-transition* is used to describe when one skill stops and another starts.
- *Concurrent skills.* Concurrent skills that operate at the same time - for example, the ability to drive the car while also being able to simultaneously plan (and replan) the best route to the shops. Concurrent skills are also represented through the use of skill-transitions.
- *Skill decompositions.* Skill decompositions are the labels we attach to hierarchical groups of skills that do something as a whole. In other words, skills are composed of *subskills*. For example, imagine a robot capable of "grocery shopping". Possible subskills could be "drive the car", "find and control a shopping trolley", and so forth. For each of these subskills, they are, in turn, also composed of subskills. For example, the ability to "drive a car" requires skills such as "perceive and obey road signs and traffic laws", while the ability to perceive road signs in turn requires the ability to interpret and discriminate certain collections of shapes and colours in the visual stream as being particular meaningful road-signs. Thus, generally, for every human behaviour there is a mountain of detail that we (designers of robot minds) either ignore or take for granted - until we try and automate such processes artificially.

Examples of skill diagrams are contained in Section 5.6.

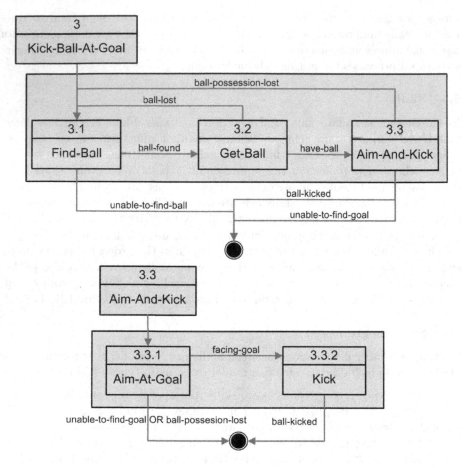

Fig. 1. A basic skill diagram representing the skills and skill transitions for `Kick-Ball-At-Goal`. This figure also illustrates decomposition, with the skill `Aim-And-Kick` decomposed.

5.6 The Design Process

Go-Design has two stages of design - "basic" and "detailed". Basic-design involves constructing a skill architecture, in which the skill abstractions are formed, and the skill-transitions between skills are identified. The product of basic-design is a skill architecture, consisting of skills, a skill hierarchy/decomposition, and skill transitions. While a skill architecture can play a role in the high-level design of an autonomous robotic system, for the purpose of implementing a software solution more detail is required. Thus, a detailed-design is needed which provides a blueprint for translating a skill architecture into software. In detailed-design, we consider how to identify the knowledge, concepts, perceptions and decision-making processes required for a robot's mind to control that robot appropriately. Detailed-design involves (and provides guidelines for):

- Identifying each skill's "type" as either an *action*, a *perception*, a *decision*, or a *behaviour*. Actions operate without perception, perceptions are decisions made about the state of the world, decisions are testable conditions, and behaviours are skills which incorporate other skills, such as other behaviours, perceptions, decisions and actions. Thus by definition, a behaviour can be decomposed.
- Designing the representation required by each skill, including the *concepts* (data structures), *percepts* (the stored memory of a perception) and *memories* (all other forms of internal state).
- Documenting the requirements of each skill, including the required level of *groundedness*.
- Identifying the dependencies between skills and knowledge representation through the use of *flows*.

Notation. An example basic-design is shown in Figure 1. Skills are represented with rectangular boxes, with the name of the skill written below a *numeric identifier*. Skill-transitions can have *transition-conditions*, which are testable conditions which signify when one skill should stop and another should start. We represent transition-conditions by adding a caption (in lowercase) to the arrows between skills on a skill decomposition diagram.

Figure 2 represents the detailed-design for a skill which enables a legged-robot to spin on the spot, while avoiding obstacles (typically used in searching for the ball). In the detailed-design process, the developer elaborates a basic-skill

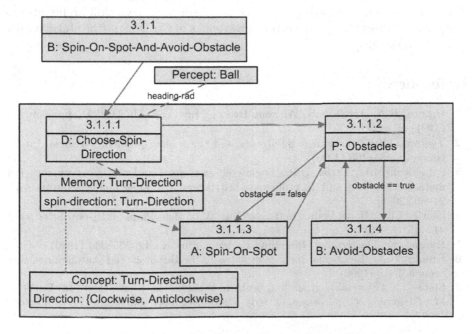

Fig. 2. Spin-On-Spot-And-Avoid-Obstacle

diagram by way of a set of guidelines and diagram integrity checks. Detailed skill-diagrams include knowledge representation requirements, such as *concepts*, *memories*, and the *knowledge flows* (represented by dotted lines) between knowledge stores and the skills which rely upon, or modify, this knowledge. Also, in a detailed-design skill diagram, each skill's "type" is identified through the use of a prefix[5].

6 Discussion and Conclusion

This paper has discussed the nature and role of grounding in designing programs for controlling autonomous mobile robots. We have argued that grounding is a problem that must be overcome for the development of scalable and robust systems that can operate in natural environments, and therefore is critical to the success of RoboCup's aspirational 2050 target. A very brief overview of a detailed grounding oriented design methodology (Go-Design) has been presented, which is currently being utilized and evaluated as a means of designing and representing systems in several domains, including RoboCup. Go-Design, while focusing on grounding by design (rather than autonomous grounding), is an initial step towards the longer-term objective of developing autonomous grounding capabilities, by assisting us (designers) understand how we ground autonomous robots. Currently, plans are in place to utilize Go-Design's skill architectures as a representation for not only designing robot systems, but implementing them - providing an efficient programming interface for grounding autonomous mobile robots. This "second phase" of development of Go-Design will also allow for dynamic (i.e. run-time) changes by the program of its own skill architecture - eliminating one of the major current drawbacks of Go-Design (the static nature of the skill designs).

References

1. Brooks, R.A.: Intelligence Without Reason. In: Proc. IJCAI 1991, pp. 569–595 (1991)
2. Tsotsos, J.K.: Behaviourist inteligence and the scaling problem. Artificial Intelligence 75, 135–160 (1995)
3. Anderson, M.L., Perlis, D.R.: Logic, self-awareness and self-improvement: The metacognitive loop and the problem of brittleness. J. of Logic and Computation 15, 21–40 (2005)
4. Searle, J.R.: Minds, brains, and programs. J. of Behavioral and Brain Sciences 3, 417–457 (1980)
5. Harnad, S.: The Symbol Grounding Problem. Physica 42, 335–346 (1990)
6. Brooks, R.A.: Elephants don't play chess. J. of Robotics and Autonomous Systems 6, 3–15 (1990)
7. Steels, L.: The symbol grounding problem is solved, so what's next? In: De Vega, M., Glennberg, G., Graesser, G. (eds.) Book Symbols, embodiment and meaning (2007)

---

[5] i.e. "B" for behavior, "P" for perception, "D" for decision, and "A" for action.

8. Taddeo, M., Floridi, L.: Solving the Symbol Grounding Problem: a Critical Review of Fifteen Years of Research. J. of Experimental and Theoretical Artificial Intelligence (2005)
9. Williams, M.-A., McCarthy, J., Gärdenfors., P., Karol, A., Stanton, C.: A framework for evaluating groundedness of representations in systems: from brains in vats to mobile robots. In: Proc. IJCAI 2005 (2005)
10. Chalup, S.K., Murch, C.L., Quinlan, M.J.: Machine Learning With AIBO Robots in the Four-Legged League of RoboCup. IEEE. Systems, Man, and Cybernetics, Part C: Applications and Reviews 37, 297–310 (2007)
11. Cangelosi, A., Greco, A., Harnad, S.: From robotic toil to symbolic theft: Grounding transfer from entry-level to higher-level categories. Connection Science 12, 143–162 (2000)
12. Ziemke, T.: Rethinking Grounding. In: Riegler, P., Stein, V. (eds.) Understanding Representation in the Cognitive Sciences, pp. 177–190 (1999)
13. This paper's author: Grounding Oriented Design. PhD Thesis (2007)
14. Mayo, M.: Symbol Grounding and its Implication for Artificial Intelligence. In: Proc. Twenty-Sixth Australian Computer Science Conference, pp. 55–60 (2003)
15. Harnad, S.: Grounding Symbolic Capacity in Robotic Capacity. In: Steels, L., Brooks, R. (eds.) The "artificial life" route to "artificial intelligence." Building Situated Embodied Agents, pp. 276–286 (1995)
16. Blank, D., Kumar, D., Meeden, L., Marshall, J.: Bringing Up Robot: Fundamental Mechanisms For Creating A Self-Motivated, Self-Organizing Architecture. J. Cybernetics and Systems 36, 125–150 (2005)

Robust Moving Object Detection from a Moving Video Camera Using Neural Network and Kalman Filter

Sanaz Taleghani[1], Siavash Aslani[2], and Saeed Shiry[3]

[1] Mechatronics Research Laboratory (MRL), Department of Computer & IT,
Qazvin Azad University, Qazvin, Iran
taleghani@mrl.ir
[2] Department of Computer & IT, Qazvin Azad University,
Qazvin, Iran
saslani@mrl.ir
[3] Computer Engineering Department, Amirkabir University,
Tehran, Iran
shiry@aut.ac.ir

Abstract. Detecting motion of objects in images, while the camera is moving, is a complicated task. In this paper, we propose a novel method to effectively solve this problem by using Neural Network and Kalman Filter. This technique uses parameters of camera motion to overcome problems caused by error in the image processing outputs. We have implemented this technique in the MRL Middle Size Soccer Robots. The experimental results show a low error rate of 2.2% which suggests that the combined approach performs significantly better than the traditional techniques.

Keywords: Motion Detection, Neural Network, Kalman Filter, Middle Size Soccer Robot.

1 Introduction

For many applications of autonomous robots it is important to detect motion of objects that are in the surroundings of the robot to avoid collisions or enable interaction. The motion detection methods that are based on image processing need high quality images as input. Since the camera used to record images for a soccer playing robot is moving, the quality of images decreases and vibration of camera produces more fatal noise. This causes the output of image processing to include considerable error which makes feature detection and object recognition a complicated task.

Various filters would help to reduce the noise to some extent. Regarding moving objects tracking, Kalman Filtering, Extended Kalman Filtering and Particle Filtering (also known as Condensation and Monte Carlo algorithms) are some of the most common used algorithms. Kalman Filter provides an efficient recursive solution which has a prediction and correction mechanism, in a way that minimizes the mean of the squared error. Due to its simplicity, Kalman filter is still being used in most of the general-purpose applications [1].

L. Iocchi et al. (Eds.): RoboCup 2008, LNAI 5399, pp. 638–648, 2009.

For localization of moving objects, Nordlund introduced a method to detect objects motion using a sequence of images which indicates the object position after two frames, but the result improves over time [2]. Such approaches are based on difference of images for motion detection, but they have some difficulties with slow-moving objects. To solve this problem, some approaches employ techniques based on estimation such as Kalman Filter and neural networks [3].

Neural networks have been implemented for image tracking applications, where they are used mostly as classifiers or measures between different types of filters [4]. Zhang and Minai [5] created a two layer pulse coupled neural network for motion detection. The two layers work in iterative fashion and find the largest matching segment between two consecutive video frames. This model adopts the image pixels as the local feature. Based on Grossberg's spreading theory and Ullman's motion decision theory [6], Guo Lei proposed a spreading and concentrating model [7] to perform motion detection. The local feature used for motion detection is the edge elements in the object's contour. The common problem of these models is the model complexity [8].

In this research we took benefits of neural networks in their learning ability and noise immunity. A MLP neural network was embedded in the Kalman Filter cycle as a corrector section. The result of implementation is quite satisfactory.

2   Problem Statement

Motion detection using image processing while the camera is moving is a difficult task. In such conditions, images are not clear and vibration of the camera decreases the quality of images significantly and causes high error in the image processing result. In the domain of soccer playing autonomous robots, it is important to have knowledge of the ball position and movement. Middle size soccer robots are a sample of systems that motion detection has an especial importance in their decision making.

As an appropriate test-bed, we have used MRL team's middle size soccer robots to implement and test different methods for detecting ball motion. These robots are

Fig. 1. MRL middle size soccer robot

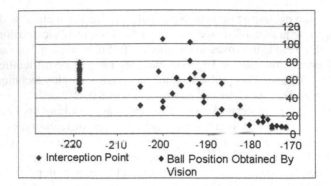

Fig. 2. The illustration of ball interception problem. Although the ball is stationary but robot supposes that the ball is moving because of the vision error in ball positioning. Robot tries to intercept the ball in the points shown in red.

equipped with omni-directional vision which consists of a camera and a hyperbolic mirror on top of each robot (Fig. 1). This kind of camera assemblies requires the lens to be mounted exactly in the focal point of the mirror while the camera has to be aligned with its symmetry axis. Due to the forces exposed to the robot during a game the camera will encounter considerable vibrations.

Movement and vibration rate of the robots during the game is significantly high which causes a considerable noise in information provided by the vision system. This noise makes the vision system to calculate different positions for a fixed ball within sequential steps. It leads the robot to try to calculate ball path and intercept it while it is not moving at all (Fig. 2).

3 Motion Detection

Having examined a number of traditional techniques of motion detection we realized they are not capable of providing acceptable result for the abovementioned test-bed. Therefore we tended to come up with a novel and optimized heuristic method.

This section explains the mentioned methods and results of their implementation in MRL middle size soccer robots.

3.1 Average and Threshold Method

This approach has a simple mathematical base. The position of objects that is observed by the vision system is recorded in some sequential steps. The difference between two successive object positions shows the movement value during the time. On the other hand each individual position of objects per steps is not precise because of the existing error in vision system. Using the discrepancy between average of some initial data calculated in (1) and average of some ultimate data calculated in (2) would be enough to resolve this problem. Thus if this discrepancy d in (3) is greater than a threshold value, we may assume that the object is moving, if otherwise then we would assume that the object is standstill.

$$\bar{X}_f = \frac{\sum\limits_{i=1}^{n} x_i}{n} \quad \bar{Y}_f = \frac{\sum\limits_{i=1}^{n} y_i}{n} \tag{1}$$

$$\bar{X}_l = \frac{\sum\limits_{i=l-n}^{l} x_i}{n} \quad \bar{Y}_l = \frac{\sum\limits_{i=l-n}^{l} y_i}{n} \tag{2}$$

$$d = \sqrt{(\bar{X}_f - \bar{X}_l)^2 + (\bar{Y}_f - \bar{Y}_l)^2} \tag{3}$$

In the above equations, X and Y are assumed as objects' coordinates in l sequential steps. In the implementation environment in middle-size robot, each time step lasts for 0.03 sec. After required information about ball position was registered in every short period of time (almost 0.5 sec) due to such data dispersal, the mentioned threshold could not be calculated for conditions which robot was moving (Fig. 3).

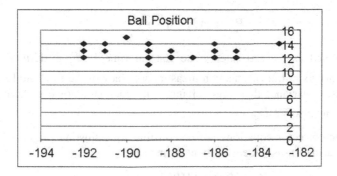

Fig. 3. Raw data observed by robot's vision. Ball position while both robot and ball are stationary.

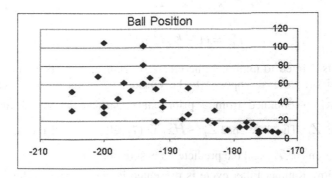

Fig. 4. Raw data observed by robot's vision. Ball position while robot is moving and the ball is stationary.

Fig. 4 presents the dispersal of ball position when ball is stationary but robot is moving. The dispersal depends on various speeds of robot which involves camera vibration and other environment error.

3.2 Kalman Filter

The Kalman Filter is a set of mathematical equations that provides an efficient computational recursive means to estimate the state of a process, in a way that minimizes the mean of the squared error [9].

The discrete Kalman Filter attempts to estimate the state of a discrete-time controlled process by using a form of feedback control. This means that merely the estimated state from the previous time step and the current actual measurement are required to compute the estimation for the current state. Thus equations for Kalman Filter can be divided into two groups: predictor equations and corrector equations:

- Predictor equations are responsible for projecting the current state estimate ahead in time.

$$\hat{x}_k^- = A\hat{x}_{k-1} + BU_{k-1} \tag{4}$$

$$P_k^- = AP_{k-1}A^T + Q \tag{5}$$

$\hat{x}_k$ is defined as the estimate of the state at time k . P_k is defined as the error covariance matrix at time k that can measure accuracy of the estimated state. Q is the process noise covariance that might change with each step however here we define it as constant and also set $U_k = 0$.

- The corrector equations are responsible for adjusting projected estimate by an actual (noisy) measurement at that time.

$$K_k = P_k^- H^T (HP_k^- H^T + R)^{-1} \tag{6}$$

$$\hat{x}_k = \hat{x}_k^- + K_k(Z_k - H\hat{x}_k^-) \tag{7}$$

$$P_k = (1 - K_k H)P_k^- \tag{8}$$

K in (6) is defined to minimize the error covariance. R in (6) is the measurement noise covariance and is supposed to be constant. At this level of Kalman Filter algorithm $\hat{x}_k$ is obtained from a priori state estimate at step k and an actual measurement Z_k. The difference $Z_k - H\hat{x}_k^-$ in (7) reflects the discrepancy between an actual measurement Z_k and the predicted measurement $H\hat{x}_k^-$.

The ongoing Kalman Filter cycle is presented in Fig. 5. Indeed the Kalman Filter contains the different parts covering the high-level operation.

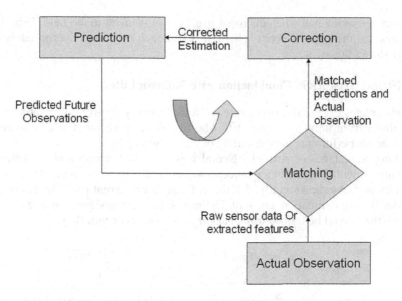

Fig. 5. The ongoing common Kalman Filter cycle

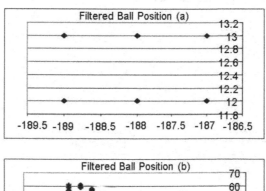

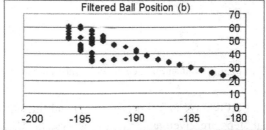

Fig. 6. Kalman Filter is not effective enough to correct ball position while the robot is moving. a) Kalman Filter reduced the noise of data of Fig 3. b) Kalman Filter could not reduce the noise of Fig 4.

In our implementation, the constant values for Kalman Filter equations are computed as $Q=0.001$ and $R=0.5$. We use ball position as output data of Kalman Filter which is provided by vision of robot. This approach is not able to recognize a fixed ball from a moving ball when the robot itself has movement in the field. Especially in the cases that the speed of robot is high, this approach may work appropriately. Fig. 6 clearly shows this issue.

3.3 Neural Network in Combination with Kalman Filter

The advantages of neural networks are twofold: learning ability and versatile mapping capabilities from input to output [10]. The multilayer perceptron is a nonparametric technique for performing a wide variety of estimation tasks [11].

Taking advantages of Artificial Neural Networks in learning and estimation, and combining it with Kalman Filter concept, we introduce an efficient technique to detect object motion. As shown in Fig. 5 Kalman Filter has different parts. In this approach, the Match and Estimation tasks of Kalman Filter are assigned to a MLP neural network that would be trained using backpropagation algorithm (Fig. 7).

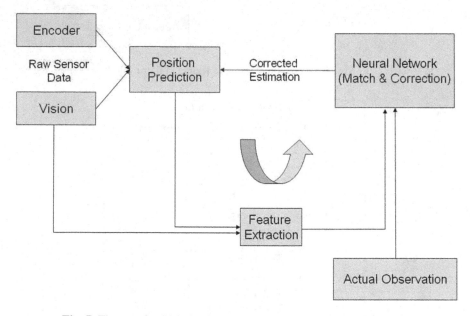

Fig. 7. The ongoing Kalman Filter cycle with embedded neural network

The location of object relative to a supervisor (A_1) and the supervisor's movement vector ΔO in serial steps are given as inputs to the system. Using these inputs, the relative location of the object in next step (A_2) can be predicted. Then the difference of the predicted value and the precept value of the vision system in next step is given to the neural network via multiplex sets. Since the training of neural network is supervised, the target value has to be supplied for every sample.

The addition of localization noise to the noise of relative ball position causes high error in global ball position. Therefore we use relative position of the ball instead of its global position. On the other hand in our vision system, the noise of Polar relative ball position is less than the noise of Cartesian global ball position. Thus, it is preferred to use Polar coordinates instead of Cartesian coordinates.

In our implementation in every step, the predictor part uses both of (r,θ) and ΔO (dx,dy,dφ). As shown in Fig. 8, at the beginning of a step the robot is positioned in the point A_1 and observes the ball in relative position B_1. During the step, robot moves as(dx,dy,dφ) and positions in the point A_2 and observes the ball in relative position B_2 in next step.

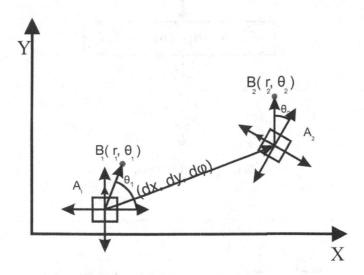

Fig. 8. At first the robot is placed at the point A_1 and observes the ball in the point B_1. Then it moves to the point A_2 and observes the ball in the point B_2 in next step.

We can compute the new relative position of the ball using movement vector of the robot according to the following equations.

$$A'_x = r_1 \cos\theta_1 - dx \tag{9}$$

$$A'_y = r_1 \sin\theta_1 - dy \tag{10}$$

$$B_x = A'_x \sin(d\phi) + A'_y \cos(d\phi) \tag{11}$$

$$B_y = A'_x \cos(d\phi) - A'_y \sin(d\phi) \tag{12}$$

$$(B_x, B_y) \xrightarrow{\text{CartToPol}} (r_2, \theta_2) \tag{13}$$

As predictor output is not accurate and the ball may also be moving, thus the predicted value differs from actual ball position observed by the vision in next step.

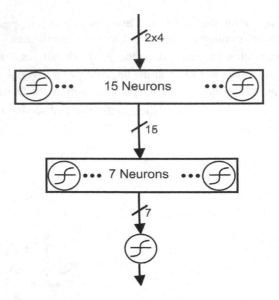

Fig. 9. Designed MLP Neural Network

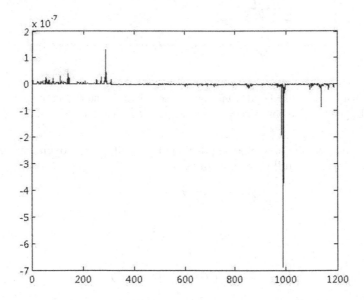

Fig. 10. Error in the training set with 1188 samples

The feature extraction section calculates this difference and after normalization passes it as input vectors to the neural network which is responsible for estimating and matching in Kalman Filter.

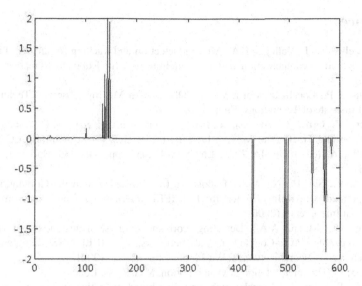

Fig. 11. Error in the validation set with 596 samples

We trained a 15-7-1 feedforward neural network by backpropagation algorithm. Neurons had *tansig* transition function (Fig 9).

The positive outputs were supposed to indicate moving ball and the negative outputs used to indicate stationary ball. The data of ball position and robot movement were logged in different states. The Robot moved in different speeds while the ball was stationary or moving in different directions. The collected data were split into a training set of 1188 vectors and a validation set of 594 vectors. During network training, it could learn all the training samples with great performance as shown in Fig 10.

Then the network was tested using the validation samples. Achieving 2.2% error shows that the trained network is not overfitted and is capable of detecting the ball motion in different states (Fig. 11).

4   Conclusion

As the movement of camera and its vibration leads to unclear and low quality noisy images, the image processing unit produces outputs with high error which makes the usual methods of motion detection unable to perform appropriately.

Having confronted with such problem we tended to design a novel technique that takes advantages of Neural Networks in learning nonlinear relations and combines it with Kalman Filter. As this technique is not directly involved with image processing procedure, it would be useful for systems where their input is provided by other resources and sensors. The result of implementation in MRL Middle Size Robots shows considerable performance improvement and suggests that the combined approach performs significantly better than traditional techniques.

References

1. Ruiz-del-Solar, J., Vallejos, P.A.: Motion Detection and tracking for an AIBO robot using camera motion compensation and Kalman Filtering. In: Robocup Symposium, Bremen (2006)
2. Nordlund, P.: Localization of a Moving Object by a Moving Observer. Technical report, Royal Institute of Technology, Stockholm (1994)
3. Herrero, E., Orrite, C., Alcolea, A., Roy, A., Guerrero, J.J., Sagues, C.: Video-Sensor for Detection and Tracking of Moving Objects. In: Perales, F.J., Campilho, A.C., Pérez, N., Sanfeliu, A. (eds.) IbPRIA 2003. LNCS, vol. 2652, pp. 346–353. Springer, Heidelberg (2003)
4. Spinko, V., Shi, D., Ng, W.S.: Endoscope tracking using Wavelet-Gravitation Network incorporated with Kalman Filter. In: 18th IEEE International Conference on Tools with Artificial Intelligence (2006)
5. Zhang, X., Minai, A.A.: Detecting corresponding segments across images using synchronizable Pulse-Coupled neural networks. In: IEEE/NNS International Joint Conference on Neural Networks, Washington, pp. 82–825 (2001)
6. Ullman, S.: The interpretation of visual motion. MIT Press, London (1979)
7. Lei, G., Baolong, G.: Visual system and distributed deduction theory. Xidan University Publisher (1995)
8. Yu, B., Zhang, L.: Pulse Coupled Neural Network for Motion Detection. In: IEEE International Joint Conference on Neural Networks, vol. 2, pp. 1179–1184 (July 2003)
9. Welch, G., Bishop, G.: An introduction to Kalman Filter. UNCChapel Hill (April 5, 2004)
10. Neji, Z., Beji, F.M.: Neural Network and time series identification and prediction. In: IEEE-INNS-ENNS International Joint Conference on Neural Networks (2000)
11. Werbos, P.J.: Beyond regression: new tools for prediction and analysis in the behavioral sciences. Ph.D. thesis, Harvard Univ., Cambridge, MA (1974)

Collaborative Localization Based Formation Control of Multiple Quadruped Robots

Qining Wang, Hua Li, Feifei Huang, Guangming Xie, and Long Wang

Intelligent Control Laboratory, College of Engineering,
Peking University, Beijing 100871, China
qiningwang@pku.edu.cn
http://www.mech.pku.edu.cn/robot/fourleg/

Abstract. In this article, we present a new formation control method based on collaborative localization. Due to the accurate position estimation, the proposed approach is computationally light and well-performing in switching between different formations of real-world multi-robot systems. Satisfactory experimental results of multi-robot formation control are obtained in the RoboCup environment.

1 Introduction

The main goal of research on sensor-based robot localization approaches is to allow autonomous mobile robots, equipped with relatively low-cost sensors and actuators, to perform tasks in complex and dynamic environments [11], [6]. These technologies have a wide range of application fields, which include global navigation in unknown environments, formation control, robots at home, and industrial automation. In this research area, global self-localization is generally regarded as the main problem [6], [16]. Many studies address the localization of a single robot based on probabilistic approaches with different sensory information [4], [22], [12]. In order to improve the effectiveness of localization, some studies focus on the problem of cooperative multi-robot localization [5], [6], [16]. However, current multi-robot localization methods have preconditions that robots should have the ability to detect and recognize each other accurately and efficiently, which are impractical in real-world tasks, especially the task like formation control that relies on the localization results. It is difficult to apply these techniques to real multi-robot systems since they cannot recognize each other with simple sensors. In this context, experience and collaboration are often adopted to overcome the difficulties of recognizing individual robot, which is applicable in formation control with only velocity and position as the input.

Formation control is an important issue in coordinated control for a group of autonomous robots, with broad applications from house security patrol to military missions. [3] proposed an architecture of formation control, analyzing the Nyquist-like stability. [7] described a discrete time model for multi-agent cooperation. Switching among proper formations intentionally will result better in RoboCup soccer competition, especially when the team needs to defense. Many

L. Iocchi et al. (Eds.): RoboCup 2008, LNAI 5399, pp. 649–659, 2009.
© Springer-Verlag Berlin Heidelberg 2009

of the current methods of formation control use positions which include absolute positions or relative positions of robots in the system and velocity as the input. Thus, precise localization is one of the essentials to perform stable formation in real-world robot systems. However, current multi-robot localization methods have preconditions that robots should have the ability to detect and recognize each other accurately and efficiently, which are impractical in real-world tasks, especially the task like formation control that relies on the localization results. It is difficult to apply these techniques to real multi-robot systems since they cannot recognize each other with simple sensors.

In this article, an approach of collaborative multi-robot localization for formation control of vision-based autonomous robots that effectively achieves accurate self-localization and dynamic formation switching with low-cost hardware equipment is implemented. We propose a protocol for formation control based on the result of self-localization. Due to the accurate position estimation, the proposed protocol is computationally light and well-performing in switching between different formations of real-world multi-robot systems.

The rest of this paper is organized as follows. Section 2 describes the details of cooperative probabilistic state estimation which is the foundation of the proposed formation control method. Section 3 presents the formation control method. Experimental results are shown in Section 4. We conclude in Section 5.

2   Multi-robot Collaborative Localization

Self-localization is one of the fundamental problems in real-world multi-robot systems on land [1], under water [8], and in the air [9]. Global positioning system (GPS) has been used to solve this problem. However, the GPS estimates are often subject to inaccuracies and unreliability, especially for multi-robot system in uncertain and unknown environments. Different approaches to multi-robot localization have been proposed [16], [6], [14]. Only a few results for multi-robot localization have been presented with no need of individual robot recognition (see [19] for instance). In this study, we used a method for global self-localization of vision-based robots in a multi-robot system, assuming that communication among robots is possible and reliable (related to [19]). Based on the Markov localization methods of a single robot [4], [22], the current position of the robot is modeled as the density of a set of particles which are seen as the prediction of the location. Initially, at time t, each location l has a belief $Bel_t(l)$. To update the belief of robot's possible location, at first, the approach uses the new odometry reading o_t:

$$Bel_t(l) \leftarrow \int P(l|o_t, l^-)Bel_t(l^-)dl^- \tag{1}$$

If robot receives new sensory information s_t, then it updates the belief with α being the normalizing constant:

$$Bel_t(l) \leftarrow \alpha P(s_t|l)Bel_t(l) \tag{2}$$

Considering the mobile robot with complex motions, let the geometric center of robot body as the location vector ϕ, which contains the x/y- global coordinates of the center point. Another vector θ is defined as the heading direction. Then every particle is updated by the motion model as follows when the robot moves:

$$\phi_t = \phi_{t-1} + \Delta_t \tag{3}$$

where Δ_t represents the displacement in x/y coordinates and heading direction.

To implement self-learning experience in self-localization, we divide the sensory update into two parts: updating position probability by landmark perception and experience retrieval. If the robot recognizes landmarks well enough, landmark based sensor model will update the belief of position with the new landmark reading s_t:

$$Bel_t(\phi_t) \leftarrow \beta P(s_t|\phi_t)Bel_t(\phi_t) \tag{4}$$

where β is a normalizing constant. It is natural that the robot may miss some landmarks with real-time recognition for a period. Thus, we set $N_1(t)$ which is the amount of lasting frames of having no landmark perception from t as a condition to activate the experience system. If $N_1(t)$ is great enough, the experience based sensor model will update the probability as follows with e_t being the new reading experience with γ being the normalizing constant different from β:

$$Bel_t(\phi_t) \leftarrow \gamma P(e_t|\phi_t)Bel_t(\phi_t) \tag{5}$$

Fig. 1 shows the steps of the method for position estimation of individual robot in the multi-robot system with self-decision (related to [15]). Fig. 2 compares the position estimation errors by using different localization approaches in the same environment.

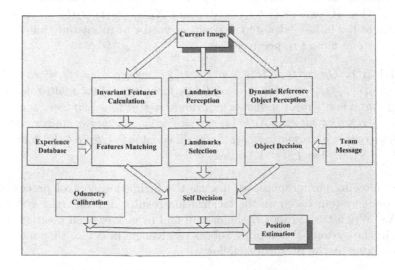

Fig. 1. An algorithm for position estimation in a multi-robot system

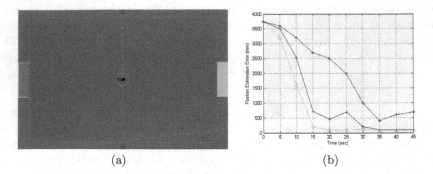

(a)   (b)

Fig. 2. (a) Standard robot soccer environment of RoboCup Four-Legged League. There are two sequenced color beacons, two goals colored in yellow and sky blue separately, and white lines. (b) Comparison of localization error between single-robot localization and multi-robot localization. The red line shows the error over time when performing single-robot localization using the approach mentioned in [4]. The blue line shows the error of single-robot localization using only landmarks and environment features as a part of the algorithm in Fig. 1. The green line is the result of four robots performing global localization simultaneously using the method as shown in Fig. 1.

3   Multi-robot Formation Control

Based on the localization results, considering a multi-robot system with M robots which are indexed as $R_1, R_2, \cdots, R_M$, we suppose that every robot of the system can localize self position through the method mentioned above, sending and receiving own position and positions of other robots through wireless network. In this section, we introduce a position feedback control protocol to implement formation control based on relatively precise self-localization.

Let $x_i(t) \in R^n$ where $i = 1, \cdots, M$, and $X_t = [x_1(t), x_2(t), \cdots, x_M(t)]^T \in R^{Mn}$ where $t = 0, 1, \cdots$ denote the position vector of robot i at time t and the system state at time t respectively.

Definition 1. *Given $f_i \in R^n$, $i = 1, \cdots, M$, where $f_i \neq f_j$, $i \neq j$. Denote $f = [f_1, f_2, \cdots, f_M]^T \in R^{Mn}$ as a candidate formation of the multi-robot system. We say that a multi-robot system is in formation f at time t, if there is a constant vector $c \in R^n$ that satisfies $X_t - 1_M \otimes c = f$. We say that a multi-robot system with convergence of formation f, if there is a constant vector $c \in R^n$ that satisfies $\lim_{t \to \infty} (X_t - 1_M \otimes c) = f$, $i = 1, \cdots, M$.*

In the following paragraphes, we discuss the formation control process of our multi-robot system based on the localization results. Initially, the system is in state X_0. When $t = 1$, robot R_1 holds still and broadcasts self position to other robots in the system. When other robots receive the position, they move to the new positions by the protocol as follows:

$$x_i(1) = x_i(0) + \varepsilon((x_1(0) - f_1) - (x_i(0) - f_i)), \ i = 1, \cdots, M. \qquad (6)$$

where ε is a feedback control parameter to tune the magnitude of the feedback. Thus, the position of the whole system is adjusted to:

$$X_1 = X_0 + \varepsilon \begin{bmatrix} 0 & 0 & 0 & \cdots & 0 \\ 1 & -1 & 0 & \cdots & 0 \\ 1 & 0 & -1 & \cdots & 0 \\ \vdots & \vdots & \vdots & \vdots & \vdots \\ 1 & 0 & 0 & \cdots & -1 \end{bmatrix} \otimes I_n (X_0 - f)$$

Denote

$$L_1 = \begin{bmatrix} 0 & 0 & 0 & \cdots & 0 \\ 1 & -1 & 0 & \cdots & 0 \\ 1 & 0 & -1 & \cdots & 0 \\ \vdots & \vdots & \vdots & \vdots & \vdots \\ 1 & 0 & 0 & \cdots & -1 \end{bmatrix} \otimes I_n$$

Then we have

$$\begin{aligned} X_1 &= X_0 + \varepsilon L_1(X_0 - f) \\ &= (I_M + \varepsilon L_1)X_0 - \varepsilon L_1 f \end{aligned} \tag{7}$$

When $t = 2$, robot R_2 holds still and broadcasts self position to other robots in the system. When other robots receive the position, they move to the new positions by the protocol as follows:

$$x_i(1) = x_i(0) + \varepsilon((x_2(0) - f_2) - (x_i(0) - f_i)), \ i = 1, \cdots, M. \tag{8}$$

Then the position of the whole system is adjusted to:

$$X_2 = X_1 + \varepsilon \begin{bmatrix} -1 & 1 & 0 & \cdots & 0 \\ 0 & 0 & 0 & \cdots & 0 \\ 0 & 1 & -1 & \cdots & 0 \\ \vdots & \vdots & \vdots & \vdots & \vdots \\ 0 & 1 & 0 & \cdots & -1 \end{bmatrix} \otimes I_n (X_1 - f)$$

Denote

$$L_2 = \begin{bmatrix} -1 & 1 & 0 & \cdots & 0 \\ 0 & 0 & 0 & \cdots & 0 \\ 0 & 1 & -1 & \cdots & 0 \\ \vdots & \vdots & \vdots & \vdots & \vdots \\ 0 & 1 & 0 & \cdots & -1 \end{bmatrix} \otimes I_n$$

Then we have

$$\begin{aligned} X_2 &= X_1 + \varepsilon L_2(X_1 - f) \\ &= (I_M + \varepsilon L_2)X_1 - \varepsilon L_2 f \end{aligned} \tag{9}$$

The rest may be deduced by analogy. Thus, when $t = M$, robot R_M holds still and broadcasts self position to other robots in the system. When other robots receive the position, they move to the new positions by the protocol as follows:

$$x_i(1) = x_i(0) + \varepsilon((x_M(0) - f_M) - (x_i(0) - f_i)), \; i = 1, \cdots, M. \qquad (10)$$

Then the position of the whole system is adjusted to:

$$\begin{aligned} X_M &= X_{M-1} + \varepsilon l_M (X_{M-1} - f) \\ &= (I_M + \varepsilon L_M) X_{M-1} - \varepsilon L_M f \end{aligned} \qquad (11)$$

where

$$L_M = \begin{bmatrix} -1 & 0 & 0 & \cdots & 1 \\ 0 & -1 & 0 & \cdots & 1 \\ 0 & 0 & -1 & \cdots & 1 \\ \vdots & \vdots & \vdots & \vdots & \vdots \\ 0 & 0 & 0 & \cdots & 0 \end{bmatrix} \otimes I_n$$

From $t = M + 1$, robots will repeat the procedures of position adjustments mentioned above. Thus, we can derive the dynamic equation of the position adjustments of the whole system as follows:

$$X_t = X_{t-1} + \varepsilon L_{r(t)} (X_{t-1} - f) \qquad (12)$$

where $L_i = (-I_M + E_i) \otimes I_n$. E_i is a $M \times M$ matrix defined as follows where the column i is 1 and others are all 0:

$$E_i = \begin{bmatrix} 0 & 0 & \cdots & 1 & \cdots & 0 \\ 0 & 0 & \cdots & 1 & \cdots & 0 \\ 0 & 0 & \cdots & 1 & \cdots & 0 \\ 0 & 0 & \cdots & 1 & \cdots & 0 \\ \vdots & \vdots & \vdots & \vdots & \vdots & \vdots \\ 0 & 0 & \cdots & 1 & \cdots & 0 \end{bmatrix}_{M \times M}$$

$r(t)$ is a periodic function defined as follows:

$$r(t) = \begin{cases} t, & 0 \le t \le M \\ t - iM, & iM \le t \le (i+1)M \end{cases} \qquad (13)$$

Definition 2. *Every robot in the multi-robot system adjusts self position by using the dynamic equation defined in Eq. (13). We say that the system passes one period, if everyone of the system has adjusted its position, namely $t = M$.*

Let

$$A_t = (\prod_{i=t}^{1}(I_m + \varepsilon L_{r(t)})) \otimes I_n \qquad (14)$$

Then we can obtain the analytical expression of Eq. (13) as follows:

$$X_t = A_t(X_0 - f) + f \qquad (15)$$

Fig. 3 shows the simulation result of generating preset diamond formation in a eight robots system. Robots' initial positions are randomly selected. In this simulation, we use the feedback control parameter $\varepsilon = 0.3$. After 25 periods, the system achieves the diamond formation.

Fig. 4 is the simulation of a fifty robots system. In this simulation, we use the feedback control parameter $\varepsilon = 0.2$. After 50 periods, the system achieves the circle formation.

The multi-robot system can change formation dynamically by using the control protocol mentioned above. Relatively precise self-localization grantee the

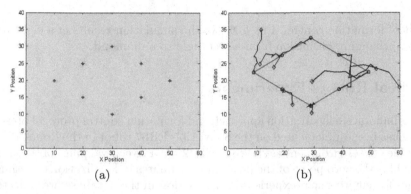

(a)   (b)

Fig. 3. (a) is the supposed final formation, where the red stars represent the position of robots. (b) shows the process of generating diamond formation from randomly selected positions, where the red "◇" and blue "○" represent the initial and final positions respectively.

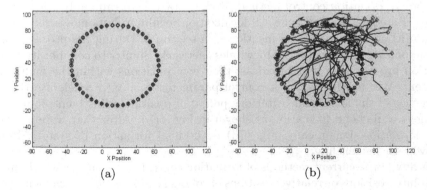

(a)   (b)

Fig. 4. (a) is the supposed final formation, where the blue "◇" represents the position of robots. (b) is the process of generating circle formation with random initial positions. The red "◇" and blue "○" represent the initial and final positions respectively.

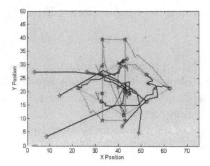

Fig. 5. Changing formation from a rectangle to a diamond. Robots start from positions marked pink "◊". First, they achieve a rectangle as shown in red stars. Then they change formation to a diamond as shown in blue "○".

result of formation control. Fig. 5 shows the simulation results of a eight robots system, changing formation from a rectangle to a diamond.

4   Real Robots Experiments

The collaborative localization approach and formation control protocol presented above has been implemented on the Sony Aibo ERS7 robot in the RoboCup four-legged league environment as shown in Fig. 2(a). The central point of the circle in the field is the zero point of the position coordinates in localization experiments. Images for environment experience construction in the database are collected in the areas of the field where the robot can not see any landmark every $100mm$ in x, $100mm$ in y and $45°$ in θ through the self-learning process. When the localization experiment is conducted, the robot performs a scanning motion with its head (pan range $[-45°, +45°]$) to search landmarks and exploit experience.

In the experiments, we used four Sony Aibo ERS7 robots to evaluate the proposed formation control method. Robots start from randomly selected positions. Based on the real-time self-localization results, robots moves as shown in Fig. 6(a) under the protocol mentioned in section 3. During the process, robots communicate with each other by wireless network. Similar to collaborative localization procedure, robots broadcast their own positions within the multi-robot system. We use the feedback control parameter $\varepsilon = 1$ to accelerate the convergence of the formation. After one period of position adjustment, four robots achieve a diamond formation as shown in Fig. 6(b). After that, robots change the formation dynamically. Fig. 6(d) is the final formation of system. Robots move into a strait line.

Many of the current methods of formation control use positions which include absolute positions or relative positions of robots in the system and velocity as the input. In this experiment, we compare the influence of different self-localization approaches on formation control. We set four Sony Aibo ERS7 robots (R_1, R_2, R_3, R_4) in the same environment as shown in Fig. 2(a). The global coordinates

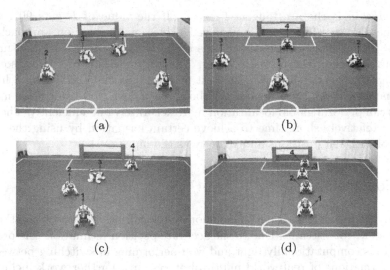

Fig. 6. Formation control of a real multi-robot system.(a) and (c) are the forming process of different formations. After adjustments, robots achieve a diamond formation as shown in (b). (d) is the final formation, a straight line.

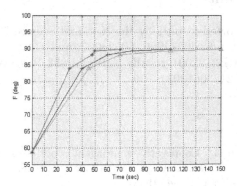

Fig. 7. Influence of different localization approaches. The red line is the result of the method proposed in this article. The blue line is the result of the multi-robot localization method mentioned in [6]. The green line is the result of the single-robot localization approach mentioned in [4].

of these four robots are notated as (x_1, y_1), (x_2, y_2), (x_3, y_3), (x_4, y_4). Robots move to certain diamond formation, where R_1 and R_4 are in the beginning and ending of one diagonal of the diamond respectively while R_2 and R_3 are on the other diagonal, under the same control protocol defined in section 3, by using three different localization approaches. To measure the influence on formation, a performance function F is defined as follows by using the characteristics of a diamond:

$$F = arctan(\frac{y_4 - y_1}{x_4 - x_1}) + arctan(\frac{y_3 - y_2}{x_3 - x_2}) \tag{16}$$

where *arctan* is the arc tangent function. If the value of F gets $90°$, we say that the diamond formation is achieved. Different localization approaches are applied to this multi-robot system. We use a computer to receive and measure the convergence of the formation. Fig. 7 shows the result of comparison. It is clear that the red line first gets the best performance that the system achieves a diamond formation. By using our collaborative multi-robot localization method, robots can perform global localization more accurately and efficiently. Therefore, it takes relatively short time to achieve certain formation by using the control protocol with self positions and velocity as the input.

5   Conclusion

This article has demonstrated a collaborative localization based formation control of multi-robot teams. Due to the accurate position estimation, the proposed protocol is computationally light and well-performing in switching between different formations of real-world multi-robot systems. Further work includes experiments of formation control in more complex environments.

Acknowledgements

This work was supported in part by the National Science Foundation of China (NSFC) under Contracts 60674050, 60528007, and 60635010, by the National 973 Program (2002CB312200), by the National 863 Program (2006AA04Z258) and 11-5 Project (A2120061303).

References

1. Arkin, R.C., Balch, T.: Cooperative multiagent robotic systems. In: Kortenkamp, D., Bonasso, R.P., Murphy, R. (eds.) Artificial Intelligence and Mobile Robots. MIT/AAAI Press, Cambridge (1998)
2. Chen, Y., Wang, Z.: Formation control: a review and a new consideration. In: Proc. of the IEEE/RSJ International Conference on Intelligent Robots and Systems, pp. 3181–3186 (2005)
3. Fax, A., Murray, R.M.: Information flow and cooperative cotnrol of vehicle formations. IEEE Transactions on Automatic Control 49(9), 1465–1476 (2004)
4. Fox, D., Burgard, W., Dellaert, F., Thrun, S.: Monte Carlo localization: Efficient position estimation for mobile robots. In: Proc. of the National Conference on Artificial Intelligence, pp. 343–349 (1999)
5. Fox, D., Burgard, W., Kruppa, H., Thrun, S.: Collaborative multi-robot localization. In: Burgard, W., Christaller, T., Cremers, A.B. (eds.) KI 1999. LNCS (LNAI), vol. 1701, pp. 255–266. Springer, Heidelberg (1999)
6. Fox, D., Burgard, W., Kruppa, H., Thrun, S.: A probabilistic approach to collaborative multi-robot localization. Autonomous Robots 8(3), 325–344 (2000)
7. Gazi, V., Passino, K.M.: Stability analysis of swarms. IEEE Transactions on Automatic Control 48(4), 692–697 (2003)

8. Kondo, H., Ura, T.: Visual observation of underwater objects by autonomous underwater vehicles. In: The 3rd International Workshop on Scientific Use of Submarine Cables and Related Technologies, pp. 145–150 (2003)
9. Merino, L., Wiklund, J., Caballero, F., Moe, A., Dios, J.R.M., Forssén, P., Nordberg, K., Ollero, A.: Vision-based multi-UAV position estimation. IEEE Robotics and Automation Magazine 13(3), 53–61 (2006)
10. Munich, M.E., Pirjanian, P., Di Bernardo, E., Goncalves, L., Karlsson, N., Lowe, D.: SIFT-ing through features with VIPR. IEEE Robotics and Automation Magazine 13(3), 72–77 (2006)
11. Murphy, R.R.: Introduction to AI Robotics. MIT Press, Cambridge (2000)
12. Nourbakhsh, I., Powers, R., Birchfield, S.: DERVISH an office-navigating robot. AI Magazine 16(2) (1995)
13. Röfer, T., et al.: GermanTeam RoboCup 2004, tech. rep. (2004), http://www.germanteam.org/GT2004.pdf
14. Roumeliotis, S.I., Bekey, G.A.: Distributed multirobot localization. IEEE Transactions on Robotics and Automation 18(5), 781–795 (2002)
15. Sala, P., Sim, R., Shokoufandeh, A., Dickinson, S.: Landmark selection for vision-based navigation. IEEE Transactions on Robotics 22(2), 334–349 (2006)
16. Schmitt, T., Hanek, R., Beetz, M., Buck, S., Radig, B.: Cooperative probabilistic state estimation for vision-based autonomous mobile robots. IEEE Transactions on Robotics and Automation 18(5), 670–684 (2002)
17. Siggelkow, S.: Feature histograms for content-based image retrieval, Ph.D. dissertation, Univ. Freiburg, Dept. Computer Sci., Freiburg, Germany (2002)
18. Vicsek, T., Czirok, A., Jacob, E.B., Cohen, I., Schochet, O.: Novel type of phase transitions in a system of self-driven particles. Physical Review Letters 75(6), 1226–1229 (1995)
19. Wang, Q., Liu, L., Xie, G., Wang, L.: Learning from human cognition: collaborative localization for vision-based autonomous robots. In: Proc. of the IEEE/RSJ International Conference on Intelligent Robots and Systems, pp. 3301–3306 (2006)
20. Wang, Q., Huang, Y., Xie, G., Wang, L.: Let robots play soccer under more natural conditions: experience-based collaborative localization in four-legged league. In: Visser, U., Ribeiro, F., Ohashi, T., Dellaert, F. (eds.) RoboCup 2007: Robot Soccer World Cup XI. LNCS (LNAI), vol. 5001, pp. 353–360. Springer, Heidelberg (2008)
21. Wang, Q., Rong, C., Liu, L., Li, H., Xie, G.: The 2006 sharPKUngfu team report. Technical report (2006), www.mech.pku.edu.cn/robot/fourleg/en/resources.htm
22. Wolf, J., Burgard, W., Burkhardt, H.: Robust vision-based localization by combining an image-retrieval system with Monte Carlo localization. IEEE Transactions on Robotics 21(2), 208–216 (2005)
23. Xie, G.: Decentralized feedback formation control of mobile robots, Intelligent control lobaratory technical report, College of Engineering, Peking University (2006)

Author Index